► Formulas for Volume and Surface Area

Figure		*Volume*	*Surface Area*

Rectangular solid

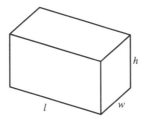

$V = lwh$

$SA = 2wh + 2hl + 2wl$

Cube

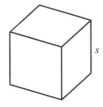

$V = s^3$

$SA = 6s^2$

Right circular cylinder

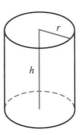

$V = \pi r^2 h$

$SA = 2\pi r^2 + 2\pi rh$

Right circular cone

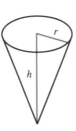

$V = \frac{1}{3}\pi r^2 h$

$SA = \pi r^2 + \pi r \sqrt{r^2 + h^2}$

Sphere

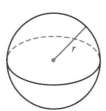

$V = \frac{4}{3}\pi r^3$

$SA = 4\pi r^2$

Intermediate Algebra

with Applications

SECOND EDITION

Linda L. Exley
Vincent K Smith

DeKalb College
Decatur, Georgia

PRENTICE HALL, Englewood Cliffs, New Jersey 07632

Library of Congress cataloging-in-publication data

Exley, Linda L.
    Intermediate algebra : with applications / Linda L. Exley, Vincent
K Smith. — 2nd ed.
       p.   cm.
    Includes index.
    ISBN 0-13-474552-3
    1. Algebra.   I. Smith, Vincent K   II. Title.
QA152.2.E96   1994
512.9—dc20   93-41474
                                    CIP

Dedicated to our spouses Lois and Charles

Editor-in-Chief: Jerome Grant
Development Editor: Bobbie Lewis
Production Editor: Edward Thomas
Marketing Manager: Melissa Acuña
Supplements Editor: Mary Hornby
Product Manager: Trudy Pisciotti
Design Director: Florence Dara Silverman
Text Designer & Page Layout: Lee Goldstein
Cover Designer: Bruce Kenselaar & William McCloskey
Photo Editor: Lori Morris-Nantz
Photo Research: Teri Stratford
Editorial Assistant: Audra Walsh
Text Composition: Interactive Composition Corporation
Copy Editor: Bill Thomas

 © 1994 by Prentice-Hall, Inc.
A Paramount Communications Company
Englewood Cliffs, New Jersey 07632

Printed in the United States of America
10 9 8 7 6 5 4 3 2 1

ISBN: 0-13-474552-3

Prentice-Hall International (UK) Limited, London
Prentice-Hall of Australia Pty. Limited, Sydney
Prentice-Hall Canada Inc., Toronto
Prentice-Hall Hispanoamericana, S. A., Mexico
Prentice-Hall of India Private Limited, New Delhi
Prentice-Hall of Japan, Inc., Tokyo
Simon & Schuster Asia Pts. Ltd., Singapore
Editora Prentice-Hall do Brazil, Ltda., Rio de Janeiro

Contents

Preface

This book is an intermediate algebra textbook with a review of beginning algebra topics. It can be used as a developmental text to prepare students for college algebra, precalculus or decision mathematics. However, it contains sufficient coverage to be used as an introductory college algebra text.

GETTING STARTED

Chapter 0 reviews such fundamental topics as sets, absolute value and the properties of the real number system. The sections in this chapter may be covered in detail, treated as a review, or omitted entirely. If the chapter is omitted, the individual topics can be covered as required with the material of the later chapters. A short review of geometry is included in Chapter 0 and geometric problems are integrated throughout the book.

PEDAGOGY

This book is written in a language that students understand. Wordy statements and mathematical jargon are kept to a minimum while methods and techniques are illustrated with worked examples whenever possible. Although simple language is used and rigor is not stressed, careful mathematics is.

The heart of the book is found in over 6500 carefully selected exercises that develop and illustrate the ideas of algebra in a modern setting. Extensive problem sets occur at the end of each section. Each set starts with a section entitled "Warmups." Warmups are a collection of problems graded in difficulty from easy to medium, which are keyed to worked examples in the text. Warmups are followed by a set of exercises called "Practice Exercises." This longer set of problems contains the drill and practice exercises so necessary for students to develop their manipulative algebra skills. The Practice Exercises are also graded in difficulty, but are mixed and not keyed to worked examples. The problem sets also have a few "Challenge Problems" that allow the student to probe into the natural extensions of the ideas presented in the text. The problem sets contain sufficient numbers of problems to provide the instructor considerable flexibility. The pedagogy is never dictated by a limited selection of problems.

The topics are developed so that instructors can tailor a course to fit their needs. The new material and all the teaching points of each section are presented with worked and annotated examples. Problem-solving steps are identified and explained. The students see what is to be done and how to do it. Most examples and all of the teaching points are reinforced with Warmup problems keyed to the examples. The topics are arranged in such a manner as to allow the student to progress through the book developing skills as they are needed. The student is not put into a position of using a concept that has not already been developed.

FEATURES OF THE TEXT

A Problem-Solving Approach A problem-solving approach is used throughout the book. Worked examples with explanatory text provide the key to an extensive collection of applied problems.

Connections Each chapter opens with an introductory paragraph putting the upcoming material in context, and showing its connections with other chapters, other courses, and historical development of mathematical ideas.

Key Problems The Warmups and Practice Exercises contain key problems marked in color. These problems illustrate the necessary teaching points of the section. The instructor can choose to work some or all of them in class as blackboard examples. They can be assigned as hand-in homework to be graded or they can be used for any purpose where the instructor wishes to be sure of a limited but representative cross-section of the material.

Let's Not Forget   The review problems in every chapter, except Chapter 0, end with a novel feature called "Let's Not Forget." This segment contains a few carefully selected problems that repeat earlier themes, particularly certain sticky ideas and problems that historically give students trouble. The Let's Not Forget problems expand with each chapter.

Check Ups   Each chapter summary ends with a section called "Checkups." These are worked examples from the text stated as problems. These can be used as a self-test for students to check their understanding of the main points in the chapter. Each problem is keyed to an example number so that students can check their work by referring back to the text.

Be Careful!   Common Student Errors are prominently flagged in the margin with the admonition, "Be Careful!" The adjoining text explains the caution.

In Your Own Words   The problem sets in each section end with a few questions marked, "In Your Own Words." These may be used to encourage good writing in mathematics and to test comprehension of certain ideas.

Calculator Boxes   Calculators are introduced where appropriate and instructional material on the proper use of calculators is included. The calculator boxes are not mere window dressing but provide detailed instructions on meaningful problems for new calculator users. However, the calculator material, including exercises, is segregated and clearly marked for those who wish to omit it. In the calculator boxes and elsewhere in the text, the distinction between exact, approximated and estimated answers is emphasized.

Chapter Summary   Each chapter contains a summary of the material and glossary of new terms.

Chapter Review   Each chapter ends with a set of review problems and a chapter test to reinforce the material.

NEW 2ND EDITION FEATURES

Problem Sets   All problem sets have been carefully edited and enhanced with more real-world applications added throughout. Problems involving geometry have been integrated throughout the text. Answers to *all* Warm-up problems are included in the back-of-book answers.

Calculators   The calculator boxes have been edited and modernized to contain information about graphing calculators. A complete introduction to graphing calculators has been added as an appendix to include keystroke examples for the TI-81, Casio fx7700, and Sharp 9200-9300 series calculators.

New Critical Thinking feature   Special problems entitled "You Decide" have been added to every chapter except Chapter 0. These are open ended problems requiring some algebra skills but also forcing some analysis of a situation. Each requires that a decision be made and a written answer provided.

Functions   The introduction to functions and graphs has been moved from Chapter 11 to Chapter 6 and has been enhanced and enlarged. A new chapter has been added on exponential and logarithmic functions.

In general   The major theme of progressing from numbers to expressions to equations to modeling has been enhanced and further emphasized in this edition. Equations have been moved from Chapter 4 to Chapter 2. A new section on problem solving has been added to Chapter 2. Coverage of the a-c method of factoring trinomials has been included. Operations with radicals has been expanded from one to three sections. More photographs have been added throughout the text.

SUPPLEMENTS

Instructors using this text may obtain the following supplements from the publisher:

Instructor's Solutions Manual   Prepared by Cheryl V. Roberts, Northern Virginia Community College, contains worked out solutions to all even-numbered problems. (0-13-475310-0)

Instructor's Solutions Manual with Tests and Syllabus   Prepared by Linda Kyle, Tarrant County Community College (with tests by ips Publishing, Inc.). Contains chapter tests, suggested teaching assignments, and teaching outlines. (0-13-475302-X)

Test Item File   Prepared by Mary Kay Schippers, Fort Hays State University. Contains hard copy of test questions from PH Test Manager and Make Test for the Macintosh. (0-13-475344-5)

PH Test Manager 2.0 (3.5" or 5.25")   Contains a bank of test questions which can be edited and manipulated to create tests: 3.5" (0-13-475443-3); 5.25" (0-13-475450-6); demo (0-13-075763-2).

Make Test for Macintosh is a state-of-the-art Macintosh test generator that makes full use of the superior MAC graphics capabilities as well as the user-friendly interface. (0-13-475468-9); demo (0-13-075771-5)

PH Algo Test (3.5" or 5.25") is a state-of-the-art IBM compatible program that allows instructors to produce exams by generating test questions from algorithms: 3.5" (0-13-475500-6); 5.25" (0-13-475518-9).

Lecture Videos A set of lecture videos which are keyed to each section of the text, and can be used in math lab. Available with an adoption of 100 copies. (0-13-475543-X); demonstration video (0-13-075722-5).

Students using this text may obtain the following supplements from the publisher:

Student's Solutions Manual Prepared by Cheryl V. Roberts, Northern Virginia Community College. Contains worked out solutions to all odd-numbered problems and to all chapter test questions. (0-13-475336-4)

Math Master Tutor IBM (3.5" or 5.25"); Math Master Tutor Macintosh carefully keyed to the text with page references, Math Master Tutor contains four modes of instruction for each section in the text: Exploration (a text-on-disk feature), Summary (a concise review), Exercises (open-ended, algorithmically generated drills with step-by-step solutions), and Quiz (timed tests which teach students test-taking strategies). Available upon adoption (with cite license) for IBM and Macintosh.

- IBM 3.5" (0-13-475385-2)
- IBM 5.25" (0-13-475393-3)
- IBM demo (0-13-075680-6)

- MAC (0-13-475401-8)
- MAC demo (0-13-075776-3)

REVIEWERS

We wish to thank the following reviewers:

Laura Clarke, Milwaukee Area Technical College (Milwaukee, WI)

Christie Gilliland, Green River Community College (Auburn, WA)

Robert L. Hoburg, Western Connecticut State University (Danbury, CT)

Joyce Huntington, Walla Walla Community College (Walla Walla, WA)

Orval LeJeune, DeKalb College-Gwinnett (Lawrenceville, GA)

Elizabeth Lundy, Linn-Benton Community College (Albany, OR)

Ruth Meyering, Grand Valley State University (Allendale, MI)

Choon J. Rhee, Wayne State University (Detroit, MI)

Daniel Richbart, Villa Maria College of Buffalo (Buffalo, NY)

Finally, we would like to thank the staff and faculty of DeKalb College for their continued support; Peggy Estes who insisted we do this work; Ed Thomas who is far too cheerful to be a production editor; designer Lee Goldstein; accuracy checker Richard Semmler (of Northern Virginia Community College); and our development editor Bobbie Lewis.

Linda L. Exley
Vincent K Smith

0 Reviewing Sets of Numbers and Geometry

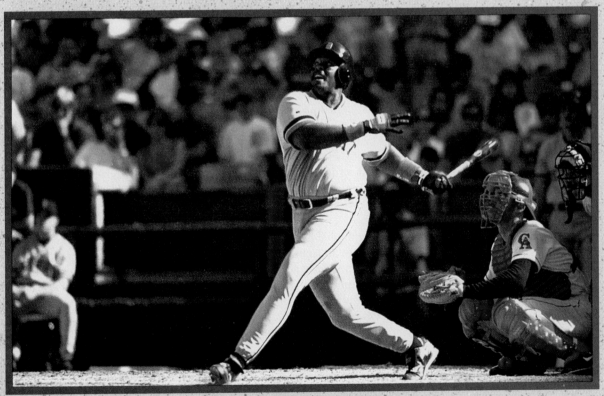

See Problem Set 0.6, Exercise 7.

► Connections

In Western Europe for nearly a thousand years after the fall of Rome, it was not fashionable to think. In fact, if caught in the act of any intellectual pursuit, one might become the featured act in the afternoon barbeque! However, the study of mathematics continued in other places, notably along the northern coast of Africa.

The Arabic word *al-jabr,* which means "to transpose," was prominent in the title of a book written by Al-Khowârizmî around A.D. 830. From this word comes our word *algebra.* During this period the Arabs developed most of what we call algebra today.

Algebra is largely arithmetic, with one or more numbers represented by letters. Therein lies its usefulness, and therein lies its difficulty. To be successful with algebra, it is necessary to do arithmetic with letters exactly as if they were numbers.

In this chapter we discuss several topics that should be a review of previous work. It would be wise to master this material before moving on.

0.1   Sets

TAPE 1

Be Careful!

A set is a collection of objects. The objects are called **elements** or **members.** We often use capital letters to name sets and list the elements in braces. For example, $A = \{a, b, c, d, e\}$ is a set containing the first five letters of the alphabet. $B = \{1, 2, 3, \ldots, 10\}$ is a set containing the counting numbers 1 through 10. (The three dots, called an *ellipsis,* mean "continue on, in the pattern that has been established.") $N = \{1, 2, 3, \ldots\}$ is a set containing all the counting numbers.

If the number of elements in a set is a counting number, the set is called a **finite set.** If there are no elements at all in a set, it is the **empty set.** Otherwise, it is called an **infinite set.** The sets A and B are finite, whereas N is infinite.

The empty set is sometimes called the **null set.** It is written as $\{\ \}$ or $\varnothing$. It would be wrong to write the empty set as $\{\varnothing\}$, as this set would contain one element.

The symbol $\in$ is used to mean **is an element of,** and the symbol $\notin$ is used to mean **is not an element of.** If $D = \{a, b, c, d, e\}$, then $a \in D$, whereas $2 \notin D$.

Sometimes, instead of listing the elements of the set, a notation called **set-builder notation** is used. $\{x \mid x \text{ is a vowel}\}$ describes the "set of all x such that x is a vowel." The bar is read "such that." It is the same set as $\{a, e, i, o, u\}$.

Set A is a **subset** of a set B if every element in A is also an element in B. The symbol $\subseteq$ means **is a subset of.** If $A = \{1, 5, 7\}$ and $B = \{1, 3, 5, 7, 9\}$, then $A \subseteq B$.

► EXAMPLE 1

List all the subsets of $\{a, b, c\}$.

Solution

The subsets are: $\{a, b, c\}$, $\{a, b\}$, $\{a, c\}$, $\{b, c\}$, $\{a\}$, $\{b\}$, $\{c\}$, $\varnothing$ ◄

Notice that the set is a subset of itself and that the empty set is a subset of every set.

Subsets of a Set

If A is any set, $A \subseteq A$ and $\emptyset \subseteq A$.

Often, sets are combined using the operations of **union** ($\cup$) and **intersection** ($\cap$).

Union and Intersection of Sets

$$A \cup B = \{x \mid x \in A \text{ or } x \in B\}$$

$$A \cap B = \{x \mid x \in A \text{ and } x \in B\}$$

That is, the union of two sets is the set of all elements that belong to **either** of the two sets (or both of them), while the intersection of two sets is the set of all elements that belong to *both* of the sets.

► **EXAMPLE 2**

If $A = \{1, 2, 7\}$ and $B = \{1, 2, 3, 4, 5,\}$, find:

(a) $A \cup B$ (b) $A \cap B$

Solutions

(a) $A \cup B = \{1, 2, 3, 4, 5, 7\}$ (b) $A \cap B = \{1, 2\}$ ◄

Algebra deals with sets of numbers. The symbols used to name the numbers are called **numerals.**

Natural Numbers

$$N = \{1, 2, 3, \ldots\}$$

Whole Numbers

$$W = \{0, 1, 2, 3, \ldots\}$$

Integers

$$Z = \{\ldots, -3, -2, -1, 0, 1, 2, 3, \ldots\}$$

Rational Numbers

$$Q = \left\{x \mid x \text{ can be written as } \frac{p}{q}, \text{ where } p \text{ and } q \text{ are integers and } q \neq 0\right\}$$

Irrational Numbers

$$I = \{x \mid x \text{ is a number but not a rational number}\}$$

Real Numbers

$$R = \{x \mid x \text{ is a rational or an irrational number}\}$$

The natural numbers are sometimes called the **counting numbers.** Some examples of rational numbers are $\frac{7}{8}$, 5.2, and -3. Rational numbers are often referred to as *fractions*. Some examples of irrational numbers are $\sqrt{2}$, $\sqrt{3}$, and π. As a matter of fact, $\sqrt{x}$, where x is a positive integer but not a perfect square, is an irrational real number.

The following diagram shows how these sets are related to each other.

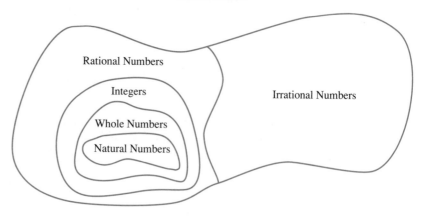

Notice that since $5 = \frac{5}{1}$, the integer 5 is a rational number. Similarly, all integers are rational numbers. As the next example shows, all rational numbers can be expressed as a repeating decimal.

► **EXAMPLE 3**

Express each rational number as a decimal.

(a) $\frac{2}{3}$ (b) $\frac{1}{2}$ (c) $\frac{3}{11}$

Solutions

(a) $\frac{2}{3} = 0.6666\ldots = 0.\overline{6}$ (b) $\frac{1}{2} = 0.5000\ldots = 0.5\overline{0}$

(c) $\frac{3}{11} = 0.272727\ldots = 0.\overline{27}$ ◄

Be Careful!   Be careful not to confuse exact values with approximations. $\frac{2}{3} = 0.\overline{6}$, which is an exact value. The number 0.67 is an approximation of $\frac{2}{3}$, as is 0.667 and 0.666667. Non-terminating decimals that do not repeat are irrational numbers.

► **EXAMPLE 4**

Classify each number as rational or irrational.

(a) $\frac{3}{7}$ (b) $0.\overline{78}$ (c) $\sqrt{7}$ (d) 1.7

Solutions

(a) $\dfrac{3}{7}$ is rational.

(b) $0.\overline{78}$ is a repeating decimal, and thus is rational.

(c) $\sqrt{7}$ is irrational, as 7 is not a perfect square.

(d) 1.7 is rational, as $1.7 = \dfrac{17}{10}$. ◀

Problem Set 0.1

Warm-Ups

In Problems 1 through 10, indicate whether each statement is true or false.

1. If $A = \{a, b, c\}$, then $a \in A$.

2. The set of natural numbers is a subset of the set of integers.

3. Every integer is a rational number.

4. $A \subseteq A$, where A is any set.

5. $-\dfrac{1}{2}$ is an integer.

6. The empty set is written as $\{\varnothing\}$.

7. 0 is a natural number.

8. If A is any set, $A \cup \varnothing = A$.

9. If A is any set, then $A \cap \varnothing = A$.

10. The set $\{2, 4, 6, \ldots, 18\}$ is an infinite set.

In Problems 11 through 13, let $A = \{2, 4, 6, 8\}$ and $B = \{-2, 0, 2\}$. See Examples 1 and 2.

11. List the element in $A \cup B$.

12. List the elements in $A \cap B$.

13. List all the subsets of B.

In Problems 14 through 17, list the elements of A.

14. $A = \{x \mid x$ is a natural number less than 10$\}$

15. $A = \{x \mid x$ is a color in the U.S. flag$\}$

16. $A = \{x \mid x$ is a natural number divisible by 5$\}$

17. $A = \{x \mid x$ is one of the first 10 letters in the alphabet$\}$

In Problems 18 through 21, write A using set-builder notation.

18. $A = \{w, x, y, z\}$

19. $A = \{2, 4, 6, 8, 10, 12\}$

20. $A = \{\ldots, -5, -3, -1, 1, 3, 5, \ldots\}$

21. $A = \{\text{Monday, Tuesday}, \ldots, \text{Sunday}\}$

In Problems 22 through 26, write each rational number as a repeating decimal. See Example 3.

22. $\dfrac{1}{3}$

23. $\dfrac{3}{4}$

24. $\dfrac{1}{6}$

25. $\dfrac{5}{8}$

26. $\dfrac{2}{7}$

In Problems 27 through 32, classify each number as rational or irrational. See Example 4.

27. $\dfrac{2}{3}$

28. 1.25

29. 2π

30. $\sqrt{17}$

31. $0.\overline{123}$

32. 0.66666

Practice Exercises

In Problems 33 through 42, indicate whether each statement is true or false.

33. If $A = \{1, 2, 3\}$, then $3 \notin A$.

34. The set of natural numbers is a subset of the set of whole numbers.

35. Every negative integer is a rational number.

36. $\varnothing \subseteq A$, where A is any set.

37. $\frac{1}{2}$ is a real number.

38. The empty set can be written as $\varnothing$.

39. 0 is a rational number.

40. If A is any set, $A \cap A = A$.

41. If A is any set, then $A \cap \varnothing = \varnothing$.

42. The set $\{2, 4, 6, \ldots\}$ is a finite set.

In Problems 43 through 45, let $A = \left\{\frac{1}{2}, 2, \frac{9}{4}, 5\right\}$ and $B = \left\{\frac{9}{4}, 5\right\}$.

43. List the elements in $A \cup B$.

44. List the elements in $A \cap B$.

45. List all the subsets of B.

In Problems 46 through 49, list the elements of A.

46. $A = \{x \mid x \text{ is a natural number less than 7}\}$

47. $A = \{x \mid x \text{ is a day of the week beginning with F}\}$

48. $A = \{x \mid x \text{ is an integer divisible by 2}\}$

49. $A = \{x \mid x \text{ is one of the last 10 letters in the alphabet}\}$

In Problems 50 through 53, write A using set-builder notation.

50. $A = \{\text{April, August}\}$

51. $A = \{-2, -1, 0, 1, 2\}$

52. $A = \{\ldots, -6, -4, -2, 0, 2, 4, 6, \ldots\}$

53. $A = \{\text{clubs, diamonds, hearts, spades}\}$

In Problems 54 through 59, classify each number as rational or irrational.

54. $\frac{2}{5}$

55. 1.2

56. 4π

57. $\sqrt{11}$

58. $0.13\overline{7}$

59. 0.1111111

In Problems 60 through 64, write each rational number as a repeating decimal.

60. $\frac{2}{3}$

61. $\frac{3}{5}$

62. $\frac{5}{6}$

63. $\frac{7}{8}$

64. $\frac{3}{7}$

Challenge Problems

65. Express 0.3 as a quotient of two integers.

66. Express 1.2 as a quotient of two integers.

67. Express 0.655 as a quotient of two integers.

In Your Own Words . . .

68. What is a real number?

69. What is a rational number?

0.2　**Properties of Real Numbers**

TAPE 1

It is much easier to visualize sets of numbers by drawing a picture of them. This is done by using a number line. Draw a line and pick a starting point called the **origin** and label it with the number 0. Move to the right of 0 and represent positive integers with a convenient unit of measure. Negative integers are represented to the left of zero. Fractions and irrational numbers are inserted in their place using the same unit of measure.

The number associated with a point is called its **coordinate.** Every real number can be located on the number line, and every point on the number line represents a real number.

To **graph** a set of numbers means to locate them on a number line. Looking at the number line, note that if a number, p, lies to the left of another number, q, then p is less than q. We write this using the symbol $<$.

Less Than

$p < q$ means p is less than q. (p is to the *left* of q.)

Similarly, if r lies to the right of s, then r is greater than s. We write this using the symbol $>$.

Greater Than

$r > s$ means r is greater than s. (r is to the *right* of s.)

Note that saying $p < q$ is the same as saying that $q > p$.

Sometimes "equals" is combined with "less than" or "greater than" by using the symbol $\leq$ or $\geq$.

Equality with Less Than or Greater Than

$p \leq q$ means p is less than or equal to q.

$r \geq s$ means r is greater than or equal to s.

The number line helps us to understand some properties of real numbers. If two real numbers p and q are graphed, one of three things can occur.

1. p is the same as q.
2. p is to the left of q.
3. p is to the right of q.

This idea is called the Trichotomy Property.

Trichotomy Property

If p and q are real numbers, exactly one of the following statements is true:

$$p = q$$

$$p < q$$

$$p > q$$

If p, q, and r are real numbers with $p < q$ and $q < r$, does p lie to the left or to the right of r? If we look at a number line, we can answer this question. Because we know that $p < q$, p lies to the left of q. Similarly, since $q < r$, we know that q lies to the left of r.

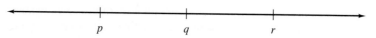

Thus we see that p is to the left of r. That is, $p < r$. This idea is called the Transitive Property.

Transitive Property

Let p, q, and r be real numbers.

If $p < q$ and $q < r$, then $p < r$.

Many properties of the real numbers deal with two numbers being equal, which are called Properties of Equality.

Properties of Equality

If p, q, and r are real numbers, then:

1. $p = p$ Reflexive
2. If $p = q$, then $q = p$.   Symmetric
3. If $p = q$ and $q = r$, then $p = r$.   Transitive

As a result of these properties of equality, we have the important Principle of Substitution.

Principle of Substitution

If p and q are real numbers with $p = q$, then we may replace p with q.

Arithmetic with real numbers uses the operations of addition, subtraction, multiplication, and division. Symbols are used to indicate which operation to perform.

Operation	Symbols	Name
Addition	$p + q$	Sum
Subtraction	$p - q$	Difference
Multiplication	$pq, p \cdot q, p(q), (p)(q)$	Product
Division	$p \div q, \dfrac{p}{q}, p/q$	Quotient

When we express sums, differences, products, and quotients in symbols, we call them **expressions** or **algebraic expressions.** In a sum or difference, the numbers being added or subtracted are called **terms.** In a product, the numbers being multiplied are called **factors.**

In doing arithmetic, we take advantage of properties of numbers without giving them much thought. For example, it is well known that $2 + 3 = 3 + 2$ and that $2 \cdot 3 = 3 \cdot 2$. These ideas are true for all real numbers and illustrate the Commutative Properties of real numbers.

Commutative Properties

If p and q are real numbers, then:

1. $p + q = q + p$ Commutative for addition

2. $pq = qp$ Commutative for multiplication

The next two properties deal with the way in which we group real numbers to add or multiply them. We know that

$$2 + (3 + 4) = (2 + 3) + 4$$

and that

$$2(3 \cdot 4) = (2 \cdot 3)4$$

These ideas are called Associative Properties.

Associative Properties

If p, q, and r are real numbers, then:

1. $p + (q + r) = (p + q) + r$ Associative for addition

2. $p(qr) = (pq)r$ Associative for multiplication

A combination of addition and multiplication such as $2(3 + 4)$ can be done in two ways. In the first way, we add 3 and 4 and then multiply 2 by 7.

$$2 \boxed{(3 + 4)} = 2 \cdot \boxed{7}$$
$$= 14$$

The second way is called the Distributive Property. We multiply each number in parentheses by 2 and then add.

$$\boxed{2} (3 + 4) = \boxed{2} \cdot 3 + \boxed{2} \cdot 4$$
$$= 6 + 8$$
$$= 14$$

Distributive Property

If p, q, and r are real numbers, then

$$p(q + r) = p \cdot q + p \cdot r$$

Two real numbers have special properties. These numbers are 0 and 1. Since the sum of 0 and any real number is the real number itself, we call 0 the identity for addition. Similarly, since 1 times any real number is the real number itself, we call 1 the multiplicative identity.

Identities

If p is a real number, then:

1. $p + 0 = 0 + p = p$ Identity for addition
2. $p \cdot 1 = 1 \cdot p = p$ Identity for multiplication

The **additive identity** is 0 and the **multiplicative identity** is 1.

Looking at a number line, notice that 3 and -3 lie on opposite sides of 0. Also, we note that each number is the same distance from 0. The numbers 3 and -3 have the property that their sum is the additive identity.

$$3 + (-3) = 0$$

The product of 2 and $\frac{1}{2}$ is the multiplicative identity. That is,

$$2 \cdot \frac{1}{2} = 1$$

Inverses

1. *Additive Inverse (Opposite).* If p is a real number, there is a real number $-p$ such that

$$p + (-p) = (-p) + p = 0$$

$-p$ is called the **opposite** or **additive inverse** of p. (We read $-p$ as negative p or the opposite of p.)

2. *Multiplicative Inverse (Reciprocal).* If p is a nonzero real number, there is a real number $\frac{1}{p}$ such that

$$p \cdot \frac{1}{p} = \frac{1}{p} \cdot p = 1$$

$\frac{1}{p}$ is called the **multiplicative inverse** or **reciprocal** of p.

► **EXAMPLE 1**

Locate each number on a number line.

(a) 5   (b) −5   (c) $-\dfrac{5}{3}$ (d) $\dfrac{7}{2}$

(e) $p; p > 0$ (f) $-p; p > 0$

(g) $q; q < 0$ (h) $-q; q < 0$

Solutions

(a) through (d) are shown below.

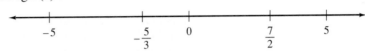

(e) If $p > 0$, then p lies to the right of 0.

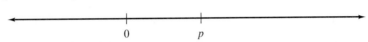

(f) If $p > 0$, then p lies to the right of 0. So $-p$ (the opposite of p) lies to the left of 0.

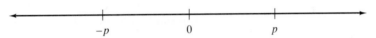

(g) If q is negative, then q lies to the left of 0.

(h) If $q < 0$, q lies to the left of 0. So $-q$ lies to the right of 0.

Using the symbol "−" can be very confusing. It has three different meanings. Consider these situations.

[a] −5   [b] 4 − 2   [c] −p

In [a] it is part of the name of a number, negative five.

In [b] it indicates subtraction, 4 minus 2 or 4 subtract 2.

In [c] it indicates the opposite of a number, the opposite of p.

We often interchange the words *minus, negative,* and *opposite* when reading problems that use the "−" symbol.

► **EXAMPLE 2**

Give the reciprocal of each real number.

(a) 2   (b) $\dfrac{3}{5}$ (c) 8   (d) $\dfrac{5}{2}$

Solutions

(a) Since $2 \cdot \dfrac{1}{2} = 1$, the reciprocal of 2 is $\dfrac{1}{2}$.

(b) Since $\dfrac{3}{5} \cdot \dfrac{5}{3} = 1$, the reciprocal of $\dfrac{3}{5}$ is $\dfrac{5}{3}$.

(c) Since $8 \cdot \dfrac{1}{8} = 1$, the reciprocal of 8 is $\dfrac{1}{8}$.

(d) Since $\dfrac{5}{2} \cdot \dfrac{2}{5} = 1$, the reciprocal of $\dfrac{5}{2}$ is $\dfrac{2}{5}$. ◄

Example 1 leads us to a discussion of the opposite of the opposite of a number. What is the opposite of the opposite of 3? The opposite of 3 is -3. So the opposite of the opposite of 3 is the opposite of -3, which is 3. Saying this in words is cumbersome; writing it in mathematical symbols makes it easier.

$$-(-3) = 3 \qquad \text{The opposite of negative three is three.}$$

This can be stated as a property for any real number.

Double Negative Property

If p is a real number, then $-(-p) = p$.

Multiplication by 0

If p is a real number, then $p \cdot 0 = 0 \cdot p = 0$.

► **EXAMPLE 3**

Indicate the property of real numbers that justifies each statement.

(a) $2 + 3 = 3 + 2$ (b) $1(3) = 3$

(c) $-5 + 0 = -5$ (d) $(4 \cdot 3) \cdot 6 = 4 \cdot (3 \cdot 6)$

Solutions

(a) $2 + 3 = 3 + 2$ Commutative Property for Addition

(b) $1(3) = 3$ Multiplicative Identity

(c) $-5 + 0 = -5$ Identity for Addition

(d) $(4 \cdot 3) \cdot 6 = 4 \cdot (3 \cdot 6)$ Associative Property for Multiplication ◄

It is easy to see that s and $-s$ are opposites because $s + (-s) = 0$. Sometimes it is more difficult to recognize opposites. Consider the following.

$$
\begin{aligned}
(p - q) + (q - p) &= p + (-q + q) - p &&\text{Associative Property} \\
&= p + 0 - p &&\text{Opposites} \\
&= p - p &&\text{Identity (Addition)} \\
&= 0 &&\text{Subtraction}
\end{aligned}
$$

Since $(p - q) + (q - p) = 0$, we see that $p - q$ and $q - p$ are opposites. These numbers come up often in algebra.

Opposites

$$p - q = -(q - p)$$

$p - q$ and $q - p$ are *opposites*.

► **EXAMPLE 4**

Find the opposite of $w - z$.

Solution

The opposite of $w - z$ is $z - w$. We say this by writing $-(w - z) = z - w$ or $z - w = -(w - z)$.   ◄

Problem Set 0.2

Warm-Ups

In Problems 1 through 7, give the opposite of each number and locate both the number and its opposite on a number line. See Example 1.

1. $\sqrt{3}$ **2.** -4 **3.** $-\dfrac{1}{3}$

4. $\dfrac{11}{3}$ **5.** 0 **6.** $p; p > 0$

7. $q; q < 0$

In Problems 8 through 13, find the reciprocal of each number. See Example 2.

8. 4 **9.** 7 **10.** $\dfrac{5}{3}$ **11.** $\dfrac{2}{3}$ **12.** $\dfrac{3}{4}$ **13.** $\dfrac{8}{5}$

In Problems 14 through 28, indicate the property of real numbers that justifies each statement. See Example 3.

14. $2 \cdot 3 = 3 \cdot 2$

15. $4 = 4$

16. If $4 = x$, then $x = 4$.

17. $(3 + 4) + \sqrt{7} = 3 + (4 + \sqrt{7})$

18. $1 + 3 = 3 + 1$

19. $9 + 0 = 9$

20. $7 + (-7) = 0$

21. $1 \cdot (-8) = -8$

22. $4 + (2 + 3) = 4 + (3 + 2)$

23. $2 \cdot \dfrac{1}{2} = 1$

24. $0 \cdot 6 = 0$

25. $4(3 + 7) = 4 \cdot 3 + 4 \cdot 7$

26. If $a < \sqrt{3}$ and $\sqrt{3} < b$, then $a < b$.

27. If $x + 2 = 5$ and $x = 3$, then $3 + 2 = 5$

28. $(2 \cdot 9) \cdot 5 = 2 \cdot (9 \cdot 5)$

29. Give a reason for each step. See Example 3.

$\boxed{1}$ $3 \cdot x + 7 \cdot x = x \cdot 3 + x \cdot 7$

$\boxed{2}$ $\qquad\qquad = x(3 + 7)$

$\boxed{3}$ $\qquad\qquad = x \cdot 10$

$\boxed{4}$ $\qquad\qquad = 10 \cdot x$

30. Find the opposite of $x - y$. See Example 4.

Practice Exercises

In Problems 31 through 45, indicate the property of real numbers that justifies each statement.

31. $2 + 3 = 3 + 2$

32. $\sqrt{6} = \sqrt{6}$

33. If $y = x$, then $x = y$.

34. $(3 \cdot 4) \cdot 7 = 3 \cdot (4 \cdot 7)$

35. $7(6 + 5) = 7 \cdot 6 + 7 \cdot 5$

36. $3(4 + \pi) = 3 \cdot 4 + 3 \cdot \pi$

37. $7 + (-7) = (-7) + 7$

38. $1 \cdot 8 = 8$

39. $5 + (-5) = 0$

40. $3 \cdot \dfrac{1}{3} = 1$

41. $0 \cdot 3 = 0$

42. $4 \cdot (3 + 7) = (3 + 7) \cdot 4$

43. If $x < 1$ and $1 < y$, then $x < y$.

44. If $x + 2 = 4$ and $x = 2$, then $2 + 2 = 4$.

45. $(2 + 9) + 5 = 2 + (9 + 5)$

In Problems 46 through 52, give the opposite of each number and locate both the number and its opposite on a number line.

46. $\sqrt{2}$

47. -2

48. $-\dfrac{2}{3}$

49. $\dfrac{2}{3} -$

50. π

51. $-p; p > 0$

52. $-q; q < 0$

In Problems 53 through 58, find the reciprocal of each number.

53. 1

54. 6

55. $\dfrac{1}{4}$

56. $\dfrac{2}{5}$

57. $\dfrac{3}{7}$

58. $\dfrac{7}{3}$

59. Give a reason for each step.

(1) $3 + (5 + 7y) = (3 + 5) + 7y$

(2) $\qquad\qquad = 8 + 7y$

60. Find the opposite of $s - t$.

Challenge Problems

61. Suppose that p and q are positive real numbers with $p > q$. Locate $p - q$ on a number line.

62. Suppose that p and q are positive real numbers with $p < q$. Locate $p - q$ on a number line.

In Your Own Words

63. Explain what opposites are.

64. Explain what the Commutative Properties say.

65. Explain what the Associative Properties say.

66. Explain what the Distributive Property says.

0.3   Absolute Value

TAPE 1

The numbers 3 and -3 share a common property. Each number is 3 units from zero on the number line. The measure of this distance is called the **absolute value** of 3 and -3 and is written $|3|$ and $|-3|$. Thus

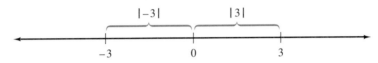

Notice that $|-3| = |3| = 3$.

A similar argument holds for 17 and -17.

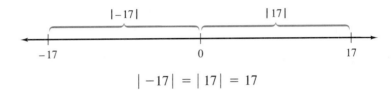

$$|-17| = |17| = 17$$

▶ **EXAMPLE 1**

Evaluate each absolute value.

(a) $|52|$ (b) $|-7|$ (c) $|\pi|$ (d) $|0|$

Solutions

(a) The number 52 is 52 units from 0 on the number line, so $|52| = 52$.

(b) Since -7 is 7 units from 0 on the number line, $|-7| = 7$.

(c) π is a positive number (approximately 3.14159), so π is π units from 0 on the number line. Therefore, $|\pi| = \pi$.

(d) As 0 is 0 units from 0 on the number line, $|0| = 0$.   ◀

Suppose that p is a positive number. Then p lies to the right of 0 on the number line.

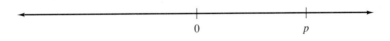

The opposite of p, $-p$, lies the same distance from 0, but to the left of 0, and, as before, $|-p| = |p| = p$.

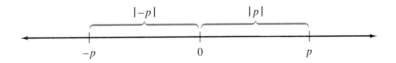

Now suppose that q is a *negative* number. Since q is *less* than 0, it lies to the left of 0 on the number line.

The opposite of q, $-q$, lies the same distance from 0, but to the right of 0.

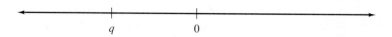

Note that, because q is negative, $-q$ is **positive!** So $-q$ is the measure of the distance and $|q| = |-q| = -q$.

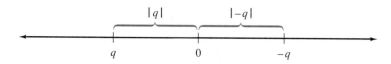

This leads to the definition of absolute value.

Absolute Value

If k is a real number,

$$|k| = \begin{cases} k, & \text{if } k \geq 0 \\ -k, & \text{if } k < 0 \end{cases}$$

Students starting the study of algebra are often troubled by statements such as "$-q$ is positive" and "$|-q| = -q$." This is understandable, as we have trained ourselves, during years of learning arithmetic, to recognize negative numbers by the minus sign that they wear. Negative numbers in the set

$$\{4, 2, -11, 81, -1, -997\}$$

are easily picked out. Unfortunately, we have taught ourselves a bad theorem!

In algebra, letters are used to stand for numbers, and a letter may be used for a negative number. Suppose that p, q, and r are real numbers. Can you pick out the negative numbers in the set $(p, q, -r)$? Try it and check your choices with the values 2 for p, -7 for q, and -1 for r.

► EXAMPLE 2

Suppose that g, h, j, and k are real numbers with $g > 0$ and $h < 0$. Simplify the following by evaluating the absolute value, *if possible.*

(a) $|g|$ (b) $|h|$ (c) $|j|$ (d) $|-k|$

Solutions

(a) Since $g > 0$, g is positive. So we can write $|g| = g$.

(b) Since $h < 0$, h is negative. So $-h$ is *positive,* and we can write $|h| = -h$.

(c) $|j|$ is *either j or $-j$,* depending on whether j is positive or negative. Without that information, $|j|$ cannot be simplified.

(d) $|-k|$ cannot be evaluated for the same reason as in part (c). However, we can write $|-k| = |k|$.   ◄

In the discussion above, notice that $|k| = |-k|$ for all values of k. Because k and $-k$ are opposites, we see that opposites have the same absolute value.

Since $p - q$ and $q - p$ are opposites, we have

$$|p - q| = |q - p|$$

► **EXAMPLE 3**

Demonstrate that $|p - q| = |q - p|$ is true for the following pairs of numbers.

(a) 14, 21   (b) −8, 26

Solutions

(a) If we let p be 14 and q be 21, then

$$|p - q| = |14 - 21| = |-7| = 7$$
$$|q - p| = |21 - 14| = |7| = 7$$

Thus $|p - q| = |q - p|$.

(b) If p is −8 and q is 26, then

$$|p - q| = |-8 - 26| = |-34| = 34$$
$$|q - p| = |26 - (-8)| = |26 + 8| = |34| = 34$$

So $|p - q| = |q - p|$.   ◄

If p and q are any two points on the number line, the distance between them is given by the value of $|p - q|$.

> **The Distance between Two Numbers on the Number Line**
>
> If p and q are *any* two numbers on the number line, the distance between them is given by
>
> $$|p - q|$$

► **EXAMPLE 4**

Find the distance on the number line between the points corresponding to each pair of numbers.

(a) 14, 21   (b) −8, 26   (c) 12, −38   (d) −37, −16

Solutions

In each case, let the first number be p and the second number be q; then calculate $|p - q|$.

(a) $|14 - 21| = |-7| = 7$
    The distance is 7 units.

(b) $|-8 - 26| = |-34| = 34$
    The distance is 34 units.

(c) $|12 - (-38)| = |12 + 38| = |50| = 50$
    The distance is 50 units.

(d) $|-37 - (-16)| = |-37 + 16| = |-21| = 21$
    The distance is 21 units.   ◄

Some properties of absolute value are very useful.

Some Properties of Absolute Value

If p and q are real numbers with $q \neq 0$, then:

1. $|p \cdot q| = |p| \cdot |q|$
2. $\left|\dfrac{p}{q}\right| = \dfrac{|p|}{|q|}$
3. $|p| = |-p|$

► **EXAMPLE 5**

Use the properties of absolute value to simplify each expression.

(a) $|2x|$　　(b) $\left|\dfrac{-3}{z}\right|$

Solutions

(a) $|2x| = |2| \cdot |x|$　　　Property 1
　　　$= 2|x|$　　　　　　2 is positive

(b) $\left|\dfrac{-3}{z}\right| = \dfrac{|-3|}{|z|}$　　　Property 2

　　　$= \dfrac{3}{|z|}$　　　　　$|-3| = 3$　◄

Problem Set 0.3

Warm-Ups

In Problems 1 through 12, evaluate each expression. See Example 1.

1. $|46|$
2. $|-46|$
3. $|7 - 11|$
4. $|54 - 16|$
5. $|0|$
6. $|7| - |11|$
7. $|54| - |16|$
8. $-|-13|$
9. $-|13|$
10. $|54| - |-16|$
11. $-|32| - |-28|$
12. $-|-17| - |-63|$

In Problems 13 through 20, assume that a, b, c, and d are real numbers with a > 0 and b < 0. Simplify by evaluating the absolute value, if possible. See Example 2.

13. $|a|$
14. $|b|$
15. $|c|$
16. $|d^2|$
17. $|-a|$
18. $|-b|$
19. $|-c|$
20. $|-d|$

In Problems 21 through 26, locate each pair of numbers on a number line and find the distance between each pair of numbers. See Examples 3 and 4.

21. 5 and -3
22. 10 and 2

23. -8 and -6
24. π and $\sqrt{7}$

25. $-\sqrt{5}$ and $-\sqrt{3}$
26. x and y

In Problems 27 through 31, use the properties of absolute value to simplify each expression. See Example 5.

27. $|5y|$ **28.** $|-14z|$ **29.** $-|-53k|$ **30.** $-|2x|$ **31.** $\left|\dfrac{6}{z}\right|$

Practice Exercises

In Problems 32 through 43, evaluate each expression.

32. $|16|$

33. $|-16|$

34. $|22-9|$

35. $|19-22|$

36. $|31-31|$

37. $|22|-|9|$

38. $|19|-|22|$

39. $-|-47|$

40. $-|47|$

41. $|19|-|-22|$

42. $-|19|-|-22|$

43. $-|-27|-|-81|$

In Problems 44 through 50, use the properties of absolute value to simplify each expression.

44. $|7z|$

45. $|-21x|$

46. $|8j|$

47. $-|81g|$

48. $-|7xy|$

49. $\left|\dfrac{u}{9}\right|$

50. $\left|\dfrac{9}{u}\right|$

In Problems 51 through 62, assume that p, q, r, and s are real numbers with $r > 0$ and $s < 0$. Simplify by evaluating the absolute value, if possible.

51. $|p|$

52. $|q|$

53. $|r|$

54. $|-p|$

55. $|s|$

56. $|-q|$

57. $|-r|$

58. $|-s|$

59. $-|r|$

60. $-|s^2|$

61. $-|-r|$

62. $-|-s|$

In Problems 63 through 68, first determine if the statement is true or false; then carefully check the result with -3 for p and 7 for q.

63. If $p < 0$, then $|-p| = p$.

64. If $p < 0$, then $|p| = -p$.

65. If $p < 0$, then $|-p| = -p$.

66. $|p - q| = |q - p|$

67. If $q > 0$, then $|q| = q$.

68. If $q > 0$, then $|-q| = q$.

In Problems 69 through 74, locate each pair of numbers on a number line and find the distance between each pair of points.

69. 7 and -1

70. 14 and 8

71. -5 and -9

72. π and $\sqrt{10}$

73. $-\sqrt{7}$ and $-\sqrt{5}$

74. $z - w$

75. Find the absolute value of the sum of -8 and -5.

76. Find the absolute value of the product of -2 and 7.

77. Find the quotient when -8 is divided by the absolute value of -4.

78. Find the sum of the absolute value of -7 and the absolute value of -19.

79. Add the absolute value of $-\sqrt{2}$ to the absolute value of π.

Challenge Problems

*In Problems 80 through 83, rewrite each expression without absolute values. (*HINT: $\pi \approx 3.14$.*)*

80. $|\pi - 3|$

81. $|3 - \pi|$

82. $|\pi + 3|$

83. $|-\pi - 3|$

In Your Own Words

84. What is the absolute value of a number?

85. Explain why we sometimes have $|x| = -x$.

86. Explain why opposites have the same absolute value.

0.4 Operations with Real Numbers

TAPE 1

The rules for addition, subtraction, multiplication, and division of real numbers are summarized next.

> ### Addition of Signed Numbers
>
> **1.** To add two real numbers with **like signs** (both positive or both negative), add their absolute values and keep the common sign.
>
> **2.** To add two real numbers with **unlike signs** (one positive and one negative), subtract the smaller absolute value from the larger absolute value and keep the sign of the number with the larger absolute value.
>
> ### Subtraction of Signed Numbers
>
> $$p - q = p + (-q)$$
>
> **1.** To subtract two real numbers, rewrite the problem as an addition problem.
>
> **2.** Follow the rules for addition.
>
> ### Multiplication and Division of Signed Numbers
>
> **1.** To multiply or divide two real numbers with **like signs** (both positive or both negative), multiply or divide their absolute values. The sign will be positive.
>
> **2.** To multiply or divide two real numbers with **unlike signs** (one positive and one negative), multiply or divide their absolute values. The sign will be negative.

▶ EXAMPLE **1**

Perform the operation indicated.

(a) $5 + (-7)$ (b) $(-6)(-5)$ (c) $-7 - (-3)$ (d) $(-8) \div 4$

Solutions

(a) $5 + (-7) = -2$ (b) $(-6)(-5) = 30$

(c) $-7 - (-3) = -7 + (+3)$ (d) $(-8) \div 4 = -2$
$\qquad\qquad\quad = -4$

◀

The following products with signed numbers are useful to know.

$$(-p)(-q) = pq$$
$$(-p)q = -pq$$
$$p(-q) = -pq$$
$$-1(p) = -p$$

Fractions are very common in algebra. Operations with fractions are reviewed here. (p, q, r, and s are real numbers.)

Fundamental Principle of Fractions

$$\frac{p}{q} = \frac{p \cdot r}{q \cdot r}; \qquad q \neq 0 \quad \text{and} \quad r \neq 0$$

Multiplication of Fractions

$$\frac{p}{q} \cdot \frac{r}{s} = \frac{p \cdot r}{q \cdot s}; \qquad q \neq 0 \quad \text{and} \quad s \neq 0$$

Division of Fractions

$$\frac{p}{q} \div \frac{r}{s} = \frac{p}{q} \cdot \frac{s}{r} = \frac{p \cdot s}{q \cdot r}; \qquad q \neq 0, \quad r \neq 0, \quad s \neq 0$$

Addition of Fractions

$$\frac{p}{q} + \frac{r}{q} = \frac{p + r}{q}; \qquad q \neq 0$$

Subtraction of Fractions

$$\frac{p}{q} - \frac{r}{q} = \frac{p - r}{q}; \qquad q \neq 0$$

Using zero in a fraction can be troublesome. The following divisions involving 0 are important.

Fractions Containing Zero

$$\frac{0}{p} = 0; \qquad p \text{ is any nonzero real number.}$$

$$\frac{p}{0} \text{ is undefined;} \qquad p \text{ is any real number.}$$

Signs in fractions must be treated carefully. In working with fractions, a fraction often must be written in another form by manipulating signs. The following is very useful in doing this.

Signs in Fractions

$$(q \neq 0)$$

$$\frac{-p}{q} = \frac{p}{-q} = -\frac{p}{q}$$

$$\frac{p}{q} = \frac{-p}{-q} = -\frac{-p}{q} = -\frac{p}{-q}$$

Mathematicians have developed a shorthand for writing products with the same number. This shorthand uses exponents. For example, instead of writing $2 \cdot 2 \cdot 2$, we write 2^3 (read "two cubed").

$$2 \cdot 2 \cdot 2 = 2^3$$

The 3 indicates how many times to use 2 as a factor.

Natural Number Exponent

If p is a real number and n is a natural number, then

$$p^n = \underbrace{p \cdot p \cdot p \cdots p}_{n \text{ factors of } p}$$

Thus

$$p^1 = p$$

$$p^2 = p \cdot p \qquad (p \text{ squared})$$

$$p^5 = p \cdot p \cdot p \cdot p \cdot p$$

In writing p^n, we call p the **base** and n the **exponent.** The exponent tells how many times to use the base as a factor. We read p^n as p to the nth power.

► **EXAMPLE 2**

Find the value of each expression.

(a) 2^4 (b) $(-3)^3$ (c) $2^2 \cdot 3^3$ (d) $(-2)^2(-5)^2$

Solutions

(a) $2^4 = 2 \cdot 2 \cdot 2 \cdot 2$
$\qquad = 16$

(b) $(-3)^3 = (-3)(-3)(-3)$
$\qquad\quad = -27$

(c) $2^2 \cdot 3^3 = 2 \cdot 2 \cdot 3 \cdot 3 \cdot 3$
$\qquad\quad = 4 \cdot 27$
$\qquad\quad = 108$

(d) $(-2)^2(-5)^2 = (-2)(-2)(-5)(-5)$
$\qquad\qquad\quad = 4 \cdot 25$
$\qquad\qquad\quad = 100$

◄

Often, more than one operation is involved in simplifying an expression. For example, $2 + 3 \cdot 4$ has multiplication and addition. We must decide which operation to perform first. Grouping symbols such as parentheses ( ), brackets [ ], or braces { } are used to make the order of operations clear.

If we write $(2 + 3) \cdot 4$, the addition is performed first and then the multiplication. So

$$(2 + 3) \cdot 4 = 5 \cdot 4 = 20$$

However, if we write $2 + (3 \cdot 4)$, the multiplication is performed first and then the addition. So

$$2 + (3 \cdot 4) = 2 + 12 = 14$$

If there are no symbols of grouping, we perform multiplications and divisions from left to right before performing additions and subtractions from left to right. So

$$2 + 3 \cdot 4 = 2 + 12 = 14$$

Fraction bars and absolute value bars are also grouping symbols. For example,

$$\frac{2 - 3}{2 + 4} = \frac{-1}{6} = -\frac{1}{6}$$

This is the same as $(2 - 3) \div (2 + 4)$. The fraction bar can be omitted because of the parentheses.

Absolute value bars are actually grouping symbols that tell where to work first. For example,

$$2|5 - 8| = 2|-3|$$
$$= 2 \cdot 3$$
$$= 6$$

Square root symbols are also grouping symbols. Operations inside the square root symbol must be performed before finding the square root. Consider

$$\sqrt{9 + 16} = \sqrt{25}$$
$$= 5$$

The following rules determine which operation to perform first.

Order of Operations

If grouping symbols are present, perform operations inside them, starting with the innermost symbol, in the following order.

1. Perform any exponentiations.
2. Perform all multiplications and divisions in order from left to right.
3. Perform all additions and subtractions from left to right.

If grouping symbols are not present, perform operations in the order given above.

Be careful to note that multiplication does not have to be done before division, nor addition before subtraction. Multiplications and divisions are performed in order as they occur working from left to right. Additions and subtractions are performed in order as they occur working from left to right.

► EXAMPLE 3

Simplify each expression.

(a) $-3(5 - 7)$ (b) $-5(-6) - (-7)$ (c) $(-5)^2 - \dfrac{8}{(-4)}$

(d) $\dfrac{(-10 - 4)(-2)}{\sqrt{28 - 12} - 11}$ (e) $-\dfrac{2}{3}\left[6(1 - 5) - |2 - 8|\right]$

Solutions

(a) $-3\,(5 - 7) = -3\,(-2)$ Work inside parentheses
$= 6$

(b) $-5(-6) - (-7) = 30 - (-7)$ Multiply first
$= 30 + 7$
$= 37$

(c) $(-5)^2 - \dfrac{8}{(-4)} = 25 - (-2)$ Multiply and divide
$= 25 + 2$
$= 27$

(d) $\dfrac{(-10 - 4)\,(-2)}{\sqrt{28 - 12} - 11} = \dfrac{(-14)\,(-2)}{\sqrt{16} - 11}$
$= \dfrac{28}{4 - 11}$
$= \dfrac{28}{-7}$
$= -4$

(e) $-\dfrac{2}{3}\left[6\,(1 - 5) - |2 - 8|\right] = -\dfrac{2}{3}[6\,(-4) - |-6|\,]$
$= -\dfrac{2}{3}[-24 - 6]$
$= -\dfrac{2}{3}[-30]$
$= 20$ ◄

► EXAMPLE 4

Subtract -8 from the sum of -5 and 2.

Solution

We must add -5 and 2, then subtract -8.

$$(-5 + 2) - (-8) = -3 - (-8) \qquad \text{Work inside parentheses}$$
$$= -3 + 8 \qquad \text{Subtract}$$
$$= 5$$

Calculator Box

Computations with a Calculator

Before starting this material, read "Using Your Calculator" in the Appendix. Calculators have the order of operations built in. That is, they will calculate $8 - 3 \cdot 2$ correctly by pressing the keys in natural order. Press $\boxed{8}$ $\boxed{-}$ $\boxed{3}$ $\boxed{\times}$ $\boxed{2}$ $\boxed{=}$ (or $\boxed{\text{ENTER}}$ or $\boxed{\text{EXE}}$) and the correct result $\boxed{2}$ appears on the screen.

However, parentheses $\boxed{(}$ and $\boxed{)}$ are needed to preserve the order of operations in a calculation like $\dfrac{9 - 5}{3 + 1}$. Since the fraction bar is a grouping symbol, we must hold the numerator and denominator together before division. Press

$\boxed{(}$ $\boxed{9}$ $\boxed{-}$ $\boxed{5}$ $\boxed{)}$ $\boxed{\div}$ $\boxed{(}$ $\boxed{3}$ $\boxed{+}$ $\boxed{1}$ $\boxed{)}$ $\boxed{=}$ (or $\boxed{\text{ENTER}}$ or

$\boxed{\text{EXE}}$) and read $\boxed{1}$ on the display.

Let's approximate $\sqrt{17}$.

Scientific Calculator:

Press $\boxed{17}$ $\boxed{\sqrt{}}$ and read $\boxed{4.123105626}$ on the display.

Graphing Calculator:

Press $\boxed{\sqrt{}}$ $\boxed{17}$ $\boxed{=}$ (or $\boxed{\text{ENTER}}$ or $\boxed{\text{EXE}}$) for the same result.

Rounding to the nearest thousandth, $\sqrt{17} \approx 4.123$. It is important to estimate calculator computations so that we can determine if our answer is reasonable. $\sqrt{17}$ should be a little larger than 4, since $\sqrt{16} = 4$. So our answer is reasonable.

We must be careful when entering a negative number on a calculator. The $\boxed{-}$ key is for *subtraction*. Find a key marked $\boxed{+/-}$, $\boxed{(-)}$, or $\boxed{\text{CHS}}$.

Let's calculate $-29 \cdot 23$.

Scientific Calculator:

Press $\boxed{29}$ $\boxed{+/-}$ $\boxed{\times}$ $\boxed{23}$ $\boxed{=}$ and read $\boxed{-667}$ on the display.

Graphing Calculator:

negation key may look like $\boxed{(-)}$. Press $\boxed{(-)}$ $\boxed{29}$ $\boxed{\times}$ $\boxed{23}$ $\boxed{\text{ENTER}}$

Calculator Exercises

Perform the following computations on a calculator and check the result with pencil and paper.

1. $93 - (6 - 8)$ **2.** $\dfrac{69 - 9}{3 \cdot 4}$ **3.** $7(28 - 11) - 6(234 - 222)$

4. $\dfrac{-17 - 13}{-3(-9 - 11)}$ **5.** $\dfrac{-3(658 - 494)}{-2\sqrt{1808} - 127}$ **6.** $-12\sqrt{\dfrac{-4 \cdot 144}{155 - 219}}$

Approximate the square roots indicated to three decimal places. Be sure to use the ≈ symbol to indicate an approximation. Estimate the answer first.

7. $\sqrt{10}$ **8.** $\sqrt{17}$ **9.** $\sqrt{79}$ **10.** $\sqrt{150}$

Answers

1. 95   **2.** 5   **3.** 47   **4.** -0.5 **5.** 6   **6.** -36

7. Estimate is a little more than 3; $\sqrt{10} \approx 3.162$

8. Estimate is about 4; $\sqrt{17} \approx 4.123$

9. Estimate is about 9; $\sqrt{79} \approx 8.888$

10. Estimate is a little more than 12; $\sqrt{150} \approx 12.247$

Problem Set 0.4

Warm-Ups

In Problems 1 through 24, perform the operation indicated. For Problems 1 through 6, see Example 1.

1. $6 - (-3)$

2. $(-7)(-7)$

3. $\dfrac{-16}{4}$

4. $-9 + (-9)$

5. $\dfrac{16}{0}$

6. $\dfrac{0}{-5}$

For Problems 7 through 18, see Examples 2 and 3.

7. $|-12 + 4| \div 4$

8. $-\dfrac{2}{3}\left(-\dfrac{1}{2} + 3\right)$

9. $\dfrac{1}{6} \div \left(-\dfrac{1}{3}\right)^2$

10. $-2 + 3 \cdot 5$

11. $(2 - 3)^2 \cdot 5$

12. $-(5 - \sqrt{18 - 2})$

13. $15 \div (-3) \cdot \dfrac{-1}{5}$

14. $3|4 - 7| + 2^2(-3)$

15. $\dfrac{2 - 7 + 4}{3(2) - (-1)}$

16. $\dfrac{-6 - 3}{4 - 7}$

17. $5[6 + 2(-5 + 1)]$

18. $3\left(\dfrac{2}{3} - 2\right) \div (-2)$

For Problems 19 through 24, see Example 4.

19. Subtract -7 from the product of -6 squared and $-\dfrac{2}{3}$.

20. Subtract 11 from the sum of -13 and -4.

21. Square the sum of -13 and 4.

22. Find the absolute value of the sum of 6 and -15.

23. Subtract -2 cubed from -2 squared.

24. Find one-half of the sum of $\dfrac{2}{5}$ and $-\dfrac{1}{3}$.

Practice Exercises

In Problems 25 through 55, perform the operation indicated.

25. $7^2 - 8^2$

26. $0 - 3^3$

27. $-7 + (-7)^2$

28. $(2)(-7) - (-3)(-6)$

29. $|-15 - (-3)| \div (-3)$

30. $-\dfrac{3}{8} - \left(\dfrac{-1}{4}\right)^2$

31. $\dfrac{\left(\dfrac{1}{8}\right)}{\left(\dfrac{-1}{4}\right)^3}$

32. $\dfrac{0}{0}$

33. $2^2 - 3 \cdot 5$

34. $2 - \dfrac{2|-1-9|}{(-2)(-3) - (-2)(7)}$

35. $\dfrac{3^2 - 4^2}{(-2)(-1) - 1}$

36. $\dfrac{0}{-1}$

37. $-2[6 - 3^2(\sqrt{5+4})]$

38. $-[\sqrt{50-1} - (3-7)]$

39. $-12 \div (-3)^3 \cdot \dfrac{-1}{3}$

40. $\dfrac{-3 - 5 - (-1)^2}{2[1 - 3(2-5) - 1]}$

41. $(2-7)^2 - 7^2$

42. $\dfrac{2 - 3(4-7)}{3}$

43. $\dfrac{3}{2} - \left(-\dfrac{1}{4} \div 4\right)$

44. $[6 - (-5) - 11] \div |-3 - 3| + 2$

45. $6\left(\dfrac{-2}{3} - 1\right) \div 3$

46. $5[-6 - 2(5-14)]$

47. $\dfrac{2\left[\dfrac{-1 - (-7)}{4 \cdot 2 - 10}\right] - 8 \cdot 3}{-(-5) - 7}$

48. $\dfrac{-3[2(3-7) - (4-9)]}{3^3 - 2 \cdot 7}$

49. $\dfrac{-4|-9-2| + 8 \cdot 5}{2\left(\dfrac{-5-9}{5+2}\right) - \dfrac{6 - 2 \cdot 3}{0 - 2}}$

50. $\dfrac{-\dfrac{1}{5}[25 - (5-15)]}{-7 - 7}$

51. Multiply $\dfrac{3}{4}$ by the sum of -24 and 16.

52. Find the reciprocal of the product of $\dfrac{5}{14}$ and $\dfrac{2}{15}$.

53. Divide the square root of the sum of -12 and 16 by -2.

54. Subtract the product of $\dfrac{1}{3}$ and -63 from the product of $-\dfrac{3}{7}$ and -21.

55. Subtract the sum of 5 and -7 from the absolute value of -11.

Challenge Problems

In Problems 56 through 58, perform the operation indicated.

56. $\dfrac{\dfrac{1}{2} - \dfrac{1}{3}}{\dfrac{1}{5}}$

57. $\dfrac{\dfrac{2}{3} - \dfrac{3}{7}}{\dfrac{1}{7} + \dfrac{1}{3}}$

58. $\dfrac{1 - \dfrac{1}{5}}{\dfrac{2}{3} - 2}$

In Your Own Words

59. State the order of operations.

0.5 Angles, Lines, and Plane Figures

TAPE 2

This section is a review of some basic ideas and terminology from geometry. Such fundamental ideas as point, line, angle, and length are left undefined. No attempt is made to use precise notation. Angles with the same measure are said to be *equal*. The symbol $\angle$ indicates the measure of an angle. Lowercase letters are used to designate the length of line segments.

Angles

- An **acute angle** is an angle of *less than* 90°.
- A **right angle** is an angle of *exactly* 90°. The symbol ⌐ is used to indicate that an angle in a figure is a right angle.
- An **obtuse angle** is an angle of *greater than* 90° but *less than* 180°.
- A **straight angle** is an angle of *exactly* 180°. Its sides form a line.
- **Complementary angles** are *two* angles whose sum is 90°. Each is called the **complement** of the other.
- **Supplementary angles** are *two* angles whose sum is 180°. Each is called the **supplement** of the other.

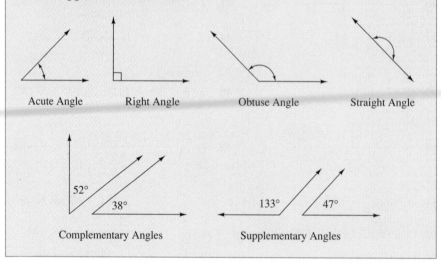

Acute Angle   Right Angle   Obtuse Angle   Straight Angle

Complementary Angles   Supplementary Angles

Two intersecting lines form the angles numbered 1, 2, 3, and 4.

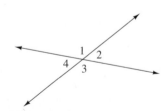

Angles 1 and 3 are **vertical angles,** as are angles 2 and 4. *Vertical angles are equal.* Angles that have a common vertex and a common side are called

adjacent angles. Angles 1 and 2 are adjacent angles, as are angles 2 and 3, angles 3 and 4, and angles 1 and 4.

► **EXAMPLE 1**

Are the following angles complementary or supplementary?

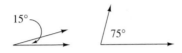

Solution

The sum of the measure of the two angles is 90°. The angles are complementary.   ◄

Lines

Two lines that lie in the same plane and do not intersect are called **parallel lines.**

Two lines that intersect at right angles are called **perpendicular.**

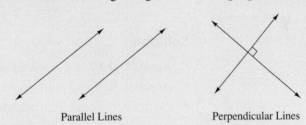

Parallel Lines   Perpendicular Lines

A **transversal** is a line that intersects two or more lines at different points. A transversal of two lines forms some angles that are given special names.

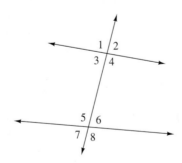

- Interior angles: 3, 4, 5, and 6
- Exterior angles: 1, 2, 7, and 8
- Corresponding angles: 1 and 5, 2 and 6, 3 and 7, 4 and 8
- Alternate interior angles: 3 and 6, 4 and 5
- Alternate exterior angles: 1 and 8, 2 and 7

> ## Parallel Lines Cut by a Transversal
>
> If two lines are cut by a transversal, the lines are parallel if any of the following statements are true:
>
> **1.** *Corresponding angles* are equal.
> **2.** *Alternate interior angles* are equal.
> **3.** *Alternate exterior angles* are equal.
>
> If two lines are parallel, all the statements are true.

► **EXAMPLE 2**

Determine the measure of each angle in the figure. Lines l_1 and l_2 are parallel. $\angle 1 = 120°$.

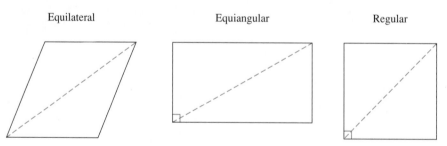

Solution

- Angles 1 and 2 are supplementary angles. So $\angle 2$ is 60°.
- Angles 1, 4, 5 and 8 are equal. So each is 120°.
- Angles 2, 3, 6 and 7 are equal. Thus each is 60°.

A **polygon** is a closed plane figure bounded by line segments. Triangles, quadrilaterals, pentagons, hexagons, and octagons are polygons with 3, 4, 5, 6 and 8 sides, respectively. If all the sides are of equal length, a polygon is called **equilateral.** If all the angles are equal, it is said to be **equiangular.** A **regular** polygon is one that is both equilateral and equiangular. Several polygons are shown next. The dashed lines are called **diagonals.**

Equilateral	Equiangular	Regular

A **triangle** is a polygon with three sides.

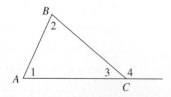

The three angles 1, 2, and 3 are the **interior angles** of the triangle, or simply the *angles* of the triangle. The three sides of a triangle and the three angles are sometimes called the **parts** of a triangle. Angle 4 is called an **exterior angle** of the triangle.

Two important relationships exist for these angles. The sum of the measures of the interior angles is 180°. The measure of an exterior angle is equal to the sum of the measures of the opposite interior angles.

$$\angle 1 + \angle 2 + \angle 3 = 180°$$

$$\angle 4 = \angle 1 + \angle 2$$

▶ **EXAMPLE 3**

Determine $\angle A$ in each figure.

(a)

30°

(b)

A

115°   20°

Solutions

(a) Since this is a right triangle, $\angle A$ must be 60°.   (b) $\angle A + 20° = 115°$

$\angle A = 95°$ ◀

Triangles

- An **isosceles triangle** has two sides of equal length. The angles opposite the equal sides are equal.
- An **equilateral triangle** has all three sides equal. It is also equiangular.
- An **acute triangle** has three acute angles.
- An **obtuse triangle** has an obtuse angle.
- A **right triangle** contains a right angle. The longest side of a right triangle (the side opposite the right angle) is called the **hypotenuse.** The other two sides are called the **legs** of a right triangle.

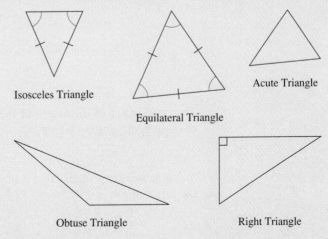

Isosceles Triangle

Equilateral Triangle

Acute Triangle

Obtuse Triangle

Right Triangle

The following result is one of the most useful relationships in all mathematics.

Pythagorean Theorem

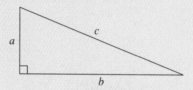

In the right triangle above, the side of length c units is the *hypotenuse* and the sides of lengths a and b units are the *legs*. The square of the length of the hypotenuse equals the sum of the squares of the lengths of the legs. That is,

$$c^2 = a^2 + b^2$$

▶ **EXAMPLE 4**

Verify the Pythagorean Theorem for a right triangle with sides of 5, 12, and 13 inches.

Solution

The hypotenuse is the longest side. So it must be 13.

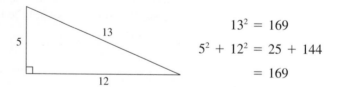

$$13^2 = 169$$
$$5^2 + 12^2 = 25 + 144$$
$$= 169$$

Thus the square of the length of the hypotenuse equals the sum of the squares of the lengths of the two legs.   ◀

A geometric property that is of great practical use involves triangles that are the same shape.

Similar Triangles

Two triangles are similar if any one of the following statements is true.

1. Two angles of one triangle equal two angles of the other.

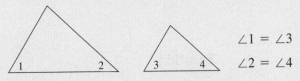

$\angle 1 = \angle 3$

$\angle 2 = \angle 4$

2. Corresponding sides are proportional.

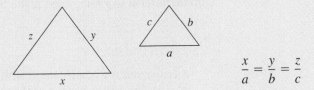

$$\frac{x}{a} = \frac{y}{b} = \frac{z}{c}$$

3. Two corresponding pairs of sides are proportional and the angle between them is equal.

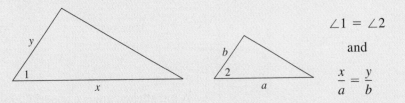

$\angle 1 = \angle 2$

and

$$\frac{x}{a} = \frac{y}{b}$$

If two triangles are similar, all these statements are true.

► EXAMPLE 5

Are the triangles in the figure similar?

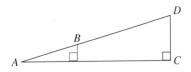

Solution

As we examine the figure we notice a large triangle and a small triangle. Both are right triangles and they share the angle at the vertex *A*. This makes the angles at *B* and *D* equal. Thus, the triangles have two equal angles. Thus, by **1.** in the Similar Triangles box, they are similar triangles.   ◄

► **EXAMPLE 6**

The given triangles are similar. Set up proportions for corresponding sides.

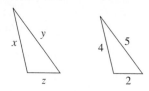

Solution

$$\frac{x}{4} = \frac{y}{5} = \frac{z}{2}$$

◄

A **quadrilateral** is a four-sided plane figure.

Quadrilaterals

* A **trapezoid** is a quadrilateral with exactly two sides parallel. The parallel sides are called **bases** and the other sides are called **legs.** A trapezoid with equal legs is called an **isosceles trapezoid.**

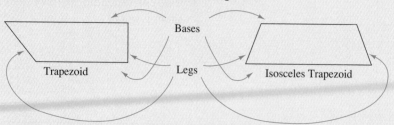

* A **parallelogram** is a quadrilateral in which *both* pairs of opposite sides are parallel. The diagonals of a parallelogram bisect each other, and any two consecutive angles are supplements of each other.
* A **rhombus** is a parallelogram with equal sides.
* A **rectangle** is a parallelogram whose angles are right angles.
* A **square** is a rectangle with equal sides.

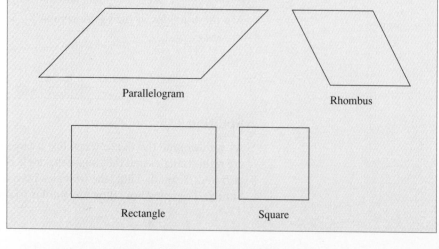

Circles

- A **circle** is the set of all points in a plane that are the same distance from a given point called the **center.**
- A **radius** of a circle is a line segment connecting the center with a point on the circle.
- A **chord** is a line segment connecting two points on the circle.
- A **diameter** is a chord that contains the center.
- A **tangent** to a circle is a line in the plane of the circle that intersects the circle in exactly one point.

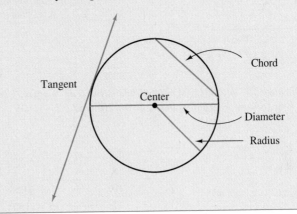

Problem Set 0.5

Warm-Ups

1. Find the measure of each angle in the figure. Lines l_1 and l_2 are parallel. Lines l_3 and l_4 are perpendicular. $\angle d = 45°$. See Examples 1 and 2.

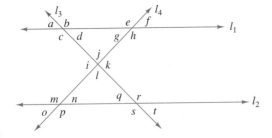

For Problems 2 through 5, find $\angle A$. See Example 3.

2.

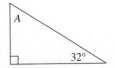

3.

4.

5.

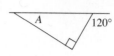

6. Verify the Pythagorean Theorem for a right triangle with sides of lengths 3, 4, and 5 units. See Example 4.

7. Why is each pair of triangles similar? See Example 5.

(a)

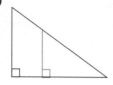

(b) $l_1 \parallel l_2$

Practice Exercises

In Problems 8 through 19, determine if each statement is true or false. Use the accompanying figure. Lines l_1 and l_2 are parallel. $\angle c = 120°$ and $\angle h = 105°$.

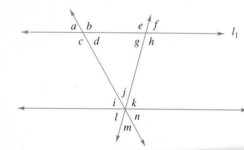

8. $\angle i = 75°$ **9.** $\angle j + \angle n = 180°$

10. $\angle g + \angle j = 105°$ **11.** $\angle g = 45°$

12. $\angle m + \angle n = 135°$ **13.** Angles h and k are supplementary.

14. $\angle l = 45°$ **15.** Angle b is an acute angle.

16. Angle f is obtuse.   **17.** Angles k and n are complementary.

18. $\angle b = 105°$ **19.** $\angle m + \angle n = 105°$

In Problems 20 through 25, find $\angle A$.

20.

21.

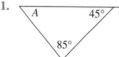

22.

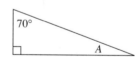

23.

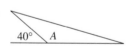

24.

25.

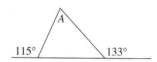

26. Find the measure of each angle in the figure. Lines l_1 and l_2 are parallel. $\angle d = 35°$. $\angle k = 130°$.

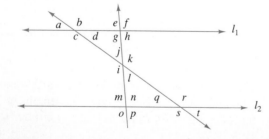

27. Verify the Pythagorean Theorem for a right triangle with sides of lengths 6, 8, and 10 units.

28. Why is the following pair of triangles similar?

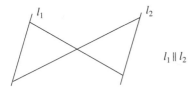

$l_1 \parallel l_2$

29. Why is the following pair of triangles similar?

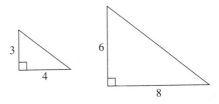

In Problems 30 through 40, determine whether each statement is true or false.

30. A rectangle is equiangular.

31. An equilateral triangle is equiangular.

32. An isosceles triangle is an acute triangle.

33. A parallelogram is a quadrilateral.

34. A square is a rhombus.

35. A pentagon is a polygon with 7 sides.

36. A square is a regular polygon.

37. A triangle is a polygon.

38. A right triangle is an acute triangle.

39. The acute angles in a right triangle are complementary.

40. A rhombus is an equilateral polygon.

Challenge Problems

41. The sum of the measures of the interior angles of a triangle is 180°. Figure out a formula for the sum of the measures of the interior angles of a polygon with n sides. (HINT: Divide a polygon into triangles.)

Quadrilateral (4 sides)   Pentagon (5 sides)

2 Triangles   3 Triangles

In Your Own Words

42. What are similar triangles?

43. What does the Pythagorean Theorem say?

0.6   Perimeter, Surface Area, Area, and Volume

TAPE 2

This section contains formulas for the perimeter, area, surface area, and volume of geometric figures. The **perimeter** of a polygon is the sum of the lengths of its sides. The perimeter of a circle is called the **circumference** of the circle.

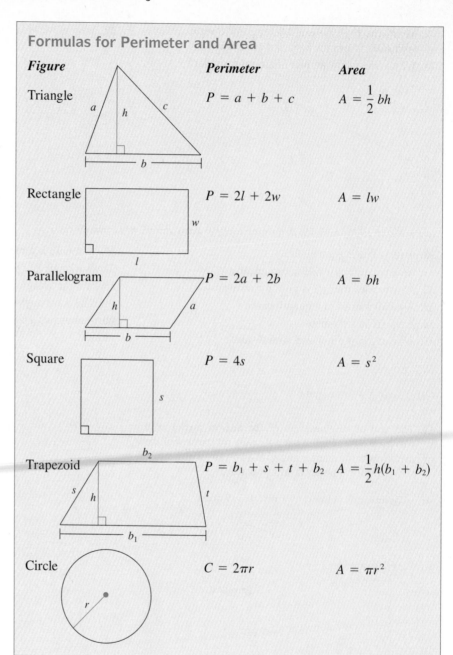

Formulas for Perimeter and Area

Figure	Perimeter	Area
Triangle	$P = a + b + c$	$A = \frac{1}{2} bh$
Rectangle	$P = 2l + 2w$	$A = lw$
Parallelogram	$P = 2a + 2b$	$A = bh$
Square	$P = 4s$	$A = s^2$
Trapezoid	$P = b_1 + s + t + b_2$	$A = \frac{1}{2} h(b_1 + b_2)$
Circle	$C = 2\pi r$	$A = \pi r^2$

► **EXAMPLE 1**

Find the perimeter of each figure.

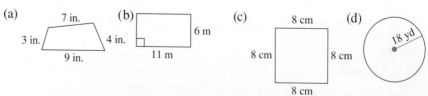

(a) 7 in. 3 in. 9 in.

(b) 6 m 4 in. 11 m

(c) 8 cm 8 cm 8 cm 8 cm

(d) 18 yd

Solutions

(a) A trapezoid is a quadrilateral, so its perimeter is given by

$$3 + 7 + 4 + 9 = 23$$

The perimeter is 23 inches (in.).

(b) The perimeter of a rectangle is twice the width plus twice the length.

$$P = 2 \cdot 6 + 2 \cdot 11$$
$$= 34$$

The perimeter is 34 meters (m).

(c) The perimeter of a square is four times the length of a side.

$$P = 4 \cdot 8$$
$$= 32$$

The perimeter is 32 centimeters (cm).

(d) The circumference of a circle is 2π times the radius.

$$C = 2 \cdot \pi \cdot 18$$
$$= 36\pi$$

The circumference is 36π yards (yd).

► EXAMPLE 2

Find the area of the following figures.

(a) A parallelogram with base 7 in. and height 21 in.
(b) A 3-foot (ft) by 8-ft rectangle
(c) A right triangle with hypotenuse of 5 cm and legs of 3 cm and 4 cm
(d) A circle of diameter 8 ft 4 in.

Solutions

(a) The formula for the area of a parallelogram is

$$A = bh$$
$$= 7 \cdot 21$$
$$= 147$$

The area is 147 square inches (in.^{2}).

(b) The area of a rectangle is

$$A = lw$$
$$= 3 \cdot 8$$
$$= 24$$

The area is 24 ft^2.

(c) In a right triangle, if the base is one leg, the height is the other leg. Therefore,

$$A = \frac{1}{2}\,bh$$

$$= \frac{1}{2} \cdot 3 \cdot 4$$

$$= 6$$

The area is 6 cm².

(d) To find the area of a circle of diameter 8 ft, 4 in., we first must write the diameter in one unit of measure. Since 8 ft is $8 \cdot 12 = 96$ in., the diameter of the circle is $96 + 4 = 100$ in. Thus the radius is 50 in. The area is given by

$$A = \pi r^2$$

$$= \pi (50)^2$$

$$= 2500\pi$$

The area is 2500π in.².

Formulas for Volume and Surface Area

Figure	*Volume*	*Surface Area*
Rectangular solid	$V = lwh$	$SA = 2wh + 2hl + 2wl$

Cube	$V = s^3$	$SA = 6s^2$

Right circular cylinder	$V = \pi r^2 h$	$SA = 2\pi r^2 + 2\pi rh$

Right circular cone	$V = \frac{1}{3}\pi r^2 h$	$SA = \pi r^2 + \pi r \sqrt{r^2 + h^2}$

Sphere	$V = \frac{4}{3}\pi r^3$	$SA = 4\pi^2$

► **EXAMPLE 3**

How many cubic meters of water will it take to fill a swimming pool 50 m long, 20 m wide, and 2 m deep?

Solution

Think of the pool as a rectangular solid.

$$V = lwh$$
$$= 50 \cdot 20 \cdot 2$$
$$= 2000$$

It will take 2000 cubic meters (m^3) of water to fill the pool.   ◄

► **EXAMPLE 4**

Approximate the number of cubic feet of silage that can be stored in a silo 80 ft high with radius of 18 ft?

Solution

If we assume that the silo is a right circular cylinder, we have

$$V = \pi r^2 h$$
$$= \pi(18)^2(80)$$

The silo can store approximately 79,451 cubic feet of silage.   ◄

► **EXAMPLE 5**

Approximate to the nearest tenth of a square inch the surface area of a can of cola that is 5 in. high with a radius of $1\frac{1}{8}$ in.

Solution

We assume that the can is a right circular cylinder.

$$SA = 2\pi r^2 + 2\pi rh$$
$$= 2\pi\left(\frac{9}{8}\right)^2 + 2\pi\left(\frac{9}{8}\right)(5)$$
$$= 2\pi \cdot \frac{81}{64} + 2\pi \cdot \frac{45}{8}$$
$$= \frac{81\pi}{32} + \frac{45\pi}{4}$$
$$= \frac{81\pi}{32} + \frac{360\pi}{32}$$
$$= \frac{441\pi}{32}$$

The surface area is approximately 43.3 square inches.   ◄

Problem Set 0.6

Warm-Ups

In Problems 1 through 6, find the perimeter of each plane figure. See Example 1.

1. A triangle with sides of 3, 4, and 5 in.

2. A trapezoid with sides of 17, 6, 19, and 5 m

3. A parallelogram with one side 13 cm and one side 4 cm

4. A rhombus with one side of 7 ft

5. A rectangle with length of 20 yd and width of 17 yd

6. A circle of radius 14 mm

7. Cecil Fielder hits a home run. How many feet must he run? (The distance from home plate to first base is 90 ft. Assume that his path around the bases forms a square.)

8. How much fencing must Tom Brown buy to enclose his backyard, pictured below?

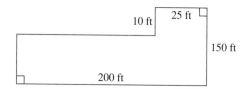

For Problems 9 through 12, see Example 2. In Problems 9 through 11, find the area of the shaded part of each figure.

9.

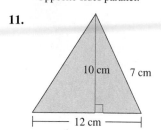

Opposite sides parallel.

10.

Quadrilateral is a square.

11.

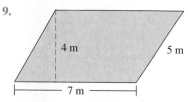

12. Find the area of a circular rug with a diameter of 8 ft, 4 in.

For Problems 13 through 18, find each volume. See Examples 3 and 4.

13. Find the volume of a cube with 5 units on a side.

14. Approximate (to the nearest hundreth) the volume of a basketball whose radius is $5\frac{1}{2}$ in.

15. How much water does it take to fill a rectangular swimming pool that measures 35 ft long, 25 ft wide, and 6 ft deep?

16. Approximately how many mm³ of cork is in a bottle cork that measures 20 mm in diameter and 45 mm in length? (Assume that the bottle cork is a right circular cylinder.)

Problem 14

17. Approximate the volume of an ice cream cone of height 3.6 in. and a base radius of 1.1 in.

18. Approximate the amount of Coca-Cola in a can 5 in. high with a radius of $1\frac{1}{8}$ in.

For Problems 19 through 22, find each surface area. See Example 5.

19. Approximate the surface area of the earth. Use 4000 miles (mi) as the radius. Approximate the answer to the nearest square mile.

20. An oil tank is a right circular cylinder with a radius of 40 ft and height of 100 ft. Approximate its surface area to the nearest square foot.

21. A box has dimensions of 14 in. by $6\frac{1}{2}$ in. by $3\frac{1}{2}$ in. Find its surface area.

22. Approximate the number of square inches of material needed to make 5000 Dole pineapple juice cans if each can is 5 in. tall and 3 in. in diameter.

Practice Exercises

In Problems 23 through 28, find the perimeter of each plane figure.

23. An equilateral triangle with sides of 2.5 in.

24. A trapezoid with sides of $17\frac{1}{2}$ m, $6\frac{1}{3}$ m, $19\frac{1}{6}$ m, and 5 m

25. A parallelogram with one side of 23 ft and one side of 8 ft

26. A rhombus with one side $6\frac{2}{3}$ ft long

27. A rectangle with length of 20.3 yd and width of 17.23 yd

28. A square with one side $5\frac{1}{4}$ ft long

29. John wants to walk around the equator. Approximately how many miles would he walk? (The diameter of the earth is 8000 miles.)

Problems 29 and 39

In Problems 30 through 32, find the area of each figure.

30.

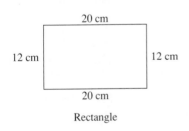

20 cm

12 cm   12 cm

20 cm

Rectangle

31.

3 m

Semicircle

32.

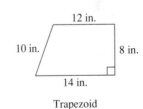

12 in.

10 in.   8 in.

14 in.

Trapezoid

33. Find the volume of a cube $2\frac{1}{3}$ units on a side.

34. Approximate the amount of air in a spherical balloon of radius 18 cm?

35. How much Clorox 2 does it take to fill a rectangular box that measures $5\frac{1}{2}$ in. tall, $3\frac{1}{2}$ in. wide, and $1\frac{1}{2}$ in. deep.

36. Approximate the amount of detergent in a cylindrical bucket that measures 10 in. in diameter and is 1 ft tall.

37. Approximate the volume of a conical sand pile of height 5 yd and a base radius of 12 yd.

38. Approximately how much trash can be put in a cylindrical garbage can that is 1 yd high and has a radius of 2 ft?

39. Approximate the volume of the earth. Use 4000 miles as the radius.

40. A water tank is a right circular cylinder with a radius of 25 ft and height of 30 ft. Approximate its surface area.

41. Suzanne is making a box from a rectangular piece of cardboard that is 3 ft by $2\frac{1}{2}$ ft by cutting out a 6-in. square from each corner. How many cubic feet of crystals will the box hold?

42. A tool box has dimensions of 3 ft by 18 in. by 16 in. Find its surface area.

43. Approximate the number of square inches of material needed to make 10,000 Campbell soup cans if each can is 4 in. tall and $2\frac{1}{2}$ in. in diameter.

44. A tennis ball has a radius of $1\frac{1}{4}$ in. If there are three tennis balls in a can $8\frac{1}{2}$ in. tall with a radius of $1\frac{1}{2}$ in., how much space is not used by the tennis balls?

45. JoAnn is making a box from a piece of cardboard that is a 3-ft square by cutting out a 6-in. square from each corner. How many cubic feet of glitter will the box hold?

46. A marble of radius $\frac{1}{2}$ in. is placed inside a cube with an edge of length 1 in. Find the volume of the part of the cube that is outside the marble.

Challenge Problems

47. Find the surface area of the trough shown at the right.

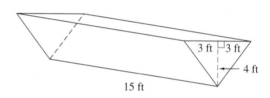

In Your Own Words

48. Make up a problem to find the volume of a sphere.

49. Make up a problem to find the surface area of a cube.

► Chapter Summary

GLOSSARY

Set   A collection of objects

Element or **member** of a set   The objects in the set. The symbol $\in$ means "is an element of."

Empty set or **null set**   A set with no elements, written as { } or $\emptyset$.

Subset of a set A A set with the property that each of its elements is in set A. A is a subset of B is written as $A \subseteq B$.

The **union** of sets A and B is the set of all members that belong to either A or B and is written $A \cup B$.

The **intersection** of sets A and B is the set of all elements that belong to both A and B and is written $A \cap B$.

Natural numbers or **counting numbers**   {1, 2, 3, . . .}

Whole numbers   {0, 1, 2, 3, . . .}

Integers   {. . . , -3, -2, -1, 0, 1, 2, 3, . . .}

Rational numbers $\left\{\dfrac{p}{q}, \text{ where } p \text{ and } q \text{ are integers and } q \neq 0\right\}$

Irrational numbers   {Numbers on a number line that are not rational}

Real numbers   {Rational and irrational numbers}

Additive identity   The number 0 is the additive identity.

Multiplicative identity   The number 1 is the multiplicative identity.

p is **less than** q The number p lies to the left of the number q on a number line. Write this as $p < q$.

r is **greater than** s The number r lies to the right of the number s on a number line. Write this as $r > s$.

$p \leq q$ The number p is less than or equal to the number q.

$r \geq s$ The number r is greater than or equal to the number s.

Reciprocal or **multiplicative inverse** of p The number that, when multiplied by p, gives a product of 1.

Opposite or **additive inverse** of p The number that, when added to p, gives a sum of 0.

Absolute value   of a real number is its distance from zero on the number line. We define the absolute value of a number as

$$|k| = \begin{cases} k, & \text{if } k \geq 0 \\ -k, & \text{if } k < 0 \end{cases}$$

Base of x^n The base is x.

Exponent of x^n The exponent is n. The exponent tells how many times to use the base as a factor.

Acute angle   An angle with measure less than $90°$.

Right angle   An angle with measure of $90°$.

Obtuse angle   An angle with measure more than $90°$ but less than $180°$.

Straight angle   An angle with measure of $180°$.

Complementary angles   Two angles whose measures total $90°$.

Supplementary angles   Two angles whose measures total $180°$.

Parallel lines   Two lines in the same plane that do not intersect.

Perpendicular lines   Two lines that intersect at right angles.

Polygon   A closed figure whose sides are line segments.

Similar triangles   Triangles that are the same shape.

Isosceles triangle   A triangle with two sides of equal length.

Equilateral triangle   A triangle with three sides of equal length.

Acute triangle   A triangle with three acute angles.

Obtuse triangle   A triangle with one obtuse angle.

Quadrilateral   A polygon with four sides.

Trapezoid   A quadrilateral with two sides parallel.

Isosceles trapezoid   A trapezoid with equal legs.

Parallelogram   A quadrilateral with both pairs of sides parallel.

Rhombus   A parallelogram with equal sides.

Rectangle   A parallelogram with right angles.

Square   A rectangle with equal sides.

Properties of Real Numbers

Properties of Addition

1. $p + q = q + p$		Commutative
2. $p + (q + r) = (p + q) + r$		Associative
3. $p + 0 = 0 + p = p$		Identity
4. $p + (-p) = 0$		Inverse

Properties of Multiplication

1. $pq = qp$		Commutative
2. $p(qr) = (pq)r$		Associative
3. $p \cdot 1 = 1 \cdot p = p$		Identity
4. $p \cdot \dfrac{1}{p} = 1; p \neq 0$		Inverse

Distributive Property

$p(q + r) = pq + pr$

Operations with Real Numbers

Addition

1. If the two numbers have the same sign, add the absolute values of the numbers and keep the common sign.

2. If the two numbers have different signs, subtract the absolute value of the smaller number from the absolute value of the larger number and keep the sign of the number with the larger absolute value.

Subtraction

1. Change the subtraction to addition.

2. Follow the rules for addition.

Multiplication and Division

1. If the signs of the two numbers are alike, multiply or divide the absolute values of the numbers. The answer will be positive.

2. If the signs of the two numbers are not alike, multiply or divide the absolute value of the numbers. The answer will be negative.

Order of Operations   If grouping symbols are present, perform operations inside them, starting with the innermost symbol, in the following order.

1. Perform any exponentiations.

2. Perform all multiplications and divisions in order from left to right.

3. Perform all additions and subtractions from left to right.

If grouping symbols are not present, perform operations in the order given above. (Fraction bars and absolute value bars act as grouping symbols.)

Properties of
Absolute Value   If p and q are real numbers, with $q \neq 0$:

1. $|p \cdot q| = |p||q|$

2. $\left|\dfrac{p}{q}\right| = \dfrac{|p|}{|q|}$

3. $|p| = |-p|$

Distance between
Two Real Numbers,
p and q $|p - q| = |q - p|$

Parallel Lines Cut by
a Transversal   If two lines are cut by a transversal, the lines are parallel if any of the following statements are true.

1. *Corresponding angles* are equal.

2. *Alternate interior angles* are equal.

3. *Alternate exterior angles* are equal.

If two lines are parallel, all the statements are true.

Pythagorean
Theorem

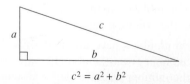

$$c^2 = a^2 + b^2$$

Perimeter and Area Formulas

Figure	Perimeter	Area
Triangle	$P = a + b + c$	$A = \frac{1}{2}bh$
Rectangle	$P = 2l + 2w$	$A = lw$
Parallelogram	$P = 2a + 2b$	$A = bh$
Square	$P = 4s$	$A = s^2$
Trapezoid	$P = b_1 + s + t + b_2$	$A = \frac{1}{2}h(b_1 + b_2)$
Circle	$C = 2\pi r$	$A = \pi r^2$

Volume and Surface Area Formulas

Figure	Volume	Surface Area
Rectangular solid	$V = lwh$	$SA = 2wh + 2hl + 2wl$
Cube	$V = s^3$	$SA = 6s^2$
Right circular cylinder	$V = \pi r^2 h$	$SA = 2\pi r^2 + 2\pi rh$
Right circular cone	$V = \frac{1}{3}\pi r^2 h$	$SA = \pi r^2 + \pi r \sqrt{r^2 + h^2}$
Sphere	$V = \frac{4}{3}\pi r^3$	$SA = 4\pi r^2$

CHECKUPS

1. Classify each number as rational or irrational.
   (a) $\frac{3}{7}$
   (b) $0.\overline{78}$
   (c) $\sqrt{7}$
   (d) 1.7 Section 0.1; Example 4

2. Indicate the property of real numbers that justifies each statement.
   (a) $2 + 3 = 3 + 2$
   (b) $1(3) = 3$
   (c) $-5 + 0 = -5$
   (d) $(4 \cdot 3) \cdot 6 = 4 \cdot (3 \cdot 6)$ Section 0.2; Example 3

3. Use the properties of absolute value to simplify.
   (a) $|2x|$
   (b) $\left| \dfrac{-3}{z} \right|$ Section 0.3; Example 5

4. Simplify $\dfrac{(-10 - 4)(-2)}{\sqrt{28 - 12} - 11}$.   Section 0.4; Example 3d

5. Find $\angle A$ in the figure.   Section 0.5; Example 3b

6. Find the perimeter of the figure.   Section 0.6; Example 1d

7. Find the area of a right triangle with hypotenuse of 5 cm and legs of 3 cm and 4 cm.   Section 0.6; Example 2c

8. Approximate the number of cubic feet of silage that can be stored in a silo 80 ft high with a radius of 18 ft.   Section 0.6; Example 4

9. Approximate to the nearest tenth of a square inch the surface area of a can of cola that is 5 in. high with a radius of $1\frac{1}{8}$ in.   Section 0.6; Example 5

► Review Problems

In Problems 1 through 14, perform the operation indicated.

1. $-(-7)[2 - (5 - 8)]$

2. $\frac{3}{7} \div \left(\frac{-9}{14} + \frac{1}{7} \right)$

3. $\frac{-6(-7)}{(-2 - 19) - \sqrt{30} - 5}$

4. $(-2 - 8) \cdot (-3)^2$

5. $\frac{(-3)(-4) - (-4)(-5)}{-5 - 2 + (-1)}$

6. $\frac{5}{0}$

7. $-|6 + (-11)| - 6^2$

8. $\frac{-2}{9}\left[-\frac{1}{2} \div \frac{1}{4} - 2(5 - 15) \right]$

9. $|(-4)(5)| - |6 - 19|$

10. $-7 + (-6) - (-4)$

11. $(6 \div 8)^2 + \frac{9}{3}$

12. $3 \cdot \frac{3 - 8}{2^2 - 3^2}$

13. $(-9)^2 - [17 - (4 - 12)]$

14. $\frac{\frac{2 - 3(5 - 9)}{(-6)}}{3 - (-6)}$

In Problems 15 through 20, indicate the property that justifies each statement.

15. $2 + 7 = 7 + 2$

16. $(3 \cdot 4) \cdot 4 = 3 \cdot (4 \cdot 4)$

17. $6(7 + 5) = 6 \cdot 7 + 6 \cdot 5$

18. $5 + (-5) = 0$

19. $9 \cdot 1 = 9$

20. $4 \cdot \frac{1}{4} = 1$

In Problems 21 through 25, classify each number as rational or irrational.

21. $\sqrt{11}$

22. $\frac{3}{8}$

23. 7.68

24. $\sqrt{14}$

25. $3.3\overline{7}$

In Problems 26 through 28, $A = \{a, b, c\}$ and $B = \{c, d, e, f\}$.

26. List the elements of $A \cup B$.

27. List the elements of $A \cap B$.

28. List the subsets of A.

29. If $C = \{x \mid x \text{ is an odd natural number less than 7}\}$, list the elements of C.

30. If $D = \{3, 6, 9, 12, 15\}$, write D using set-builder notation.

In Problems 31 through 35, evaluate each expression.

31. $|-5| - |25|$

32. $|3x^2y|$

33. $-|-25|$

34. $-|6p^2|$

35. $-|-4| - |-21|$

In Problems 36 through 40, assume that a, b, and c are real numbers with a > 0 and b < 0.

36. $|a|$

37. $|b|$

38. $|c|$

39. $|-a|$

40. $|-b|$

In Problems 41 through 44, find the perimeter and area of each figure.

41.
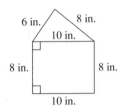

42.
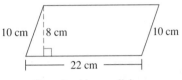
Opposite sides parallel.

43.

44.

In Problems 45 through 47, find the volume and surface area of each figure.

45.

Cone

46.
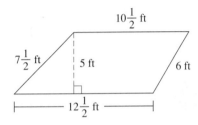
Cylinder

47.

48. Find the measure of each angle. l_1 and l_2 are parallel.
$\angle b = 65°.$
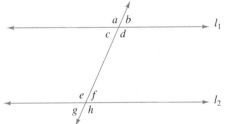

In Problems 49 through 50, indicate why the pair of triangles are similar.

49. A pair of equilateral triangles

50. A pair of right triangles with an acute angle of $33°$

► Chapter Test

In Problems 1 through 9, choose the correct answer.

1. If A is a real number and $A < 0$, which of the following is true?
   A. $|A| = A$
   B. $|A| = -A$
   C. $|A - B| = |A + B|$
   D. All are true

2. The set of irrational numbers is a subset of (?)
   A. the set of rational numbers
   B. the set of integers
   C. the set of natural numbers
   D. the set of real numbers

3. The rational number $\frac{1}{6}$ expressed as a decimal is (?)
   A. 0.16 **B.** $0.1\overline{6}$ **C.** $0.\overline{16}$ **D.** 0.2

4. If $A = \{1, 2, 5\}$, which statement is true?
   A. $1 \notin A$ **B.** $1 \subseteq A$ **C.** $\emptyset \subseteq A$ **D.** $a \in A$

5. The statement $5 + 4 = 4 + 5$ illustrates which property?
   A. Distributive   **C.** Commutative
   B. Associative   **D.** Closure

6. The distance between -5 and p on the number line is (?)
   A. $p - 5$ **B.** $p + 5$ **C.** $|p - 5|$
   D. $|p + 5|$

7. The opposite of $2 - x$ is (?)
   A. $2 + x$
   B. $-2 - x$
   C. $x - 2$
   D. None of the above

8. If p is a negative real number, which statement is true?
   A. p lies to the right of 0 on a number line.
   B. $-p$ lies to the left of 0 on a number line.
   C. $-p$ lies to the right of 0 on a number line.
   D. None of the above

9. The reciprocal of $\frac{3}{8}$ is (?)
   A. $\frac{3}{8}$ **B.** $\frac{8}{3}$ **C.** 1 **D.** Does not exist

In Problems 10 through 16, perform the operations indicated.

10. $(-7)^2 - (-13)$

11. $\frac{-5}{9} \div (-15)$

12. $-5^2 + 4 \cdot 6$

13. $3\{2 - 3[5 + 2(8 - 9)]\}$

14. $\frac{16 - 4^2}{8}$

15. $-\frac{2}{3}\left(3 - \frac{9}{2}\right)$

16. $3|11 - 17|$

In Problems 17 through 20, rewrite the expression using the property listed.

17. $6 + (-7)$ Commutative Property of Addition

18. $x(6 + y)$ Distributive Property

19. $-5 + 5$ Additive Inverse

20. $2(x + t)$ Commutative Property of Multiplication

21. Find the measures of each angle. $\angle a = 130°$ and $\angle d = 30°$.

22. Find the volume of a cylindrical tank 12 ft high that has a radius of 4 ft.

23. Find the amount of fencing needed to enclose a rectangular garden with dimensions 28 m by 22 m.

24. Find the surface area of a circular ball of radius $3\frac{1}{2}$ cm.

25. Find the area of the parallelogram below.

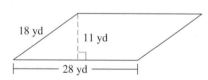

1 Integer Exponents and Polynomials

See Section 1.1, Exercises 94 and 95.

► **Connections**

For centuries, mathematicians have sought to approximate the value of π. Near the end of the seventeenth century, James Gregory showed that $\dfrac{\pi}{4}$ is approximately the value of the expression

$$x - \frac{1}{3}x^3 + \frac{1}{5}x^5 - \frac{1}{7}x^7$$

when x is 1. That is,

$$\frac{\pi}{4} \approx 1 - \frac{1}{3} + \frac{1}{5} - \frac{1}{7}$$

By extending this expression to hundreds of terms and doing all the calculations by hand, Gregory approximated π to 100 decimal places. Expressions such as $x - \dfrac{1}{3}x^3 - \dfrac{1}{5}x^5 + \dfrac{1}{7}x^7$ are called polynomials.

Today, polynomials are among the most useful expressions in mathematics. For example, they may be used by a business manager to find the cost of manufacturing calculators, by a biologist to find the volume of a spherical cell, or by a physicist to find the distance traveled by a free-falling body.

This chapter deals with polynomials. We learn how to add, subtract, multiply, and divide polynomials. In addition, we learn the important skill of factoring, or unmultiplying, polynomials. Algebra is full of uses of factoring. In fact, the inherent difficulty in factoring very large numbers is the basis of a revolutionary new breakthrough in developing secret codes.

1.1   Integer Exponents

TAPE 2

In this section, we define integer exponents. First, we examine natural number exponents and develop properties of exponents. Next, we define zero exponents and then negative integer exponents so that the properties of exponents are valid.

In Chapter 0 we learned that natural number exponents were developed as shorthand for multiplications with the same number, such as $x^3 = x \cdot x \cdot x$ or $a^6 = a \cdot a \cdot a \cdot a \cdot a \cdot a$. This leads to the definition of natural number exponents.

Natural Number Exponents

If x is a real number and n is a natural number, then

$$x^n = \underbrace{x \cdot x \cdot x \ldots x.}_{n \text{ factors of } x}$$

In writing x^n, we call x the **base** and n the **exponent.** The exponent gives the number of times the base is used as a factor. Sometimes the exponent is called the *power.* Read x^n as x to the nth power.

Be Careful!

Note: x^n does NOT mean n times x.

► **EXAMPLE 1**

Identify the base and simplify each of the following:

(a) 3^2 (b) $(-3)^2$ (c) -3^2 (d) $-2x^3$ (e) $(-2x)^3$

Solutions

(a) 3^2 is 3 squared. The base is 3.

$$3^2 = 3 \cdot 3 = 9.$$

(b) $(-3)^2$ is -3 squared. The base is -3.

$$(-3)^2 = (-3)(-3) = 9.$$

(c) -3^2 is the opposite of 3 squared. The base is 3.

$$-3^2 = -3 \cdot 3 = -9.$$

(d) $-2x^3$ is -2 times the cube of x. The base is x.
$-2x^3$ is simplified.

(e) $(-2x)^3$ is the cube of $-2x$. The base is $-2x$.

$$(-2x)^3 = (-2x)(-2x)(-2x)$$
$$= (-2)(-2)(-2)x \cdot x \cdot x \qquad \text{Commutative and Associative Properties}$$
$$= -8x^3$$
◄

Example 1 points out that, when working with exponents, it is very important to identify the base for each exponent. Negative signs can be troublesome. Notice that $-p^n$ and $(-p)^n$ are not necessarily the same.

Identifying the Base

$-p^n$	$(-p)^n$
The base is p.	The base is $-p$.
$-p^n = -p \cdot p \cdot p \cdots p$	$(-p)^n = (-p)(-p) \cdots (-p)$
$-4^2 = -4 \cdot 4 = -16$	$(-4)^2 = (-4)(-4) = 16$

Looking more closely at $(-p)^n$, we see that if n is odd then

$$(-p)^n = (-p)(-p) \cdots (-p) \qquad (n \text{ factors of } -p, \text{ and } n \text{ is odd})$$
$$= -p^n$$

However, if n is even,

$$(-p)^n = (-p)(-p) \cdots (-p) \qquad (n \text{ factors of } -p, \text{ and } n \text{ is even})$$
$$= p^n$$

This discussion leads to the following result for exponents.

Odd and Even Powers

If p is a real number and n is a natural number, then

$$(-p)^n = \begin{cases} -p^n & \text{if } n \text{ is odd} \\ p^n & \text{if } n \text{ is even} \end{cases}$$

Not only is the exponent notation convenient to use, but it has some very nice properties. Suppose we want to multiply $x^3 \cdot x^2$. Use the definition of exponent and write

$$x^3 \cdot x^2 = (x \cdot x \cdot x)(x \cdot x) = x \cdot x \cdot x \cdot x \cdot x = x^5$$

This leads to the first property of exponents.

Product with Same Base

$$x^m \cdot x^n = x^{m+n}$$

To multiply two expressions with the same base, keep the base and add the exponents.

Look at $(x^2)^3$. Use the definition of exponents and write

$$(x^2)^3 = x^2 \cdot x^2 \cdot x^2$$

$$= x^{2+2+2} \qquad \text{Product with Same Base Property}$$

$$= x^6$$

This leads to the second property of exponents:

Power of a Power

$$(x^m)^n = x^{mn}$$

To raise a power to a power, keep the base and multiply the exponents.

Now consider a product raised to a power, such as $(xy^3)^2$. Again we use the definition of exponent and write

$$(xy^3)^2 = (xy^3)(xy^3)$$

$$= x \cdot x \cdot y^3 \cdot y^3 \qquad \text{Commutative and Associative Properties}$$

$$= x^2 y^6 \qquad \text{Product with Same Base Property}$$

This leads to the third property of exponents:

Product to a Power

$$(xy)^n = x^n y^n$$

To raise a product to a power, raise each factor in the product to the power.

The next property is just like the previous one except that we use a quotient instead of a product. Consider $\left(\dfrac{x^2}{y}\right)^2$. We apply the definition of exponent and write

$$\left(\frac{x^2}{y}\right)^2 = \frac{x^2}{y} \cdot \frac{x^2}{y}$$

$$= \frac{x^4}{y^2}$$

This leads to the fourth property of exponents:

Quotient to a Power

$$\left(\frac{x}{y}\right)^n = \frac{x^n}{y^n}, \qquad y \neq 0$$

To raise a quotient to a power, raise both the numerator and the deonominator to the power.

For the final property of exponents, let's look at some examples of quotients with the same base. Consider $\dfrac{x^5}{x^3}$, $\dfrac{x^3}{x^5}$, and $\dfrac{x^4}{x^4}$. Again using the definition of exponent, we write

$$\frac{x^5}{x^3} = \frac{x \cdot x \cdot x \cdot x \cdot x}{x \cdot x \cdot x} = x^2 \qquad \text{Divide out common factors}$$

$$\frac{x^3}{x^5} = \frac{x \cdot x \cdot x}{x \cdot x \cdot x \cdot x \cdot x} = \frac{1}{x^2} \qquad \text{Divide out common factors}$$

$$\frac{x^4}{x^4} = \frac{x \cdot x \cdot x \cdot x}{x \cdot x \cdot x \cdot x} = 1 \qquad \text{Divide out common factors}$$

So we have the fifth property of exponents:

Quotient with Same Base

$$\frac{x^m}{x^n} = \begin{cases} x^{m-n}, & \text{if } m > n \\ 1, & \text{if } m = n, \qquad x \neq 0 \\ \dfrac{1}{x^{n-m}}, & \text{if } n > m \end{cases}$$

The Quotient with Same Base Property helps us to understand how we can define exponents of 0. Consider the quotient $\dfrac{2^5}{2^5}$. This is a number divided by itself and has the value 1. However, if we subtract exponents, we get,

$$\frac{2^5}{2^5} = 2^{5-5} = 2^0$$

For the quotient with same base to make sense in both situations,

$$2^0 = 1$$

This leads to our definition of an exponent of 0.

Zero Exponent

If x is any nonzero real number, then $x^0 = 1$.

► **EXAMPLE 2**

Use the definition of zero exponent to simplify each expression.

(a) $3x^0$ (b) $(-5)^0$ (c) -5^0 (d) $(2a + 3b)^0$

Solutions

(a) $3x^0 = 3 \cdot 1 = 3$ The base is x.
(b) $(-5)^0 = 1$ The base is -5.
(c) $-5^0 = -1$ The base is 5.
(d) $(2a + 3b)^0 = 1$ The base is $2a + 3b$.   ◄

   Now we have a definition for whole-number exponents. To define negative integer exponents, we want the five properties of exponents to be valid. So the following must be true.

$$3^{-4} \cdot 3^4 = 3^{-4+4}$$ Product with same base

$$= 3^0$$

$$= 1$$ Definition of zero exponent

So we see that

$$3^{-4} \cdot 3^4 = 1$$

$$3^{-4} = \frac{1}{3^4}$$ Divide both sides by 3^4

Since this is true in general, we make the following definition:

Negative Integer Exponent

For x any nonzero real number and n a natural number,

$$x^{-n} = \frac{1}{x^n}$$

Thus we see that x^n and x^{-n} are *reciprocals*.

► **EXAMPLE 3**

Rewrite each of the following expressions without negative exponents and simplify.

(a) 5^{-3} (b) $2^4 x^{-5}$ (c) $-3^{-2} z^4$ (d) $(-5)^{-2} xy^{-1}$ (e) $7x^{-3}$

Solutions

(a) $5^{-3} = \dfrac{1}{5^3} = \dfrac{1}{125}$

Notice that the negative exponent did not make the answer negative.

(b) $2^4 x^{-5} = 2^4 \cdot \dfrac{1}{x^5}$

$\qquad\quad = \dfrac{2^4}{x^5} = \dfrac{16}{x^5}$

(c) $-3^{-2} z^4 = -\dfrac{1}{3^2} z^4$

Notice that the base for the exponent, -2, is 3, and not -3.

$\qquad\qquad = -\dfrac{z^4}{9}$

(d) $(-5)^{-2} xy^{-1} = \dfrac{1}{(-5)^2} \cdot x \cdot \dfrac{1}{y}$

$\qquad\qquad\quad = \dfrac{x}{25y}$

The next example is similar, but is even more treacherous as there is a tendency to let the 7 stick to the variable, particularly when working in a hurry.

(e) $7x^{-3} = 7 \cdot \dfrac{1}{x^3}$

$\qquad\quad = \dfrac{7}{x^3}$

Notice that the base for the exponent, -3, is x, not $7x$. ◄

Now that we have defined integer exponents, we summarize the properties of exponents. Notice how negative exponents allow us to write property 5 in a more compact form.

Properties of Exponents

If m and n are natural numbers and x and y are real numbers, then:

1. $x^m x^n = x^{m+n}$ \qquad\qquad Product with same base

2. $(x^m)^n = x^{mn}$ \qquad\qquad Power of a power

3. $(xy)^n = x^n y^n$ \qquad\qquad Product to a power

4. $\left(\dfrac{x}{y}\right)^n = \dfrac{x^n}{y^n}, \quad y \neq 0$ \qquad Quotient to a power

5. $\dfrac{x^m}{x^n} = x^{m-n}, \quad x \neq 0$ \qquad Quotient with same base

► EXAMPLE **4**

Use the properties of exponents to rewrite each expression. The answers should not contain zero or negative exponents.

(a) $5^5 \cdot 5^{-3}$ (b) $(a^{-2})^{-3}$ (c) $\left(\dfrac{c^{-2}}{d}\right)^{-2}$ (d) $(3x^{-4})^{-2}$

(e) $\dfrac{(x + 2)^2}{(x + 2)^8}$ (f) $(-2x^0y^3)^2$

Solutions

(a) $5^5 \cdot 5^{-3} = 5^{5+(-3)} = 5^2$ Product with same base

(b) $(a^{-2})^{-3} = a^{(-2)(-3)} = a^6$ Power of a power

(c) $\left(\dfrac{c^{-2}}{d}\right)^{-2} = \dfrac{(c^{-2})^{-2}}{d^{-2}}$ Quotient to a power

$\qquad = \dfrac{c^4}{d^{-2}}$ Power of a power

$\qquad = \dfrac{c^4}{\dfrac{1}{d^2}}$ Definition of negative exponent

$\qquad = c^4d^2$

(d) $(3x^{-4})^{-2} = 3^{-2}(x^{-4})^{-2}$ Product to a power

$\qquad = 3^{-2}x^8$ Power of a power

$\qquad = \dfrac{x^8}{9}$ Definition of negative exponent

(e) $\dfrac{(x + 2)^2}{(x + 2)^8} = \dfrac{1}{(x + 2)^6}$ Quotient with same base

(f) $(-2x^0y^3)^2 = (-2y^3)^2$ Definition of zero exponent

$\qquad = (-2)^2y^6$ Product to a power

$\qquad = 4y^6$ ◄

Properties 2, 3, and 4 together are sometimes called *power rules* and are used in problems in the form

$$(\text{EXPRESSION})^{\text{POWER}}$$

We must be very careful in simplifying problems of this type. The following two examples illustrate an important difference in simplifying an expression raised to a power.

► EXAMPLE **5**

Use the properties of exponents to simplify each of the following:

(a) $(2^{-1} \cdot 3^3)^2$ (b) $(2^{-1} + 3^3)^2$

Solutions

(a) $(2^{-1} \cdot 3^3)^2$ is an expression raised to a power. Notice that the expression is a product. It can be simplified by multiplying exponents.

$$(2^{-1} \cdot 3^3)^2 = (2^{-1})^2 \, (3^3)^2 \qquad \text{Product to a power}$$

$$= 2^{-2}3^6 \qquad\qquad \text{Power of a power}$$

$$= \frac{1}{2^2} \cdot 3^6 \qquad\qquad \text{Definition of negative exponent}$$

$$= \frac{3^6}{2^2} = \frac{729}{4}$$

(b) $(2^{-1} + 3^3)^2$ is also an expression raised to a power. Notice that the expression is a sum of two terms. It cannot be rewritten by multiplying exponents.

$$(2^{-1} + 3^3)^2 = (2^{-1} + 3^3)(2^{-1} + 3^3) \qquad \text{Base is } (2^{-1} + 3^3). \text{ Exponent is } 2$$

$$= \left(\frac{1}{2} + 27\right)\left(\frac{1}{2} + 27\right) = 756.25$$

Notice that this is NOT $2^{-2} + 3^6$ (which is 729.25), obtained by multiplying exponents. ◄

Many problems with negative exponents can be simplified by applying some consequences of the definition. First, we note the following idea:

$$\frac{1}{x^{-n}} = \frac{1}{\dfrac{1}{x^n}} = x^n$$

Notice the result of raising a fraction to a negative integer exponent.

$$\left(\frac{p}{q}\right)^{-n} = \frac{1}{\left(\dfrac{p}{q}\right)^n} = \frac{1}{\dfrac{p^n}{q^n}} = \left(\frac{q}{p}\right)^n$$

$$\left(\frac{p}{q}\right)^{-n} = \left(\frac{q}{p}\right)^n$$

Let's simplify $\dfrac{a^{-2}}{bc^{-3}}$.

$$\frac{a^{-2}}{bc^{-3}} = \frac{\dfrac{1}{a^2}}{b \cdot \dfrac{1}{c^3}} = \frac{\dfrac{1}{a^2}}{\dfrac{b}{c^3}} = \frac{1}{a^2} \cdot \frac{c^3}{b} = \frac{c^3}{a^2 b}$$

Notice that a^{-2} was a factor of the numerator and that a^2 ends up as a factor of the denominator. Also, c^{-3} was a factor of the denominator and c^3 ends up as a factor of the numerator. This leads directly to the following results:

---

1. If P^{-n} is a factor of the numerator of a fraction, the fraction may be rewritten with P^n as a factor of the denominator.

2. If Q^{-m} is a factor of the denominator of a fraction, the fraction may be rewritten with Q^m as a factor of the numerator.

Be Careful!

In other words, a factor of the numerator (denominator) of a fraction may be moved to the denominator (numerator) of the fraction by changing the sign of its exponent. It is very important to remember that these results apply only to factors!

► **EXAMPLE 6**

Simplify the following. The answers should not contain negative exponents.

(a) $\dfrac{x^{-2}}{yz^{-3}}$ (b) $\left(-\dfrac{2}{5}\right)^{-2}$ (c) $\dfrac{3^{-1} + 3^{-2}}{1 - 3^{-2}}$

Solutions

(a) $\dfrac{x^{-2}}{yz^{-3}} = \dfrac{z^3}{x^2 y}$ Both x^{-2} and z^{-3} are *factors*

(b) $\left(-\dfrac{2}{5}\right)^{-2} = \left(\dfrac{2}{5}\right)^{-2}$ Even power

$\qquad = \left(\dfrac{5}{2}\right)^2 = \dfrac{25}{4}$

(c) We must be very careful in this problem. The two boxed rules for simplifying negative exponents do not apply here. These rules apply only to factors. The 3^{-1} and 3^{-2} in the numerator are *terms, not factors,* as are the two numbers in the denominator.

$$\frac{3^{-1} + 3^{-2}}{1 - 3^{-2}} = \frac{\dfrac{1}{3} + \dfrac{1}{3^2}}{1 - \dfrac{1}{3^2}} \qquad \text{Definition of negative exponent}$$

$$= \frac{\dfrac{1}{3} + \dfrac{1}{9}}{1 - \dfrac{1}{9}}$$

$$= \frac{\dfrac{4}{9}}{\dfrac{8}{9}} = \frac{4}{9} \cdot \frac{9}{8} = \frac{1}{2}$$

◄

Calculator Box

Exponents on a Calculator

Calculators have a special key for squaring. It is marked $\boxed{x^2}$ and is often a second function.

Let's find $(17.302 - 13.877)^2$.

Scientific Calculator:
Press $\boxed{17.302}$ $\boxed{-}$ $\boxed{13.877}$ $\boxed{=}$ $\boxed{x^2}$ and read $\boxed{11.730625}$ on the display.

Graphing Calculator:
Press $\boxed{(}$ $\boxed{17.302}$ $\boxed{-}$ $\boxed{13.877}$ $\boxed{)}$ $\boxed{x^2}$ $\boxed{\text{ENTER}}$
The parentheses, $\boxed{(}$ $\boxed{)}$, are *important!*

To find powers other than 2, we use the exponent key. This key is usually marked $\boxed{y^x}$ or $\boxed{x^y}$ and sometimes $\boxed{\wedge}$ or $\boxed{a^b}$. To find 4^7 with either type calculator,

Press $\boxed{4}$ $\boxed{y^x}$ $\boxed{7}$ $\boxed{=}$ (or $\boxed{\text{ENTER}}$) and read $\boxed{16384}$ on the display

For negative exponents, we use the change-sign key or the negation key. To approximate 7^{-3}:

Scientific Calculator:
Press $\boxed{7}$ $\boxed{y^x}$ $\boxed{3}$ $\boxed{+/-}$ $\boxed{=}$ and read $\boxed{0.00291545}$ on the display.

Graphing Calculator:
Press $\boxed{7}$ $\boxed{y^x}$ $\boxed{(-)}$ $\boxed{3}$ $\boxed{\text{ENTER}}$ and read the same result.

Calculator Exercises

In Problems 1 through 6, find the value of each expression. Approximate to three decimal places if necessary.

1. 17^2 **2.** 7^3 **3.** $(-3)^{10}$ **4.** $(7.24)^{-2}$ **5.** $(-\sqrt{6})^3$ **6.** $(1.33 - 0.84)^2$

In Problems 7 through 11, estimate the value of each expression; then use a calculator to approximate the value to five decimal places.

7. $(2.14)^3$ **8.** $(-9.1234)^2$ **9.** $(1.01)^5$ **10.** $\dfrac{\pi^2}{3}$ **11.** π^{-2}

Answers

1. 289   **2.** 343   **3.** 59,049   **4.** ≈ 0.019 **5.** ≈ -14.697 **6.** ≈ 0.240

7. 8; 9.80034   **8.** 81; 83.23643   **9.** 1; 1.05101   **10.** 3; 3.28987

11. 0.1; 0.10132

Scientific Notation

An application of integer exponents is scientific notation, a compact way of writing very large and very small numbers.

Multiply 123,456 by 123,456 on a calculator and it may display an answer like

$$1.524138394 \ 10 \quad \text{or} \quad 1.52413839 \ E \ 10$$

Either of these is short for $1.52413839 \times 10^{10}$, a large number written in scientific notation.

Divide 1 by 123,456, and the calculator may display an answer like

$$8.10005184 \ -06 \quad \text{or} \quad 8.10005184 \ E \ -6$$

This is short for $8.10005184 \times 10^{-6}$, a small number written in scientific notation. Each number is written as a number between 1 and 10 times a power of 10. As is usually the case in calculator computations, these are both approximations.

Scientific notation is a convenient way to express very large and very small numbers. Two examples that occur in physics are Planck's constant,

$$h \approx 0.0000000000000000000000000006625 \text{ erg-sec}$$

and the speed of light in a vacuum,

$$c \approx 29,900,000,000 \text{ cm/sec.}$$

Written in scientific notation, these become,

$$h \approx 6.625 \times 10^{-27} \text{ erg-sec} \quad \text{and} \quad c \approx 2.99 \times 10^{10} \text{ cm/sec}$$

These are both approximations, as the $\approx$ symbol indicates.

To Write a Number in Scientific Notation

1. Starting with the number in decimal format, move the decimal point until the number is between 1 and 10, including 1. Count the number of places moved.

2. Multiply the number formed in step 1 by 10 to the power equal to the number of decimal places moved. If the original number was between 0 and 1, the power of 10 is negative. If the original number was greater than 10, the power is positive.

3. This procedure is for a positive number. If the original number is negative, perform steps 1 and 2 on the absolute value of the original number and then attach a negative sign to the result.

► EXAMPLE **7**

Write the following in scientific notation.

(a) 93,000,000   (b) 0.000001554   (c) −254,000

Solutions

(a) Think of 93,000,000 as 93000000.0; then move the decimal point until the number is between 1 and 10.

$$9.3\,0\,0\,0\,0\,0\,0.0 \qquad \text{Move the decimal point}$$

The decimal point was moved seven places. Now we write the number in scientific notation.

$$9.3 \times 10^7$$

The exponent is positive because 93,000,000 is greater than 10.

(b) First move the decimal point.

$$0.0\,0\,0\,0\,0\,1.554$$

We moved the decimal point six places. We write the number in scientific notation.

$$1.554 \times 10^{-6}$$

The exponent is negative because 0.000001554 is between 0 and 1.

(c) −254,000 is negative, so we work with its absolute value.

$$2.5\,4\,0\,0\,0.0$$

We moved the decimal point five places. Thus, in scientific notation, the number is

$$2.54 \times 10^5$$

The exponent is positive because 254,000 is greater than 10. The original number was negative. So we write

$$-2.54 \times 10^5 \qquad \blacktriangleleft$$

► **EXAMPLE 8**

Write the following numbers without exponents.

(a) 8.771×10^{14} (b) 3.2×10^{-13} (c) -9.99231×10^{-3}

Solutions

To change scientific notation to standard decimal format, reverse the steps given above.

(a) $8.7\,7\,1\,0\,0\,0\,0\,0\,0\,0\,0\,0\,0.0$

$$877,100,000,000,000$$

Note that the decimal point was moved 14 places to the right because the exponent 14 is positive.

(b) $0.0\,0\,0\,0\,0\,0\,0\,0\,0\,0\,0\,3.2$

$$0.00000000000032$$

This time we moved the decimal point 13 places to the left (to make a number between 0 and 1), because the exponent −13 is negative.

(c) Because the given number is negative, we work with its absolute value.

$$0.0\,0\,9.9\,9\,2\,3\,1$$
$$0.00999231$$

We moved the decimal point three places to the left because the exponent of 10, -3, is negative. Now we append a negative sign because the original number was negative.

$$-0.00999231$$

◀

Calculator Box

Scientific Notation on a Calculator

As noted above, a calculator will often give answers in scientific notation. The exact value of 500^5 is given by

$$500^5 = 31,250,000,000,000 = 3.125 \times 10^{13}$$

Compute 500^5 on a calculator $\boxed{500}$ $\boxed{x^y}$ $\boxed{5}$ $\boxed{=}$ and note how the answer is displayed.

The exact value of $\dfrac{1}{8,000,000}$ is given by

$$\frac{1}{8,000,000} = 0.000000125 = 1.25 \times 10^{-7}$$

Divide 1 by 8,000,000 on a calculator and note how this answer is displayed.

It is sometimes convenient to *enter* numbers into a calculator in scientific notation. A scientific calculator should have a key marked $\boxed{\text{EE}}$ or $\boxed{\text{EXP}}$. (If you cannot find such a key, consult an instruction manual.) To enter 7.2×10^8, press $\boxed{7.2}$ $\boxed{\text{EE}}$ $\boxed{8}$. The display should show 7.2^{08} or 7.2 E 08 or something similar. To enter 1.25×10^{-7}, press

Scientific Calculator:

$\boxed{1.25}$ $\boxed{\text{EE}}$ $\boxed{7}$ $\boxed{+/-}$

Graphing Calculator:

$\boxed{1.25}$ $\boxed{\text{EE}}$ $\boxed{(-)}$ $\boxed{7}$

The display should show 1.25 E −7 or something similar.

As an example, we can use a calculator to find the product of Planck's constant and the speed of light in a vacuum. Recall that Planck's constant in scientific notation is $h \approx 6.625 \times 10^{-27}$ erg-sec, and the speed of light in a vacuum is $c \approx 2.99 \times 10^{10}$ cm/sec. Therefore, we are to calculate

$$hc \approx (6.625 \times 10^{-27})(2.99 \times 10^{10})$$

Scientific Calculator:

| 6.625 | EE | 27 | +/− | x | 2.99 | EE | 10 | = |

Graphing Calculator:

| 6.625 | EE | (−) | 27 | x | 2.99 | EE | 10 |

| ENTER |

and read 1.980875^{-16} on the display. So

$$hc \approx 1.980875 \times 10^{-16}$$

or

$$hc \approx 0.000000000000001980875 \text{ erg-cm}$$

Calculator Exercises

Perform the calculations indicated using the following constants:

Plank's constant: $h \approx 6.625 \times 10^{-27}$

Speed of light: $c \approx 2.99 \times 10^{10}$

Avogadro's number: $N \approx 6.02217 \times 10^{23}$

Write answers in scientific notation. (Remember to indicate approximations.)

1. c^2 **2.** $\sqrt{N}$ **3.** $\dfrac{1}{h}$

4. $\dfrac{2c^2}{h}$ **5.** $(5 \times 10^7)\sqrt{h}$ **6.** $\sqrt{\dfrac{2Nh}{\pi c}}$

Answers

1. $c^2 \approx 8.9401 \times 10^{20}$ **2.** $\sqrt{N} \approx 7.7603 \times 10^{11}$ **3.** $\dfrac{1}{h} \approx 1.5094 \times 10^{26}$

4. $\dfrac{2c^2}{h} \approx 2.6989 \times 10^{47}$ **5.** $(5 \times 10^7)\sqrt{h} \approx 4.0697 \times 10^{-6}$

6. $\sqrt{\dfrac{2Nh}{\pi c}} \approx 2.9146 \times 10^{-7}$

Problem Set 1.1

Warm-Ups

In Problems 1 through 8, identify the base and simplify each expression. See Example 1.

1. 2^4 **2.** $(-4)^2$ **3.** -4^2 **4.** $(-4)^3$

5. -4^3 **6.** $3\left(-\dfrac{1}{5}\right)^2$ **7.** $-3(-2)^2$ **8.** $-(-2)^3$

In Problems 9 through 12, use the definition of zero exponent to simplify each expression.
See Example 2.

9. $\left(\dfrac{2xy^2}{z^{11}}\right)^0$ 10. $-2^0\left(-\dfrac{1}{2}\right)^4$ 11. $\dfrac{(-5x^0)^2}{(-5x^2)^0}$ 12. $2x^0$

In Problems 13 through 18, rewrite each expression without negative exponents and simplify. See Example 3.

13. 2^{-3} 14. 6^{-1} 15. $(-2)^{-4}$
16. -2^{-4} 17. -2^{-3} 18. $-2x^{-3}$

In Problems 19 through 28, use the properties of exponents to rewrite each expression.
The answers should not contain zero or negative exponents. See Example 4.

19. $2^3 \cdot 2^{-2}$ 20. $3^{-3} \cdot 3^3$ 21. $[(x+4)^2]^3$ 22. $(2^{-2})^{-3}$

23. $\left(\dfrac{2x}{y^2}\right)^3$ 24. $\left(\dfrac{x^3}{y^{-2}}\right)^{-1}$ 25. $(-2t^2)^3$ 26. $(xy)^{-1}$

27. $\dfrac{(c-a)^3}{(c-a)^5}$ 28. $\dfrac{(2-y)^{-2}}{(2-y)^4}$

In Problems 29 through 32, simplify each expression using the properties of exponents.
See Example 5.

29. $(x^{-3}y^5)^3$ 30. $(x^3+y^5)^3$ 31. $(2\cdot 3^2)^2$ 32. $(2-3^2)^2$

In Problems 33 through 40, simplify each expression. The answers should not contain negative exponents. See Example 6.

33. $\dfrac{2^{-3}}{3}$ 34. $\dfrac{3}{5^{-2}}$ 35. $\dfrac{6^{-1}}{5^{-2}}$ 36. $\dfrac{-2x^{-2}}{y}$

37. $\left(\dfrac{2}{3}\right)^{-2}$ 38. $\left(-\dfrac{4}{3}\right)^{-1}$ 39. $\dfrac{2^{-1}+2^{-2}}{2^{-1}}$ 40. $\dfrac{2^{-2}+3^{-2}}{2^{-2}-3^{-2}}$

In Problems 41 through 44, write each number in scientific notation. See Example 7.

41. 0.000000021367   42. -0.0000012345
43. $-32{,}000$ 44. 77,722,000,000,000

In Problems 45 through 48, write each number without exponents. See Example 8.

45. 1.609×10^5 46. 5.43×10^{-5} 47. 1.1×10^{-9}
48. -8.0×10^{11}

Practice Exercises

In Problems 49 through 92, use definitions and properties to rewrite each expression.
The answers should not contain zero or negative exponents.

49. $(-3)^2$ 50. -3^2 51. $(-3)^{-4}$ 52. $3^0\left(\dfrac{1}{2}\right)^2$

53. $\left(\dfrac{1}{2}\right)^{-3}$ 54. $\left(\dfrac{2}{3}\right)^{-3}$ 55. $\left(-\dfrac{1}{10}\right)^{-1}$ 56. $-\left(\dfrac{3}{5}\right)^{-1}$

57. $-2^2(-3^3)$ 58. $(-2)^2(-3)^3$ 59. $\left(-\dfrac{1}{2}\cdot\dfrac{2}{3}\right)^3$ 60. $\left(-\dfrac{1}{2}\right)^2\left(\dfrac{1}{2}\right)^3$

61. $3x^{-2}$ 62. $2^{-3}x$ 63. $\dfrac{-2x^{-2}}{y^{-3}}$ 64. $\dfrac{3x^{-1}}{y^{-2}}$

65. $\dfrac{(-5)^4}{5^5}$

66. $\dfrac{(-7)^9}{-7^8}$

67. $(x-2)^3(x-2)^5$

68. $(z+1)^7(z+1)$

69. $4^{-1}+4^{-2}$

70. $2^{-1}-3^{-1}$

71. $\dfrac{3^{-1}+3^{-2}}{3^{-1}}$

72. $\dfrac{2^{-2}-3^{-2}}{2^{-2}+3^{-2}}$

73. $\dfrac{1}{3^{-1}}+\left(\dfrac{1}{3}\right)^{-1}$

74. $(2^{-1}-2^{-3})^{-1}$

75. $(-4x)^3$

76. $(-2x)^4$

77. $-(3x^{-2}y)^{-3}$

78. $(-3^{-2}x)^{-1}$

79. $\left(\dfrac{3x}{y^3}\right)^2$

80. $\left(\dfrac{-2x}{y^2}\right)^5$

81. $\left(\dfrac{x^{-1}}{3y}\right)^2$

82. $\left(-\dfrac{3x^{-3}}{y^3}\right)^2$

83. $\dfrac{4^3}{4^6}$

84. $\dfrac{3^2}{3^{-3}}$

85. $\dfrac{(-y)^{-4}}{(-y)^{-2}}$

86. $\dfrac{(3t)^{-3}}{(3t)^{-6}}$

87. $\dfrac{(16xy^0z)^3}{(8x^2yz)^2}$

88. $\dfrac{(14xz)^2}{(7x^2z)^3}$

89. $\left(\dfrac{-30x^5yz^8}{45x^5yz^3}\right)^2$

90. $\left(\dfrac{-x^4y^7z^8}{xy^6z^3}\right)^3$

91. $\dfrac{(x^2y^2)^{-2}}{(2xy^2)^{-1}}$

92. $\dfrac{(-2x^2z^{-1})^{-3}}{(2x^{-2}z^{-1})^2}$

93. The U.S. Environmental Protection Agency estimates for household and industrial wastes in the United States are listed next.

Year	Total in Metric Tons
1960	8.0×10^7
1965	9.3×10^7
1970	1.1×10^8
1975	1.12×10^8
1980	1.35×10^8
1985	1.42×10^8
1988	1.63×10^8

Find the total number of metric tons of household and industrial waste in the years listed.

94. Use the following pie graph to estimate how many met-

Waste in 1988

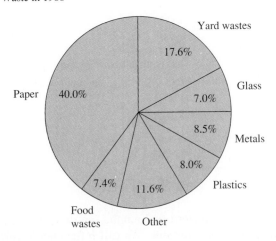

ric tons each of paper, yard wastes, glass, metals, plastics, and food wastes were in the 1.63×10^8 metric tons of household and industrial wastes in 1988.

95. For each category in the following table, estimate the number of metric tons of garbage that were recycled in 1988 when household and industrial wastes totaled 1.63×10^8 metric tons. Use graph in problem 94.

Percentage of
Household and Industrial
Wastes Recycled in 1988

Paper	25.6%
Metals	24.6%
Glass	12%
Yard wastes	1.6%
Plastics	1.1%
Food wastes	0.0%

Source: U.S. Environmental Protection Agency.

96. If 100 million tin cans are used each day in the United States, how many tin cans are used in a year? If 15% of these are recycled, how many tin cans are recycled?

97. Nanoelectronics is a field that creates electronic devices that are measured in nanometers or billionths of a meter. Using a powerful microscope, scientists have made the world's tiniest battery that is $\frac{1}{100}$ the diameter of a red blood cell. The diameter of a red blood cell is about 7 microns. A micron is $\frac{1}{1000}$ of a millimeter. Find the size in nanometers of the battery.

98. Light travels at the rate of 2.99×10^{10} cm/sec. A light-year is the distance light travels in a year. The nearest star, other than the sun, is 4.3 light-years away. How far away is the nearest star in centimeters?

99. Astronomers believe the Milky Way galaxy consists of a great disk with spiral arm segments. The sun is thought to be about 23,000 light-years from the center. If a light-year is about 5.9×10^{12} miles, how many miles is the sun from the center of the galaxy?

100. Find the square of $3x^2y^3$.

101. Find the cube of $-5x^2z^4$.

102. Square the quotient when $11x^5z$ is divided by $7y^3$.

103. Cube the quotient when $12x^2y$ is divided by $8uv^3$.

104. What is the area of a square whose side has length $3x$ feet?

105. What is the area of a square whose side has length $3x^3$ meters?

106. What is the volume of a cube whose side has length $4t$ inches?

107. What is the volume of a cube whose side has length $2t^2$ centimeters?

Challenge Problems

In Problems 108 through 115, simplify each expression. Assume that n is a natural number.

108. $6^n \cdot 6^2$

109. $\dfrac{3^{3n}}{3^n}$

110. $(8^n)^2$

111. $(5^2)^n$

112. $2 \cdot 2^n$

113. $\dfrac{9^{n+3}}{9^{n+1}}$

114. $x^n \cdot x^{2n}$

115. $x^2 \cdot x^n$

We stated that the definition of negative exponents is consistent with the five properties of exponents. In Problems 116 through 120, demonstrate the truth of the properties for the given values of m and n.

116. Using $m = -4$ and $n = -1$, demonstrate that the definition

$$x^{-k} = \frac{1}{x^k}$$

satisfies the property $x^m \cdot x^n = x^{m+n}$.

117. Using $m = -3$ and $n = 2$, demonstrate the property

$$(x^m)^n = x^{m \cdot n}$$

118. Using $n = -5$, demonstrate the property

$$(x \cdot y)^n = x^n \cdot y^n$$

119. Using $n = -1$, demonstrate the property

$$\left(\frac{x}{y}\right)^n = \frac{x^n}{y^n}$$

120. Using $m = 3$ and $n = -2$, demonstrate that

$$\frac{x^m}{x^n} = x^{m-n}$$

In Your Own Words

121. Explain what is squared in each expression.
(a) $x^2 + y^2$ (b) $(x + y)^2$ (c) $(xy)^2$

122. Explain when the situation (expression)^{power} tells us to multiply exponents.

123. Explain what is meant by -7^2 and $(-7)^2$.

124. Explain why $(x^{-1} + y^{-1})^{-1}$ is not the same as $x + y$.

1.2 Polynomial Definition, Evaluation, Addition, and Subtraction

TAPE

The expression bx^n, where b is a nonzero real number and n is a nonnegative integer, is called a **monomial** in one variable. We call b the **coefficient,** x the **variable,** and n the **degree.** The following are monomials in one variable.

$$\boxed{1}\ 2x^2 \qquad \boxed{2}\ \frac{1}{2}y^3 \qquad \boxed{3}\ \sqrt{3}x^5 \qquad \boxed{4}\ -5z \qquad \boxed{5}\ -7$$

	Coefficient	**Variable**	**Degree**
$\boxed{1}$	2	x	2
$\boxed{2}$	$\dfrac{1}{2}$	y	3
$\boxed{3}$	$\sqrt{3}$	x	5
$\boxed{4}$	-5	z	1
$\boxed{5}$	-7	any letter	0

(Note in $\boxed{5}$ that $-7 = -7x^0$.) Monomials such as -7 are called **constants.**

The expression $bx^m y^n$, where b is a nonzero real number and m and n are nonnegative integers, is called a **monomial in two variables.** Its degree is $m + n$. For example, $-12x^2y^4$ is a monomial in the two variables x and y. Its degree is 6 since $2 + 4 = 6$.

A **polynomial** is a finite sum of monomials. Its degree is the highest degree of the monomials that make it up. Consider $4x^5 + 3x^3 + (-2x^2) + (-6)$. This is a polynomial in one variable since it is the sum of four monomials in the same variable. However, we would write it in the form

$$4x^5 + 3x^3 - 2x^2 - 6$$

Its degree is 5. This polynomial is written so that the monomial of highest degree comes first and the degrees of the other monomials decrease in order. This is called **standard form.** The coefficient of the first term is called the **leading coefficient.** The monomials are most often called **terms.** The last term, -6, is called the **constant term.**

Consider the polynomial in two variables:

$$4x^5 - 6x^4y + 7x^2y^3 - 3x + y^4$$

Its degree is 5. Notice that the powers of x are in decreasing order. This is considered standard form with respect to x for a polynomial in two variables.

► EXAMPLE 1

Put each polynomial in standard form with respect to x. Give the degree of each.

(a) $5 - 3x^2 + 2x$ (b) $2x + 3$

(c) $x^5y - x^6 + 2y^6$ (d) $5 - 2x^2y^3 + 3xy + 3y^5$

Solutions

(a) Standard form is $-3x^2 + 2x + 5$.
   The degree is 2.

(b) Standard form is $2x + 3$.
   The degree is 1.

(c) Standard form is $-x^6 + x^5y + 2y^6$.
   The degree is 6.

(d) Standard form is $-2x^2y^3 + 3xy + 3y^5 + 5$.
   The degree is 5.

◄

Polynomials that have one term are called **monomials.** Those that have two terms are called **binomials,** and those that have three terms are called **trinomials.**

Following are examples of each.

Monomials	Binomials	Trinomials
x	$x + y$	$x + y + z$
$5y^4$	$z^3 - 7$	$w^2 - 8w + 6$

Polynomials have many different values, depending on the value of each variable. To **evaluate** a polynomial means to find its value. Use the order of operations discussed in Section 0.4.

► **EXAMPLE 2**

Evaluate each polynomial for the values given for each variable.

(a) $3x^2 - x + 4$ if x is -2 (b) $3x^2y^4 - xy + y^5$ if x is 2 and y is -1

Solutions

(a) $3x^2 - x + 4 = 3(-2)^2 - (-2) + 4$
$$= 3(4) + 2 + 4$$
$$= 12 + 2 + 4$$
$$= 18$$

(b) $3x^2y^4 - xy + y^5 = 3(2)^2(-1)^4 - (2)(-1) + (-1)^5$
$$= 3(4)(1) - (-2) + (-1)$$
$$= 12 + 2 - 1$$
$$= 13$$ ◄

The Distributive, Associative, and Commutative Properties apply to polynomials just as they do to real numbers.

Properties of Polynomials

If A, B, and C are polynomials, then:

1. $AB = BA$ Commutative for Multiplication
2. $A + B = B + A$ Commutative for Addition
3. $A(BC) = (AB)C$ Associative for Multiplication
4. $A + (B + C) = (A + B) + C$ Associative for Addition
5. $A(B + C) = AB + AC$ Distributive

These properties enable us to simplify many polynomials. For example, the binomial $7x + 3x$ can be simplified.

$$7x + 3x = x(7 + 3)$$ Distributive Property
$$= x(10)$$
$$= 10x$$ Commutative Property

This would not work with $7x + 3x^2$. We say that $7x$ and $3x$ are **like terms.** $5x^3$ and $-6x^3$ are like terms, while $2x$ and $5x^2$ are **unlike terms. Like terms** are "alike" everywhere except the coefficient.

Addition of polynomials is just a matter of adding up like terms. For example, to add the polynomials $5x^2 - 7x + 6$ and $4x^2 + 4x - 5$, we use the Associative and Commutative Properties to rearrange the terms:

$$(5x^2 - 7x + 6) + (4x^2 + 4x - 5)$$
$$= (5x^2 + 4x^2) + (-7x + 4x) + (6 - 5)$$

Then we add the like terms.

$$= 9x^2 - 3x + 1$$

Addition of Polynomials

To add two polynomials, combine like terms.

► ## EXAMPLE 3

Perform the addition in each.

(a) $(4x^3 + 4x^2 - x + 2) + (-7x^2 + x - 5)$

(b) $(3xy^2 + 4x^3y) + (xy^2 - 2x^3y)$

Solutions

(a) $(4x^3 + 4x^2 - x + 2) + (-7x^2 + x - 5)$
$$= 4x^3 + (4x^2 - 7x^2) + (-x + x) + (2 - 5)$$
$$= 4x^3 - 3x^2 - 3$$

(b) $(3xy^2 + 4x^3y) + (xy^2 - 2x^3y) = (3xy^2 + xy^2) + (4x^3y - 2x^3y)$
$$= 4xy^2 + 2x^3y$$ ◄

The Opposite of a Polynomial

If A is a polynomial, $-A$ is called the **opposite** of A and

$$A + (-A) = 0$$

► ## EXAMPLE 4

Find the opposite of each polynomial.

(a) $3x^2$ (b) $x^2 - x - 2$

Solutions

(a) $-(3x^2) = -3x^2$, since $3x^2 + (-3x^2) = 0$

(b) $-(x^2 - x - 2) = -x^2 + x + 2$, since
$$(x^2 - x - 2) + (-x^2 + x + 2) = 0$$ ◄

Example 4 suggests that to find the opposite of a polynomial we change the sign of each term in the polynomial.

We define subtraction as we did for real numbers.

Subtraction of Polynomials

If A and B are polynomials, then

$$A - B = A + (-B)$$

To subtract B from A, we add the opposite of B to A.

► EXAMPLE 5

Perform the subtraction indicated.

(a) $(3x^3 - 7x + 2) - (x^3 + x - 8)$

(b) $(5x^4y^2 + 6xy^3 - y^4) - (-4x^4y^2 + 2xy^3 - 3y^4)$

Solutions

(a) $(3x^3 - 7x + 2) - (x^3 + x - 8) = (3x^3 - 7x + 2) + (-x^3 - x + 8)$
$$= 2x^3 - 8x + 10$$

(b) $(5x^4y^2 + 6xy^3 - y^4) - (-4x^4y^2 + 2xy^3 - 3y^4)$
$$= (5x^4y^2 + 6xy^3 - y^4) + (4x^4y^2 - 2xy^3 + 3y^4)$$
$$= 9x^4y^2 + 4xy^3 + 2y^4 \qquad \blacktriangleleft$$

We can combine the operations of addition and subtraction together by using grouping symbols.

► EXAMPLE 6

Perform the operations indicated in the expression

$$[(x^2 - x) - (2x^2 + 3x + 2)] + (6x^2 - x + 1)$$

Solution

Start inside the brackets.

$$[(x^2 - x) - (2x^2 + 3x + 2)] + (6x^2 - x + 1)$$
$$= [(x^2 - x) + (-2x^2 - 3x - 2)] + (6x^2 - x + 1)$$
$$= [-x^2 - 4x - 2] + (6x^2 - x + 1)$$
$$= 5x^2 - 5x - 1 \qquad \blacktriangleleft$$

When doing additions and subtractions of polynomials, think of them in terms of parentheses preceded by a positive or a negative sign.

Look at an addition.

$$(5x^2 + x - 8) + (3x^2 + 4x + 3) = 5x^2 + x - 8 + 3x^2 + 4x + 3$$
$$= 8x^2 + 5x - 5$$

Parentheses preceded by a positive sign can be removed without changing the problem.

Look at a subtraction.

$$(4x^2 - 3x - 7) - (x^2 + 5x - 2) = (4x^2 - 3x - 7) + (-x^2 - 5x + 2)$$
$$= 4x^2 - 3x - 7 - x^2 - 5x + 2$$
$$= 3x^2 - 8x - 5$$

Thus parentheses preceded by a negative sign can be removed if the sign of each term inside the parentheses is changed.

Calculator Box

Evaluating Polynomials with a Calculator

Let's evaluate the polynomial $5x^4 - 3x^2 + x - 6$ when x is 2.1. We enter the polynomial with 2.1 substituted for x.

| 5 | × | 2.1 | y^x | 4 | − | 3 | × | 2.1 | x^2 | + | 2.1 | − | 6 | = |

We read 80.1105 on the display. Therefore, the value of the polynomial is 80.1105 when x is 2.1. Notice that we *did not* write down any intermediate results as we went through the evaluation. The calculator kept the running total. This is an important part of good calculator technique.

Next, let's estimate the value of the polynomial $5t^7 - t^3 + 6$ when t is 1.0286, then approximate the value to three decimal places with a calculator. Since 1.0286 is close to 1, we estimate the value using 1 for t.

$$5 \cdot 1^7 - 1^3 + 6 = 5 - 1 + 6 = 10$$

So our estimate is 10. Now, with our calculator we press

| 5 | × | 1.0286 | y^x | 7 | − | 1.0286 | y^x | 3 | + | 6 | = |

and see 11.00282153 on the display. Therefore, the value of the polynomial is approximately 11.003 to three decimal places when x is 1.0286. This approximation is reasonably close to our estimate and thus is a reasonable answer.

Calculator Exercises

Approximate the value of each polynomial to the nearest thousandth when x is 2.171 and y is 0.964. Make an estimate to see if the answer is reasonable.

1. $8x^5 - 11x^4 + 2x + 3$ **2.** $13y^{11} + 25y^6 - 9y + 1$

3. $2x^2 + 3xy + y^2 - 11$ **4.** $12x^5 - 10x^2y + 4y^3 - 2$

Answers

1. 148.805 (estimate: 87)   **2.** 21.073 (estimate: 30)

3. 5.634 (estimate: 4)   **4.** 534.884 (estimate: 346)

Problem Set 1.2

Warm-Ups

In Problems 1 through 6, write each polynomial in standard form with respect to x, and give its degree. See Example 1.

1. $x + 6$

2. $-3x^2y^2 + 4xy^3 - 5 + 2x^3y$

3. $4 - 3x + x^2$

4. $9 - x^3$

5. $\frac{1}{2}x - 3x^3 + 4x^2 - 7$

6. $4x^3 + \frac{2}{3}x - 17x^7 + 89$

In Problems 7 through 10, evaluate each polynomial when x is −2. See Example 2.

7. $3x + 7$

8. $4x^2 - 2x + 1$

9. $2x^3 - x - 2$

10. $x^2 - x$

In Problems 11 through 14, evaluate each polynomial when x is 1 and y is −1. See Example 2.

11. $xy + 6$

12. $x^2y^2 - xy + 1$

13. $2x^3y - 5x^2y^2 + xy + y^2$

14. $x^3 - y^3$

In Problems 15 through 18, perform the operations indicated. See Example 3.

15. $(4x - 5) + (3x - 9)$

16. $(y^2 + 6y - 5) + (2y^2 - 7)$

17. $(x^2 + 4xy + y^2) + (4x^2 - 7xy - 4y^2)$

18. $(5x^3 - 6x^2 + 5x - 7) + (10x^3 + 7x^2 - x - 1)$

In Problems 19 through 22, find the opposite of each polynomial. See Example 4.

19. $5x^6 - 3x^4 + 3$

20. $-4x^5 + 7x^4 - x^3 - 4$

21. $15x^4y - 12x^3y^2 + 7x^2y^2 - 12$

22. $3x^3y^2 - 5xy^2 + 6$

In Problems 23 through 26, perform the subtractions indicated. See Example 5.

23. $(5x - 8) - (7x + 4)$

24. $(z^2 - 6z) - (2z^2 - z + 8)$

25. $(2x^2 - 5xy + 3y^2) - (6x^2 + 3xy - 8y^2)$

26. $(6x^4 + 5x^3 - x^2 + x - 4) - (7x^4 - 4x^3 - 4x^2 + 7x - 9)$

In Problems 27 through 30, perform the operations indicated. See Example 6.

27. $[(x^2 + 2x + 5) + (4x^2 - 8)] - (5x^2 - x)$

28. $(v^4 + 4v^2 - 8) - [(2v^4 + v - 7) + (3v^3 - 5v^2 + 6v + 5)]$

29. $[(5x^3y - x^2y^2 + 7xy + 5) - (3x^3y + 2x^2y^2 - 4xy)] + (2x^3y + 7)$

30. $(t^3 - 27) - [(t^3 + 27) - (27 - t^3)]$

Practice Exercises

In Problems 31 through 40, write each polynomial in standard form with respect to x, give its degree, and identify those that are monomials, binomials, or trinomials.

31. $x - 4$

32. $-5x^2y^2 + 7x^4y^3 - 8 + 3x^3y$

33. $1 - x + x^2$

34. x

35. $\frac{2}{5}x - 2x^5 + 7x^4 - 9$

36. $4x^5 + \frac{1}{2}x^2 - 11x^8 + 19$

37. $x^5 + y^5$

38. $6xy + 2x^2 + 3y^2$

39. $-x^4y^3 + x^7y^2 + y^7$

40. $4 - 2x^6$

In Problems 41 through 46, evaluate each polynomial when x is −3.

41. $3x + 7$

42. $4x^2 - 2x + 1$

43. $8 - x$

44. $x^2 - x$

45. $1 - x^2$

46. $2x^3 - x - 2$

In Problems 47 through 52, evaluate each polynomial when x is −2 and y is 3.

47. $xy + 6$

48. $x^2y^2 - xy + 1$

49. $x^2 + 2xy + y^2$

50. $x^3 - y^3$

51. $x^2 - y^2$

52. $2x^3y - 5x^2y^2 + xy + y^2$

In Problems 53 through 70, perform the indicated operations in each expression.

53. $(x - 5) + (23x + 9)$

54. $(7x + 6) + (x + 4)$

55. $(3y^2 + 4y - 8) - (3y^2 - 9)$

56. $(z^2 - 3z) - (4z^2 + z - 5)$

57. $(x^2 - 4xy - y^2) - (3x^2 + 7xy - 5y^2)$

58. $(5x^2 - 2xy + 4y^2) - (7x^2 - 5xy - 11y^2)$

59. $(3x^3 - 4x^2 + 3x - 9) + (12x^3 + 6x^2 + x + 1)$

60. $(2x^4 + 5x^3 + x^2 - x + 4) - (3x^4 - 8x^3 + 5x^2 - 2x - 1)$

61. $(8x^5 - 4x^3 + 7x) - (3x^4 - 17x^3 + x^2 - 12)$

62. $(4x^6 - 2x^4 + 13) - (9x^5 + 4x^4 - x^3 + 3)$

63. $(13x^5 - 4x^3 - 7x) + (11x^5 + 14x^4 - 3x)$

64. $(x^2 - 7x - 5) - (x^2 + 6x - 4) + (x^2 - 2x + 11)$

65. $(9y + 8) + (10y + 6) - (y - 4)$

66. $(z^2 - 3) - (z^2 + 3) + (z^2 - 2z - 1)$

67. $(w^3 - 1) - (w^3 - 2w^2 + 2w - 2) - (w^2 + 4w - 4)$

68. $[(3x^2 - 4x + 7) + (2x^2 + 10)] - (7x^2 - x)$

69. $(t^3 + 27) - [(t^3 - 27) - (27 - t^3)]$

70. $[(z^2 - 2z) - (2z^2 + 8)] - (6z^2 - 2z - 7)$

71. Find the sum of $6x^2 - 5x + 1$ and $x + 7$.

72. Subtract $3s + 2$ from $s^2 + 7s - 6$.

73. Find the sum of $3x^2 - 7x + 5$ and $3x + 4$.

74. Subtract $x^2 + x + 1$ from the sum of $x^2 + 1$ and $x^2 + 2x + 1$.

75. Subtract $x^2 - x - 1$ from the sum of $x^2 - 1$ and $x^2 - 2x - 1$.

76. Subtract the sum of $3x + y$ and $4x - 8y$ from the sum of $6x - 5y$ and $3x - 13y$.

77. Subtract $x - y$ from $y - x$.

78. Find the sum of $2x^2 + 2x + 2$, $-x^2 - 4x - 4$, and $6x^2 + 7x - 3$.

79. Subtract $-5t^3 - t^2 + t + 3$ from $t^3 + 7t^2 - t - 5$.

80. Find a polynomial that gives the perimeter of a rectangle if the width is given by W units and the length by $2W - 3$ units. What is the perimeter of such a rectangle with a width of 7 ft?

81. Find a polynomial that gives the perimeter of a triangle if the lengths of the sides are $s + 1$ units, $s^2 - 2s + 2$ units, and $s^3 + 2s^2 + s - 9$ units.

Problem 82

82. Two Amtrak passenger trains leave Philadelphia, one traveling north and the other south. If the northbound train travels $65x$ miles while the southbound train travels $54x + 27$ miles, how far apart are the trains?

83. A barbed-wire fence is cut into three sections of lengths $3t$ yards, $2t - 1$ yards, and $t^2 + 4t - 5$ yards. Write a polynomial that gives the original length of the fence.

84. Alex has a collection of 42 Wilson and Penn tennis balls. If x of them are Wilson, how many are Penn?

85. A stone is thrown into the air in a manner such that its height in feet after t seconds is given by the polynomial $64t - 16t^2$. How high is the stone after 3 seconds? After 4 seconds?

86. A ball is thrown down from the top of the Mile High skyscraper. Its height from the ground in feet after t seconds is given by the polynomial $5280 - 74t - 16t^2$. How high is the ball after 5 seconds? After 10 seconds? After 15 seconds and 16 seconds?

87. Two angles of a triangle have measure $x + 30$ degrees and $3x - 80$ degrees, respectively. What is the measure of the third angle?

88. Charles and Kitty have saved $10x + 4$ dollars for their summer trip. If Charles saved $6x - 22$ dollars, how much has Kitty saved?

89. A section of cast-iron pipe $4s$ feet long is cut into three pieces. If one piece is 16 ft long and another is $2s - 7$ feet long, what is the length of the third piece?

90. An angle has measure $x^2 - 2x + 30$ degrees. Find the measure of its supplement.

91. One acute angle of a right triangle has measure $x - 30$ degrees. What is the measure of the other acute angle?

92. The sum of the squares of the first n natural numbers is given by the expression $\frac{1}{3}n^3 + \frac{1}{2}n^2 + \frac{1}{6}n$. What is the sum of the squares of the first five natural numbers? The first six?

Challenge Problems

In Problems 93 through 96, evaluate each polynomial when x is -1. Assume that n is a natural number.

93. $x^n - 1$

94. $x^{2n} + 1$

95. $x^{2n+1} + 1$

96. $x^{2n-1} - 1$

In Problems 97 through 100, perform the operations indicated. Assume that m and n are natural numbers.

97. $(2x^{2n} - x^n + 1) + (5x^{2n} + 2x^n + 3)$

99. $(5x^{2n} - 3x^n - 2) - (3x^{2n} - x^n - 4)$

98. $(x^m - 5x^m y^n + y^n) + (2x^m + x^m y^n - 5y^n)$

100. $(x^{mn} - y^{mn}) - (2x^{mn} - 2y^{mn})$

In Your Own Words

101. What is a polynomial?

102. How do we evaluate a polynomial?

103. Explain how to combine like terms.

104. Explain the meaning of an exact answer, an approximation, and an estimate.

105. How do we subtract polynomials?

1.3   Multiplication of Polynomials

TAPE 3

We multiply polynomials by using the Distributive, Commutative, and Associative Properties for polynomials given in Section 1.2 and the laws of exponents from Section 1.1. To multiply monomials, multiply the coefficients and use the laws of exponents. For example,

$$(5x^2)(-3x^3)(4x) = (5)(-3)(4)x^2 \cdot x^3 \cdot x$$

$$= -60x^6$$

Use the Distributive Property to multiply a monomial and a binomial.

$$a(b + c) = ab + ac$$

The product of two binomials is one of the most important products that we learn. Let's look at $(a + b)(c + d)$. Using the Distributive and Commutative Properties, write

$$(a + b)(c + d) = a(c + d) + b(c + d)$$
$$= ac + ad + bc + bd$$

Note that there are four terms in the product found by multiplying each term in the first binomial by each term in the second binomial.

Product of Two Binomials

$$(a + b)(c + d) = ac + ad + bc + bd$$

A common way to remember this is:

- ac represents the product of the "first" terms,
- ad represents the product of the "outside" terms,
- bc represents the product of the "inside" terms, and
- bd represents the product of the "last" terms.

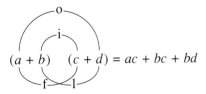

$$(a + b)\quad(c + d) = ac + bc + bd$$

This is often called the **foil** method.

► ### EXAMPLE 1

Find the product of $(2x + 3)$ and $(x - 4)$.

Solution

- The product of the **first** is $(2x)(x)$ or $2x^2$.
- The product of the **outside** is $(2x)(-4)$ or $-8x$.
- The product of the **inside** is $(3)(x)$ or $3x$.
- The product of the **last** is $(3)(-4)$ or -12.

Thus

$$(2x + 3)(x - 4) = 2x^2 + (-8x) + 3x + (-12)$$
$$= 2x^2 - 5x - 12 \qquad ◄$$

The distributive property allows other multiplications, such as

$$(a + b)(c + d + e) = a(c + d + e) + b(c + d + e)$$
$$= ac + ad + ae + bc + bd + be$$

Again, each term in the first polynomial is multiplied by each term in the second polynomial. It is important to note that the *foil* method works only for the product of two binomials.

► EXAMPLE 2

Find each product.

(a) $xyz^2(xy^2z)$ (b) $x^3(x - 2y)$

(c) $(x + 2)(x + 1)$ (d) $(x - y)(x^2 - xy + 1)$

Solutions

(a) $xyz^2(xy^2z) = x^2y^3z^3$

(b) $x^3(x - 2y) = x^3 \cdot x - x^3 \cdot 2y$
$$= x^4 - 2x^3y$$

(c) $(x + 2)(x + 1) = x \cdot x + 2 \cdot x + 1 \cdot x + 2 \cdot 1$
$$= x^2 + 3x + 2$$

(d) $(x - y)(x^2 - xy + 1)$
$$= x \cdot x^2 - x \cdot xy + x \cdot 1 - y \cdot x^2 + y \cdot xy - y \cdot 1$$
$$= x^3 - x^2y + x - x^2y + xy^2 - y$$
$$= x^3 - 2x^2y + xy^2 + x - y$$

◄

The next several examples develop some multiplications that occur very often in algebra. They are **special products** and should be memorized. The first product is called the **square of a binomial.**

► EXAMPLE 3

Find each product.

(a) $(p + q)^2$ (b) $(p - q)^2$

Solutions

(a) $(p + q)^2 = (p + q)(p + q)$
$$= p^2 + pq + pq + q^2$$
$$= p^2 + 2pq + q^2$$

(b) $(p - q)^2 = (p - q)(p - q)$
$$= p^2 - pq - pq + q^2$$
$$= p^2 - 2pq + q^2$$

◄

Square of a Binomial

$$(p + q)^2 = p^2 + 2pq + q^2$$
$$(p - q)^2 = p^2 - 2pq + q^2$$

These should be memorized as a pattern and remembered in words rather than symbols. The square of a binomial is the first term squared plus (or minus) twice the product of the two terms plus the last term squared. Notice that these patterns are alike everywhere except for the sign in the middle term.

► EXAMPLE 4

Find each product.

(a) $(x + 4)^2$ (b) $(2x - 3y)^2$

Solutions

(a) $(x + 4)^2$

$$(p + q)^2 = p^2 + 2pq + q^2 \qquad \text{Square of binomial}$$

We replace p with x and q with 4.

$$(x + 4)^2 = x^2 + 2x \cdot 4 + 4^2$$

(the first term squared plus twice the product of the two terms plus the last term squared).

$$= x^2 + 8x + 16$$

(b) $(2x - 3y)^2$

$$(p - q)^2 = p^2 - 2pq + q^2 \qquad \text{Square of binomial}$$

We replace p with $2x$ and q with $3y$.

$$(2x - 3y)^2 = (2x)^2 - 2(2x)(3y) + (3y)^2$$
$$= 4x^2 - 12xy + 9y^2 \qquad ◄$$

Another special product comes from multiplying the sum of two numbers by the difference of the same two numbers. It is called the **difference of two squares.**

► EXAMPLE 5

Find the product of $(p - q)$ and $(p + q)$.

Solution

$$(p - q)(p + q) = p^2 + pq - pq - q^2$$
$$= p^2 - q^2$$

Note that $(p - q)(p + q) = (p + q)(p - q)$. ◄

Difference of Two Squares
$$(p - q)(p + q) = p^2 - q^2$$

The product of the sum of two numbers and the difference of the same two numbers is the first number squared minus the second number squared. Also, the commutative property tells us that $(p - q)(p + q) = (p + q)(p - q)$.

► **EXAMPLE 6**

Find each product.

(a) $(x + 2)(x - 2)$ (b) $(4r + 7)(4r - 7)$

Solutions

(a) $(x + 2)(x - 2)$

$$(p + q)(p - q) = p^2 - q^2 \qquad \text{Difference of two squares}$$

We replace p with x and q with 2.

$$(x + 2)(x - 2) = x^2 - 2^2$$
$$= x^2 - 4$$

(the first number squared minus the second number squared).

(b) $(4r + 7)(4r - 7) = (4r)^2 - 7^2$
$$= 16r^2 - 49 \qquad ◄$$

The next special product is called the **cube of a binomial.**

► **EXAMPLE 7**

Find each product.

(a) $(p + q)^3$ (b) $(p - q)^3$

Solutions

(a) $(p + q)^3 = (p + q)(p + q)(p + q)$
$$= (p + q)(p^2 + 2pq + q^2)$$
$$= p^3 + 2p^2q + pq^2 + p^2q + 2pq^2 + q^3$$
$$= p^3 + 3p^2q + 3pq^2 + q^3$$

(b) $(p - q)^3 = (p - q)(p - q)(p - q)$
$$= (p - q)(p^2 - 2pq + q^2)$$
$$= p^3 - 2p^2q + pq^2 - p^2q + 2pq^2 - q^3$$
$$= p^3 - 3p^2q + 3pq^2 - q^3 \qquad ◄$$

Cube of a Binomial

$$(p + q)^3 = p^3 + 3p^2q + 3pq^2 + q^3$$
$$(p - q)^3 = p^3 - 3p^2q + 3pq^2 - q^3$$

Notice that the powers of p decrease in each term while the powers of q increase in each term. The coefficients of the two middle terms are both 3.

► **EXAMPLE 8**

Find each product.

(a) $(x + 2y)^3$ (b) $(2x - 3y)^3$

Solutions

(a) Since $x + 2y$ is a binomial, $(x + 2y)^3$ is the cube of a binomial. Replace p with x and q with $2y$ in the formula

$$(p + q)^3 = p^3 + 3p^2q + 3pq^2 + q^3$$

and get

$$(x + 2y)^3 = x^3 + 3x^2(2y) + 3x(2y)^2 + (2y)^3$$
$$= x^3 + 6x^2y + 12xy^2 + 8y^3$$

(b) Replace p with $2x$ and q with $3y$ in the formula

$$(p - q)^3 = p^3 - 3p^2q + 3pq^2 - q^3$$

and get

$$(2x - 3y)^3 = (2x)^3 - 3(2x)^2(3y) + 3(2x)(3y)^2 - (3y)^3$$
$$= 8x^3 - 36x^2y + 54xy^2 - 27y^3 \qquad \blacktriangleleft$$

Summary of Special Products

Square of a Binomial

$$(p + q)^2 = p^2 + 2pq + q^2$$
$$(p - q)^2 = p^2 - 2pq + q^2$$

Difference of Two Squares

$$(p + q)(p - q) = p^2 - q^2$$

Cube of a Binomial

$$(p + q)^3 = p^3 + 3p^2q + 3pq^2 + q^3$$
$$(p - q)^3 = p^3 - 3p^2q + 3pq^2 - q^3$$

We can combine the operations of multiplication, addition, and subtraction along with symbols of grouping. The order of operations from Section 0.4 also still applies.

► **EXAMPLE 9**

Simplify each expression.

(a) $6 - 2[(x + 2)(x - 3) - x(x + 1)]$ (b) $(3x - 2)^2 - (x - 7)(2x + 3)$

Solutions

(a) Begin on the inside of the grouping symbols.

$$6 - 2[(x + 2)(x - 3) - x(x + 1)] = 6 - 2[x^2 - x - 6 - x^2 - x]$$
$$= 6 - 2(-2x - 6) \qquad \text{Combine like terms}$$
$$= 6 + 4x + 12 \qquad \text{Distributive Property}$$
$$= 4x + 18$$

(b) Begin by performing the indicated multiplications.

Be Careful!

$$(3x - 2)^2 - (x - 7)(2x + 3) = 9x^2 - 12x + 4 - (2x^2 - 11x - 21)$$

Notice that when we multiplied $(x - 7)$ and $(2x + 3)$ we enclosed the product in parentheses. Care must be taken with the minus sign in front of these parentheses. The problem indicates subtraction of the entire product. Each sign in the parentheses must be changed.

$$= 9x^2 - 12x + 4 - 2x^2 + 11x + 21$$
$$= 7x^2 - x + 25 \qquad \blacktriangleleft$$

Problem Set 1.3

Warm-Ups

In problems 1 through 8, find each product. See Example 1.

1. $(x + 2)(x + 1)$ **2.** $(x + 4)(x + 5)$

3. $\left[\dfrac{1}{2}x + 2\right]\left[\dfrac{1}{2}x + 1\right]$ **4.** $(x + 7)(x + 1)$

5. $(x - 6)(x - 7)$ **6.** $(x - 5)(x - 8)$

7. $(x + 7)(x - 5)$ **8.** $(x - 9)(x + 7)$

In Problems 9 through 18, find each product. See Example 2.

9. $(4ab)(-3a^2b)$ **10.** $(2x^3)(x^2y^3)$

11. $(rs)s^3$ **12.** $3x(x + 4)$

13. $-3x^2(1 - x)$ **14.** $x(2x + 5)$

15. $(5x - 7)(3x - 8)$ **16.** $(3x - 1)(4x + 5)$

17. $(x - y)(x^2 - 2xy^2 + y^3)$ **18.** $(x + 2y)(x^2 + 3xy + y^2)$

In Problems 19 through 22, find each product. See Examples 3 and 4.

19. $(x + 5)^2$ **20.** $(5x + 12)^2$

21. $(3x - 4y)^2$ **22.** $(6x + 7)^2$

In Problems 23 through 26, find each product. See Examples 5 and 6.

23. $(x - 2)(x + 2)$ **24.** $(1 + 4y)(1 - 4y)$

25. $(5x + 7)(5x - 7)$ **26.** $(8 - 9z)(8 + 9z)$

In problems 29 through 30, find each product. See Examples 7 and 8.

27. $(x + 2)^3$ **28.** $(x - 4)^3$

29. $(3a - b)^3$ **30.** $(3w + 4x)^3$

In Problems 31 through 34, simplify each expression. See Example 9.

31. $(x + 3)(x - 1) + 2x(x - 7)$ **32.** $(x + 3)^2 - (x - 8)(2x + 5)$

33. $5 - 2[3(x^2 - x) + (x + 1)(x + 3)]$ **34.** $12 - [5 - (x + 4)^2 + (2x - 1)(3x + 2)]$

Practice Exercises

In Problems 35 through 80, find each product.

35. $(2ab)(-4a^3b)$

36. $(4x^5)(x^3y^6)$

37. $(2rs)s^4$

38. $x^3(-7x^4y)$

39. $-2^3xy^3(x^4y)$

40. $\dfrac{1}{3}a^4b^5(-3^2ab)$

41. $3b^5(-3^2abc)^3$

42. $(-ab)^2$

43. $(-r^4t)^5$

44. $(-3cd^6)^3$

45. $(rst^5)^3(-3r^7)^2$

46. $(-u^3)(-u)^3(-u)^4$

47. $(abc^5)(a^3bc)^7(-ab^2c)^4$

48. $3x(2x - 5)$

49. $a^3bc(3a + 5ab^3c^6 - 4b)$

50. $x(2x - 1)$

51. $-\dfrac{3}{5}x^2(5x - 3y - 1)$

52. $3a^4(a^5 + a^3 - a)$

53. $\dfrac{1}{3}x^4(3x^7 - 9x^6 + x^3 - 3)$

54. $(x + 3)(x + 5)$

55. $(x + 1)(x + 9)$

56. $(x - 5)(x + 3)$

57. $\left(\dfrac{1}{2}x - 1\right)\left(\dfrac{1}{2}x + 1\right)$

58. $\left[\dfrac{1}{2}x + 3\right]\left[\dfrac{1}{2}x + 4\right]$

59. $(5 + x)(7 + x)$

60. $(x - 5)(x - 7)$

61. $(x^2 - 4)(x^2 + 6)$

62. $(x^2 + 2)(x^2 - 5)$

63. $\left[x - \dfrac{1}{3}\right]\left[x + \dfrac{2}{3}\right]$

64. $\left[x - \dfrac{2}{5}\right]\left[x + \dfrac{3}{5}\right]$

65. $(3 - x)(2 - x)$

66. $(x - 4)(x - 9)$

67. $(7 - x^2)(4 + x^2)$

68. $(5x^2 + 7)(4x^2 + 9)$

69. $(4x + 5)(2x + 5)$

70. $(2x + 7)(3x + 1)$

71. $(11x + 5)(3x + 1)$

72. $(2x - 3)(x - 5)$

73. $(4 - x)(5 - 4x)$

74. $(6 - x)(7 - 2x)$

75. $(5x^2 - 6)(x^2 - 7)$

76. $(2x^2 + 1)(2x^2 - 1)$

77. $(c - d)(c^2 - 6cd + d^3)$

78. $(2x - 3y)(x^2 - 2xy + 3y^2)$

79. $(x + 3)(x - 5)(x + 7)$

80. $(2x + 7)(x - 7)(3x + 1)$

In Problems 81 through 96, use special products to find each product.

81. $(x - 3)(x + 3)$

82. $(x + 4)(x - 4)$

83. $(x + 6)^2$

84. $(11x + 12)^2$

85. $(9x - 8)^2$

86. $(5x - 4)^2$

87. $(7x + 5)(7x - 5)$

88. $(x - 5y)(x + 5y)$

89. $(5x - 3y)^2$

90. $(7x + 6)^2$

91. $(4x + 7y)^2$

92. $(2x + 5)^2$

93. $(x + 3)^3$

94. $(x + 2y)^3$

95. $(a - 2b)^3$

96. $(4s - 3t)^3$

In Problems 97 through 100, simplify each expression.

97. $(x + 4)^2 - (x - 7)(2x - 3)$

98. $(x + 6)(x - 3) - 3x(x - 5)$

99. $22 - [7 - (x + 3)^2 - (2x - 5)(3x - 2)]$

100. $7 - 3[3(x^2 + x) - (x - 1)(x + 3)]$

101. If N is an *even* integer, find the product of N and the next *even* integer.

102. If N is an *odd* integer, find the product of the next two *odd* integers.

103. Find a polynomial that gives the area of a circle whose radius is $2x - 5$ feet.

104. Suppose N is an integer. Find the product of the next two consecutive integers.

105. Cube the integer just before the integer N.

106. Find a polynomial that gives the volume of a sphere whose radius is $x - 5$ meters.

107. Barbara Phipps keeps her favorite mare on a pasture in the shape of a triangle whose base is $2W$ yards and whose height is $3W^2$ yards. How many square yards are there in the pasture?

108. Donald Crankshaw planes a block of wood that is W inches wide and $3W^2$ inches long until it is uniformly $\dfrac{1}{2}$ in. thick. How many cubic inches are there in the block of wood?

Problem 107

Challenge Problems

In Problems 109 through 116, find each product. Assume that m and n are natural numbers.

109. $x^n(x^2 - x + 1)$

110. $2x^2(x^{2n} + 3x^n + 1)$

111. $(x^n + 1)^2$

112. $(x^n + 2)(x^n - 2)$

113. $(x^n + y^n)^2$

114. $(x^n + 1)^3$

115. $[(x - y) + 7][(x - y) - 7]$

116. $[(s + t) + 3]^2$

In Your Own Words

117. Explain the special products.

118. Explain how to multiply a binomial by a trinomial.

119. Perform the multiplication indicated.
(a) $(p - q)^2$ (b) $(q - p)^2$
Why are these products the same?

1.4 Pascal's Triangle (Optional)

TAPE 3

Consider $(p + q)^n$, where n is a whole number.

$$(p + q)^0 = 1$$
$$(p + q)^1 = p + q$$
$$(p + q)^2 = p^2 + 2pq + q^2$$
$$(p + q)^3 = p^3 + 3p^2q + 3pq^2 + q^3$$
$$(p + q)^4 = \;?$$

What will be the result of raising $p + q$ to the fourth power? There will be five terms. The first will be p^4, and the last will be q^4.

$$p^4 + \_\,\_\,\_\,\_\,\_\,\_ + \_\,\_\,\_\,\_\,\_\,\_ + \_\,\_\,\_\,\_\,\_\,\_ + q^4$$

Note that in $(p + q)^3$ the powers of p decrease while the powers of q increase, and the exponents total three in each term. So if we follow this pattern, we will have everything but the coefficients.

$$p^4 + \_\,\_\,\_ p^3q + \_\,\_\,\_ p^2q^2 + \_\,\_\,\_ pq^3 + q^4$$

However, we can figure out the coefficients by a particular pattern known as **Pascal's Triangle.** It is named for the famous French mathematician Blaise Pascal.

Removing everything but the coefficients in the pattern

$$
\begin{array}{ll}
1 & n = 0 \\
p \;+\; q & n = 1 \\
p^2 \;+\; 2pq \;+\; q^2 & n = 2 \\
p^3 + 3p^2q + 3pq^2 + q^3 & n = 3
\end{array}
$$

yields the following pattern:

$$
\begin{array}{ll}
1 & n = 0 \\
1 \quad 1 & n = 1 \\
1 \quad 2 \quad 1 & n = 2 \\
1 \quad 3 \quad 3 \quad 1 & n = 3
\end{array}
$$

If we could find the next row in the triangle, we would have the coefficients for the expansion of $(p + q)^4$. An interesting and uncomplicated pattern is present. The triangle is made by placing 1's on the outsides. The numbers inside are found by adding the two numbers directly above.

$$
\begin{array}{ll}
1 & n = 0 \\
1 \quad 1 & n = 1 \\
1 \quad 2 \quad 1 & n = 2 \\
1 \quad 3 \quad 3 \quad 1 & n = 3 \\
1 \quad 4 \quad 6 \quad 4 \quad 1 & n = 4
\end{array}
$$

Thus the coefficients we need are 1, 4, 6, 4, and 1.

$$(p + q)^4 = p^4 + 4p^3q + 6p^2q^2 + 4pq^3 + q^4$$

Pascal's Triangle can be used to find the coefficients in expanding $(p + q)^n$ when n is a nonnegative integer. In Chapter 10 we look at a statement of the general result of expanding a binomial, the Binomial Theorem. It is clear that using the triangle would be difficult if the power were very large.

We can expand $(p - q)^n$ in the same manner, but the signs will alternate from positive to negative.

► **EXAMPLE 1**

Expand each binomial.

(a) $(x + 2)^4$ (b) $(2x + 3)^5$ (c) $(x - 3)^6$ (d) $(2x - 3)^4$

Solutions

(a) There will be five terms. We make the powers of x decrease and the powers of 2 increase.

$$(x + 2)^4 = x^4 + \_\_\_x^3(2)^1 + \_\_\_x^2(2)^2 + \_\_\_x^1(2)^3 + 2^4$$

Now we use the triangle to find the coefficients.

$$
\begin{array}{ccccccc}
 & & & 1 & & & \\
 & & 1 & & 1 & & \\
 & 1 & & 2 & & 1 & \\
1 & & 3 & & 3 & & 1 \\
\end{array}
$$

$\longrightarrow$ 1   4   6   4   1 $n = 4$

   1   5   10   10   5   1

Notice that we can identify the row we need in the triangle by looking at the second number in the row.

The coefficients we need are 1, 4, 6, 4, and 1.

$$(x + 2)^4 = x^4 + 4x^3(2) + 6x^2(2)^2 + 4x(2)^3 + 2^4$$
$$= x^4 + 8x^3 + 24x^2 + 32x + 16$$

(b) There will be six terms. We make the powers of $2x$ decrease and the powers of 3 increase. The exponents will total 5 in each term.

$$(2x + 3)^5 = (2x)^5 + \_\_\_(2x)^4(3)^1 + \_\_\_(2x)^3(3)^2$$
$$+ \_\_\_(2x)^2(3)^3 + \_\_\_(2x)^1(3)^4 + (3)^5$$

We use the triangle to find the coefficients.

$$
\begin{array}{ccccccc}
 & & & 1 & & & \\
 & & 1 & & 1 & & \\
 & 1 & & 2 & & 1 & \\
1 & & 3 & & 3 & & 1 \\
\end{array}
$$

   1   4   6   4   1

$\longrightarrow$ 1   5   10   10   5   1

The coefficients are 1, 5, 10, 10, 5, and 1.

$$(2x + 3)^5 = (2x)^5 + 5(2x)^4(3)^1 + 10(2x)^3(3)^2$$
$$+ 10(2x)^2(3)^3 + 5(2x)^1(3)^4 + (3)^5$$
$$= 32x^5 + 240x^4 + 720x^3 + 1080x^2 + 810x + 243$$

(c) There will be seven terms. The powers of x will decrease and the powers of

3 will increase. The exponents will total six in each term. The signs will alternate, beginning with positive.

$$(x - 3)^6 = x^6 - \_\_\_ x^5(3)^1 + \_\_\_ x^4(3)^2 - \_\_\_ x^3(3)^3$$
$$+ \_\_\_ x^2(3)^4 - \_\_\_ x(3)^5 + \_\_\_ (3)^6$$

We look for the coefficients in the triangle.

$$
\begin{array}{ccccccccccccc}
 & & & & & & 1 & & & & & & \\
 & & & & & 1 & & 1 & & & & & \\
 & & & & 1 & & 2 & & 1 & & & & \\
 & & & 1 & & 3 & & 3 & & 1 & & & \\
 & & 1 & & 4 & & 6 & & 4 & & 1 & & \\
 & 1 & & 5 & & 10 & & 10 & & 5 & & 1 & \\
\longrightarrow 1 & & 6 & & 15 & & 20 & & 15 & & 6 & & 1
\end{array}
$$

The coefficients are 1, 6, 15, 20, 15, 6, and 1.

$$(x - 3)^6 = x^6 - 6x^5(3)^1 + 15x^4(3)^2 - 20x^3(3)^3$$
$$+ 15x^2(3)^4 - 6x(3)^5 + (3)^6$$
$$= x^6 - 18x^5 + 135x^4 - 540x^3 + 1215x^2 - 1458x + 729$$

(d) There will be five terms. The powers of $2x$ will decrease and the powers of 3 will increase. The exponents will total four in each term. The signs will alternate, beginning with positive.

$$(2x - 3)^4 = (2x)^4 - \_\_\_ (2x)^3(3)^1 + \_\_\_ (2x)^2(3)^2 - \_\_\_ (2x)^1(3)^3 + 3^4$$

We find the coefficients.

$$
\begin{array}{ccccccccc}
 & & & & 1 & & & & \\
 & & & 1 & & 1 & & & \\
 & & 1 & & 2 & & 1 & & \\
 & 1 & & 3 & & 3 & & 1 & \\
\longrightarrow 1 & & 4 & & 6 & & 4 & & 1 \\
1 & & 5 & & 10 & & 10 & & 5 & 1
\end{array}
$$

They are 1, 4, 6, 4, and 1.

$$(2x - 3)^4 = (2x)^4 - 4(2x)^3(3) + 6(2x)^2(3)^2 - 4(2x)^1(3)^3 + 3^4$$
$$= 16x^4 - 96x^3 + 216x^2 - 216x + 81 \qquad \blacktriangleleft$$

Problem Set 1.4

Warm-Ups

In Problems 1 through 9, expand each binomial. See Example 1.

1. $(x + 1)^3$ **2.** $(x - 2)^3$ **3.** $(x + 2)^4$

4. $(3x - 2)^5$

5. $(2x + 3y)^6$

6. $(x + y)^6$

7. $(u - 2w)^6$

8. $(x + 2)^8$

9. $(x - 2)^{10}$

Practice Exercises

In Problems 10 through 30, expand each binomial.

10. $(x + 2)^3$

11. $(x - 1)^3$

12. $(x + 3)^4$

13. $(y - 2)^4$

14. $(2x + 1)^4$

15. $(b - 3)^5$

16. $(y - 1)^4$

17. $(2x - 1)^4$

18. $(b - 2)^5$

19. $(a + 1)^5$

20. $(3x - 1)^3$

21. $(2x + 3)^4$

22. $(a + 2)^5$

23. $(3x - 2)^3$

24. $(2x + 5)^4$

25. $(3x + 2)^5$

26. $(2x - 3y)^6$

27. $(x - y)^6$

28. $(u - 3w)^6$

29. $(x - 2)^8$

30. $(x + 2)^{10}$

31. Find the fourth term in the expansion of $(3x - y)^{11}$.

32. Find the third term in the expansion of $(3x + 7)^8$.

33. Find the middle term in the expansion of $(5z - 2)^8$.

In Your Own Words

34. Explain how to raise a binomial to the nth power if n is a whole number.

35. How is Pascal's Triangle formed?

36. Explain why $(x + y)^n \neq x^n + y^n$.

1.5   Greatest Common Factor and Factoring by Grouping

TAPE 3

Just as we can multiply 2 and 3 to get 6, we can *unmultiply* 6 to obtain 2 and 3. We call such unmultiplication factoring. To **factor** a number means to write the number as a product of numbers. Factoring often deals with natural numbers. A natural number is **prime** if the only way it can be written as a product of natural numbers is as the number itself times 1. For example, 5 is prime because $5 = 5 \cdot 1$. This is the only way to write 5 as a product of natural numbers. Similarly, 2, 3, 7, and 11 are prime. The number 1 is not considered to be a prime number. The smallest prime number is 2.

There are usually several ways to factor a natural number that is not prime. For example,

$$12 = 2^2 \cdot 3, \qquad 12 = 2 \cdot 6, \qquad 12 = 3 \cdot 4, \qquad 12 = 12 \cdot 1$$

Frequently, the most useful of these is $2^2 \cdot 3$ because it is a product of prime numbers. There is only one way to factor a natural number using prime factors. When a number is written as a product of prime numbers, it is **factored completely** or it is written in its **prime factorization.** The numbers in the product are called **factors.**

In factoring large numbers, it is helpful to know when a number is divisible by a prime.

Some Tests for Divisibility

1. A natural number is divisible by 2 if it is even, that is, if its last digit is either 0, 2, 4, 6, or 8.

2. A natural number is divisible by 3 if the sum of the digits of the number is divisible by 3.

3. A natural number is divisible by 5 if its last digit is 0 or 5.

To factor a number into prime factors, start with the smallest prime that will divide into the number and continue until all prime factors are found.

► EXAMPLE **1**

Factor each number into prime factors.

(a) 45   (b) 108   (c) 392

Solutions

(a) Notice that 45 is not even. So it is not divisible by 2. However, it is divisible by 3. So

$$45 = 3 \cdot 15$$
$$= 3 \cdot 3 \cdot 5$$
$$= 3^2 \cdot 5$$

(b) Since 108 is even, it is divisible by 2.

$$108 = 2 \cdot 54$$

Now 54 is also divisible by 2. So

$$= 2 \cdot 2 \cdot 27$$
$$= 2 \cdot 2 \cdot 3 \cdot 3 \cdot 3$$
$$= 2^2 \cdot 3^3$$

(c) Since 392 is even, divide by 2.

$$392 = 2 \cdot 196$$
$$= 2 \cdot 2 \cdot 98$$
$$= 2 \cdot 2 \cdot 2 \cdot 49$$
$$= 2^3 \cdot 7^2$$ ◄

Let's consider finding the largest factor that 15 and 25 have in common. The only factor of both is 5. The largest common factor is 5. If the numbers are too large to do this by inspection, the prime factorization of the numbers will help us. Such a factor is called the **greatest common factor.**

To Find the Greatest Common Factor (GCF)

1. Write the prime factorization of each number.

2. List the prime factors that are common to all the numbers.

3. The GCF is the product of the factors in step 2 with each factor raised to the smallest power that occurs in any number.

► EXAMPLE 2

Find the GCF for each pair of numbers.

(a) 24 and 80   (b) 72 and 378

Solutions

(a) $24 = 2 \cdot 12$ $80 = 2 \cdot 40$
$\quad\quad = 2 \cdot 2 \cdot 6$ $\quad\quad = 2 \cdot 2 \cdot 20$
$\quad\quad = 2 \cdot 2 \cdot 2 \cdot 3$ $\quad\quad = 2 \cdot 2 \cdot 2 \cdot 10$
$\quad\quad = 2^3 \cdot 3$ $\quad\quad = 2 \cdot 2 \cdot 2 \cdot 2 \cdot 5$
$\quad\quad\quad\quad\quad\quad\quad\quad\quad\quad = 2^4 \cdot 5$

The different prime factors are 2, 3, and 5. The only factor common to both numbers is 2. Since 2 occurs three times in 24 and four times in 80, the greatest common factor is 2^3.

The GCF is 2^3, or 8.

(b) $72 = 2 \cdot 36$ $378 = 2 \cdot 189$
$\quad\quad = 2 \cdot 2 \cdot 18$ $\quad\quad = 2 \cdot 3 \cdot 63$
$\quad\quad = 2 \cdot 2 \cdot 2 \cdot 9$ $\quad\quad = 2 \cdot 3 \cdot 3 \cdot 21$
$\quad\quad = 2 \cdot 2 \cdot 2 \cdot 3 \cdot 3$ $\quad\quad = 2 \cdot 3 \cdot 3 \cdot 3 \cdot 7$
$\quad\quad = 2^3 \cdot 3^2$ $\quad\quad = 2^1 \cdot 3^3 \cdot 7$

The common prime factors are 2 and 3. Since 2 occurs three times in 72 and once in 378, the GCF will contain 2. Since 3 occurs two times in 72 and three times in 378, the GCF will contain 3^2.
The GCF is $2 \cdot 3^2$, or 18. ◄

▶ **EXAMPLE 3**

Find the GCF of $108x^3y^6$ and $24x^4y^2$.

Solution

$$108x^3y^6 = 2^2 \cdot 3^3 \cdot x^3y^6$$

$$24x^4y^2 = 2^3 \cdot 3 \cdot x^4y^2$$

The common prime factors are 2, 3, x, and y. The GCF is $2^2 \cdot 3 \cdot x^3y^2$, or $12 \cdot x^3y^2$.

◀

Let's consider writing a polynomial as a product of polynomials. In other words, let's examine factoring a polynomial. The polynomials that we factor in this chapter will use only integers as coefficients.

The first type of polynomial factoring that we will study is called **factoring out common factors.** It is done by using the Distributive Property to factor out the greatest common factor of the terms in the polynomial. When we multiplied polynomials in Section 1.4, we used the Distributive Property and wrote

$$a(b + c) = ab + ac$$

To factor, turn this around and write

$$ab + ac = a(b + c)$$

Notice that $ab + ac$ is a sum of two terms and *is not factored,* whereas $a(b + c)$ is a product and is *factored.*

We say that we have *taken out* a common factor. If no more factoring can be done, the polynomial is **factored completely.** If no factoring can be done, the polynomial is **prime.** Monomials are considered to be prime polynomials.

To factor out common factors, we must identify the *factors* in each *term.* In the polynomial

$$3x^2 + 5x$$

there are two terms. In the first term, 3, x, and x^2 are factors. In the second term, 5 and x are factors. So x is a common factor. Thus

$$3x^2 + 5x = x(3x + 5)$$

Be Careful!

Notice that $3x^2 + 5x$ is a *sum,* while $x(3x + 5)$ is a *product.*

Let's consider $x(a + y) - 2b(a + y)$. If we think of it as two terms, the first term is $x(a + y)$ and the second term is $2b(a + y)$. Naming the factors of each term, we have

- x and $(a + y)$ are factors of $x(a + y)$.
- 2, b, and $(a + y)$ are factors of $2b(a + y)$.

Thus $(a + y)$ is the only common factor. So we factor

$$x(a + y) - 2b(a + y) = (a + y)(x - 2b)$$

Notice again that $x(a + y) - 2b(a + y)$ is a *difference,* whereas $(a + y)(x - 2b)$ is a *product.*

Consider the polynomial

$$4(x + y)^3 + 2(x + y)^2 + 6(x + y)$$

If we think of it as a trinomial, then $2(x + y)$ is a factor of each term. So $2(x + y)$ is a common factor and

$$4(x + y)^3 + 2(x + y)^2 + 6(x + y) = 2(x + y)[2(x + y)^2 + (x + y) + 3]$$

We must distinguish between polynomials that are factored and those that are not factored. A polynomial is *factored* if it is written as a *product*.

▶ **EXAMPLE 4**

Factor each polynomial completely.

(a) $ax^2 + ax^3$ (b) $54a^2bc^3 - 72ab^2c^2 + 144abc^2$

Solutions

(a) $ax^2 + ax^3 = ax^2(1 + x)$ Common factor ax^2
    Check by multiplying.
(b) $54a^2bc^3 - 72ab^2c^2 + 144abc^2$

$$= 2 \cdot 3^3 a^2 bc^3 - 3^2 \cdot 2^3 ab^2c^2 + 2^4 \cdot 3^2 abc^2$$
$$= 2 \cdot 3^2 abc^2(3ac - 2^2b + 2^3)$$
$$= 18abc^2(3ac - 4b + 8)$$ ◀

▶ **EXAMPLE 5**

Factor $-15x^2y + 5xy^2$ completely.

Solution

$$-15x^2y + 5xy^2 = -5xy(3x - y)$$

We could have factored out 5 instead of -5.

$$-15x^2y + 5xy^2 = 5xy(-3x + y)$$

Sometimes it is convenient to factor out a negative coefficient.   ◀

▶ **EXAMPLE 6**

Factor each polynomial completely.

(a) $x(a + b) + y(a + b)$ (b) $3a(m - n)^2 + 2a(m - n)^3$

Solutions

(a) $x(a + b) + y(a + b) = (a + b)(x + y)$
    Think of this as two terms with a common factor of $(a + b)$.
(b) $3a(m - n)^2 + 2a(m - n)^3 = a(m - n)^2[3 + 2(m - n)]$

$$= a(m - n)^2(3 + 2m - 2n)$$
$$= a(m - n)^2(2m - 2n + 3)$$ ◀

The next example contains opposites. Remember that

$$p - q = -(q - p)$$

To handle the opposites, we must look at how to multiply three numbers. Multiplying $(3)(y)(-z)$ gives $-3yz$. The negative sign is usually written in front of the product. Similarly, $3y[-(a - b)]$ is the product of three numbers, 3, y, and $-(a - b)$. So $3y[-(a - b)] = -3y(a - b)$.

► **EXAMPLE 7**

Factor $2x(a - b) + 3y(b - a)$.

Solution

We will replace $b - a$ with $-(a - b)$.

$$2x(a - b) + 3y\ (b - a)\ = 2x(a - b) + 3y\ [-(a - b)]$$

$$= 2x(a - b) - 3y(a - b) \qquad *$$

Be Careful!

Now a common factor of $(a - b)$ appears.

$$= (a - b)(2x - 3y) \qquad ◄$$

Compare the original problem, $2x(a - b) + 3y(b - a)$, with the step marked with *, $2x(a - b) - 3y(a - b)$. This maneuver occurs over and over again in working with polynomials. After doing several problems such as Example 7, it is easy to skip some steps and go directly from the original problem to the * step.

Factoring out the common factors should be done before any other type of factoring is done. Check for a common factors *first*.

After a little rearranging, we can factor

$$ax + bx + ay + by$$

Notice that there is no common factor. However, grouping the terms will allow us to factor.

$$ax + bx + ay + by = (ax + bx) + (ay + by)$$

$$= x(a + b) + y(a + b)$$

$$= (a + b)(x + y)$$

This type of factoring is called **grouping.** It should be used when factoring a polynomial with four or more terms.

► **EXAMPLE 8**

Factor each polynomial completely.

(a) $ax - by + ay - bx$ (b) $rm - rn - sm + sn$ (c) $rx - r + x - 1$

Solutions

(a) $ax - by + ay - bx = (ax + ay) - (by + bx)$
Notice the negative in front of the parentheses.

$$= a(x + y) - b(x + y)$$

$$= (x + y)(a - b)$$

Often, there is more than one way to group.

$$ax - by + ay - bx = (ax - bx) + (ay - by)$$
$$= x(a - b) + y(a - b)$$
$$= (a - b)(x + y)$$

It does not matter which we use.

(b) $rm - rn - sm + sn = (rm - rn) - (sm - sn)$
$$= r(m - n) - s(m - n)$$
$$= (m - n)(r - s)$$

(c) $rx - r + x - 1 = (rx - r) + (x - 1)$
$$= r(x - 1) + (x - 1)$$
$$= (x - 1)(r + 1)$$

The 1 in $(r + 1)$ is necessary. Why?

[The coefficient of $(x - 1)$ is 1.]   ◄

Grouping is often used to factor completely polynomials with four or more terms. Be careful to look for a common factor before grouping.

► **EXAMPLE 9**

Factor for each polynomial completely.

(a) $a^2x + ax + abx + bx$ (b) $y^7 - y^6z + y^5 - y^4z$

Solutions

(a) There is a common factor of x. Factor this out first.

$$a^2x + ax + abx + bx = x(a^2 + a + ab + b)$$

Now use grouping.

$$= x[(a^2 + a) + (ab + b)]$$
$$= x[a(a + 1) + b(a + 1)]$$
$$= x(a + 1)(a + b)$$

(b) $y^7 - y^6z + y^5 - y^4z = y^4(y^3 - y^2z + y - z)$
$$= y^4[(y^3 - y^2z) + (y - z)]$$
$$= y^4[y^2(y - z) + (y - z)]$$
$$= y^4(y - z)(y^2 + 1)$$

In using grouping to factor, we get expressions such as

$$m(x + y) - n(x + y)$$ ◄

Be Careful!

It is very important to recognize that this is *not* factored. *It is a difference, not a product.* However, $(x + y)(m - n)$ is factored because it is a *product.*
Factoring should always be checked by multiplying the factors to see if we get the original polynomial.

Problem Set 1.5

Warm-Ups

In Problems 1 through 10, factor each number into prime factors. See Example 1.

1. 64   **2.** 56   **3.** 45   **4.** 196   **5.** 162
6. 1000   **7.** 243   **8.** 900   **9.** 216   **10.** 180

In Problems 11 through 14, find the GCF. See Example 2.

11. 24; 48   **12.** 125; 75   **13.** 36; 54   **14.** 162; 72

In Problems 15 through 21, find the GCF. See Example 3.

15. x^3y^2; x^4y^6 **16.** a^4b^2; ab^3 **17.** s^5t^5; st^2 **18.** $30x^3y^2$; $21x^2y^4$
19. $96x^2y$; $80x$ **20.** $100x^2yz^4$; $40y^3z^4$ **21.** $192x^2y^3$; $48x^2y^4$; $72x^2y^3$

In Problems 22 through 30, factor out all common factors in each polynomial.
See Example 4.

22. $x^2y - xy^2$ **23.** $48a^2b^3c + 64a^3b^2c$
24. $21a^2b - 14ab^2 + 7ab$ **25.** $20p^2q^7 - 28p^5q^5 + 36p^8q^3$
26. $15 - 5a$ **27.** $15t^3 - 15t^2$
28. $18x^2y^3z^4 - 16x^2y^4z^2 - 14x^6y^6z^6$ **29.** $15m^2n^3 - 9m^4n^4 + 6m^3n^5$
30. $12p^8 + 14p^7 - 10p^9$

In Problems 31 through 33, factor each polynomial. See Example 5.

31. $-6x^2 + 3x$ **32.** $-18x^5 - 27x^4$ **33.** $-6uv + 3v$

In Problems 34 through 40, factor each polynomial. See Example 6.

34. $5(a + 2) + b(a + 2)$ **35.** $m(p - q) + n(p - q)$
36. $(x + 2)(x + 3) - x(x + 3)$ **37.** $(x + y)^2 + 3(x + y)$
38. $a(z + 2)^3 - b(z + 2)^2$ **39.** $5(1 - r)^4 + 3(1 - r)^3 - (1 - r)^2$
40. $x(5 + 2t)^2 - (2t + 5)$

In Problems 41 through 44, factor each polynomial. See Example 7.

41. $r(s - t) - u(t - s)$ **42.** $2x(1 - x) - 3y(x - 1)$
43. $w(3u - v)^3 - 3u(3u - v)^2 + v(v - 3u)$ **44.** $16(y - x)^3 + 24(y - x)^2 - 32(x - y)$

In Problems 45 through 54, factor each polynomial by grouping. See Example 8.

45. $7x + 7y + ax + ay$ **46.** $xy + x + by + b$
47. $a^2 - ab + 2a - 2b$ **48.** $6r - 6t - sr + st$
49. $a^2 + ab + a + b$ **50.** $y^2 - xy - y + x$
51. $5r + s^2 - 5s - rs$ **52.** $5a + 10 + ab + 2b$
53. $uv + wv - u - w$ **54.** $z^3 - z^2 + z - 1$

In Problems 55 through 60, factor each polynomial completely. See Example 9.

55. $a^3 - a^2c + a^2b - abc$ **56.** $x^6 + x^5y + x^4 + x^3y$
57. $xz^2 - z^2 + xz - z$.   **58.** $30st^2 + 60t^2 - 24st - 48t$
59. $15abx - 15aby - 25acx + 25acy$ **60.** $y^3 - xy^2 - y^2 + xy$

Practice Exercises

In Problems 61 through 87, factor out all common factors in each polynomial.

61. $42a^4b^2c + 54a^2b^5c$

62. $-8x^3 + 4x$

63. $24r^4s^4 - 48r^3s^3$

64. $12s^6 - 18s^3$

65. $27a^4b - 18ab^5 + 9ab$

66. $25p^5q^9 - 35p^3q^7 + 45p^6q^2$

67. $-16x^3 - 24x^6$

68. $18 - 6a$

69. $12t^4 - 12t^3$

70. $-9uv - 3v$

71. $2\sqrt{3}x - \sqrt{3}x^2$

72. $3\sqrt{2}x^2 + \sqrt{2}x^4$

73. $2\sqrt{5}x + 4\sqrt{5}x^3$

74. $6\sqrt{7}x^2 - 2\sqrt{7}x$

75. $c(d - 2) + b(d - 2)$

76. $4(r - t) - n(r - t)$

77. $a(s - t) + u(t - s)$

78. $a^4(x - b) + a^3(x - b)$

79. $(x + y)^3 - 5(x + y)^2$

80. $a(z + 2)^3 - b(z + 2)$

81. $3(1 - r)^3 - 3(1 - r)^4 - 4(1 - r)^2$

82. $z(5 + 2x)^2 - (2x + 5)$

83. $2t(a - x) - t(x - a)$

84. $z(4u - 3v)^3 - 3u(4u - 3v)^2 + v(3v - 4u)$

85. $c^2(5 + b)^4 - d^2(5 + b)^5 - 2(b + 5)^2$

86. $-x(3 - 2x)^3 - 2x^2(3 - 2x)^2 - (3 - 2x)^2$

87. $15(a - b)^4 + 45(a - b)^2 - 35(b - a)$

In Problems 88 through 99, factor each polynomial by grouping.

88. $4x + 4y + zx + zy$

89. $xy + 6y + x + 6$

90. $z^2 + 2z - yz - 2y$

91. $ar - br - as + bs$

92. $t + b + at + ab$

93. $a + b^2 - b - ab$

94. $3x + y^2 - 3y - xy$

95. $4s + 12 - 3b - bs$

96. $2zu + 2zv^2 - 3wu - 3wv^2$

97. $t^5 + t^4 + s^4t + s^4$

98. $am + an - m - n$

99. $q^4 - q^2t + q^2 - t$

In Problems 100 through 105, factor each polynomial completely.

100. $b^4z - b^4 - b^2 + b^2z$

101. $ax^2 + b^2x^2 - ax - b^2x$

102. $uz - vz + uz^2 - vz^2$

103. $16c^2r + 32c^2s - 8cr - 16cs$

104. $36a + 12ax - 18a^2 - 6a^2x$

105. $w^3 - w^2t + w^3t - w^4$

Challenge Problems

In Problems 106 through 109, factor each polynomial completely. Assume that m and n are natural numbers.

106. $x^{n+2} + x^2$

107. $x^{n+2} - x^{n+3}$

108. $3x^{4n} - 2x^{3n} + x^{2n}$

109. $x^{2mn} + x^{4mn}$

In Your Own Words

110. Explain how to factor by grouping.

111. Make up a polynomial that can be factored by grouping.

1.6 Factoring Binomials and Trinomials

TAPE 3

Factoring Binomials

Some binomials can be factored using a special product, the difference of squares, that we learned in Section 1.3.

$$p^2 - q^2 = (p - q)(p + q)$$

However, $p^2 + q^2$, the sum of squares, will not factor using real coefficients. The sum and difference of cubes can also be factored. Notice the following:

$$(p - q)(p^2 + pq + q^2) = p^3 + p^2q + pq^2 - p^2q - pq^2 - q^3$$
$$= p^3 - q^3$$
$$(p + q)(p^2 - pq + q^2) = p^3 - p^2q + pq^2 + p^2q - pq^2 + q^3$$
$$= p^3 + q^3$$

Thus, we can factor $p^3 - q^3$ and $p^3 + q^3$.

Factoring Binomials

$$p^2 - q^2 = (p + q)(p - q) \qquad \text{Difference of squares}$$
$$p^3 - q^3 = (p - q)(p^2 + pq + q^2) \qquad \text{Difference of cubes}$$
$$p^3 + q^3 = (p + q)(p^2 - pq + q^2) \qquad \text{Sum of cubes}$$

These rules tell how to factor the difference of squares and the difference of cubes as well as the sum of cubes.

To factor the difference of squares, we must recognize the squares.

▶ **EXAMPLE 1**

Factor $4x^2 - 9y^2$ completely.

Solution

This is the difference of squares.

$$4x^2 - 9y^2 = (2x)^2 - (3y)^2$$
$$= (2x - 3y)(2x + 3y) \qquad ◀$$

To factor the sum or difference of cubes, we must recognize the cubes.

▶ **EXAMPLE 2**

Factor each expression completely.

(a) $x^3 - 8$ (b) $8x^3 + 27$

Solutions

(a) This is the difference of cubes. Think of $x^3 - 8$ as $x^3 - 2^3$.

$$x^3 - 8 = x^3 - 2^3$$
$$= (x - 2)(x^2 + 2x + 4)$$

(b) This is the sum of cubes.

$$8x^3 + 27 = (2x)^3 + 3^3$$
$$= (2x + 3)[(2x)^2 - (2x)(3) + 3^2]$$
$$= (2x + 3)(4x^2 - 6x + 9) \quad \blacktriangleleft$$

Remember always to look for common factors. This should be done before using the difference of squares or sum or difference of two cubes.

▶ **EXAMPLE 3**

Factor each completely.

(a) $x^4 - 16$ (b) $8x^3 - 64$

Solutions

(a) This is the difference of squares. Think of $x^4 - 16$ as $(x^2)^2 - 4^2$.

$$x^4 - 16 = (x^2)^2 - 4^2$$
$$= (x^2 + 4)(x^2 - 4)$$

Be Careful!

Since $x^2 - 4$ will factor, continue factoring.

$$= (x^2 + 4)(x + 2)(x - 2)$$

(b) This one is tricky. It is the difference of cubes, but unless we look for a common factor, we may not factor it completely.

$$8x^3 - 64 = 8(x^3 - 8)$$
$$= 8(x^3 - 2^3)$$
$$= 8(x - 2)(x^2 + 2x + 4) \quad \blacktriangleleft$$

Factoring Trinomials

The product of two binomials sometimes gives a trinomial. To factor a trinomial, we look for two binomials. That is why multiplication of binomials is such an important skill.

There are some guidelines that we can use to help us to factor trinomials. Generally, it is easier to factor a trinomial if it is in standard form.

If a trinomial is in standard form with a positive leading coefficient, the following guidelines are useful.

Guidelines for Factoring Trinomials

1. If the last term is positive, the factors will have the form

$$(\_\_ + \_\_)(\_\_ + \_\_) \quad \text{or} \quad (\_\_ - \_\_)(\_\_ - \_\_)$$

The + or − sign is determined by the coefficient of the middle term.

2. If the last term is negative, the factors will have the form

$$(\_\_ + \_\_)(\_\_ - \_\_) \quad \text{or} \quad (\_\_ - \_\_)(\_\_ + \_\_)$$

Practice makes factoring trinomials easier. The more practice, the less trial and error. Always check an answer by multiplying the factors.

► **EXAMPLE 4**

Factor each trinomial completely.

(a) $x^2 + 3x + 2$ (b) $-6x + x^2 + 5$

Solutions

(a) $x^2 + 3x + 2$. The first guideline shows what the factors must look like.

$$(x + \text{\_\_})(x + \text{\_\_})$$

The missing numbers have a product of 2 and a sum of 3. We need 2 and 1.

$$x^2 + 3x + 2 = (x + 2)(x + 1)$$

Check by multiplying.

(b) $-6x + x^2 + 5$. Write the trinomial in standard form first, $x^2 - 6x + 5$. Since the last term is positive and the middle term is negative, the first guideline gives factors of the following type:

$$(x - \text{\_\_})(x - \text{\_\_})$$

The missing numbers have a product of 5 and a sum of -6. We need -5 and -1.

$$x^2 - 6x + 5 = (x - 5)(x - 1)$$

Check by multiplying.   ◄

► **EXAMPLE 5**

Factor each trinomial completely.

(a) $x^2 + 3x - 18$ (b) $x^2 - 3x - 18$

Solutions

(a) $x^2 + 3x - 18$. The second guideline gives factors in the form

$$(x + \text{\_\_})(x - \text{\_\_})$$

The missing terms have a product of 18. They could be

6 and 3   or

9 and 2   or

18 and 1

Since the middle term is $+3x$, the only possibility is 6 and 3. The 6 must go in the factor with the "+" and the 3 in the factor with the "−". So

$$x^2 + 3x - 18 = (x + 6)(x - 3)$$

Check by multiplying.

(b) $x^2 - 3x - 18$. By the second guideline, the factors must look as follows:

$$(x + \text{\_\_})(x - \text{\_\_})$$

Note that this trinomial is the same as the one in part (a) except that the sign of the middle term is negative. Just swap the 6 and the 3 so that the middle term will be $-3x$.

$$x^2 - 3x - 18 = (x + 3)(x - 6)$$

Check by multiplying.   ◀

► **EXAMPLE 6**

Factor each trinomial completely.

(a) $6x^2 - 13x + 6$ (b) $3 + 5x - 2x^2$

Solutions

(a) $6x^2 - 13x + 6$. Notice that

$$6x^2 = (2x)(3x) \quad \text{and} \quad 6x^2 = (6x)(x)$$

So the factors must look as follows:

$$(2x - \text{\_\_})(3x - \text{\_\_}) \quad \text{or} \quad (6x - \text{\_\_})(x - \text{\_\_})$$

We can write 18 as $6 \cdot 3$ or $9 \cdot 2$ or $18 \cdot 1$. There are several possible combinations to try. Some possible factors are:

$\boxed{1}$ $(6x - 2)(x - 3)$		$\boxed{2}$ $(6x - 3)(x - 2)$
$\boxed{3}$ $(2x - 3)(3x - 2)$		$\boxed{4}$ $(2x - 2)(3x - 3)$
$\boxed{5}$ $(6x - 6)(x - 1)$		$\boxed{6}$ $(6x - 1)(x - 6)$
$\boxed{7}$ $(2x - 6)(x - 1)$		$\boxed{8}$ $(2x - 1)(3x - 6)$

After multiplying each of these, note that $\boxed{3}$ is the correct answer. Notice that $\boxed{1}$ could not be the correct answer because $(6x - 2)$ has a factor of 2, that is, $(6x - 2) = 2(3x - 1)$. This would mean that there was a common factor of 2 in $6x^2 - 13x + 6$. The same idea applies to $\boxed{2}$, $\boxed{4}$, $\boxed{5}$, $\boxed{7}$, and $\boxed{8}$ as well. This idea can help cut down on trial and error. So

$$6x^2 - 13x + 6 = (2x - 3)(3x - 2)$$

(b) $3 + 5x - 2x^2$. This one is tricky! Do not try to rearrange the terms. It is easier to factor as it is, because the coefficient of x^2 is negative.

$$3 + 5x - 2x^2 = (3 - x)(1 + 2x)$$

Check by multiplying.   ◀

Alternative Approach to Factor Trinomials:
The ac Method

An alternative way to factor trinomials that does not require as much trial and error is called the **ac** method. To factor $ax^2 + bx + c$ by the **ac** method, we look for two numbers whose product is ac and whose sum is b. Then we factor by grouping. The procedure is illustrated in the next examples.

► **EXAMPLE 7**

Factor $3x^2 + 11x + 10$.

Solution

Notice that $a = 3$ and that $c = 10$. The product ac is $3(10)$ or 30. Now we look for factors of 30 whose sum is 11.

Factors of 30	Sum of Factors
1(30)	$1 + 30 = 31$
2(15)	$2 + 15 = 17$
3(10)	$3 + 10 = 13$
5(6)	$5 + 6 = 11$

The factors we are looking for are 5 and 6. Write $11x$ as $5x + 6x$ and use grouping to factor.

$$3x^2 + 11x + 10 = 3x^2 + 5x + 6x + 10$$
$$= x(3x + 5) + 2(3x + 5)$$
$$= (3x + 5)(x + 2) \qquad ◄$$

► **EXAMPLE 8**

Factor $6a^2 - 7a - 3$.

Solution

The product of 6 and -3 is -18. We look for factors of -18 whose sum is -7.

Factors of -18	Sum of Factors
1(-18)	$1 + (-18) = -17$
2(-9)	$2 + (-9) = -7$
3(-6)	$3 + (-6) = -3$
6(-3)	$6 + (-3) = 3$
9(-2)	$9 + (-2) = 7$
18(-1)	$18 + (-1) = 17$

The factors we are looking for are 2 and -9. Now we write $-7a$ as $2a - 9a$ and use grouping.

$$6a^2 - 7a - 3 = 6a^2 + 2a - 9a - 3$$
$$= 2a(3a + 1) - 3(3a + 1)$$
$$= (3a + 1)(2a - 3) \qquad ◄$$

Before factoring a trinomial, look for common factors.

► **EXAMPLE 9**

Factor each trinomial completely.

(a) $t^4 - 2t^3 + t^2$ (b) $6x^2 - 6xy - 36y^2$

Be Careful!   **Solutions**

(a) $t^4 - 2t^3 + t^2 = t^2(t^2 - 2t + 1)$ Common factor
$$= t^2(t - 1)(t - 1)$$
$$= t^2(t - 1)^2$$

(b) $6x^2 - 6xy - 36y^2 = 6(x^2 - xy - 6y^2)$ Common factor
$$= 6(x - 3y)(x + 2y)$$ ◄

Sometimes recognizing the square of a binomial special product can speed up factoring a trinomial. It gives us these factoring rules. If a trinomial can be factored by one of these rules, the trinomial is called a **perfect trinomial square.**

Factoring Perfect Squares

$$p^2 + 2pq + q^2 = (p + q)^2$$
$$p^2 - 2pq + q^2 = (p - q)^2$$

► **EXAMPLE 10**

Determine if each trinomial is a perfect square.

(a) $x^2 - 6x + 9$ (b) $x^2 + 10x + 25$ (c) $x^2 - 12x - 36$
(d) $x^2 + 8x + 15$ (e) $x^2 - x + 4$

Solutions

(a) $x^2 - 6x + 9 = x^2 - 2 \cdot x \cdot 3 + 3^2$. So the first and last terms are squares and the middle term is twice the product of x and 3.
$x^2 - 6x + 9 = (x - 3)^2$, so it is a perfect square.

(b) $x^2 + 10x + 25 = x^2 + 2 \cdot x \cdot 5 + 5^2$. This is also a perfect square since
$$x^2 + 10x + 25 = (x + 5)^2$$

(c) The last term must be positive. So
$x^2 - 12x - 36$ is *not* a perfect square.

(d) The last term is not a square. So
$x^2 + 8x + 15$ is *not* a perfect square.

(e) $x^2 - x + 4 \neq (x - 2)^2$

The first and last terms are squares, but the middle term is not twice the product of the x and 2.
$x^2 - x + 4$ is *not* a perfect square.   ◄

Problem Set 1.6

Warm-Ups

In Problems 1 through 4, factor each polynomial completely. See Example 1.

1. $x^2 - 9$

2. $4x^2 - 1$

3. $9x^2 - 4$

4. $64y^2 - 25$

In Problems 5 through 8, factor each polynomial completely. See Example 2.

5. $x^3 - 1$

6. $8 + y^3$

7. $64 + t^3$

8. $27x^3 - 8y^3$

In Problems 9 through 12, factor each polynomial completely. See Example 3.

9. $4x^2 - 64$

10. $12x^2 - 3$

11. $3t^3 + 24s^3$

12. $2x^4 - 16x$

In Problems 13 through 16, factor each polynomial completely. See Example 4.

13. $x^2 + 3x + 2$

14. $s^2 + 7s + 6$

15. $-5x + x^2 + 6$

16. $x^2 - 10x + 9$

In Problems 17 through 20, factor each polynomial completely. See Example 5.

17. $z^2 + z - 12$

18. $x^2 - 5x - 24$

19. $x^2 - 3x - 10$

20. $-20y^2 + x^2 + xy$

In Problems 21 through 28, factor each polynomial completely. See Examples 6, 7, and 8.

21. $2x^2 + 3x + 1$

22. $6x^2 - 11x + 4$

23. $8s^2 - 22st + 15t^2$

24. $2 - 13z + 21z^2$

25. $2x^2 - 7x + 6$

26. $2t^2 + t - 6$

27. $6a^2 + 7a - 3$

28. $4y^2 - 16y + 15$

In Problems 29 through 32, factor each polynomial completely. See Example 9.

29. $4x^2 - 4x - 8$

30. $5x^2 + 15x + 10$

31. $-6y^3 + 6y^2 + 12y$

32. $6s^3 + 15s^2 - 9s$

Practice Exercises

In Problems 33 through 74, factor each polynomial completely.

33. $x^2 - 81$

34. $x^2 - 49$

35. $x^3 - 8$

36. $x^3 - 64$

37. $x^2 + 9x + 8$

38. $s^2 + 6x + 8$

39. $t^2 + t - 20$

40. $x^2 - x - 30$

41. $z^3 + 27$

42. $125 + r^3$

43. $3x^2 - x - 10$

44. $3t^2 + t - 44$

45. $16 - x^2$

46. $81 - x^2$

47. $27 - 8v^3$

48. $64 - 125k^3$

49. $6y^2 + 25yz + 4z^2$

50. $4x^2 + 16xy + 15y^2$

51. $5t^2 - 28t + 15$

52. $6s^2 - 31s + 35$

53. $x^3 - 8y^3$

54. $t^3 + 64s^3$

55. $8x^3 - 2x$

56. $P - P^3$

57. $6y^3 + 6y^2 - 12y$

58. $x^4 - 4x^3 - 60x^2$

59. $2x^2 + 6xy + 4y^2$

60. $12w^2 + 10wz - 50z^2$

61. $81x^4 + 24x$

62. $40t^2 + 5t^5$

63. $7t^2 + 42t + 63$

64. $14x^4 - 11x^3 - 15x^2$

65. $x^4 - 2x^2 - 8$

66. $x^4 - 10x^2 + 9$

67. $z^4 - 3z^2 - 4$

68. $1 + 3k^2 + 2k^4$

69. $x^4 - 1$

70. $81t^4 - 16$

71. $x^6 - 8x^3 + 15$

72. $x^6 + 6x^3 + 9$

73. $4x^6 - 4x^3 - 8$

74. $x^6 + 9x^3 + 8$

In Problems 75 through 84, determine if the polynomial is factored or not. If not factored completely, factor it completely, if possible.

75. $16x^2 - 16$

76. $81 - 16x^4$

77. $2ab + 2ax - by - xy$

78. $y^2 - yz - z + y$

79. $x(x^2y - 1)$

80. $(x^2 + 2)^2$

81. $6s^2 + st - 15t^2$

82. $x^{10} + x^5 - 6$

83. $3x^2 + x + 2$

84. $a^2b - cd^2$

In Problems 85 through 90, find the value of k.

85. Find a value of k so that $z + 4$ will be a factor of $z^2 - k$.

86. Find a value of k so that $3y + 1$ is a factor of $ky^3 + 1$.

87. Find all values of k so that $t^2 + kt - 7$ will factor.

88. Find all values of k so that $x^2 + kx - 3$ will factor.

89. Find all values of k so that $s - 7$ is a factor of $s^2 - s + k$.

90. Find all values of k so that $y + 4$ is a factor of $y^2 + ky - 8$.

In Problems 91 and 92, express the volume of the shaded figure as a polynomial in completely factored form.

91.

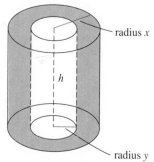

radius x

h

radius y

92.

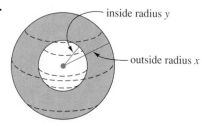

inside radius y

outside radius x

In Problems 93 and 94, express the area of the shaded region as a polynomial in completely factored form.

93.

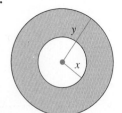

y

x

94.

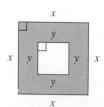

x

y

x y y x

y

x

Challenge Problems

In Problems 95 through 110, factor each polynomial completely.

95. $(x + y)^2 - 4$

96. $(a + b)^3 + 8$

97. $125 - (s - t)^3$

98. $144 - (x + y)^2$

99. $1 - x^9$

100. $a^9 + b^9$

In Problems 101 through 104, assume that p and q are natural numbers.

101. $x^{2p} - 1$

102. $x^{3q} - 1$

103. $x^{3p} + 1$

104. $x^p y^3 - x^p$

In Problems 105 and 106, factor each as the difference of two squares and then factor each as the difference of two cubes.

105. $x^6 - 1$

106. $x^6 - 64$

In Problems 107 through 110, assume that n is a natural number.

107. $x^{2n} + 2x^n - 8$

108. $x^{2n} + 3x^n - 10$

109. $x^{2n} + 2x^n y^m + y^{2m}$

110. $x^{2n} + x^n y^m - 6y^{2m}$

111. Consider the number $2^{32} - 1$.
   (a) Using a calculator, find the value of $2^{32} - 1$ and then show that 3 and 17 are factors.
   (b) Factor $2^{32} - 1$ as the difference of two squares and prove that 3 and 17 are factors.

112. Show that the product of $x^2 + x - 2$, $x^2 - x + 6$, and $x^2 - 4x + 3$ is a perfect square.

In Your Own Words

113. Explain the guidelines used to determine the signs in the factors of a trinomial.

114. Explain how to factor the sum or difference in two cubes.

1.7   Summary of Factoring

TAPE 4

Now that we have looked at several types of factoring, here is a procedure to summarize how to factor a polynomial completely.

> ### Procedure to Factor a Polynomial Completely
>
> **1.** Factor out the greatest common factor (GCF) if there is one. This should be done *first*.
>
> **2.** Count the number of terms.
>
> **3.** If the polynomial is a binomial, check for difference of squares or cubes, or sum of cubes. $p^2 + q^2$ **does not factor** *with real coefficients*.
>
> **4.** If the polynomial is a trinomial, look for two binomial factors.
>
> **5.** If the polynomial has four or more terms, try grouping.
>
> **6.** Make sure that each factor is prime.
>
> **7.** Check to see if the product of the factors is the original polynomial.

► **EXAMPLE 1**

Factor each polynomial completely.

(a) $s^2 - 81t^2$ (b) $4x^4 + 32x$ (c) $2x^4 - x^2 - 1$

Solutions

(a) There is no common factor. We count two terms. This is the difference of squares.
$$s^2 - 81t^2 = (s - 9t)(s + 9t)$$

(b) This polynomial has a common factor of $4x$.
$$4x^4 + 32x = 4x(x^3 + 8)$$

Now $x^3 + 8$ has two terms, and we have the sum of cubes.
$$= 4x(x + 2)(x^2 - 2x + 4)$$

(c) This polynomial has no common factor. There are three terms.
$$2x^4 - x^2 - 1 = (2x^2 + 1)(x^2 - 1)$$

We continue to factor $x^2 - 1$.
$$= (2x^2 + 1)(x + 1)(x - 1)$$ ◄

► **EXAMPLE 2**

Factor each polynomial completely.

(a) $x^4 + x^3 - 2x^2$ (b) $6r^2t - 2rt^2 + 3rst - st^2$

Solutions

(a) There is a common factor.
$$x^4 + x^3 - 2x^2 = x^2(x^2 + x - 2)$$
$$= x^2(x + 2)(x - 1)$$

Check by multiplying.

(b) There is a common factor.
$$6r^2t - 2rt^2 + 3rst - st^2 = t(6r^2 - 2rt + 3rs - st)$$

Use grouping on the polynomial in the parentheses.
$$t[(6r^2 - 2rt) + (3rs - st)] = t[2r(3r - t) + s(3r - t)]$$
$$= t[(3r - t)(2r + s)]$$
$$= t(3r - t)(2r + s)$$ ◄

Often, it is necessary to determine if a polynomial is factored. Which of the following polynomials is factored?

$$a^3 - b^3 \qquad (a - b)^3$$

The polynomial $a^3 - b^3$ is not factored, whereas the polynomial $(a - b)^3$ is factored. To factor a polynomial means to write it as a *product* of polynomials. Thus $a^3 - b^3$ is a difference, *not* a product, but $(a - b)^3$ is a product, as

Be Careful!

$$(a - b)^3 = (a - b)(a - b)(a - b)$$

Factored Form	**Not Factored**
• $(a - b)^3$	$a^3 - 3a^2b + 3ab^2 - b^3$
• $(a + b)^3$	$a^3 + 3a^2b + 3ab^2 + b^3$
• $(a - b)(a^2 + ab + b^2)$	$a^3 - b^3$
• $(a + b)(a^2 - ab + b^2)$	$a^3 + b^3$

► **EXAMPLE 3**

Determine if each polynomial is factored. Factor each polynomial completely if possible.

(a) $50 - 2x^2$ (b) $x(2 + t) - y(2 + t)$ (c) $x^2 + 5$ (d) $4(x^2 - 1)$

Solutions

(a) This polynomial is not factored. It is a binomial with a common factor of 2.

$$50 - 2x^2 = 2(25 - x^2) \qquad \text{Common factor}$$
$$= 2(5 - x)(5 + x) \qquad \text{Difference of two squares}$$

(b) At first glance, this polynomial looks like it is factored. However, it is not factored. There are two terms with a common factor of $2 + t$.

$$x(2 + t) - y(2 + t) = (2 + t)(x - y)$$

(c) $x^2 + 5$ is prime.

(d) This polynomial is factored, *but* it is not factored completely.

$$4(x^2 - 1) = 4(x - 1)(x + 1)$$

◄

Problem Set 1.7

Warm-Ups

In Problems 1 through 8, factor each polynomial completely. See Examples 1 and 2.

1. $25x^2 - 25$

2. $x^3 - x$

3. $2w^3 + 250x^3$

4. $3x^4 + x^2 - 4$

5. $x^4 + x^3 - 2x^2$

6. $2x^3 - 4x^2 - 6x$

7. $2yz - 8y + 8 - 2z$

8. $az^2 - az^2y - 4ay + 4a$

In Problems 9 through 24 determine if each polynomial is factored. Factor each completely if possible. See Example 3.

9. $x^2 + x + 1$

10. $2ab + 2ax - by - xy$

11. $16x^2 + 8x + 1$

12. $16 + 15y^2 - y^4$

13. $(3x - 5)^3$

14. $2x^2 - 5x - 7$

15. $18a^3 + 36b^3c^3 + 27b^2c^2$

16. $x^6 + x^3 - 2$

17. $x^2 - 10x + 25$

18. $x^6 + x^5 + x^4 + x^3$

19. $a^2z + a^2x - z - x$

20. $x^2(x + y)$

21. $y^2 - yz - z + y$

22. $9r^2 - 23r - 8$

23. $21x^2 + 24x - 36$

24. $2x^2 + x + 1$

Practice Exercises

In Problems 25 through 60, determine whether or not each polynomial is factored. Factor it completely, if possible.

25. $x^2 + 9x + 14$

26. $9 - 16x^8$

27. $135r^3t^5 + 225r^2$

28. $x^4 - 8x^2 + 16$

29. $1 - 27a^3b^3$

30. $x^4 - 64$

31. $x^2 - y^2 + x + y$

32. $a^3b(x + 1) + ab^2(1 + x)$

33. $s^3t(2st + u)$

34. $x^2 + x$

35. $32 - 2x^2$

36. $r^3t(s - 1) + rt(1 - s)$

37. $r^3(x - t) + (t - x)$

38. $6s^2 + st - 15t^2$

39. $27r^3 + 64$

40. $a^2b + ab^2 + a + b$

41. $x(r + s) - y(s + r)$

42. $x^4 + 3x^2 + 2$

43. $x^8 - 16$

44. $8x^9 + y^3$

45. $a(x - y) - b(y - x)$

46. $x^2 + 12x + 36$

47. $x^2 - 6x - 5$

48. $3x^2 + x + 1$

49. $x^4 - x^3 + x^2 - x$

50. $2 - 2x^2$

51. $x^4 + x^2 - 20$

52. $x^{10} + x^5 - 6$

53. $81 - 18x + x^2$

54. $z(a + b)$

55. $a^9 + 8$

56. $(a + b)^2$

Challenge Problems

In Problems 57 through 60, factor each polynomial completely. Assume that n is a natural number.

57. $x^{4n} - 2x^{2n} + 1$

58. $x^{5n} + 8x^{2n}$

59. $x^{4n} - x^{2n}$

60. $x^{6n} - 8$

61. We can factor certain trinomials by writing the polynomial as the difference of two squares. Consider $x^4 + x^2 + 1$. This will not factor as it is. If the middle term were $2x^2$, it would factor. Let's make the middle term $2x^2$ by adding x^2 and subtracting x^2.

Factor each polynomial.

(a) $x^4 + 3x^2 + 4$

(b) $x^4 + 2x^2 + 9$

(c) $x^4 + 7x^2 + 16$

(d) $x^4 - 8x^2 + 4$

(e) $x^4 - 11x^2 + 25$

$$x^4 + x^2 + 1 = (x^4 + x^2 + 1) + (x^2 - x^2)$$

$$= (x^4 + 2x^2 + 1) - x^2$$

$$= (x^2 + 1)^2 - x^2$$

$$= [(x^2 + 1) - x][(x^2 + 1) + x]$$

$$= (x^2 - x + 1)(x^2 + x + 1)$$

In Your Own Words

62. Explain the steps to follow in factoring a polynomial.

63. What do we mean when we say that a polynomial is factored?

64. What do we mean when we say that a polynomial is factored completely?

1.8 Division of Polynomials

TAPE 4

Division of polynomials can be done like division of real numbers. Recall that there are several ways of writing a division problem. If P and D are polynomials, the following statements have the same meaning.

$$P \text{ divided by } D, \qquad P \div D, \qquad \frac{P}{D}, \qquad P/D, \qquad D\overline{)P}$$

We call P the **dividend** and D the **divisor.**

When we divide real numbers, the divisor must be nonzero. For polynomials, the divisor must not have a value of zero.

We divided monomials by monomials in Section 1.1 using the laws of exponents. For example,

$$\frac{4x^5}{2x^3} = 2x^2 \qquad \text{and} \qquad \frac{3a^2}{6a} = \frac{a}{2}$$

The laws of exponents do not tell how to do other division, such as $\frac{6x^3 + 3x^2}{2x}$. The next result follows from the *properties of real numbers,* found in Section 0.2.

Division by a Monomial

If A, B, and C are monomials, then

$$\frac{A + B}{C} = \frac{A}{C} + \frac{B}{C}$$

(C must not have a value of 0.)

► **EXAMPLE 1**

Perform the division $\dfrac{6x^3 + 3x^2}{2x}$.

Solution

$$\frac{6x^3 + 3x^2}{2x} = \frac{6x^3}{2x} + \frac{3x^2}{2x}$$

$$= 3x^2 + \frac{3x}{2}$$

$$= 3x^2 + \frac{3}{2}x \qquad \blacktriangleleft$$

If the divisor is not a monomial, we use long division. It is much like long division of numbers. The steps are illustrated in this example.

► **EXAMPLE 2**

Divide $(2x^3 - x^2 + 5x + 4)$ by $(2x + 1)$.

Solution

STEP 1   Put each polynomial in standard form and write

$$2x + 1 \overline{)\, 2x^3 - x^2 + 5x + 4}$$

STEP 2   What is $\dfrac{2x^3}{2x}$? It is x^2. So write

$$\begin{array}{r} x^2 \\ 2x + 1 \overline{)\, 2x^3 - x^2 + 5x + 3} \end{array}$$

STEP 3   Multiply x^2 by $2x + 1$ and write

$$\begin{array}{r} x^2 \\ 2x + 1 \overline{)\, 2x^3 - x^2 + 5x + 4} \\ 2x^3 + x^2 \end{array}$$

Be Careful!   STEP 4   **Subtract** $2x^3 + x^2$ from $2x^3 - x^2$.

$$\begin{array}{r} x^2 \\ 2x + 1 \overline{)\, 2x^3 - x^2 + 5x + 4} \\ 2x^3 + x^2 \\ \hline -2x^2 \end{array}$$

STEP 5   Bring down the next term.

$$\begin{array}{r} x^2 \\ 2x + 1 \overline{)\, 2x^3 - x^2 + 5x + 4} \\ 2x^3 + x^2 \\ \hline -\,2x^2 + 5x \end{array}$$

Now we start the process over. What is $\dfrac{-2x^2}{2x}$? It is $-x$. So write

$$\begin{array}{r} x^2 \;-\; x \\ 2x + 1 \overline{)\, 2x^3 - x^2 + 5x + 4} \\ 2x^3 + x^2 \\ \hline -\,2x^2 + 5x \end{array}$$

Multiply $-x$ by $2x + 1$.

$$
\begin{array}{r}
x^2 - x \\
2x + 1 \overline{\smash{\big)}\, 2x^3 - x^2 + 5x + 4} \\
\underline{2x^3 + x^2 } \\
-2x^2 + 5x \\
-2x^2 - x
\end{array}
$$

Subtract $-2x^2 - x$ from $-2x^2 + 5x$.

$$
\begin{array}{r}
x^2 - x \\
2x + 1 \overline{\smash{\big)}\, 2x^3 - x^2 + 5x + 4} \\
\underline{2x^3 + x^2 } \\
-2x^2 + 5x \\
\underline{-2x^2 - x } \\
6x
\end{array}
$$

Now we begin again. Bring down the 4. Since $\dfrac{6x}{2x} = 3$, write

$$
\begin{array}{r}
x^2 - x + 3 \\
2x + 1 \overline{\smash{\big)}\, 2x^3 - x^2 + 5x + 4} \\
\underline{2x^3 + x^2 } \\
-2x^2 + 5x \\
\underline{-2x^2 - x } \\
6x + 4
\end{array}
$$

Multiply 3 by $2x + 1$.

$$
\begin{array}{r}
x^2 - x + 3 \\
2x + 1 \overline{\smash{\big)}\, 2x^3 - x^2 + 5x + 4} \\
\underline{2x^3 + x^2 } \\
-2x^2 + 5x \\
\underline{-2x^2 - x } \\
6x + 4 \\
6x + 3
\end{array}
$$

Be Careful!

Subtract $6x + 3$ from $6x + 4$ and write

$$
\begin{array}{r}
\text{Quotient} \longleftarrow \quad x^2 - x + 3 \\
\text{Divisor} \longrightarrow \quad 2x + 1 \overline{\smash{\big)}\, 2x^3 - x^2 + 5x + 4} \longleftarrow \text{Dividend} \\
\underline{2x^3 + x^2 } \\
-2x^2 + 5x \\
\underline{-2x^2 - x } \\
6x + 4 \\
\underline{6x + 3} \\
1 \longleftarrow \text{Remainder}
\end{array}
$$

The degree of 1 is less than the degree of $2x + 1$, so the procedure stops. Thus

$$
(2x^3 - x^2 + 5x + 4) \div (2x + 1) = x^2 - x + 3 + \frac{1}{2x + 1}
$$

Be Careful!   Be careful when carrying out the subtraction step. Most of the errors made in long division result from not subtracting correctly.

- The polynomial $2x + 1$ is the **divisor.**
- The polynomial $2x^3 - x^2 + 5x + 4$ is the **dividend.**
- The polynomial $x^2 - x + 3$ is the **quotient.**
- The polynomial 1 is the **remainder.**

To check the division, we multiply the divisor by the quotient and add the remainder. This should give the dividend.

$$\text{Divisor} \cdot \text{Quotient} + \text{Remainder} = \text{Dividend}$$

$$(2x + 1)(x^2 - x + 3) + 1 = (2x^3 - x^2 + 5x + 3) + 1$$

$$= 2x^3 - x^2 + 5x + 4$$

► **EXAMPLE 3**

Perform the division $\dfrac{x^2 + x - 6}{x + 3}$.

Solution

Both polynomials are in standard form.

$$x + 3 \,\overline{\smash{\big)}\, x^2 + x - 6}$$

$$
\begin{array}{r}
x \\
x + 3 \overline{\smash{\big)}\, x^2 + x - 6} \\
\underline{x^2 + 3x} \\
- 2x
\end{array}
$$
  $\longleftarrow$ $x^2 \div x = x$
  $\longleftarrow$ $x(x + 3)$
  $\longleftarrow$ $x^2 + x - (x^2 + 3x)$

$$
\begin{array}{r}
x - 2 \\
x + 3 \overline{\smash{\big)}\, x^2 + x - 6} \\
\underline{x^2 + 3x} \\
- 2x - 6 \\
\underline{- 2x - 6} \\
0
\end{array}
$$
  $\longleftarrow$ $-2x \div x = -2$
  $\longleftarrow$ Bring down -6
  $\longleftarrow$ $-2(x + 3)$

$$\frac{x^2 + x - 6}{x + 3} = x - 2$$

To check the answer,

$$(x - 2)(x + 3) = x^2 + x - 6$$ ◄

► **EXAMPLE 4**

Divide $2 - 2x + x^3$ by $x + 1$.

Solution

Write each polynomial in standard form first. Notice that there is no x^2 term in the dividend. Indicate this by writing $0x^2$ in its proper place.

$$
\begin{array}{r}
x^2 - x - 1 \\
x + 1 \overline{\smash{\big)}\ x^3 + 0x^2 - 2x + 2} \\
\underline{x^3 + x^2} \\
-x^2 - 2x \\
\underline{-x^2 - x} \\
-x + 2 \\
\underline{-x - 1} \\
3
\end{array}
$$

$$(2 - 2x + x^3) \div (x + 1) = x^2 - x - 1 + \frac{3}{x + 1}$$ ◀

► **EXAMPLE 5**

Divide $(2x^3 - 3x^2y + 2xy^2 + y^3)$ by $(2x + y)$.

Solution

$$
\begin{array}{r}
x^2 - 2xy + 2y^2 \\
2x + y \overline{\smash{\big)}\ 2x^3 - 3x^2y + 2xy^2 + y^3} \\
\underline{2x^3 + x^2y} \\
-4x^2y + 2xy^2 \\
\underline{-4x^2y - 2xy^2} \\
4xy^2 + y^3 \\
\underline{4xy^2 + 2y^3} \\
-y^3
\end{array}
$$

$$(2x^3 - 3x^2y + 2xy^2 + y^3) \div (2x + y)$$

$$= x^2 - 2xy + 2y^2 + \frac{-y^3}{2x + y}$$ ◀

► **EXAMPLE 6**

Perform the division $\dfrac{x^3 - y^3}{x - y}$.

Solution

$$
\begin{array}{r}
x^2 + xy + y^2 \\
x - y \overline{\smash{\big)}\ x^3 + 0x^2y + 0xy^2 - y^3} \\
\underline{x^3 - x^2y} \\
x^2y + 0xy^2 \\
\underline{x^2y - xy^2} \\
xy^2 - y^3 \\
\underline{xy^2 - y^3}
\end{array}
$$

$$\frac{x^3 - y^3}{x - y} = x^2 + xy + y^2$$ ◀

Problem Set 1.8

Warm-Ups

In Problems 1 through 6, perform the division indicated. See Example 1.

1. $\dfrac{a^4 b^2}{b}$

2. $\dfrac{xy^3}{x^3 y}$

3. $\dfrac{(x - y)^4 z^2}{(x - y)^2 z}$

4. $\dfrac{4a + 2b}{2b}$

5. $\dfrac{6x^2 + 12y}{6x}$

6. $\dfrac{x^2 - x}{x}$

In Problems 7 through 13, perform the division indicated. See Examples 2 and 3.

7. $\dfrac{z^2 - 4z - 12}{z + 2}$

8. $\dfrac{x^2 - x - 2}{x + 1}$

9. $\dfrac{2x^2 - 5x - 3}{2x + 1}$

10. $\dfrac{x^2 + 3x + 4}{x + 2}$

11. $\dfrac{x^2 + 2x - 4}{x + 3}$

12. $(4x^4 - 3x^2 + 1 + 2x) \div (x + 2x^2 - 2)$

13. $\dfrac{x^5 - 3x^4 + x^3 - 2x^2 + 5x + 4}{x^2 - 2x + 3}$

In Problems 14 through 17, perform the division indicated. See Examples 4 and 5.

14. $\dfrac{x^2 - 2xy + y^2}{x - 3y}$

15. $\dfrac{x^3 + x^2 y - xy^2 + 2y^3}{x + 2y}$

16. $\dfrac{x^3 + y^3}{x + y}$

17. $(x^4 + x^2 - x^3 + 1)/(x^2 - 1)$

Practice Exercises

In Problems 18 through 46, perform the division indicated.

18. $\dfrac{-16a^2 c}{8ac}$

19. $\dfrac{27ab}{18b^2}$

20. $\dfrac{(a + b)^3 t}{(a + b)^2 t^3}$

21. $\dfrac{5x^2 + 10x - 15}{10y}$

22. $\dfrac{a^2 b - ac}{ab}$

23. $\dfrac{x^2 - xy + y^2}{xy}$

24. $\dfrac{x^2 - 4x + 4}{x - 2}$

25. $\dfrac{6x^2 + 7x - 3}{2x + 3}$

26. $\dfrac{x^3 + 3x + 3x^2 + 1}{x + 1}$

27. $\dfrac{y^2 + 2y - 3}{y - 1}$

28. $\dfrac{6x^2 - x - 2}{3x + 1}$

29. $(2x^3 - x^2 - 3x + 3) \div (2x + 1)$

30. $\dfrac{2x^3 - x^2 - 5x + 2}{2x - 3}$

31. $\dfrac{x^2 + 6x^3 + x - 5}{3x + 2}$

32. $\dfrac{x^3 - 6x^2 + 12x - 8}{x - 2}$

33. $\dfrac{2x^2 - 5x - 3}{x - 3}$

34. $\dfrac{3x + x^2 - 5}{4 + x}$

35. $(6x^2 + 5xy + y^2)/(2x + y)$

36. $\dfrac{12x^2 + 17x + 12}{4x + 3}$

37. $\dfrac{3x^4 - 2x^3 - 2x + x^2 + 4}{x + 3x^2 - 1}$

38. $\dfrac{6x^3 + 4x^2 + x - 2}{3x^2 - x - 2}$

39. $\dfrac{21x^3 + 5x^2 + 3x + 8}{7x + 4}$

40. $(2x^4 - x^3 + 5x + 3x^2 + 7) \div (x^2 + 4 + 2x)$

41. $\dfrac{x^4 - x + 2}{3 + x}$

42. $\dfrac{x^4 - 2x^2 + 1}{x + 1}$

43. $\dfrac{x^3 + x^6 - 3}{x^3 - 1}$

44. $\dfrac{x^4 - y^4}{x - y}$

45. $\dfrac{4x^5 - 6x^3 + 2x^2 + 1}{2x^2 + 1}$

46. $\dfrac{x^5 + x^4 - 4x + 1}{x^2 - 2}$

47. Divide $x^3 + x^2 - 10x + 8$ by $x - 2$.

48. Find the remainder when $4x^3 - 12x^2 - x + 7$ is divided by $2x - 1$.

49. Find the remainder when $6t^3 - 13t^2 + 21t + 17$ is divided by $3t - 2$.

Challenge Problems

In Problems 50 through 52, perform the division indicated.

50. $\dfrac{x^2 - 2ax + 2a^2}{x - a}$

51. $\dfrac{x^{2n} - 2x^n + 4}{x^n + 1}$; n is a natural number.

52. $\dfrac{x^n - 1}{x - 1}$; n is a natural number.

53. Find the value of k so that, if $x^3 + 2x^2 + x + k$ is divided by $x + 3$, the remainder will be zero.

In Your Own Words

54. Explain the procedure for long division.

55. When do we use long division?

1.9 Synthetic Division (Optional)

TAPE 4

The long-division procedure that we learned in Section 1.8 can be shortened if the divisor is of the form $x - r$, where r is a constant. This process is called **synthetic division.**

Let's divide $x^2 + 3x - 5$ by $x - 1$ using long division.

$$
\begin{array}{r}
x + 4 \\
x - 1 \overline{\smash{\big)}\, x^2 + 3x - 5} \\
\underline{x^2 - x} \\
4x - 5 \\
\underline{4x - 4} \\
-1
\end{array}
$$

Removing all the variables but leaving their coefficients, we have the display

$$
\begin{array}{r}
1\quad\ \ 4 \\
1\quad-1\,|\,\overline{1\quad\ \ 3\quad-5} \\
1\quad-1 \\
\overline{4\quad-5} \\
4\quad-4 \\
\overline{-1}
\end{array}
$$

Many of the numbers in this display are repetitions of the numbers above them. Removing them and condensing further, we obtain

$$
\begin{array}{r}
1\quad\ \ 4 \\
1\quad-1\,|\,\overline{1\quad\ \ 3\quad-5} \\
-1\quad-4 \\
\overline{4\quad-1}
\end{array}
$$

If the first number in the dividend is copied below the line, the quotient appears there and we can eliminate the top line.

$$
\begin{array}{r}
1\quad-1\,|\,1\quad\ 3\quad-5 \\
-1\quad-4 \\
\overline{1\quad\ 4\quad-1}
\end{array}
$$

Finally, if we eliminate the leading 1 in the divisor and change the sign of the remaining number and all the signs in the second row (so that we can add instead of subtract), we get the following display:

$$
\begin{array}{r}
1\,|\,1\quad\ 3\quad-5 \\
1\quad\ \ 4 \\
\overline{1\quad\ 4\quad-1}
\end{array}
$$

All the information in the original long-division problem is contained in this condensed form. We only need to know how to make it and read the result.

Suppose that we wish to divide $3x^2 - 2x + 10$ by $x - 2$. **Remember,** we use synthetic division **only** when we are dividing by a binomial of the form $x - r$. Here r is 2. We make the first line of the display by writing r and the coefficients of the dividend.

$$
2\,|\,3\quad-2\quad10
$$

Next, we draw a line and bring down the 3.

$$
\begin{array}{r}
2\,|\,3\quad-2\quad10 \\
\overline{} \\
3
\end{array}
$$

Next we multiply 2 by 3 and write

$$
\begin{array}{r}
2\,|\,3\quad-2\quad10 \\
6 \\
\overline{3}
\end{array}
$$

Now we *add* -2 and 6.

$$
\begin{array}{r}
2\,|\,3\quad-2\quad10 \\
6 \\
\overline{3\quad\ \ 4}
\end{array}
$$

Multiply 2 by 4 and write

$$
\begin{array}{r|rrr}
2 & 3 & -2 & 10 \\
  &   & 6 & 8 \\
\hline
  & 3 & 4 &
\end{array}
$$

Add 10 and 8.

$$
\begin{array}{r|rrr}
2 & 3 & -2 & 10 \\
  &   & 6 & 8 \\
\hline
  & 3 & 4 & 18
\end{array}
$$

Remainder
Constant
Coefficient of x

The last line contains the coefficients of the quotient and the remainder. Therefore, $3x^2 - 2x + 10$ divided by $x - 2$ is $3x + 4$ with a remainder of 18.

Synthetic division can be used only when the divisor is of the form $x - r$, that is, a variable to the first power minus a constant. The dividend **must** be in standard form.

► EXAMPLE **1**

Use synthetic division to divide

$$4x^4 - 6x^3 - 3x^2 - 4x + 5 \quad \text{by} \quad x - 2$$

Solution

We note that both polynomials are in standard form.

$$
\begin{array}{r|rrrrr}
2 & 4 & -6 & -3 & -4 & 5 \\
  &   & 8 & 4 & 2 & -4 \\
\hline
  & 4 & 2 & 1 & -2 & 1
\end{array}
$$

Remainder
Constant
Coefficient of x
Coefficient of x^2
Coefficient of x^3

$$\frac{4x^4 - 6x^3 - 3x^2 - 4x + 5}{x - 2} = 4x^3 + 2x^2 + x - 2 + \frac{1}{x - 2}$$

◄

► EXAMPLE **2**

Use synthetic division to perform the division

$$(x^3 - 8x - 1)/(x + 3)$$

Solution

We notice that the x^2 term is missing in the dividend. This means that the coefficient of x^2 is 0. Also, the divisor must be written in the form $x - r$, so

$$x + 3 = x - (-3).$$

$$\begin{array}{r|rrr} -3 & 1 & 0 & -8 & -1 \\ & & -3 & 9 & -3 \\ \hline & 1 & -3 & 1 & -4 \end{array}$$

$$(x^3 - 8x - 1)/(x + 3) = x^2 - 3x + 1 + \frac{-4}{x + 3} \quad \blacktriangleleft$$

► EXAMPLE 3

Find the remainder when $x^3 - x^2 - 5x + 6$ is divided by $x - 1$.

Solution

The easiest way to find the remainder in such a situation is by synthetic division.

$$\begin{array}{r|rrrr} 1 & 1 & -1 & -5 & 6 \\ & & 1 & 0 & -5 \\ \hline & 1 & 0 & -5 & 1 \end{array}$$

The remainder is 1. ◄

Problem Set 1.9

Warm-Ups

In Problems 1 through 4, use synthetic division to perform the division indicated. See Examples 1 and 2.

1. $\dfrac{2x^2 - 5x - 3}{x - 3}$

2. $\dfrac{y^2 + 2y - 3}{y - 1}$

3. $\dfrac{x^3 + 3x^2 + 3x + 1}{x + 1}$

4. $\dfrac{x^3 + 8}{x + 2}$

In Problems 5 and 6, use synthetic division to find the remainder when the given polynomial is divided by $x - 4$. See Example 3.

5. $x^3 - 64$

6. $x^3 - 2x^2 - 9x + 3$

Practice Exercises

In Problems 7 through 22, use synthetic division to perform the division indicated.

7. $\dfrac{x^4 - 2x^2 + 1}{x - 1}$

8. $\dfrac{3x + x^2 - 5}{4 + x}$

9. $\dfrac{7x^2 + 4x^3 + x + 5}{x + 2}$

10. $\dfrac{3x^3 - 12x^2 - 14x - 6}{x - 5}$

11. $(2x^2 - 9x + 9) \div (x - 3)$

12. $\dfrac{2x + 1 + 3x^3 + 4x^2}{x + 1}$

13. $\dfrac{x^4 - 2x^2 - 5x + 3}{x - 2}$

14. $\dfrac{x^4 + x^2 + 1}{x + 1}$

15. $(2x^3 - x^2 - 3x - 4)/(x - 2)$

16. $(z^5 - 1) \div (z - 1)$

17. $\dfrac{y^4 + 1}{y + 1}$

18. $(x^4 - 16) \div (x - 2)$

19. $(x^4 + x^2 - 2) \div (x - 1)$

20. $\dfrac{x^5 - 1}{x - 1}$

21. $\dfrac{x^4 + x^3 + x^2 + x + 1}{x - 1}$

22. $\dfrac{x^4 - x^3 + x^2 - x + 1}{x - 1}$

In Problems 23 and 24, use synthetic division to find the remainder when the polynomial is divided by $x - 2$.

23. $2x^4 - 4x^3 + x^2 - x - 2$

24. $x^3 + 3x^2 - 3x + 4$

In Your Own Words

25. Explain how to divide $x^2 - x - 6$ by $x - 3$ using synthetic division.

► Chapter Summary

GLOSSARY

Base in the expression x^n The base is x.

Exponent in x^n The exponent is n.

Monomial   An expression of the form bx^n, where n is a whole number and b is a real number.

Coefficient in the expression bx^n The coefficient is b.

Polynomial   A sum of monomials.

A polynomial written in **standard form**   The monomial of highest degree is written first and other monomials are written in decreasing order of degree.

To **factor** a polynomial   To write the polynomial as a product of polynomials.

To **factor** a polynomial **completely**   To write the polynomial as a product of prime polynomials.

Binomial   A polynomial with two terms.

Trinomial   A polynomial with three terms.

To **evaluate** a polynomial   To find the value of a polynomial given the value of its variable(s).

Greatest common factor of a set of numbers   The largest factor that is a factor of each number in the set.

Exponent Definitions   If n is a natural number:

1. $x^n = x \cdot x \cdot x \ldots x$ (n x's)

2. $x^0 = 1; \quad x \neq 0$

3. $x^{-n} = 1/x^n, \qquad x \neq 0$

Properties of Exponents

1. $x^m \cdot x^n = x^{m+n}$ Product with same base

2. $(x^m)^n = x^{mn}$ Power of a power

3. $(xy)^n = x^n y^n$ Product to a power

4. $\left(\dfrac{x}{y}\right)^n = \dfrac{x^n}{y^n}, \quad y \neq 0$ Quotient to a power

5. $\dfrac{x^m}{x^n} = x^{m-n}, \quad x \neq 0$ Quotient with same base

Special Products

Square of a Binomial

1. $(p + q)^2 = p^2 + 2pq + q^2$

2. $(p - q)^2 = p^2 - 2pq + q^2$

Difference of Two Squares

3. $(p - q)(p + q) = p^2 - q^2$

Cube of a Binomial

4. $(p + q)^3 = p^3 + 3p^2q + 3pq^2 + q^3$

5. $(p - q)^3 = p^3 - 3p^2q + 3pq^2 - q^3$

Division of Polynomials

To divide polynomials, use long division if the divisor is not a monomial. If the divisor is a monomial, divide each term in the dividend by the divisor.

Procedure to Factor a Polynomial Completely

1. Factor out the GCF if there is one. Do this *first*.

2. Count the number of terms.

3. If two terms, try difference of squares, difference of cubes, or sum of cubes. NOTE: $p^2 + q^2$ will not factor with real coefficients.

4. If three terms, look for two binomial factors.

5. If four or more terms, try grouping.

6. Make sure that each factor is prime.

7. Check to see if the product of the factors is the original polynomial.

Factoring Patterns

1. $p^2 - q^2 = (p - q)(p + q)$ Difference of two squares

2. $p^3 - q^3 = (p - q)(p^2 + pq + q^2)$ Difference of two cubes

3. $p^3 + q^3 = (p + q)(p^2 - pq + q^2)$ Sum of two cubes

4. $p^2 + 2pq + q^2 = (p + q)^2$ Perfect square

5. $p^2 - 2pq + q^2 = (p - q)^2$ Perfect square

CHECKUPS

1. Simplify $(3x^{-4})^{-2}$.   Section 1.1; Example 4d

2. Evaluate the following polynomial when x is 2 and y is -1.

$3x^2y^4 - xy + y^5$ Section 1.2; Example 2b

3. Subtract: $(3x^3 - 7x + 2) - (x^3 + x - 8)$ Section 1.2; Example 5a

4. Multiply: $(x - y)(x^2 - xy + 1)$ Section 1.3; Example 2d

5. Factor $-15x^2y + 5xy^2$ completely.   Section 1.5; Example 5

6. Factor $8x^3 + 27$ completely.   Section 1.6; Example 2b

7. Factor $6x^2 - 13x + 6$ completely.   Section 1.6; Example 6a

8. Divide $2 - 2x + x^3$ by $x + 1$.   Section 1.8; Example 4

► Review Problems

In Problems 1 and 2, evaluate each polynomial when x is 1 and y is −3.

1. $x^3 - xy^2 + y^3$ **2.** $x^2 - y^2$

In Problems 3 through 17, use the properties of exponents to simplify each expression.

3. $(-5)^2$

4. -5^2

5. $(-2)^2(-3)^{-3}$

6. $\dfrac{(x - t)^4}{(x - t)^2}$

7. $(x^{-2}y^3)^4$

8. $(-xy)^{-3}$

9. $(a + b)^4(a + b)$

10. $\left(\dfrac{x^4}{y^{-7}z^8}\right)^{-5}$

11. $\left(\dfrac{-2x}{y}\right)^2$

12. $\left(\dfrac{xz}{-y^2}\right)^3$

13. $\dfrac{-10^0}{10}$

14. $(-2x^2z^3)^3$

15. $\dfrac{(-6)^4}{(-6)^5}$

16. $\left(\dfrac{2x^5}{x^4z^7}\right)^0$

17. $\dfrac{-5^3}{-2^4}$

In Problems 18 through 49, perform the operations indicated.

18. $(x^2 - 3x + 8) + (3x^2 - x - 6)$

19. $abc(a^2b^3c - ac^4)$

20. $\dfrac{7t^2 - 7t}{7t}$

21. $(7p^3 - 4p^2 - 7) - (p^3 + 6p - 5)$

22. $(2x - 3)(x + 4)$

23. $\dfrac{xy^3z^2}{x^2yz}$

24. $(3r - 7) + (r - 1)$

25. $(r^3s^2t)(r^3st^4)$

26. $\dfrac{x^2 + 2x - 3}{x + 3}$

27. $(x^4 - x^3y + 3x^2y^2 - 5xy + y^5) - (2x^4 + 2x^3y - x^2y^2 + 7xy)$

28. $(2x - 3)^2$

29. $\dfrac{x^3 + 5x^2 - 10x + 20}{x + 7}$

30. $(x + 2)(x^2 - 2x + 4)$

31. $(t - 5)^3$

32. $(8p^5 - 6p^4 + 3p^2 - 3) + (6p^4 - 5p^2 + 6p - 2)$

33. $(7r - 2)(2r - 1)$

34. $(x + 3)(x^2 - 2xy + y^2)$

35. $(c + 3d)(c - 3d)$

36. $\dfrac{18r^2t^2 - 12t + 14rt}{6rt}$

37. $(x - y) - (x + y)$

38. $(s + 3)^3$

39. $xy^2(xy - 1)$

40. $(2x - y)(x^2 + 2xy + y^2)$

41. $(8s^3 + 27) \div (2s + 3)$

42. $2x + x(x - 2)$

43. $(x + 3)^2 - (x - 2)^2$

44. $(3x - 7)(5x + 3)$

45. $\dfrac{x^4 + 2x^3 - x^2 + 6}{x^2 - x + 2}$

46. $3a^2b(3ab^3)$

47. $r(r - t) + (2r - t)(2r + t)$

48. $(x^4 - x^3 - x^2 - x + 1) \div (x + 1)$

49. $6 - [x(x - 1) - 2(x + 3)]$

In Problems 50 through 79, factor each polynomial completely, if possible.

50. $2x^2 - 162$

51. $x^2 - 7x - 18$

52. $at - 4z^2 - 2az + 2zt$

53. $64y^3 + 125$

54. $4z^2 - 4$

55. $2x^2 + x + 1$

56. $12x^2 - 38x - 14$

57. $2x^2 + x - 1$

58. $12a^2b^2c - 96a^2b^3c + 54abc$

59. $4z^2(s - t) - 2z(t - s)$

60. $x^4 - 15x^2 - 16$

61. $8 + r^2$

62. $x^6 + 28x^3 + 27$

63. $144 - x^2$

64. $2a^2b + 4a^2y - 3abz - 6ayz$

65. $30(u - v)^2 + 42(v - u)$

66. $2ax + 4x + 2a + 4$

67. $x^2 + 4$

68. $x^2 + 18x + 81$

69. $3y^2 - 11y + 8$

70. $w^2 - x^2 - 4x - 4$

71. $a^6 - 125$

72. $x^8 - 81$

73. $15 - x - 2x^2$

74. $3x^2 + 4x + 5$

75. $16x^2 + 8x + 1$

76. $27a^3 - 1$

77. $27a^3 - 3a$

78. $x^2 + 4x + 4 - z^2$

79. $x^2 + z^2 - 2xz - y^2$

80. If a stone is thrown into the air in such a manner that its height in feet after t seconds is given by the polynomial $5 + 100t - 16t^2$, how high is the stone after 3 seconds? After 6 seconds?

81. Bernard's porch has a triangular roof whose base is twice its height. What is the area of Bernard's porch roof if its height is x feet?

In Problems 82 through 85, write each number in scientific notation.

82. 31,640,000

83. 0.000998

84. 0.10059

85. 123.456

In Problems 86 through 89, write each number without exponents.

86. 6.338×10^{-4}

87. 1.001×10^{11}

88. 5×10^7

89. 2.006×10^{-7}

Let's Not Forget . . .

How many terms are in each polynomial?

90. $y^2 - 49$

91. $y^2 + 6y - 7$

92. $15x^3y^2z$

93. $2x^4 + 3x^2 - 5x - 6$

Find each product.

94. $(2x)^2$

95. $2x(x + 2)$

96. $(x - 1)(x + 2)$

97. $(x - 2)^2$

98. $(x + 1)^3$

Simplify each expression.

99. -3^2

100. $(-3)^2$

101. $(-3)^3$

102. $(-b)^2$

103. $(-b)^3$

Identify the expressions that are in factored form. Factor those that are not factored.

104. $(2x + 5)^3$

105. $8x^3 + 125$

106. $(s + t)^3 - a(s + t)^2$

107. $5x(x^2 + y^2)$

108. $12ab^2$

109. $a(1 - b) + (b - 1)$

How many terms are in each expression? Which expressions have y + 7 as a factor?

110. $y^2 - 49$

111. $(y + 7)^4$

112. $y^2 - 6y - 7$

113. $y^2 + 14y + 49$

114. $4(y + 7) - (7 + y)$

► You Decide

The Fulton County health code for commercial pools requires a sand filter to be large enough to filter all the water in a pool every 6 hours. The county says that 1 sq ft of filter bed can filter 15 gallons per minute. Sand filters are sold in two sizes—30 in. with 4.9 sq ft of filter bed and 36 in. with 7.02 sq ft of filter bed.

You work for *Metro Pool and Chemical Company*, and a customer with a pool 25 ft by 45 ft needs a new filter system. The customer's pool is 3 ft deep in the shallow end and 8.5 ft deep in the deep end.

To calculate the number of gallons of water in the pool, you can use $7.5 \cdot length \cdot width \cdot average\ depth$.

Figure out what size filter(s) you would recommend that the customer install to meet minimum county standards. Design and write a report for the customer explaining how you made your decision.

► Chapter Test

In Problems 1 through 6, choose the correct answer.

1. $\dfrac{-2x^{-1}}{y^{-2}} = (?)$

   A. $\dfrac{-2y^2}{x}$ **B.** $\dfrac{y^2}{-2x}$

   C. $\dfrac{-2x}{y^2}$ **D.** $-2xy^2$

2. The value of the polynomial $3x^3 - 2x^2y^2$ when x is -1 and y is 3 is (?)

   A. 15   **C.** 21

   B. -15 **D.** -21

3. $\left(\dfrac{-2a^5b}{b^4}\right)^2 = (?)$

   A. $\dfrac{4a^{10}}{b^6}$ **C.** $\dfrac{4a^7}{b}$

   B. $\dfrac{-2a^{10}}{b^6}$ **D.** $\dfrac{-4a^7}{b^8}$

4. $\dfrac{(-3)^4}{(-3)^6} = (?)$

   A. $\dfrac{1}{6}$ **C.** $-\dfrac{1}{6}$

   B. $\dfrac{1}{9}$ **D.** $-\dfrac{1}{9}$

5. $-3^2ab^4(2ab)(-a^4) = (?)$

   A. $-18a^6b^5$ **C.** $12a^4b^4$

   B. $18a^6b^5$ **D.** $-12a^6b^5$

6. Which polynomial is factored?

   A. $x^2 + 2x$ **C.** $x^3 - 8$

   B. $a(r - s) + b(r - s)$ **D.** $(x + 2)^3$

In Problems 7 through 13, perform the operations indicated.

7. $\dfrac{6x^2 - 4x^3}{x^2}$

8. $(2x - 3)(5x + 7)$

9. $(5x^3y^2 - 4x^2y + 3xy) + (-3x^3y^2 + x^2y - 7xy)$

10. $(4x - 7)^2$

11. $(6x^3 - 2x^2 + x - 2) \div (x + 1)$

12. $(5x^3 - x^2 + 4) - (x^3 - 2x - 7)$ **13.** $7 - [2(x + 4) - (x + 2)(x - 2)]$

In Problems 14 through 20, factor each polynomial completely.

14. $8a^3 - 27$ **15.** $4x^2 + 18x - 10$

16. $ax - bx + 2a - 2b$ **17.** $x(p - q) - y(q - p)$

18. $a^4 - 2a^2 - 8$ **19.** $4y^2 - 25$

20. $27a^2b^3 - 36a^3b^2 + 45a^4b^4$

In Problems 21 and 22, write each number in scientific notation.

21. 0.00001076 **22.** $9{,}009{,}000$

In Problems 23 and 24, write each number without exponents.

23. 5.22×10^4 **24.** 7.009×10^{-4}

2 Using Linear Equations and Inequalities in One Variable

See Problem Set 2.5, Exercise 32

Connections

During the Renaissance, the Hindu-Arabic notation for numbers became popular throughout Western Europe. This compact number system was soon followed by the development of the common symbols we use today for addition, multiplication, and equality. These symbols became the key that unlocked algebra for widespread use.

With the use of these Hindu-Arabic numerals and new symbols, complicated arithmetical questions became equivalent to stating the equality of two number representations. We call such statements *equations*. Equations are one of the central ideas of algebra.

Just as equations may arise when the language of mathematics is used to model real-life situations, inequalities may arise when one quantity is larger or smaller than another quantity. For example, charges on a credit card must be less than the credit limit and the load limit on a bridge must be greater than the weight of the vehicles using it.

A large class of equations of the form $Ax + B = 0$ are called *linear equations*. In this chapter we study linear equations and inequalities and their applications. We also examine techniques for solving other types of equations and inequalities and a general way to approach problem solving.

2.1   Solving Linear Equations in One Variable

TAPE 4

An **equation** is a statement that two numbers are equal. This statement may or may not be true. For example;

$$5 = 5$$

$$17 = 31$$

are two such statements. The first is true and the second false.

Equations are interesting when they contain letters, such as x, that stand for numbers whose value is unknown. For example, the equation

$$x + 4 = 11$$

is true if x is 7 and false otherwise. We call x a **variable.** Any number that replaces the variable and makes the equation true is called a **solution** or **root.** The set of all solutions is called the **solution set.** To **solve** an equation means to find the solution set.

Two equations that have the same solution set are called **equivalent.** The three equations

$$2(x - 3) = x - 12$$

$$2x - 6 = x - 12$$

$$x = -6$$

are equivalent because each has $\{-6\}$ as its solution set.

In Problems 9 and 10, solve for x. See Example 4.

9. $a(x - b) = b(x + b)$

10. $3a - (a - x) = 2b(x + 3)$

In Problems 11 and 12, solve for x. See Example 5.

11. $\dfrac{x}{2} = \dfrac{x - a}{3}$

12. $\dfrac{ax + b}{5} = \dfrac{b - ax}{2}$

Practice Exercises

In Problems 13 through 28, solve for the variable indicated.

13. $A = lw$, for l (area of rectangle)

14. $A = \dfrac{1}{2}bh$, for b (area of triangle)

15. $V = lwh$, for w (volume of box)

16. $d = rt$, for r (distance, rate, time)

17. $I = Prt$, for t (simple interest)

18. $V = \dfrac{1}{3}Bh$, for B (volume of pyramid)

19. $C = \pi d$, for d (circumference of circle)

20. $A = P + Prt$, for t (amount accumulated with interest)

21. $A = P + Prt$, for P (amount accumulated with interest)

22. $C = \dfrac{5}{9}(F - 32)$, for F (Celsius–Fahrenheit)

23. $P = 2l + 2w$, for l (perimeter of rectangle)

24. $V = \dfrac{1}{3}\pi r^2 h$, for h (volume of cone)

25. $S = 2\pi rh$, for h (area of side of cylinder)

26. $E = IR$, for R (voltage)

27. $S = \dfrac{n}{2}[2a + (n - 1)d]$, for a (sum of sequence)

28. $S = 2(lh + hw + lw)$, for h (surface area of box)

In Problems 29 through 37, use the appropriate formula to find each of the following.

29. The area of a rectangle is 46.7 ft² and its width is 6.6 ft. Find its length.

30. The area of a triangle is 6 yd² and its height is 4 yd. Find the length of its base.

31. A jogger runs 10 km in 55 minutes. What is his rate?

32. The circumference of a circle is 21.2 m. Find the length of the diameter of the circle correct to two decimal places.

33. The perimeter of a rectangle is $110\dfrac{1}{4}$ ft and its width is $27\dfrac{1}{2}$ ft. Find its length.

34. The Kentucky Derby is a $1\dfrac{1}{4}$-mile horse race. Strike the Gold won the 117th Derby in May 1991 with a time of 2 minutes and 3 seconds. Approximated to tenths, what was his speed in mph?

35. The following table gives the low temperature for certain cities on a day in December. Complete the chart with the corresponding Fahrenheit or Celsius degrees.

City	Fahrenheit	Celsius
Washington	32°	
Paris		5°
Fairbanks	−31°	
Nassau		20°
Moscow		−10°
Miami	77°	

36. Juanita has inherited some money from her aunt who had deposited $1000 in each of three savings accounts drawing simple interest. The total interest earned in each account and the number of years that the accounts were active are shown below.

	Interest Earned	Number of Years
Account 1	$750	15
Account 2	$540	12
Account 3	$300	5

Determine which account had the highest rate of interest.

37. For shipping fruit, the Pierpont Manufacturing Company makes rectangularly shaped wire-bound boxes in different sizes. A typical box, shown below, has length l, width w, and height h. The formula for the surface area for such a box is $S = 2wh + 2lh + 2wl$. The surface area, length, and width of three types at boxes are listed below. The trucking company that ships the fruit has asked Pierpont to label each box with its height. Find the height of each box.

Type of Box	Surface Area (in.²)	Length (in.)	Width (in.)
1	1152	24	12
2	1440	24	12
3	1512	24	12

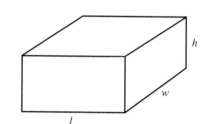

In Problems 38 through 57, solve each equation for x.

38. $x - b = 6$

39. $3x + 5 = a$

40. $4bx + a = 7$

41. $ax + ab = b$

42. $\dfrac{x + a}{3} = 1$

43. $\dfrac{x - a}{4} = \dfrac{3}{2}$

44. $b(2 - x) = a$

45. $3(5 + ax) = 1$

46. $2ax + 3 = 4ax$

47. $ax - 2 = bx$

48. $a(x + b) = b(x - a)$

49. $2b - (x - a) = a(x - 2)$

50. $(x + a)^2 - (x - a)^2 = 1$

51. $bx + 4 - a^2 = ax$

52. $a(x + a - 1) = x - a$

53. $(x - b)(x + a) = x^2 + ab$

54. $\dfrac{1}{3}bx - a = ax + b$

55. $\dfrac{2}{3}bx + a + b = \dfrac{2}{5}x$

56. $\dfrac{4}{7}(ax - b) = 2$

57. $\dfrac{5}{6}(bx + 2a) = \dfrac{1}{6}ax + b^2$

Challenge Problems

58. Sometimes we must be careful to note restrictions when reducing. Consider

$$x = \frac{(b - 2)(b + 2)}{b - 2}$$

Notice that, if we reduce by dividing out the common factor $(b - 2)$, we will have *divided by zero if b* has the value 2. In cases like this, we alert the formula user with the restriction $b \neq 2$.

$$x = b + 2, \qquad b \neq 2$$

Solve each equation for x, and note any restrictions.
(a) $a(a - x) = x + 1$ (b) $2(x + 2) = b(b + x)$
(c) $(b + 2)(x - b + 3) = 0$
(d) $a(x - a) = x - 1$

59. Solve $\dfrac{x + 2}{x - a} = \dfrac{x + a}{x - 1}$ for x. (Be careful to note where denominators are zero.)

In Your Own Words

60. Describe how to solve the equation $Ax + bx = C$ for x.

61. What does solving a formula for a variable mean?

2.3   Linear Inequalities in One Variable

TAPE 5

An **inequality** is a statement that says one number is greater than another. For example,

$$5 > 3$$
$$7 > -9$$
$$-2 > -5$$

are such statements. Sometimes these statements are written the other way around and say that one number is less than another. For example,

$$3 < 5$$
$$-9 < 7$$
$$-5 < -2$$

Often, we wish to allow equality along with inequality. For example, x **less than or equal** to 5 can be written

$$x \leq 5$$

We write x **greater than or equal** to 3 as

$$x \geq 3$$

Like equations, inequalities are interesting when they contain letters, such as x, that stand for numbers whose value is unknown. For example, the inequality,

$$x + 4 > 11$$

is true if x is 9 and false if x is 3. Note that the statement is true if x is any number greater than 7 and false if x is 7 or less.

The collection of numbers that, when substituted for x, make the statement true is called the **solution set** of the inequality. To **solve** an inequality is to find its solution set. Usually, the solution set for an inequality cannot be written by listing its elements. We use set-builder notation. For example, to write the solution set for the inequality above, we need to write the set of all numbers greater than 7. In set-builder notation, that is

$$\{x \mid x > 7\}$$

Two inequalities with the same solution set are said to be **equivalent.**

It is often convenient and instructive to draw a picture of the solution set of inequalities. This picture of the solution is called the **graph** of the inequality.

To graph the set $\{x \mid x < 1\}$, we construct a copy of the real number line and locate all numbers less than 1 on it. They lie to the left of 1. We darken that portion of the line containing the solution set and indicate by an arrowhead that it continues indefinitely. We indicate with a **parenthesis** that the number 1 does *not* belong to the solution set.

There are other notations that are used for this graph. A common notation uses a small, hollow circle instead of the parenthesis to indicate that a number is not included in the graph. Such a graph is shown below:

To graph the set $\left\{ x \mid x \geq -\dfrac{1}{2} \right\}$, we draw a number line and shade the numbers to the right of $-\dfrac{1}{2}$. We use a **bracket** to show that $-\dfrac{1}{2}$ is included.

The alternate notation for this situation is a small, darkened circle instead of the bracket to indicate that a number is included in the graph. Such a graph is shown below:

Linear, or first-degree, inequalities in one variable can be solved in a manner very similar to the method used to solve first-degree equations. Our general approach is to use addition and multiplication properties in a series of steps. Our strategy is to form equivalent inequalities until we have an inequality of the form:

$$x < a$$

$$x \leq a$$

$$x \geq a$$

$$x \geq a$$

We can then write the solution set.

The addition property of inequality works exactly like the Addition Property of Equality. That is, the same number can be added to both sides of an inequality without changing its meaning. Also, both sides of an inequality can be multiplied by the same *positive* number. However, notice what happens when we multiply both sides of the true statement $-3 < 3$ by a *negative* number.

$$-3 < 3 \qquad \text{A \textit{true} statement}$$

$$(-1)(-3) < (-1)(3) \qquad \text{Multiply both sides by \textit{negative} 1}$$

$$3 < -3 \qquad \text{A \textit{false} statement}$$

Notice that the last statement becomes *true* if the direction of the inequality is *reversed*.

$$3 > -3 \qquad \text{A } \textit{true} \text{ statement}$$

This idea leads to the following property: If both sides of an equality are multiplied or divided by the same negative number and the inequality symbol is reversed, the resulting inequality is equivalent to the first inequality.

Here are three tools used in solving inequalities.

Tools Used to Solve Inequalities

Addition Property

If the same number is added to (or subtracted from) both sides of an inequality, the resulting inequality is equivalent to the first inequality.

Multiplication by Positive Number

If both sides of an inequality are multiplied or divided by the same **positive** number, the resulting inequality is equivalent to the first inequality.

Multiplication by Negative Number

If both sides of an inequality are multiplied or divided by the same **negative** number and the inequality symbol is **reversed,** the resulting inequality is equivalent to the first inequality.

► ## EXAMPLE 1

Solve $x + 5 < 7$ and graph the solution set.

Solution

$$x + 5 < 7$$

We add -5 to both sides,

$$x + 5 - 5 < 7 - 5$$
$$x < 2$$

and then write the solution set:

$$\{x \mid x < 2\}$$

To graph the solution set, we shade the portion of the number line to the left of 2. We indicate with a parenthesis that 2 *does not* belong to the solution set.

► **EXAMPLE 2**

Solve $2x - 1 \geq 5$ and graph the solution set.

Solution

$$2x - 1 \geq 5$$

We add 1 to both sides.

$$2x - 1 + 1 \geq 5 + 1$$
$$2x \geq 6$$

Divide both by 2 $\left(\text{multiply by } \dfrac{1}{2}\right)$.

$$\frac{2x}{2} \geq \frac{6}{2}$$
$$x \geq 3$$
$$\{x \mid x \geq 3\}$$

We graph the solution set on a number line. In this case, we shade that portion of the line to the right of 3. We show with a bracket that 3 belongs to the solution set.

Sometimes we find ourselves in a situation where it is convenient to multiply or divide by a negative number.

► **EXAMPLE 3**

Solve $-2x + 1 < 7$.

Solution

$$-2x + 1 < 7$$

Subtract 1 from both sides,

$$-2x < 6$$

Be Careful!   Divide both sides by -2.

$$x > -3$$
$$\{x \mid x > -3\}$$

Note that the inequality symbol is reversed ◄

The distributive property is often very useful when solving inequalities.

► **EXAMPLE 4**

Solve $3(t - 1) > (t + 1)$.

Solution

$$3(t - 1) > 7(t + 1)$$

$$3t - 3 > 7t + 7 \qquad \text{Distributive Property}$$

$$3t - 10 > 7t \qquad \text{Subtract 7 from both sides}$$

$$-10 > 4t \qquad \text{Subtract } 3t \text{ from both sides}$$

$$-\frac{5}{2} > t \qquad \text{Divide both sides by 4}$$

$$\left\{ t \mid t < -\frac{5}{2} \right\} \qquad \blacktriangleleft$$

In Example 4, notice that we wrote $t < -\dfrac{5}{2}$ rather than $-\dfrac{5}{2} > t$. Either is correct. The variable is usually written on the left.

Like equations, inequalities may contain fractions in decimal format.

► EXAMPLE 5

Solve $x + 0.6 < 0.4(x + 2)$.

Solution

Multiply both sides by 10 to clear fractions.

$$10x + 6 < 4(x + 2)$$

$$10x + 6 < 4x + 8 \qquad \text{Distributive Property}$$

$$10x - 4x < 8 - 6$$

$$6x < 2$$

$$x < \frac{1}{3}$$

Using decimals, $\dfrac{1}{3}$ is written as $0.\overline{3}$.

$$\{ x \mid x < 0.\overline{3} \} \qquad \blacktriangleleft$$

The inequality of Example 5 is stated with decimal numbers rather than fractions; thus it is good form to state the solution set with decimals. It is **important** to note that the number $0.\overline{3}$ is **not** the same as *approximations* of $\dfrac{1}{3}$ such as 0.3 or 0.333.

In the next example, notice how the inequality symbol is unaffected when we multiply by a *positive* number, but is reversed when we multiply by a *negative* number.

▶ **EXAMPLE 6**

Solve $\dfrac{x+1}{3} - \dfrac{x+2}{2} < \dfrac{1}{6}$.

Solution

Multiply both sides by 6 to clear fractions.

$$2(x+1) - 3(x+2) < 1 \qquad \text{Inequality symbol does not change}$$
$$2x + 2 - 3x - 6 < 1 \qquad \text{Distributive Property}$$
$$-x - 4 < 1 \qquad \text{Combine like terms}$$
$$-x < 5 \qquad \text{Add 4 to both sides}$$

Be Careful!   Next we multiply both sides by -1.

$$x > -5 \qquad \text{Note inequality change!}$$
$$\{x \mid x > -5\} \qquad \qquad \blacktriangleleft$$

We have seen when solving equations that sometimes the variable falls out, leaving a statement that is always true, or a statement that is always false. The same thing may occur when solving inequalities. The next two Examples illustrate the situation.

▶ **EXAMPLE 7**

Solve $2x + 2 < 3(x + 1) - x + 3$.

Solution

$$2x + 2 < 3(x + 1) - x + 3$$
$$2x + 2 < 3x + 3 - x + 3 \qquad \text{Distributive Property}$$
$$2x + 2 < 2x + 6 \qquad \text{Combine like terms}$$
$$2 < 6 \qquad \text{Subtract } 2x$$

Because the last statement is *always* true, the original inequality is true for *any* value of x. Thus the solution set is the set of all real numbers

$$\{x \mid x \text{ is any real number}\} \qquad \qquad \blacktriangleleft$$

Compare this result with that of the following example.

▶ **EXAMPLE 8**

Solve $x + 3 \leq 2(x - 3) - x + 3$.

Solution

$$x + 3 \leq 2(x - 3) - x + 3$$
$$x + 3 \leq 2x - 6 - x + 3$$
$$x + 3 \leq x - 3$$
$$3 \leq -3$$

As the last statement is *never* true, the original inequality is false for *any* value of x. Thus the solution set is the empty set.

$$\varnothing$$

◄

Compound Inequalities

A compound inequality is a statement made by combining two inequalities with the word **and** or **or.** For example, the statement

$$x < 1 \quad \text{or} \quad x > 4$$

is a compound inequality. It is true if x is less than 1 *or* if x is greater than 4. Thus it is true when x is to the left of 1 on the number line *or* x is to the right of 4.

The statement

$$x < 10 \quad \text{and} \quad x > 3$$

is a compound inequality that is true if x is less than 10 *and* greater than 3. Thus it is true when x is to the left of 10 *and* to the right of 3.

Notice from the graph of the inequality that, if x is in the solution set, x must be *between* 3 and 10. Therefore, we usually write inequalities such as

$$x < 10 \quad \text{and} \quad x > 3$$

in the **compact form**

$$3 < x < 10$$

This inequality could be written the other way around us

$$10 > x > 3$$

We prefer to write such inequalities using *less than* symbols as this puts the smaller number on the left.

The compact form will **never** be written using both < and > symbols. Such statements as $2 < x > 5$ are nonsense and not acceptable.

► **EXAMPLE 9**

Sketch the graph of each set.

(a) $\{x \mid x \leq -0.7 \text{ or } x > 2.3\}$ (b) $\left\{x \mid \dfrac{1}{2} \leq x < \dfrac{3}{2}\right\}$

Solutions

(a) $\{x \mid x \leq -0.7 \text{ or } x > 2.3\}$

For x to belong to this set, x must be to the left of (or equal to) -0.7, or x must be to the right of 2.3. Sketching this, we have

(b) $\left\{x \mid \dfrac{1}{2} \leq x < \dfrac{3}{2}\right\}$

This set contains all numbers that are greater than (or equal to) $\dfrac{1}{2}$ and less than $\dfrac{3}{2}$, that is, all numbers between $\dfrac{1}{2}$ and $\dfrac{3}{2}$, including $\dfrac{1}{2}$.

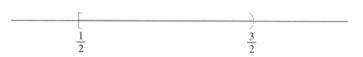

To solve a compound inequality written in the compact form, work on all sides at the same time until the variable is isolated in the middle.

► **EXAMPLE 10**

Graph the solution set for $1 < 2x - 3 < 5$.

Solution

Because $1 < 2x - 3 < 5$ is shorthand for two inequalities, we can work on both at the same time. First, we add 3 to all three sides and get

$$1 + 3 < 2x - 3 + 3 < 5 + 3$$

$$4 < 2x < 8$$

Divide all three sides by 2:

$$\frac{4}{2} < \frac{2x}{2} < \frac{8}{2}$$

$$2 < x < 4$$

which is the graph of the set

$$\{x \mid 2 < x < 4\}$$

► **EXAMPLE 11**

Find the solution set for $-3 \le 1 - 2x < 5$.

Solution

$$-3 \le 1 - 2x < 5$$

Work on all three sides at one time to isolate x in the middle. First, subtract 1.

$$-4 \le -2x < 4$$

Then divide by -2 carefully, because -2 is *negative*.

*Be
Careful!*

$$2 \ge x > -2 \qquad \text{Note inequality change}$$

We usually turn these around so that the smaller number is on the left when we write the solution set.

$$\{x \mid -2 < x \le 2\} \qquad ◄$$

Interval Notation

Another common way to write sets is with **interval notation.** We use parentheses and brackets just as we did on the graphs. Now we have *three* ways to indicate the solution set of an inequality: set-builder notation, interval notation, and the graph. We summarize these methods in the following table.

Set-builder Notation	Graph	Interval Notation
$\{x \mid p < x < q\}$		(p, q)
$\{x \mid p \le x \le q\}$		$[p, q]$
$\{x \mid p \le x < q\}$		$[p, q)$
$\{x \mid p < x \le q\}$		$(p, q]$
$\{x \mid x > p\}$		$(p, +\infty)$
$\{x \mid x \ge p\}$		$[p, +\infty)$
$\{x \mid x < p\}$		$(-\infty, p)$
$\{x \mid x \le p\}$		$(-\infty, p]$
All real numbers ($\mathbb{R}$)		$(-\infty, +\infty)$

► **EXAMPLE 12**

Write a set in interval notation that represents each graph.

(a) 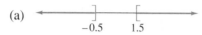   (b)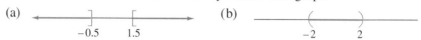

Solutions

(a)

The graph shown is the set of numbers less than or equal to -0.5 together with the numbers greater than or equal to 1.5. In set-builder notation, this would be $\{x \mid x \le -0.5 \text{ or } x \ge 1.5\}$. Interval notation almost copies the symbols on the graph, reading from left to right. To connect the two parts, we use the symbol for the union of two sets, $\cup$, instead of the word *or*. In interval notation, this set is $(-\infty, -0.5] \cup [1.5, +\infty)$.

(b)

The graph shown is the set of numbers between -2 and 2. In set-builder notation, this set is $\{x \mid -2 < x < 2\}$. In interval notation, this set is $(-2, 2)$.   ◄

Inequalities arise in solving word problems just as equations do. The following phrases are used frequently.

Words	Symbol
is more than	$>$
is less than	$<$
is not less than	$\ge$
is not more than	$\le$
is at least	$\ge$
is at most	$\le$

► **EXAMPLE 13**

Three times a number increased by 7 is at most 31. Write an appropriate mathematical model and then find all such numbers.

Solution

Let n be the number.

We can write a mathematical model by using the language of algebra and inserting $\le$ for "is at most"

$$3n + 7 \le 31$$

Solving gives us

$$3n \le 24$$

$$n \le 8$$

The numbers must be less than or equal to 8.   ◄

Problem Set 2.3

Warm-Ups

In Problems 1 through 6 solve each inequality and graph the solution set. See Examples 1 and 2.

1. $x - 3 \leq 4$

2. $x + 1 > -5$

3. $x - 7 \geq 3$

4. $x + 8 < 22$

5. $2x + 1 < -5$

6. $3x - 2 \geq 3$

In Problems 7 and 8, solve each inequality. See Example 3.

7. $-4x \geq 2$

8. $-\dfrac{1}{3}x \leq 1$

In Problems 9 and 10, solve each inequality. See Example 4.

9. $6(x - 2) \geq 2(x - 1)$

10. $3(x + 3) + 2(x - 1) \leq 0$

In Problems 11 and 12, solve each inequality. See Example 5.

11. $1.1x - 0.5 > 0.7x$

12. $2.5t + 1.5 \leq 3.0t - 2.5$

In Problems 13 and 14, solve each inequality. See Example 6.

13. $\dfrac{2}{3}(2x - 1) < \dfrac{1}{6}x + 1$

14. $\dfrac{x - 3}{5} - \dfrac{x - 1}{2} \leq 1$

In Problems 15 and 16, solve each inequality. See Examples 7 and 8.

15. $x + 3 \leq x + 4$

16. $2(x + 2) > 2(x + 1) + 2$

In Problems 17 through 22, sketch the graph of each set. See Example 9.

17. $\{x \mid 1 < x < 2\}$

18. $\{x \mid 2 \leq x \leq 10\}$

19. $\{x \mid -1 < x \leq 1\}$

20. $\{x \mid 0 < x < 2\}$

21. $\{x \mid x \leq -7 \text{ or } x > -2\}$

22. $\{x \mid x \leq 0 \text{ or } x > 3\}$

In Problems 23 and 24, graph each solution set. See Example 10.

23. $2 < x + 3 \leq 7$

24. $-3 \leq \dfrac{1}{2}x - 1 < 0$

In Problems 25 and 26, solve each inequality. See Example 11.

25. $1 < 2 - x < 5$

26. $-7 \leq 1 - 4x \leq -3$

In Problems 27 through 34, write a set in interval notation to represent each graph. See Example 12.

27.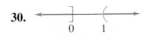

28.

29.

30.

31.

    5   10

32.

    -2   2

33.

    5

34.

    -3

Problems 35 through 38 involve inequalities. Write a mathematical model for each statement; then solve and write a sentence that describes the solution. See Example 13.

35. Twice a number plus 1 is at most 7. Find all such numbers.

36. Jim has grades of 90, 75, and 82 on his first three tests. What must he make on his fourth test in order to have an average of at least 80?

37. The perimeter of a rectangle is to be no more than 72 ft. If the length is 4 ft more than the width, find the measure of the largest possible width.

38. Two times 3 more than a number is no less than 12. Find all such numbers.

Practice Exercises

In Problems 39 through 82, solve the inequality and graph the solution set.

39. $2 - x \geq 1$

40. $5 - x > 1$

41. $4x < 2$

42. $2x \leq 1$

43. $\frac{1}{2}x > -5$

44. $\frac{1}{4}x \leq 0$

45. $6x + 3 \leq 2x - 7$

46. $12x + 17 \leq 11 - 3x$

47. $\frac{1}{5}x - \frac{1}{3} \leq \frac{1}{5} + 2x$

48. $\frac{1}{2}y - \frac{1}{3} > \frac{1}{3}y + \frac{1}{6}$

49. $5(x + 1) < 2(1 - x)$

50. $3(x + 7) - (x + 2) \geq 0$

51. $\frac{1}{3}(x + 2) > \frac{1}{4}(x - 2)$

52. $-\frac{1}{2}(x - 6) > \frac{2}{3}(x + 6)$

53. $\frac{x}{7} - \frac{1}{3} < \frac{x}{3}$

54. $\frac{x + 2}{5} - \frac{x - 3}{4} < 2$

55. $\frac{x - 3}{6} \leq \frac{x}{6} - 1$

56. $\frac{3x + 2}{2} \leq \frac{6x - 1}{4}$

57. $\frac{1 - x}{3} > \frac{2}{3}(5 - x)$

58. $\frac{1}{8}x - 2(x + 3) < \frac{1}{4}$

59. $6(y + 3) > 2(y - 1)$

60. $3(t - 7) > -5(t + 3)$

61. $5(x - 3) + 3 \le -2(2x - 4)$

62. $11(2x - 1) - 3(3x + 2) < 12x$

63. $3[5 - (s + 2)] - 2s \ge 6s - 7$

64. $6 - 7z - 3(3z - 4) > -8(2z - 7)$

65. $\frac{3}{4}(y + 1) < \frac{1}{8}y$

66. $\frac{1}{6}y - \frac{1}{2}(y - 7) \ge -\frac{3}{2}$

67. $\frac{x}{2} - \frac{1}{7} > \frac{1}{4}(2x - 9)$

68. $\frac{2x}{3} + \frac{x + 3}{2} < -1 + \frac{7x}{6}$

69. $\frac{2(x - 7)}{3} - \frac{3(x - 3)}{2} \ge 1 - x$

70. $\frac{1}{4}x - \frac{3}{4}(1 - x) > \frac{1}{2}(x + 1)$

71. $0 < \frac{x - 7}{4}$

72. $\frac{3t + 4}{16} \ge 0$

73. $-5 < \frac{3(r - 1)}{-2}$

74. $1 - 3(2x + 5) \le -\frac{1}{2}$

75. $0.4x - 20.8 \le 8.0 - 0.8x$

76. $5.3 - 1.2x > 0.4x - 58.7$

77. $0.5 - 10.3x > 4.0x - 28.1$

78. $0.02u - 2 \ge 0.01u$

79. $2x + 5 > 2(x + 3)$

80. $5(2x - 1) < 10(x + 1) - 15$

81. $3(1 - x) \le 3(x + 1) - 6x$

82. $2(x + 3) \ge 3(2 - x) + 5x$

In Problems 83 through 90, sketch the graph of each set.

83. $\{x \mid 2 < x < 3\}$

84. $\{x \mid -9 < x < 0\}$

85. $\{x \mid x < 6 \text{ or } x > 11\}$

86. $\{x \mid x < -3 \text{ or } x > 3\}$

87. $\{x \mid x \le -3 \text{ or } x > 3\}$

88. $\{x \mid x < -11 \text{ or } x \ge -8\}$

89. $\{x \mid 0.5 < x \le 1.5\}$

90. $\{x \mid 1.3 \le x < 1.4\}$

In Problems 91 through 96, which inequalities are nonsense?

91. $-3 < x < -5$

92. $-1 > t \ge 7$

93. $0 \le y \le 3$

94. $4 < x < -6$

95. $1 > x > 0$

96. $-2 < x < 2$

In Problems 97 through 104, write a set in interval notation to represent each graph.

97.

98.

99.

100.

101.
-11 -8

102.
6 9

103.
-3 3

104.
-2 0

In Problems 105 through 112, find the solution set for each compound inequality.

105. $3 \leq 2x + 1 \leq 5$

106. $-10 \leq 5x + 5 < 5$

107. $4 \leq 1 - x \leq 10$

108. $-6 < 2 - 8x < 18$

109. $5 < \dfrac{1}{2}x - 2 \leq 8$

110. $3 < \dfrac{2}{3}x + \dfrac{1}{3} \leq \dfrac{15}{3}$

111. $\dfrac{1}{2} < 2x - 3 < 3$

112. $x < \dfrac{1}{2}$ and $x > -\dfrac{1}{2}$

Problems 113 through 120 involve inequalities. Write a mathematical model for each statement. Then solve and write a sentence that describes the solution.

113. Four times a number minus 2 is at most 10. Find all such numbers.

114. The perimeter of a rectangle is to be no more than 82 m. If the length is 7 m more than the width, find the measure of the largest possible width.

115. Gloria has grades of 82, 65, and 73 on her first three tests. What must she score on her fourth test to have an average of at least 80?

116. Five less than a number, multiplied by 3, is no less than 30. Find all such numbers.

117. A 150-pound man can burn 600 calories per hour when exercising on a rowing machine. How long would it take to burn at least enough calories to compensate for eating a small pepperoni pizza containing 450 calories?

118. Revenue earned from the sale of doll houses is the product of the number sold and the price. Each doll house sells for $229.95. If the company needs a monthly revenue of at least $37,500, how many must be sold each month?

119. A basketball team that plays 18 games per season is striving for a winning season (more games won than lost). So far, the team has played 9 games and has won twice as many as they have lost. How many more games must be won to ensure a winning season?

120. During 1990, Disneyland in Anaheim, California, had about 4 times as many visitors as Six Flags Magic Mountain in Valencia, California. How many visitors would Six Flags Magic Mountain have to have if the attendance numbers of these two California attractions were to surpass those of Walt Disney World in Lake Buena Vista, Florida, which enjoyed crowds numbering 28,500,000?

Challenge Problems

In Problems 121 through 126, solve for x; a, b, and c are constants.

121. $x - 5 < b$

122. $x + a \geq 6$

123. $3x - 2a > b$

124. $5x + 3b \leq a$

125. $ax < 5; a > 0$

126. $ax < 5; a < 0$

In Problems 127 through 130, find the solution set.

127. $x \leq a$ or $x \geq b; a < b$

128. $1 < ax < 3; a > 0$

129. $1 < ax < 3; a < 0$

130. $a \leq ax + b \leq b; a < 0; a < b$

In Problems 131 and 132, write a set to represent each graph.

131.

$$\xrightarrow{\hspace{2cm}}$$

$-1 \quad 1 \quad 2$

132.

$-8 \ -6 \qquad -1 \quad 2$

In Your Own Words

133. What numbers are in the set $\{x \mid x < 1\}$?

134. What numbers are in the set $\{t \mid t \geq 1\}$?

135. What numbers are in the set $\{y \mid y < -7 \text{ or } y \geq -1\}$?

136. What numbers are in the set $\left\{x \mid \dfrac{2}{3} \leq x < 4\right\}$?

137. Explain how to solve a linear inequality.

138. What is a compound inequality?

2.4   Problem Solving

TAPE 5

The usefulness of mathematics lies in solving problems. However, the problem of *problem solving* is not easily solved! Let's look at problem solving as a carpenter might look at a new project. As a first step, he must decide what he is trying to build. That is, he must make sure that *he understands the problem.* Next, he *plans his strategy.* Has he built something like this before? If so, will the old approach work here? If not, does he know of another approach that will work? Once his plan is formulated, he *applies the tools and materials of his trade* to complete the project.

Solving problems in mathematics uses steps much like the ones used by the carpenter. We must

1. understand the problem,

2. plan a strategy, and

3. apply our mathematical tools.

Each of these steps has its difficulties. In step 1, we must, like the carpenter, be sure that we understand what it is we are trying to do! For step 3 we use the tools and ideas developed over the ages and select the ones appropriate to our problem. (In fact, this book deals almost exclusively with the development of such tools.) In this section, however, we will focus on step 2, *planning a strategy.*

Several strategies are used in problem solving. We will look at three of them: forming a mathematical model, estimating an answer, and looking for a pattern.

Forming a Mathematical Model

We have seen in earlier sections of this chapter how variables can be used to make an equation or an inequality that describes some given information mathematically. The idea is to form this mathematical model and thus solve the problem. This approach to problem solving is called *modeling.*

We have seen that models can be formed by using a known formula. When we use a known formula to form the mathematical model, we often use the variables from the formula and just substitute the given data into the formula. The formula is actually the model.

► **EXAMPLE 1**

The formula $I = PRT$ calculates simple interest, where I is the interest, P is the principal, R is the rate of interest, and T is time. Jason wants to know how much money he must invest for 2 years at 8% annual interest to earn $192. Form an appropriate mathematical model and answer Jason's question.

Solution

The given information tells us that I is $192, R is 8% (0.08), and T is 2 years. Since the principal is what we want to find, we will call it P. Now we substitute this information into the formula $I = PRT$ and we have a mathematical model for the problem.

$$I = PRT$$
$$192 = P(0.08)(2)$$
$$192 = 0.16P$$
$$1200 = P$$

Now we can answer the question. Jason must invest $1200 at 8% to earn $192 in 2 years.   ◄

Simple interest is calculated only on the principal. Interest that is calculated on the principal plus any interest earned is called *compound interest*.

► **EXAMPLE 2**

If P dollars are invested at an annual interest rate of r compounded n times a year, then A, the amount of dollars in the account after t years, is given by $A = P\left(1 + \dfrac{r}{n}\right)^{nt}$. Using this formula as a model, approximate the amount that would accumulate in an account if $2500 were invested for 2 years at 9% when the interest is compounded:
(a) Annually (b) Semiannually (c) Quarterly (d) Monthly

Solution

The information tells us that the formula $A = P\left(1 + \dfrac{r}{n}\right)^{nt}$ is the model and that we wish to find A when P is $2500, r is 9%, and t is 2. If we substitute these numbers into the formula it becomes

$$A = 2500\left(1 + \frac{0.09}{n}\right)^{2n}.$$

To approximate each amount in parts (a) through (d), we substitute the different values for n into the formula and use a calculator. In (c) $n = 4$ and thus the amount in the account is

$$A = 2500\left(1 + \frac{0.09}{4}\right)^{2(4)}.$$

With a calculator, enter

The answers are listed in the following table.

Interest Period	Amount A in Dollars in Account after 2 Years	
(a) Annually ($n = 1$)	$2500(1 + 0.09)^{2(1)}$	$= 2970.25$
(b) Semiannually ($n = 2$)	$2500\left(1 + \dfrac{0.09}{2}\right)^{2(2)}$	≈ 2981.30
(c) Quarterly ($n = 4$)	$2500\left(1 + \dfrac{0.09}{4}\right)^{2(4)}$	≈ 2987.08
(d) Monthly ($n = 12$)	$2500\left(1 + \dfrac{0.09}{12}\right)^{2(12)}$	≈ 2991.03

◄

Sometimes we use a formula, but we must make up the model from the problem situation. The model does not come directly from the formula.

► **EXAMPLE 3**

Denise Hairston invested part of her savings at 7% simple interest and the remainder at 8% simple interest. In one year she receives $1415 in interest. If her total savings were $19,000 set up an appropriate mathematical model and determine how much money she invested at each rate.

Solution

Let p be the amount invested at 7%. Then she has the rest of her savings ($19,000 - p$) invested at 8%. The total interest on the $19,000 is the sum of the interest on each of the two parts.

Total interest = Interest earned at 7% + Interest earned at 8%

So, using the formula $I = PRT$, we form the mathematical model for the problem.

$$1415 = p \cdot 0.07 \cdot 1 + (19,000 - p) \cdot 0.08 \cdot 1$$
$$1415 = 0.07p + 1520 - 0.08p$$
$$1415 = -0.01p + 1520$$
$$-105 = -0.01p$$
$$10,500 = p$$

Thus $19,000 - p = 8500$.

Denise Hairston invested $10,500 at 7% and $8,500 at 8%.   ◄

Sometimes a picture or diagram is helpful in forming the mathematical model.

► **EXAMPLE 4**

In a right triangle, one leg is 3 cm longer than the other leg. If the shorter leg is increased by 6 cm, the area is increased by 24 cm². Find a model that can be used to find the length of each leg in the original triangle.

Solution

We are to find the length of the two legs of the original triangle.

Let l be the length in centimeters (cm) of the shorter leg. Then $l + 3$ is the length in centimeters of the other leg.

The problem mentions another triangle. The other triangle is made by increasing the shorter leg (l) of the original triangle by 6 cm. Thus one leg is $l + 3$ and the other leg is $l + 6$.

The information in the problem is about the area of these triangles. We use the formula for the area of a triangle $\left(A = \dfrac{1}{2}bh \right)$ and determine that the area of a right triangle is one-half the product of the legs. We can write an expression for the area of each triangle.

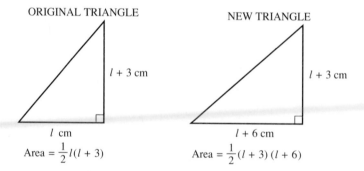

ORIGINAL TRIANGLE

$l + 3$ cm

l cm

Area $= \dfrac{1}{2}l(l + 3)$

NEW TRIANGLE

$l + 3$ cm

$l + 6$ cm

Area $= \dfrac{1}{2}(l + 3)(l + 6)$

To form a model, we use the fact that the area of the new triangle is 24 cm² more than the area of the original triangle and form an equation. If we add 24 to $\dfrac{1}{2}l(l + 3)$, the smaller area, we will have the larger area.

Area of a smaller triangle $+ 24 =$ Area of larger triangle

$$\frac{1}{2}l(l + 3) + 24 = \frac{1}{2}(l + 3)(l + 6)$$ ◄

Estimating an Answer

Another approach to problem solving is to use trial and error to guess the answer. This approach may be time consuming and may not give the exact answer. Sometimes, experience and reasoning allow us to narrow the possibilities.

► **EXAMPLE 5**

Estimate how many years it would take to double an investment of $1000 at 8% interest compounded quarterly.

Solution

We use the formula for computing compound interest and substitute 1000 for P, 8% for r, and 4 for n. Since we want to double 1000, A is 2000. The unknown quantity is time, so we use t as a variable.

$$A = P\left(1 + \frac{r}{n}\right)^{nt}$$

$$2000 = 1000\left(1 + \frac{0.08}{4}\right)^{4t}$$

$$= 1000(1.02)^{4t}$$

We do not know how to solve this equation. We will try different numbers for t to see which numbers are "close" to making the equation true.

t	$1000\,(1.02)^{4t}$
1	$1000(1.02)^{4} \approx 1082$
2	$1000(1.02)^{4(2)} \approx 1172$
5	$1000(1.02)^{4(5)} \approx 1486$
8	$1000(1.02)^{4(8)} \approx 1885$
9	$1000(1.02)^{4(9)} \approx 2040$

So we estimate that it will take between 8 and 9 years for the money to double.

◄

Looking for a Pattern

Another approach to solve problems is to look for a pattern. An interesting pattern was discovered by Fibonacci, an Italian mathematician who lived during the thirteenth century. He presented the pattern as the answer to the following problem.

If a pair of rabbits produces another pair of rabbits at the end of two months and after that produces another pair of rabbits each month, and if each pair of rabbits produced follows this same pattern, how many pairs of rabbits will there be at the beginning of each month if all the rabbits live?

Beginning of Month	Number of Pairs of Rabbits
1	1
2	1
3	2
4	3
5	5
6	8

This pattern is called the *Fibonacci sequence.* The pattern for the numbers is: the first two numbers are 1, and every number thereafter is the sum of the two preceding numbers. These numbers have turned up in many natural phenomena, such as in the spirals in a pine cone or the bottom of a pineapple.

► EXAMPLE 6

Fill in the missing number in each row by looking for a pattern.

	1st Place	2nd Place	3rd Place	4th Place	5th Place
(a)	1	3	5	☐	9
(b)	1	4	☐	16	25
(c)	14	11	8	5	☐
(d)	2	4	8	☐	32

Solutions

(a) The given numbers are consecutive odd integers. So the missing number is 7.

(b) Each of the given numbers is the square of its place number. So the missing number is 3^2, or 9.

(c) Each number from 2nd place on is 3 less than the preceding number. The missing number is $5 - 3$ or 2.

(d) Each number is a power of 2. The power is the place number. Thus the missing number is 2^4, or 16.   ◄

► EXAMPLE 7

Fill in the missing number in each row by looking for a pattern.

	1st Place	2nd Place	3rd Place	4th Place	nth Place
(a)	1	8	27	☐	n^3
(b)	1	4	9	16	☐

Solutions

(a) Each number is the cube of its place number. The missing number is 4^3, or 64.

(b) Each number is the square of the place number. So the missing number is n^2.   ◄

► EXAMPLE 8

The *half-life* of a substance is the amount of time it takes for half of the substance to decay. If 100 grams (g) of a substance with a half-life of 10 years is present initially, how many grams will be present after 10 years, 20 years, 30 years, and

40 years? Find an expression for the amount of the substance remaining after t years.

Solution

A half-life of 10 years means that *half* of the substance decays every 10 years, and thus *half* of it remains. So we will let t be 10, 20, 30, and 40 years and calculate how much of the substance remains by finding half of each previous amount. We will write in all the details involved in figuring each amount so that we can look for a pattern as we work.

Number of Years	Grams of Substance Remaining
10	$100\left(\dfrac{1}{2}\right)$ $= 50$
20	$100\left(\dfrac{1}{2}\right) \cdot \dfrac{1}{2} = 100\left(\dfrac{1}{2}\right)^2 = 25$
30	$100\left(\dfrac{1}{2}\right)^2 \cdot \dfrac{1}{2} = 100\left(\dfrac{1}{2}\right)^3 = 12.5$
40	$100\left(\dfrac{1}{2}\right)^3 \cdot \dfrac{1}{2} = 100\left(\dfrac{1}{2}\right)^4 = 6.25$

To find an expression for the amount remaining after t years, we must look for a pattern. If we look at the calculations we have just made, we see that each amount is 100 times a power of $\dfrac{1}{2}$ and that the power is the number of years divided by 10. We feel fairly sure that $100\left(\dfrac{1}{2}\right)^{t/10}$ gives the amount of the radioactive substance remaining after t years. If we wish, we can use negative exponents and write this as $100 \cdot 2^{-t/10}$ ◀

Problem Set 2.4

Warm-Ups

In Problems 1 and 2 form an appropriate mathematical model and answer the question. See Example 1.

1. J. R. made $300 in simple interest by loaning Matthew $1000 for three years. What rate did J. R. charge Matthew?

2. Sadie Roden invested $100,000 at $8\frac{1}{2}\%$ per year simple interest. If her account earned $42,500 in interest, how many years was the money invested?

For Problems 3 and 4, see Example 2.

3. Joe Begay invested $10,000 in an account that pays 6.5% compounded quarterly. How much will Joe have in his account after two years?

4. Mary Grayhills invested $17,424 in a savings account that pays $7\frac{1}{4}\%$ interest compounded semiannually. What will her account be worth in 5 years?

In Problems 5 through 8 set up an appropriate mathematical model and answer the question. See Example 3.

5. Peggy's Card Shop invests $25,000 for 1 year. Part of it is to be invested at 5.5% and the rest at 6.5%. If a total of $1470 in interest is earned, how much money is invested at each rate?

6. Jane Batten wishes to invest $10,000 for 1 year. Part of it is to be invested at 5% and the rest at 6%. If $560 in interest is earned, how much money is invested at each rate?

7. Alonzo has $6500 invested at 4.5%. How much should he invest at 5.25% to yield 5% on his investments?

8. Damon needs a return of 6% on his investments. He has $14,000 in bonds that yield 4% and $16,000 more to invest. What yield should he get on the $16,000?

In Problems 9 through 12, form an appropriate mathematical model and answer the question. See Example 4.

9. If the sides of a square are increased by 4 cm, the area is increased by 40 cm². Find the length of a side of the original square.

10. In a right triangle, the length of one leg is twice the length of the other leg. If each leg is increased by 2 m, the resulting triangle has an area of 17 m² more than the area of the original triangle. Find the length of each leg in the original triangle.

11. The length of a rectangle is twice its width. If its length and width are each decreased by 2 cm, its area is decreased by 38 cm². Find the original dimensions of the rectangle.

12. The length of a rectangle is 5 ft. If the length and width are each increased by 2 ft, the area is increased by 24 ft². Was the original rectangle a square?

For Problems 13 and 14, see Example 5.

13. Estimate how long it would take to double an investment of $1000 at 6% interest compounded monthly.

14. Five-hundred grams of cobalt 60 decays according to the relationship $A = 500\left(\dfrac{1}{2}\right)^{0.2t}$, where t is the elapsed time in years, and A is the amount remaining after t years. If 150 g of cobalt 60 remains, estimate how many years have elapsed.

In Problems 15 through 18, fill in the missing number. See Example 6.

	1st Place	2nd Place	3rd Place	4th Place	5th Place
15.	3	7	11	15	☐
16.	400	200	100	☐	25
17.	3	9	27	81	☐
18.	7	4	1	☐	−5

In Problems 19 and 20, fill in the missing number. See Example 7.

	1st Place	2nd Place	3rd Place	4th Place	nth Place
19.	1	3	5	☐	$2n - 1$
20.	2	4	6	8	☐

For Problems 21 and 22, see Example 8.

21. The number of a certain kind of bacteria doubles every 24 hours. There are 1000 bacteria present to begin with. Complete the following table and find a pattern for the number of bacteria present after t days.

Number of Days	Number of Bacteria Present
1	
2	
3	
4	
5	
t	

22. A sheet of a certain kind of paper is 0.001 inch in thickness. Fold the sheet in half to have 2 pieces in thickness, fold this in half again to have 4 pieces in thickness, and again to have 8 pieces in thickness. Complete the following chart and find a pattern to determine the thickness after n folds. If the paper is folded in this manner 30 times, what would its thickness be?

Number of Folds	Thickness (in.)
1	
2	
3	
4	
5	
n	

Practice Exercises

23. If $1000 is deposited in an account today, how much will the account be worth in 25 years at:
   (a) 7% simple interest
   (b) 7% compound interest, compounded annually
   (c) 7% compound interest, compounded quarterly

24. If $1000 is deposited in an account today, how much will the account be worth in 5 years at
   (a) 7% simple interest
   (b) 7% compound interest, compounded annually
   (c) 7% compounded interest, compounded quarterly

25. To finance his college education, Manuel's parents invested $10,000 in a cerfiticate of deposit on his 6th birthday. The certificate of deposit pays interest of 5% compounded annually.
   (a) Fill in the following table. What is the value of the CD on Manuel's 11th birthday?

Birthday	Principal	Interest Earned during Previous Year	Value
7th	$10,000	$500	$10,500
8th	10,500		
9th			
10th			
11th			

   (b) Use the compound interest formula $A = P\left(1 + \dfrac{r}{n}\right)^{nt}$ to find the value of the CD on Manuel's 11th birthday.
   (c) Use the compound interest formula to find the value of the CD on Manuel's 16th birthday.

26. To finance college, Maria's parents invested $10,000 in a certificate of deposit on her 6th birthday with 6.5% interest compounded annually. What is the value of the CD on Maria's 11th birthday? Her 16th birthday?

27. Leonard White wishes to invest $12,000 for 1 year, part of it at 5.5% and the rest at 6.5%. If a total of $700 in interest is earned, how much money has Leonard invested at each rate?

28. A total of $33,500 is to be invested, some of it at 6.5% and the remainder at 7.3%. If a total interest of $2273.50 is earned in 1 year, how much money is invested at each rate?

29. Ken and Margaret won a lottery prize of $100,000, after taxes. They invest part of it in a second mortgage paying 10% and the rest in utility stock paying 8%. If their total yearly income from these investments is $9400, how much was invested at each rate?

30. Francesca wants to get a return of 6% on her total investments. If she has $15,000 invested at 5%, how much must she invest at 7.5% to realize her goal?

In Problems 31 through 42, find an appropriate model and answer the question.

31. Tony wishes to divide $100,000 between AAAA bonds that yield 4.25% and AA bonds that yield 5.25%. How should he divide his investment to obtain an annual return of $4500?

32. Lauretta has $12,000 invested at 5%. How much of this investment should she convert to another that yields 5.5% to increase her yearly return by $10?

33. Gerald invested $240,000 in two real estate deals last year. He made a profit of 20% on one deal, but lost 5% on the second. If Gerald's net profit was 13.75%, how much did he invest in each deal?

34. Barbara has invested $25,000 in two accounts. One is a money market account paying 5.5% annual interest and the other is a CD that pays 6% yearly. If Barbara receives $1440 from these two investments, how much has she invested in each?

35. If the sides of a square are increased by 2 in., the area is increased by 48 in.^{2}. Find the length of a side of the square.

36. One leg in a right triangle is twice the other leg. If each leg is decreased by 4 ft, the area of the resulting triangle is 28 ft^2 less than the area of the original triangle. Find the length of each leg in the original triangle.

37. Glenn gave away tickets to the Redskins game. Bonnie took half of them and Jackie took one-fifth of them. If Bonnie has three more tickets than Jackie, how many tickets did Glenn give away?

38. Find three consecutive even integers such that one-third of the first is 2 more than the second subtracted from the third.

39. The sum of one-half a number and two-thirds of the number is 3 more than the number. Find the number.

40. Mom left a plate of brownies on the kitchen table. Amanda ate one-third of them and Chuck ate one-half of them. If Chuck ate one more than Amanda, how many brownies were on the table?

41. When one-fifth of a number is subtracted from one-third of the same number, the result is 1 more than one-tenth of the number. Find the number.

42. A 1-acre field is in the shape of a right triangle. If the longest leg measures 110 yd, how many yards of fencing are needed to fence the two legs? (*Hint:* 1 acre = 4840 yd^2.)

43. Find a pattern in the following display.

$$1 = 1$$
$$1 + 3 = 4$$
$$1 + 3 + 5 = 9$$
$$1 + 3 + 5 + 7 = 16$$

What are the next two rows in the display?

44. Find a pattern in the following display.

$$1 = 1$$
$$3 + 5 = 8$$
$$7 + 9 + 11 = 27$$
$$13 + 15 + 17 + 19 = 64$$

What are the next two rows in the display?

45. A bactericide is injected into a bacteria culture containing 50,000 bacteria. The number of bacteria present is given by the expression $50{,}000 \cdot 3^{-0.05t}$, where t is the number of hours after the bactericide is administered. Estimate when the culture will contain 25,000 bacteria and when it will contain 10,000 bacteria.

46. A well-known soft drink bottler has determined that profit for a new clear cola is given by the formula $P = 5x^2\left(\frac{1}{2}\right)^x$, where P is the total profit in millions of dollars and x is the advertising budget in millions of dollars. Estimate the advertising budget that will give the largest profit.

47. In 1991 the world population was 5.4 billion. Suppose the world population is increasing at 1.5% each year.
(a) Fill in the following table and find the world population in 1992, 1993, 1994, and 1995.

Number of Years after 1991	World Population
1	
2	
3	
4	

(b) Find a formula for calculating world population t years after 1991.

48. A new latex composition ball is dropped from a height of 12 feet. Each time it strikes the ground, it bounces back to a height of $\frac{9}{10}$ the distance from which it fell.
(a) Fill in the following table and find the height of the ball on the bounces indicated.

Number of Bounces	Return Height
1	10.8
2	
3	
4	

(b) Find a formula for the height of the ball on the nth bounce.

(c) Use the formula to estimate the number of the first bounce for which the height is less than 6 feet.

Challenge Problem

49. In 1992 a small country with a population of 100,000 people can produce food for exactly 200,000 in a given year. Every year after 1992, the food-production capac- ity is predicted to increase by 50,000 and the population is predicted to increase by 5%.
(a) Fill in the following table and find a pattern for population and food-production capacity after 1992.

Years after 1992	Population	Food-production Capacity
1		
2		
3		
4		
5		
6		
t		

(b) Estimate when the population will overtake the food-production capacity.

50. Suppose a job paid one penny for the first day of the month, 2 cents for the second day, 4 cents for the third and so on, each day's salary doubling that of the previous day. What would the pay be for the 30th day? What would be the total pay for a 30-day month? (Hint: Make a table and look for a pattern.)

In Your Own Words

51. Discuss three strategies for using mathematics to solve problems.

2.5   Linear Equations as Models for Applications

TAPE 5

In this section we continue to look at problem solving. We focus on three types of problems: mixture, work, and distance. The strategy that we use to solve the problems in this section is to make a model. Following a systematic procedure will help to organize our work.

A Procedure to Model Word Problems

1. Read the problem to determine what must be found.

2. Assign a variable, such as *x*, to represent one of the quantities to be found. (This is the "let" statement.)

3. Express all other quantities to be found in terms of *x* (or the variable chosen).

4. Draw a figure or picture, if possible. Label it.

5. Reread the problem and form a mathematical model (equation).

6. Solve the equation and find the values of all the quantities to be found.

7. Check the values in the original word problem. They should answer the question and make sense.

8. Write an answer to the original question.

Mixture

These problems contain percents. Remember that percents represent hundredths. To do arithmetic with them, we replace them with their value. That is, we would replace 7% with 0.07 or $\dfrac{7}{100}$.

► **EXAMPLE 1**

Jack Glover wishes to add enough 50% antifreeze solution to 16 gallons of a 5% antifreeze solution to obtain a 20% antifreeze solution. Find an appropriate model and determine how much of the 50% solution Jack should add.

Solution

STEP 1   Read the problem and determine what is to be found.

The problem asks for the number of gallons of 50% antifreeze solution to be added.

STEP 2   Assign a variable to represent the quantity to be found.

Let x be the number of gallons of 50% solution added.

STEP 4   Draw a figure.

Mixture problems become clearer with a picture.

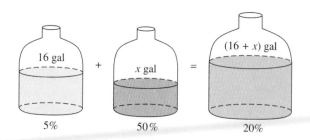

The amount of antifreeze in the first solution plus the amount of antifreeze in the solution added must equal the amount of antifreeze in the final solution. (The amount of antifreeze in 16 gallons of a 5% solution is 16(0.05) gallons.)

STEP 5   Read the problem again and form an equation as a model.

$$16 \cdot 0.05 \quad + \quad x \cdot 0.5 \quad = (16 + x) \cdot 0.2$$

Antifreeze in + Antifreeze in = Antifreeze in
5% solution   50% solution   20% solution

STEP 6   Solve the equation.
Multiply both sides by 100 to clear decimals.

$$16 \cdot 5 + x \cdot 50 = (16 + x) \cdot 20$$

$$80 + 50x = 320 + 20x$$

$$30x = 240$$

$$x = 8$$

STEP 7   Check this value in the original word problem to see if it makes sense and answers the question.

Adding 8 gallons of antifreeze answers the question and makes sense.

STEP 8   Write an answer to the question.

Jack should add 8 gallons of the 50% solution.   ◀

► EXAMPLE 2

A chemist has 10 gallons of a 3% alcohol solution. Find an appropriate model and determine how many gallons of pure alcohol she should add to the 10 gallons to obtain a new solution that is 4% alcohol.

Solution

Determine what is to be found and assign a variable to represent this quantity.

Let x be the number of gallons of pure alcohol added. Draw a figure.

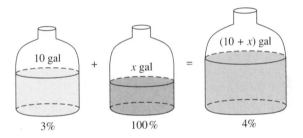

Read the problem again and form an equation as a model.

The amount of alcohol in the original mixture plus the amount of alcohol added must equal the amount of alcohol in the final solution.

$$10 \cdot \frac{3}{100} + x \cdot \frac{100}{100} = (10 + x) \cdot \frac{4}{100}$$

Solve the equation.

We multiply both sides by 100 to clear fractions.

$$10 \cdot 3 + x \cdot 100 = (10 + x) \cdot 4$$

$$30 + 100x = 40 + 4x$$

$$96x = 10$$

$$x = \frac{10}{96} = \frac{5}{48}$$

Check this value in the original word problem to see if it makes sense and answers the question.

Adding $\frac{5}{48}$ of a gallon makes sense and answers the question.

Write an answer to the question.

The chemist should add $\frac{5}{48}$ gallon of pure alcohol, or about $\frac{1}{10}$ gal.   ◀

Work

▶ EXAMPLE 3

Pipe A alone can fill a tank in 2 hours, and pipe B alone can fill the same tank in 3 hours. If both pipes run at the same time, find an appropriate model and determine how long will it take to fill the tank.

Solution

Let x be the number of hours it takes both pipes, working together, to fill the tank.

As pipe A alone fills the tank in 2 hours, it must fill $\frac{1}{2}$ of a tank in 1 hour. So in x hours it must fill $x \cdot \frac{1}{2}$ of a tank. Similarly, pipe B must full $\frac{1}{3}$ of a tank in 1 hour or $x \cdot \frac{1}{3}$ of a tank in x hours. Both pipes working together for x hours fill one tank, so we have the equation

$$\frac{x}{2} + \frac{x}{3} = 1$$

We multiply both sides by 6 to clear fractions.

$$3x + 2x = 6$$
$$5x = 6$$
$$x = \frac{6}{5}$$

Because $\frac{6}{5}$ of an hour answers the question and makes sense, we write the solution:

It takes $\frac{6}{5}$ of an hour for both pipes to fill the tank.   ◀

Distance

The next example is a distance problem. We must remember the formula

$$\text{Distance} = \text{Rate} \times \text{Time}$$

which we usually write $D = RT$. Often it is useful to solve this equation for rate or time, giving us the three forms of the distance formula:

$$D = RT, \qquad R = \frac{D}{T}, \qquad T = \frac{D}{R}$$

▶ EXAMPLE 4

Two trains, 780 miles apart, are heading toward each other at rates of 85 mph and 110 mph. Form an appropriate model and determine when they will crash.

Solution

Let x be the number of hours until they crash.

A picture or diagram is very helpful in solving this type of problem. Usually, there are two pieces in a distance problem. In this problem there are the two trains.

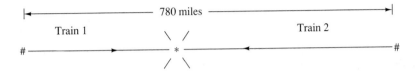

Next we find D, R, and T for each train. One of these should be found from the "let statement" and one should be given in the problem statement. In this problem the let statement gives the time for each train, because they each travel x hours before they crash. The problem statement gives both rates.

	Distance	Rate	Time
Train 1		85	x
Train 2		110	x

We find the third member of the DRT triple using the appropriate form of the distance formula. In this case, $D = RT$.

	Distance	Rate	Time
Train 1	$85x$	85	x
Train 2	$110x$	110	x

With D, R, and T for each train, we must find an equation. In this problem the trains started out 780 miles apart and one went $85x$ miles and the other $110x$ miles. Thus

$$85x + 110x = 780$$
$$195x = 780$$
$$x = 4$$

The trains will crash in 4 hours; this answers the question and makes sense. ◄

► EXAMPLE 5

Frank McComb rides his bike from home to the stadium at a constant rate of 6 mph. He returns along the same route at a rate of 10 mph. If the round trip took $1\frac{3}{5}$ hours, form an appropriate model and determine how far the stadium is from his house.

Solution

Let x be the distance from Frank's house to the stadium (in miles).

There are two pieces in this problem, going and returning.

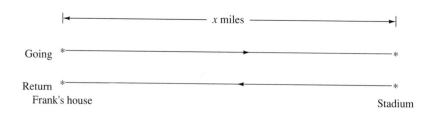

Find D, R, and T for each piece. The distance comes from the let statement and each rate is given in the problem itself.

	Distance	Rate	Time
Going	x	6	
Returning	x	10	

The missing part of the chart can be found using the formula $T = \dfrac{D}{R}$.

	Distance	Rate	Time
Going	x	6	$\dfrac{x}{6}$
Returning	x	10	$\dfrac{x}{10}$

To find an equation, we note that the round trip took $1\frac{3}{5}$ hours. So the time going added to the time returning is $1\frac{3}{5}$ hours or $\frac{8}{5}$ hours.

$$\frac{x}{6} + \frac{x}{10} = \frac{8}{5}$$

$$30\left(\frac{x}{6} + \frac{x}{10}\right) = 30 \cdot \frac{8}{5} \qquad \text{Clear fractions}$$

$$5x + 3x = 48$$

$$8x = 48$$

$$x = 6$$

The stadium is 6 miles from Frank's house.

Problem Set 2.5

For each problem, set up an appropriate mathematical model and answer the question.

Warm-Ups

For Problems 1 through 4, see Examples 1 and 2.

1. How many gallons of a 60% solution should Christine add to 30 gallons of a 10% solution to produce a 40% solution?

2. How many gallons of 50% lemon juice should Orval add to 20 gallons of a 5% lemonade punch to produce a 10% lemonade punch?

For Problems 5 through 7, see Example 3.

5. Pipe A alone can fill the "Little Squirt" pool in 2 hours, and pipe B alone can fill the pool in 4 hours. If both pipes run at the same time, how long will it take to fill the pool?

6. Kimberly leaves the hot and cold water faucets on her kitchen sink turned on full blast. The hot water faucet alone can fill the sink in 2 minutes and the cold faucet

For Problems 8 through 11, see Examples 4 and 5.

8. A TWA 747 flying 720 km/hr travels 161 km farther than a Sabina 737 flying 650 km/hr. If departure and arrival times were the same for both planes, how far did each plane fly?

9. Frank, who lives in College Park, starts walking toward Chuck's house in Stone Mountain at a rate of 2 mph. Chuck starts at the same time toward Frank's house at a rate of 3 mph. If their houses are 20 miles apart, in how many hours will they meet? How far has each walked?

3. A chemist has 10 gallons of a 30% alcohol solution. How many gallons of pure alcohol should he add to obtain a mixture of 40% alcohol?

4. Linda has 10 gallons of a 5% iodine solution, but wishes to lower the concentration to 2.5%. How many gallons of water should she add?

alone can fill it in 3 minutes. The drain is open and it can drain the sink in 4 minutes. How long will it take for the sink to fill up?

7. Rowland can mow his yard in 21 minutes. It takes Don 28 minutes to mow the same lawn. How long should it take both of them, working together, to mow the yard?

10. Mr. Singh sets off for Tampa, a distance of 244 miles, in his Trihull Special at a constant rate. After 4 hours he has to reduce his speed by 12 mph due to a malfunctioning spider valve. If he finishes the trip in 3 hours, what was his original speed?

11. Lois and Joan walk in opposite directions from the same point, Lois walking at the rate of $6\frac{2}{3}$ mph and Joan walking 6 mph. In how many hours will they be 38 miles apart?

Practice Exercises

12. How many liters of a 65% solution should be added to 35 liters of a 12% solution to produce a 15% solution?

13. Virginia Carson has 30 ounces of 15% silver alloy that she wishes to upgrade to 20% silver. How many ounces of 50% silver alloy should she add?

14. How many gallons of pure alcohol should be added to 20 gal of an 8% alcohol solution to produce a 10% solution?

15. Jasinia Parks has 10 milliliters (mL) of a 6% iodine solution, but wishes to lower the concentration to 3%. How many milliliters of water should she add?

16. How many grams (g) of an amalgam containing 20% silver should be added to 100 g of an amalgam contain-

ing 32% silver to obtain an amalgam containing 28% silver?

17. How much pure lead should a glass manufacturer add to 532 kg of a molten mixture of 20% lead crystal to make a batch of 24% lead crystal?

18. A vat contains 20 liters of a 40% detergent solution. How much should be drained and replaced by pure detergent to obtain 20 liters of a 64% solution?

19. Pipe A alone can fill a tank in 5 hours and pipe B alone can fill the tank in 6 hours. If both pipes run at the same time, how long will it take to fill the tank?

20. Pipes A and B fill a tank. Pipe A alone can fill it in 2 hours and pipe B alone can fill it in 5 hours. The drain can empty it in 4 hours. If both pipes and the drain are open, how long will it take to fill the tank?

21. Jose can fill a sound tire with air in 3 minutes. An irate motorist brings Jose a tire that will go flat in 5 minutes if filled with air. How many minutes will it take Jose to fill the faulty tire with air so he can locate the leak?

22. J.R. can dig a certain length of ditch in 2 hours working alone, while John takes 3 hours to dig the same length of ditch if he is working alone. How long does it take John and J.R. to dig the length of ditch working together?

23. The cold water tap can fill a sink to the overflow level in 3 minutes. The hot water tap can fill the sink to the same level in 4 minutes. How long will it take both taps, running together, to fill a half-full sink to the overflow level?

24. Pipe A alone can fill a tank in 2 hours, pipe B alone can fill the same tank in 3 hours, and pipe C alone can fill it in 6 hours. How long will it take all three pipes working together to fill the tank?

25. The left valve can fill a tank in 5 hours. The right valve can fill the same tank in 10 hours. The drain can empty a full tank in $2\frac{1}{2}$ hours. If the tank is full and both valves are opened and the drain is opened, how long will it take to empty the tank?

26. If Barney's boat were full of water, it would take the bilge pump 2 hours to pump it dry. Barney could bail the water out in 3 hours if the pump were not working. A leak develops that takes 12 hours to fill the boat. How long does it take Barney and the bilge pump working together to empty the leaking boat so that Barney can repair it?

27. John can chop a batch of slaw in 30 minutes working alone, while JoAnne can chop the same batch in 42 minutes if she works alone. If they work together, they each work half as fast, due to the intensity of their conversation. How long does it take John and JoAnne, working together, to chop a batch of slaw?

28. Two planes leave Daytona at the same time traveling in opposite directions. If they are 2480 km apart after 2 hours and one plane travels 40 km/hr faster than the other plane, what are their average rates?

29. If Shu drives his old truck from home to Clarks Hill at 50 mph and returns immediately at 60 mph, his total elapsed driving time is 5.5 hours. How far is it from Shu's home to Clarks Hill?

30. Charlotte sets off on her bicycle at a rate of 15 mph to the local K-mart, a distance of 5 miles. After traveling part way, a low tire forces her to reduce her speed. She pedals at a rate of 10 mph for the rest of the trip. If the total trip takes 24 minutes $\left(\frac{2}{5}\text{ of an hour}\right)$, how far did Charlotte pedal at 15 mph?

31. The Midnight Express leaves the station at midnight traveling east. At 2 A.M. a slow freight train leaves the same station traveling west. At 4 A.M. the distance traveled by the Midnight Express is 210 miles more than the distance traveled by the slow freight. If the average speed of the Midnight Express is twice the average speed of the slow freight, how far has the slow freight traveled at 4 A.M.?

32. Carolyn and Irma jog the same track. Carolyn jogs at a rate of 0.15 mile/min and Irma at 0.10 mile/min. If Carolyn starts 1.5 minutes after Irma starts, how long will it take Carolyn to overtake Irma?

33. John and Jim run the same track. John runs at the rate of $\frac{1}{6}$ mile/min and Jim at $\frac{1}{8}$ mile/min. Jim starts first, and 3 minutes later John starts. How long will it take John to catch Jim?

34. Mary and Susan jog together. Mary jogs at the rate of $\frac{1}{5}$ mile/min and Susan at the rate of $\frac{1}{6}$ mile/min. If Mary starts 1 minute after Susan starts, how long will it take Mary to overtake Susan?

35. Jonathan rented a bicycle for 4 hours so that he could look at wildflowers along the road to the state park. He estimates that his average speed going will be 5 mph and returning will be 4 mph. How far can he ride and still return the bicycle within the 4-hour rental time? How far can he ride if he doubled his rental time?

Problem 35.

36. Two Japanese bullet trains start at noon from two cities that are 550 km apart and head toward one another. One train travels at 210 km/hr and the other at 175 km/hr. How far has each train traveled when they meet? At what time will they meet?

Solution

(a) Cara can be 3° from Laurie's heading. So Cara's heading can be either 148° or 154°.

(b) A number line will help us to see the use of absolute value. We locate 151 on the line and then move 3 units right and left of 151.

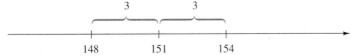

If h is Cara's heading, then the distance between Cara's heading and Laurie's heading is equal to 3.

The absolute value of the difference in the two headings must be equal to 3.

Thus $|h - 151| = 3$.   ◀

Problem Set 2.6

Warm-Ups

In Problems 1 through 6, solve each equation. See Example 1.

1. $|x| = 3$

2. $|x + 5| = 5$

3. $|2x| = 5$

4. $\left|\dfrac{1}{2}x\right| = \dfrac{1}{4}$

5. $|1 - 2x| = 4$

6. $\left|\dfrac{3x}{2} - \dfrac{1}{2}\right| = \dfrac{1}{4}$

In Problems 7 and 8, find the solution set. See Example 2.

7. $|x| = -2$

8. $|x + 2| = -39$

In Problems 9 through 12, find the solution set. See Example 3.

9. $|x| = 0$

10. $|x - 7| = 0$

11. $|2x - 5| = 0$

12. $\left|\dfrac{1}{3} - \dfrac{2}{3}x\right| = 0$

In Problems 13 through 16, find the solution set. See Example 4.

13. $|2x - 3| - 5 = 0$

14. $|4z + 1| + 6 = 0$

15. $|5 - 2x| + 2 = 5$

16. $3 - |x + 4| = 3$

In Problems 17 through 20, find the solution set. See Example 5.

17. $|x + 2| = |x|$

18. $|x + 1| - |x - 3| = 0$

19. $|2x - 1| = |x + 3|$

20. $\left|\dfrac{x + 1}{3}\right| - |x| = 0$

In Problems 21 and 22, answer the equation and find a model using absolute value for each problem. See Example 6.

21. Eleanor and Toby are on a diet of 2000 calories a day. They agreed that they could deviate 500 calories a day from this diet. What is the smallest and largest number of calories each can eat in a day?

22. The Twist Drill and Tool Company makes blue steel bolts. If the tolerance level for the length of their 3-inch bolt is $\dfrac{1}{16}$ of an inch, what is the longest and shortest acceptable bolt?

Practice Exercises

In Problems 23 through 58, find the solution set.

23. $|x| = 6$

24. $|x| = -1$

25. $|2x| = 0$

26. $|y + 3| = 1$

27. $|x - 4| = 2$

28. $|x + 11| = 0$

29. $|x - 6| = -7$

30. $|x - 7| = 7$

31. $|x + 13| = 7$

32. $|2x + 7| = 3$

33. $|4x + 5| - 4 = 4$

34. $|5x + 1| = 9$

35. $|10x - 3| = 0$

36. $|12t + 15| + 5 = 1$

37. $|5x + 11| = 9$

38. $\left|\dfrac{1}{4}x + 3\right| - 2 = 0$

39. $6 - |2.5x - 4| = 0$

40. $\left|\dfrac{2}{7}v - \dfrac{1}{3}\right| = \dfrac{1}{7}$

41. $|2.3 - x| = 4.5$

42. $\left|\dfrac{3}{4} - \dfrac{1}{2}x\right| = \dfrac{1}{8}$

43. $|2s| = |s + 3|$

44. $|x| = |x - 4|$

45. $|x + 7| = |x - 4|$

46. $|x - 8| = |x - 1|$

47. $|3x + 4| = |x - 2|$

48. $|x + 6| - |2x - 3| = 0$

49. $|x + 1| = |2x + 1|$

50. $|x - 9| = |x + 9|$

51. $|2x - 5| = |4x + 3|$

52. $|x - 3| - |3 - x| = 0$

53. $|2x + 9| = |7 - x|$

54. $|6 + x| = |6 - x|$

55. $|9x - 5| = |x + 3|$

56. $|2x + 5| = |7 - 2x|$

57. $|x + 1| = \left|\dfrac{1}{2}x\right|$

58. $\left|\dfrac{1}{3}x\right| = |x + 2|$

In Problems 59 through 64, answer the question and find a model using absolute value.

59. Find all numbers such that the absolute value of the number is 3.

60. Find all numbers such that the absolute value of 1 less than the number is $\dfrac{1}{2}$.

61. Find all numbers such that the absolute value of 3 less than four times the number is 1.

62. Find all numbers such that the absolute value of 3 minus $\dfrac{1}{5}$ of the number is 2.

63. The normal setting on a Wesson camp refrigerator keeps the inside temperature at 40°F. If the error in the ther-

mostat is 2.5%, what is the lowest and highest inside temperature?

64. In a poll conducted just before a recent election, Rosita Cheves had 42% of the votes. If the error in the poll itself was 4%, what was the smallest and largest percentage of votes for Rosita Cheves?

65. Kelli was a contestant on "Jeopardy" and she selected the category "Absolute Value Equations" for $500. The answer was "{−2, −8}." Kelli stated the question, "What is the solution set of $|x + 5| = 3$?" Should Kelli be awarded the $500? Why?

Challenge Problems

In Problems 66 through 70, find the solution set.

66. $|x - 5| = a, a > 0$

67. $|x - a| = 1$

68. $|x - a| = b, b > 0$

69. $|x + a| = |x|$

70. $|2x - a| = |x|$

In Problems 71 through 74, use properties 1 and 2 to solve each equation. Be sure to check all solutions. Why must these equations be checked?

71. $|2x + 1| = x$

72. $|2x - 3| = x + 1$

73. $|x - 2| = x + 3$

74. $|x + 5| = x + 2$

In Your Own Words

75. What do we mean by the absolute value of a number?

76. Describe how to solve equations of the form $|\text{variable}| = \text{a constant}$.

2.7   Boundary Numbers and Absolute Value Inequalities

TAPE 6

In Section 2.3 we solved linear inequalities in one variable. The tools that we used there are not sufficient to solve many other kinds of inequalities. We now examine another technique that applies to the linear inequalities, as well as to most other types of inequalities.

Let's solve the inequality $2x + 1 < 7$ by the method developed in Section 2.3.

$$2x + 1 < 7$$
$$2x < 6$$
$$x < 3$$
$$\{x \mid x < 3\}$$

The number 3 plays an important role in this solution set. It divides the number line into two regions, labeled A and B.

Region A	Region B

3

The solution set proved to be the region A. That is, the statement $2x + 1 < 7$ is true for **every** number in region A and false for **every** number in region B.

Is there a simple way to find this number that has the property of so dividing the number line? Note that the number 3 is the solution to the equation $2x + 1 = 7$. We call 3 a **boundary number** for the inequality $2x + 1 < 7$.

> **Boundary Numbers**
>
> Solutions to the equation formed by replacing the inequality symbol of an inequality with an equals symbol are *boundary numbers* for that inequality.

Let's see how to use this technique to solve an inequality.

► **EXAMPLE 1**

Solve $3x - 7 < 5$.

Solution

First, we solve the associated equation in order to find the boundary number(s).

$$3x - 7 = 5$$
$$3x = 12$$
$$x = 4$$

4 is the only boundary number.

We locate 4 on a number line and note that it divides the line into two regions, A and B.

We select *any* number from region A and check to see if it is in the solution set for $3x - 7 < 5$. A particularly convenient number in region A is 0. Since $3(0) - 7 < 5$ is a *true* statement, 0 must be in the solution set for the inequality. Thus *all* of region A is in the solution set.

Next we test *any* number from region B, say 5. Since $3(5) - 7 < 5$ is *false,* we conclude that 5 is not in the solution set. So *none* of region B is in the solution set.

The original inequality did not include equality, so the number 4 is *not* included in the solution set.

Therefore, the solution set is $\{x \mid x < 4\}$.   ◄

A Procedure for Solving Inequalities

1. Find the boundary numbers.

2. Locate the boundary numbers on a number line.

3. Determine which regions formed by the boundary numbers make the inequality true by testing with one number inside each region.

4. Shade only the regions that test true.

5. Check the boundary numbers themselves.

6. Write the solution set.

► **EXAMPLE 2**

Solve the inequality $|x - 2| > 5$.

Solution

STEP 1 Find the boundary numbers.

We solve the associated equation.

$$|x - 2| = 5 \quad \text{(an absolute value equation)}$$
$$x - 2 = 5 \quad \text{or} \quad x - 2 = -5$$
$$x = 7 \quad \text{or} \quad x = -3$$

The boundary numbers are -3 and 7.

STEP 2 Locate the boundary numbers on a number line.

STEP 3 Determine in which regions the inequality is true by testing each region.

A table is particularly convenient for this step.

Region	Number in Region	Statement $\|x - 2\| > 5$	Truth of Statement	Region in Solution Set?
A	-4	$\|-4 - 2\| > 5$	True	Yes
B	0	$\|0 - 2\| > 5$	False	No
C	8	$\|8 - 2\| > 5$	True	Yes

STEP 4 Shade the regions that test true.

We shade region A and region C on the number line.

STEP 5 Check the boundary numbers themselves.

The statement $|x - 2| > 5$ does not include equality, so the boundary numbers are not in the solution set.

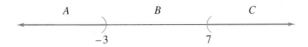

STEP 6 Write the solution set.

$$\{x \mid x < -3 \quad \text{or} \quad x > 7\}$$ ◄

► **EXAMPLE 3**

Solve $|1 - 2x| \leq 3$.

Solution

First we find the boundary numbers.

$$|1 - 2x| = 3$$
$$1 - 2x = 3 \quad \text{or} \quad 1 - 2x = -3$$
$$-2x = 2 \quad \text{or} \quad -2x = -4$$
$$x = -1 \quad \text{or} \quad x = 2$$

The boundary numbers are -1 and 2.

Next, we locate the boundary numbers on a number line.

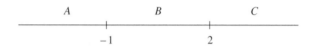

The next step is to test regions A, B, and C. A table such as the one that follows can be made or the regions can be shaded as each region is tested.

Region	Number in Region	Statement $\mid 1 - 2x \mid \leq 3$	Truth of Statement	Region in Solution Set?
A	-2	$\mid 1 - 2(-2) \mid \leq 3$	False	No
B	0	$\mid 1 - 2(0) \mid \leq 3$	True	Yes
C	3	$\mid 1 - 2(3) \mid \leq 3$	False	No

Region B is the only region in the solution set.

The statement $|1 - 2x| \leq 3$ includes equality, so the boundary numbers *are* in the solution set.

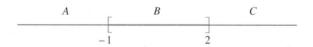

The last step is to write the solution set.

$$\{x \mid -1 \leq x \leq 2\}$$

◄

► **EXAMPLE 4**

Solve $|x - 6| > |2 - 3x|$. Write the solution set in interval notation.

Solution

We find the boundary numbers by solving the equation

$$|x - 6| = |2 - 3x|$$

We recall that we must solve the two equations

$$x - 6 = 2 - 3x \quad \text{or} \quad x - 6 = -(2 - 3x)$$
$$4x - 6 = 2 \qquad\qquad \text{or} \quad x - 6 = -2 + 3x$$
$$4x = 8 \qquad\qquad \text{or} \qquad -4 = 2x$$
$$x = 2 \qquad\qquad \text{or} \qquad -2 = x$$

Thus we get the boundary numbers -2 and 2.

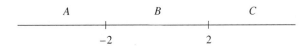

Next, we test each region. Region A contains -4. Testing with -4 gives

$$|-4 - 6| > |2 - 3(-4)|$$
$$|-10| > |2 + 12|$$
$$10 > 14$$

which is false, so region A is not shaded.

Region B contains 0, a convenient number.

$$|0 - 6| > |2 - 3(0)|$$
$$6 > 2$$

Because this is a true statement, we shade region B.

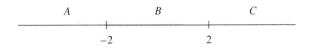

Region C contains 3. Testing with 3 gives

$$|3 - 6| > |2 - 3(3)|$$
$$|-3| > |2 - 9|$$
$$3 > 7$$

which is not true. We do not shade region C.

The statement $|x - 6| > |2 - 3x|$ does not allow equality, so the boundary numbers are not in the solution set.

Write the solution set in interval notation.

$$(-2, 2)$$

► **EXAMPLE 5**

Solve $|x - 5| \geq 0$.

Solution

First we find the boundary numbers.

$$|x - 5| = 0$$
$$x - 5 = 0$$
$$x = 5$$

The only boundary number is 5.

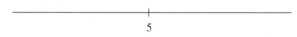

A convenient number less than 5 is 0.

$$|0 - 5| \geq 0$$
$$5 \geq 0$$

is a true statement, so the left region is in the solution set. A number to the right of 5 is 7, and

$$|7 - 5| \geq 0$$
$$2 \geq 0$$

is also a true statement, so the right region also belongs to the solution set.

Since equality is included in the original inequality, the boundary number is included in the solution set.

$$\{x \mid x \text{ is any real number}\}$$

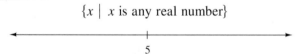

► **EXAMPLE 6**

Solve $|x| < -5$.

Solution

To find the boundary numbers, we solve the equation

$$|x| = -5$$

Since there is no solution to this equation, there are no boundary numbers. Thus there is just one region and it is the entire real line.

If we test any number in the region, say 10, we find that the statement is false.

$$|10| < -5$$
$$10 < -5$$

Since there are no boundary numbers to include, we get the empty set as our solution set:

$$\emptyset$$

NOTE: We could have seen that the solution set for Example 6 was the empty set because the absolute value of a number is never negative.

Absolute value inequalities such as

$$|X| < p, \qquad |X| > p$$
$$|X| \le p, \qquad |X| \ge p$$

with $p > 0$, arise frequently in mathematics. Often the solution set is not needed, but rather an equivalent compound inequality without the absolute value symbol. It is useful to remember the following result.

Absolute Value Inequalities

If $p > 0$ and X is an expression, then

$$|X| < p \quad \text{is equivalent to} \quad -p < X < p$$

and

$$|X| > p \quad \text{is equivalent to} \quad X < -p \text{ or } X > p$$

The results hold if the $<$ and $>$ are replaced by $\le$ and $\ge$.

► **EXAMPLE 7**

Write $|x + 1| < 5$ as an equivalent compound inequality.

Solution

We let $x + 1$ be the X in the statement

$$|X| < p \quad \text{is equivalent to} \quad -p < X < p$$

and we have

$$|x + 1| < 5 \quad \text{is equivalent to} \quad -5 < x + 1 < 5$$

Subtracting 1 from all sides yields

$$-6 < x < 4 \qquad \blacktriangleleft$$

► **EXAMPLE 8**

Write $|x - 3| \ge 7$ as an equivalent compound inequality.

Solution

We take $x - 3$ as the X and have that

$$|x - 3| \ge 7 \quad \text{is equivalent to} \quad x - 3 \le -7 \text{ or } x - 3 \ge 7$$

Adding 3 to both sides of each part, we get the compound inequality

$$x \le -4 \quad \text{or} \quad x \ge 10 \qquad \blacktriangleleft$$

We learned in Section 0.3 that the distance between two numbers on the number line is always given by the absolute value of their difference. This idea leads to a very important inequality.

► **EXAMPLE 9**

Find all numbers that are less than $\frac{1}{7}$ unit from 4 on the number line. Write an absolute value inequality to model the statement.

Solution

We let x be such a number. The distance between 4 and x is given by $|x - 4|$. As this is less than $\frac{1}{7}$, we get the inequality

$$|x - 4| < \frac{1}{7}$$

$$-\frac{1}{7} < x - 4 < \frac{1}{7}$$

$$-\frac{1}{7} + 4 < x < \frac{1}{7} + 4$$

$$\frac{27}{7} < x < \frac{29}{7}$$

So $\left\{ x \mid \dfrac{27}{7} < x < \dfrac{29}{7} \right\}$ is the set of all such numbers.　◄

Problem Set 2.7

Warm-Ups

In Problems 1 through 6, find the solution set. See Examples 2 and 3.

1. $|x - 1| < 5$

2. $|x + 3| > 1$

3. $|x + 2| \geq 8$

4. $|7 - x| \leq 9$

5. $|5x - 7| - 2 < 0$

6. $|4x + 1| - 4 \geq 0$

In Problems 7 through 10, find the solution set. Write the solution set in interval notation. See Example 4.

7. $|x + 1| > |2x + 3|$

8. $|2x| \geq |x - 2|$

9. $|2x + 3| < |5x - 3|$

10. $|3 - x| \geq |2x + 5|$

In Problems 11 through 14, solve each inequality. See Example 5.

11. $|x - 7| \geq 0$

12. $|x + 8| > 0$

13. $|x - 5| < -7$

14. $|x + 1| > -1$

In Problems 15 through 18, write each inequality as an equivalent compound inequality. See Examples 7 and 8.

15. $|x - 7| \leq 5$

16. $|2x - 5| > 6$

17. $\left|2x + \dfrac{1}{2}\right| \geq 7$

18. $\left|\dfrac{1}{2}x - 1\right| < 3$

For Problems 19 through 26, answer the question and find an absolute value inequality to model each problem. See Example 9.

19. Find all numbers that are within 2 units of -4 on the number line.

20. Find all numbers that are less than 1 unit from 5 on the number line.

21. Find all numbers that are less than $\dfrac{1}{3}$ unit from $\dfrac{1}{2}$ on the number line.

22. Find all numbers whose absolute value is less than or equal to 5.

23. Find all numbers whose absolute value is greater than 7.

24. Find all numbers whose absolute value is at least 10.

25. Find all numbers whose absolute value is at most 3.

26. If 6 is added to three times a number, the absolute value of the result is at least 1. Find all such numbers.

Practice Exercises

In Problems 27 through 44, find the solution set. (Use set builder notation.)

27. $|x| \geq 8$

28. $|x| < 2$

29. $|x - 3| < 9$

30. $|x + 1| \geq 16$

31. $|x + 5| \leq 3$

32. $|x - 7| > 4$

33. $|2x + 5| > 7$

34. $|3x - 1| \leq 10$

35. $|2x - 9| - 1 \leq 0$

36. $|3x + 7| - 7 > 0$

37. $|5x - 8| > 9$

38. $|4x + 3| \leq 7$

39. $|10y + 3| - 1 \geq 0$

40. $|5 - 2z| - 3 < 0$

41. $3(|x - 1| + 3) < 15$

42. $2(3 + |2x - 1|) \geq 5$

43. $10 \leq 3(4 - |x + 3|)$

44. $0 > 5(|s| - 4)$

In Problems 45 through 62, find the solution set. (Use interval notation.)

45. $|t - 2.5| < 6.3$

46. $\left|\dfrac{2}{3}x - \dfrac{3}{4}\right| < 6.3$

47. $|x - 2.73| > 6.72$

48. $|x - 2.4| > 7.8$

49. $\left|\dfrac{2x - 1}{3}\right| \leq 4$

50. $\left|\dfrac{3x}{5} - 2\right| \geq \dfrac{1}{5}$

51. $|2 - x| + 3 \geq 4$

52. $\left|3 - \dfrac{1}{4}x\right| + \dfrac{3}{8} > 1$

53. $|x + 1| \geq 0$

54. $|x - 7| > 0$

55. $|z - 3| < -5$

56. $|y + 2| > -4$

57. $|2x + 3| \leq -1$

58. $|3x + 2| \geq -4$

59. $|3x| \geq |x - 4|$

60. $|2x + 7| < |3x - 2|$

61. $|4x + 3| \leq |3x + 18|$

62. $|-3x + 2| > |x - 2|$

In Problems 63 through 70, write each inequality as an equivalent compound inequality.

63. $|x + 1| \leq 3$

64. $|3x + 4| > 5$

65. $|x| > 4.3$

66. $|x| \leq 1.5$

67. $|x + 3| < 2$

68. $|2x + 3| \geq 7$

69. $\left|2x - \dfrac{1}{3}\right| < 2$

70. $\left|\dfrac{1}{3}x + 1\right| \leq 2$

In Problems 71 through 76, answer the question and find an absolute value inequality to model each problem.

71. Find all numbers whose absolute value is less than 4.

72. Find all numbers whose absolute value is greater than or equal to $\dfrac{3}{2}$.

73. Find all numbers whose absolute value is at most 8.

74. If 4 is subtracted from twice a number, the absolute value of the result is at most 5. Find all such numbers.

75. Find all numbers that are less than 2 units from 1 on the number line.

76. Find all numbers that are less than 0.05 unit from 3 on the number line.

77. Richard is flying from Manchester, New Hampshire, on a heading of 270°. Richard's fight plan calls for veering no more than 4° from his original heading. At what headings will he fly if he stays on course?

78. Recent research revealed that normal body temperature may be different from 98.6°F. If the deviation from 98.6° may be 0.6° for a healthy person, what is the range of normal body temperatures?

79. A company has budgeted 22% of its total budget for advertising a new product. Miscellaneous expenses could cause this percentage to be off by 5%. What percentage of the total budget might be spent on advertising?

Challenge Problems

80. Prove that $|X| < a \ (a > 0)$ is equivalent to $-a < X < a$ using boundary numbers.

In Problems 81 through 90, find the solution set. Assume that $a > 0$.

81. $|x - a| \leq 4$

82. $|x - 1| > d; \ d > 0$

83. $|x - a| < d; \ d > 0$

84. $|x - a| > d; \ d < 0$

85. $|x - a| < a$

86. $|x - a| \geq a$

87. $|x - a| \leq 0$

88. $|x - a| > 0$

89. $|x + a| < a$

90. $|x + a| \geq a$

91. Find all numbers that are within s units of a.

In Your Own Words

92. What is a boundary number for an inequality?

93. What do we do if an inequality has no boundary numbers?

► Chapter Summary

GLOSSARY

Equation   A statement that two numbers are equal.

Variable   A letter representing a number.

Solution or **root**   A number that makes an equation a true statement when it replaces the variable.

Solution set   The set of all solutions of an equation.

Solve an equation   To find the solution set.

Equivalent equations   Equations that have the same solution set.

Inequality A statement that says one number is greater or less than another number.

Solution of an inequality A number that makes the statement true when substituted for the variable.

Solution set of an inequality The set of all solutions to the inequality.

Solve an inequality To find the solution set.

Linear equation An equation that can be written in the form

$$Ax + B = 0; \quad A \neq 0$$

Contradiction An equation that is always false.

Identity An equation that is always true.

Literal equation An equation that contains more than one letter.

Equivalent inequalities Inequalities that have the same solution set.

The **graph** of an inequality A picture of the solution set on a number line.

Boundary numbers Numbers that border the graph of the solution set of an inequality.

Properties of Equality

1. Add or subtract the same number to both sides.
2. Multiply or divide both sides by the same nonzero number.

Absolute Value Equations

Equations with absolute values can be solved by using these properties:

1. If $q \geq 0$, then $|X| = q$ is equivalent to the two equations $X = q$ or $X = -q$.
2. If $q < 0$, there are no solutions of the equation $|X| = q$.
3. $|U| = |V|$ is equivalent to the two equations $U = V$ or $U = -V$.

Linear Inequalities

Linear inequalities may be solved using the following tools.

Addition Property
The same number may be added or subtracted on both sides.

Multiplication by Positive Number
Both sides may be multiplied or divided by the same positive number.

Multiplication by Negative Number
Both sides may be multiplied or divided by the same negative number, *but* the inequality symbol must be *reversed*.

The Method of Boundary Numbers

An excellent technique for solving nonlinear inequalities is the method of boundary numbers.

1. Form an equation by replacing the inequality symbol with an equals sign. The solution(s) to this equation is the boundary number(s).
2. Locate and label the boundary numbers on a real number line.
3. Test the original inequality with one number from the interior of each region formed by the boundary numbers. The entire region is in the solution set if the test statement is true, or none of the region is in the solution set if the test statement is false.
4. Check the boundary numbers themselves in the original inequality to see if they are included in the solution set.

Problem Solving

1. Understand the problem.
2. Plan a strategy.
3. Apply mathematical tools.

Common strategies include making a mathematical model, estimating an answer, and looking for a pattern.

Modeling   Word problems can be modeled by the following steps:

1. Determine what is to be found.
2. Assign a variable, such as x, to represent one of the quantities to be found.
3. Express all other quantities to be found in terms of x (or the variable chosen).
4. Draw a figure or picture, if possible. Label it.
5. Reread the problem and form an equation or inequality (the mathematical model).
6. Solve the equation or inequality and find the values of all the quantities to be found.
7. Check the values in the original word problem. They should answer the question and make sense.
8. Write an answer to the original question.

CHECKUPS

In Problems 1 through 5, solve each equation.

1. $5(2x + 3) = 3 - 2(3x - 5)$ — Section 2.1; Example 3
2. $ax - b(x + a) = 6 + 2bx$ for x. — Section 2.2; Example 4
3. Solve $\dfrac{x + 1}{3} - \dfrac{x + 2}{2} < \dfrac{1}{6}$. — Section 2.3; Example 6
4. Solve $-3 \le 1 - 2x < 5$. — Section 2.3; Example 11
5. Three times a number increased by 7 is at most 31. Write an appropriate model and then find all such numbers. — Section 2.3; Example 13
6. In a right triangle, one leg is 3 cm longer than the other leg. If the shorter leg is increased by 6 cm, the area is increased by 24 cm². Find a model that can be used to find the length of each leg in the original triangle. — Section 2.4; Example 4
7. Jack Glover wishes to add enough 50% antifreeze solution to 16 gallons of a 5% antifreeze solution to obtain a 20% antifreeze solution. Find an appropriate model and determine how much of the 50% solution should Jack add? — Section 2.5; Example 1
8. Two trains, 780 miles apart, are heading toward each other at rates of 85 mph and 110 mph. When will they crash? — Section 2.5; Example 4
9. $\left| \dfrac{3 - 2x}{5} \right| - 7 = 0$ — Section 2.6; Example 4
10. Solve $|x - 2| > 5$. — Section 2.7; Example 2

► Review Problems

In Problems 1 through 42, find the solution set. Graph solution sets for inequalities and express the sets in interval notation.

1. $6(x - 5) - (2 - x) = 2x + 1$

2. $2[3 - 2(x - 2)] = 6(x - 4)$

3. $|2z - 3| = 4$

4. $2x - 3 \leq 5$

5. $1 < 2x < 4$

6. $|1 - x| = 0$

7. $|x + 16| > 8$

8. $\frac{1}{2}x - \frac{1}{3} = \frac{1}{4}$

9. $2(x - 7) + 5(3 - x) \leq 0$

10. $x \geq 5$ or $2x - 1 \leq 3$

11. $\left|\frac{3 + y}{2}\right| = 5$

12. $|2x - 3| \leq 3$

13. $\frac{3}{2}(x + 2) = \frac{1}{2}(x - 2)$

14. $|3x - 1| = |2x + 1|$

15. $\frac{1}{2}x - \frac{3}{4} < \frac{1}{4}x + 1$

16. $|3x + 4| = \frac{1}{4}$

17. $\frac{1}{2} - \frac{1}{4}x = \frac{1}{6}$

18. $6x + 3 = 2(3x + 1)$

19. $|6x - 7| \geq 0$

20. $|6 - x| = 4$

22. $|x - 1.5| < 4.6$

23. $2x + 1 < 5$ and $x + 2 > 5$

24. $3(2x - 1) = 2(3x + 1) - 5$

25. $\frac{1}{2}x - 2 \leq \frac{1}{4}(x - 3)$

26. $-3 \leq 2x + 1 < 7$

27. $\left|\frac{3x + 1}{2}\right| \leq 2$

28. $\left|\frac{1}{2}x + 3\right| = \frac{1}{4}$

29. $6x + 5 = 2x - 3$

30. $|x| = |2x + 1|$

31. $|3x + 5| = 6$

32. $\left|\frac{1}{3} - x\right| > \frac{2}{3}$

33. $2(x + 1) + 3 < 2x - 1$

34. $3x - (1 - x) = 0$

35. $\left|\frac{1 - x}{4}\right| = \frac{1}{2}$

36. $\frac{1}{2}x + \frac{1}{3}(x + 1) = \frac{1}{6}$

37. $|3x - 5| = 0$

38. $3 - (2 + x) = 6x - 7$

39. $3\left(x - \frac{2}{3}\right) + 4 > 1 + 3x$

40. $2x + 1 < 7$ or $3x - 5 > 25$

41. $|2x - 7| = -3$

42. $3[2 - (6 - x) + 3(x + 2)] = 3(x - 2)$

43. What are the dimensions of the largest picture frame that can be constructed from 60 in. of molding if the width is to be two-thirds of the length?

44. The length of a rectangle is 8 ft more than the width. If the perimeter is 104 ft, find the dimensions of the rectangle.

45. The sum of two consecutive even integers is 50. Find the integers.

46. The sum of three natural numbers is 29. The second number is 5 more than the first number and the third number is twice the first. Find the three numbers.

47. Find three consecutive odd integers such that the first

plus twice the second plus three times the third is 94.

48. Martin can mow the lawn in 2 hours and Tom can mow the lawn in 6 hours. If they both mow at the same time, how long will it take to mow the lawn?

49. Angelita has 15 gallons of 8% Clorox solution, but she wishes to lower the concentration to 5%. How many gallons of water should she add?

50. Asha has $14,000 invested at 4.75% and 5.5%. If her yearly interest is $702.50, how much does she have at each rate?

51. The perimeter of a rectangle is 86 m. If the length is two less than twice the width, find the dimensions of the rectangle.

52. Two cars are 760 km apart. They are approaching each other at speeds of 110 km/hr and 80 km/hr. In how many hours will they meet?

53. When 5 is subtracted from two times a number, the result is at most 17. Find all such numbers that fit this statement.

54. If one-third of a number is added to 15, the result is less than 25. Find all such numbers.

55. The perimeter of a square must be less than or equal to 100 ft. What could the length of a side be?

56. Amanda must score at least 30 points to qualify for the state gymnastics meet. If she has 9.2 points on the floor exercise, 8.7 points on the beam, and 7.1 points on the vault, how many points must she score on the uneven parallel bars to qualify?

57. Jo has test scores of 75, 86, 88, and 62. What must she score (out of 100 points) on her fifth test to have an average of at least 80?

58. Sharon has a pickle barrel and she sells pickles for 15 cents each. It costs her 3 cents for each pickle and she spent a total of $1.80 on wrappers. How many pickles must she sell to make a profit of more than $3.00?

Let's Not Forget . . .

Identify the expressions that are in factored form. Factor those that are not factored.

59. $(x - 3)^3$

60. $x^3 - 27$

61. $(x + y) - 4(x + y)^2$

62. $2xy(x^2 + y^2)$

63. x^2yz

How many terms are in each expression? Which expressions have $x + 2$ as a factor?

64. $x^2 - 4$

65. $(x + 2)^2$

66. $x^3 + 8$

67. $xy(x + 2)$

68. $(x + 2)^2 - y(x + 2)$

69. $x^2 + 4x + 4$

Simplify each expression.

70. -8^2

71. $(-8)^2$

Find each product.

72. $(x - 2y)^2$

73. $(a + 2b)^3$

74. $(x^2y)^2$

75. $x^3(x^2y - xy)$

Label each as an equation, inequality, or an expression. Solve the equations and inequalities, and perform the operations indicated on the expressions.

76. $2x - 5 = x - 3(4 - x)$

77. $2x - 5 < x - 3(4 - x)$

78. $|3 - 6| + 1$

79. $|3x + 2| \le 6$

80. $3(x - 4) - (x - 1)$

▶ You Decide

You are planning to drive from Los Angeles to Chicago and to look at some of the sights along the way. You have two possible possible routes.

A. The *Northern Route.* I-15 to Salt Lake City to I-80 to Chicago with possible side trips to Las Vegas (10 miles) and Yellowstone National Park (350 miles). You figure the interstate mileage to be 2100 miles and the side-trip mileage as given.

B. The *Southern Route.* I-40 to Memphis then I-55 to Chicago with possible side trips to the Grand

Canyon (100 miles), the Petrified Forest (20 miles), White Sands and Carlsbad Caverns (240 miles), and Graceland (30 miles). You figure the interstate mileage to be 2400 miles and the side-trip mileage as given.

You expect to drive 8 hours each day and to average 60 mph on the interstates and 30 mph on the side trips. Which route do you decide to take? What side trips will you include and why? If you leave Los Angeles at 8:00 A.M. on May 10, when do you expect to arrive in Chicago?

► Chapter Test

In Problems 1 through 7, choose the correct answer.

1. The solution set for the equation
$3(2x - 1) - 4x = 3(5 - x)$ is (?)
   A. $\left\{\dfrac{14}{5}\right\}$ **B.** $\{-12\}$ **C.** $\left\{\dfrac{18}{5}\right\}$ **D.** $\{6\}$

2. The graph of the set $\{x \mid 1 < x < 3\}$ is (?)

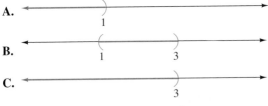

A.

B.

C.

D.

3. When solved for x, the literal equation $b(1 - 6x) = 2ax$ becomes (?)
   A. $x = \dfrac{b(1 - 6x)}{2a}$ **B.** $x = \dfrac{b}{2a + 6b}$
   C. $x = \dfrac{b}{2a - 3b}$ **D.** $x = \dfrac{1}{2a + 6}$

4. The solution set for $\dfrac{x - 3}{6} \le \dfrac{x}{2} - 1$ is (?)
   A. $\left\{x \mid x \ge -\dfrac{9}{2}\right\}$ **B.** $\left\{x \mid x \le -\dfrac{9}{2}\right\}$
   C. $\left\{x \mid x \ge \dfrac{3}{2}\right\}$ **D.** $\left\{x \mid x \le \dfrac{3}{2}\right\}$

5. If x is the smallest of three consecutive odd integers, which of the following is the sum of the first plus twice the second plus four times the third?
   A. $x + 2x + 2 + 4x + 4$ **B.** $x + 2x + 4x$
   C. $x + 2(x + 2) + 4(x + 4)$ **D.** $x + 2(2x) + 4(4x)$

6. Which set represents the following graph?

   A. $\{x \mid x < -5$ or $0 < x \le 3\}$
   B. $\{x \mid x \le -5$ or $x > 0$ or $x \le 3\}$
   C. $\{x \mid x \le -5$ or $x \le 3\}$
   D. $\{x \mid x < -5$ or $0 < x < 3\}$

7. The solution set for $|x + 3| < 0$ is (?)
   A. $(-\infty, -3)$
   B. $(-3, +\infty)$
   C. $\varnothing$
   D. $(-\infty, +\infty)$

In Problems 8 through 12, find the solution set.

8. $|2x + 3| = 7$

9. $1 - 2x + 2(x - 7) < 3x$

10. $\dfrac{2}{3}(x - 1) = \dfrac{1}{6} + x$

11. $|1 - 2x| = -3$

12. $|3x + 1| > 10$

13. Little Red Riding Hood is going from her house to her grandmother's house, which is a distance of 600 miles. She rides her Honda for 5 hours and then rides 9 hours on a bus the rest of the way. If the bus went 20 mph faster than the Honda, how fast did the Honda go?

14. Katura has 134 in. of picture-frame material. What are the dimensions of the largest frame that she can make if the length is to be 5 in. more than the width?

15. Shirley Ann Exum has 7 quarts of a 5% vinegar solution. How many quarts of a 20% vinegar solution should she add to obtain a 13% vinegar solution?

3 Rational Expressions

See Problem Set 3.4, Exercise 32.

► **Connections**

To find the capacitance of a spherical capacitor, an electrical engineer might use the formula, $C = \dfrac{4kR_1R_2}{R_2 - R_1}$. Expressions such as the right side of this formula are common in science and engineering. They are called rational expressions. Rational expressions are simply polynomials transformed into fractions.

Just as fractions form an important part of arithmetic, rational expressions play an important role in algebra. Rational expressions, like the fractions of arithmetic, indicate a division operation. We will learn how to reduce rational expressions, in much the same way as we simplify common fractions.

In this chapter we define rational expressions; then we learn how to simplify, add, subtract, multiply, and divide them. Complex fractions (fractions containing fractions) are introduced and we learn how to simplify them. The factoring skills we acquired in Chapter 1 are a necessary prerequisite to this study, and are essential to our understanding of it. We will use the skills developed in this chapter to solve rational equations and inequalities together with the problem-solving skills from Chapter 2.

3.1 Fundamental Principle of Rational Expressions

TAPE 6

In Section 0.1, a rational number was defined as the quotient of two integers, a and b, written as $\dfrac{a}{b}$, with $b \neq 0$, and in Section 1.8 this idea was extended to the division of polynomials. Next we examine quotients of polynomials in more detail.

A **rational expression** is the quotient of two polynomials. Rational expressions are often called *algebraic fractions* or simply *fractions*. We use the terms *numerator* and *denominator* just as we would with fractions in arithmetic. The following are examples of rational expressions.

$$\boxed{1}\ \frac{3}{x} \qquad \boxed{2}\ \frac{5x^2 + 2x - 1}{x^3 + 3x^2 - x + 1} \qquad \boxed{3}\ \frac{x}{(x + 1)(x - 2)}$$

Each of these is a quotient of two polynomials and each has many different values, depending on the number x.

$\boxed{1}\ \dfrac{3}{x}$ has the value $\dfrac{3}{2}$ when x is 2 and $\dfrac{3}{7}$ when x is 7.

$\boxed{2}\ \dfrac{5x^2 + 2x - 1}{x^3 + 3x^2 - x + 1}$ has the value of -1 when x is 0 and $\dfrac{23}{19}$ when x is 2.

$\boxed{3}\ \dfrac{x}{(x + 1)(x - 2)}$ has the value $-\dfrac{3}{10}$ when x is -3.

Let's consider the rational expression $\dfrac{1}{x - 2}$. If x is 4, its value is $\dfrac{1}{2}$.

However, if x is 2, the expression takes the form $\dfrac{1}{0}$. This is not a real number. We say that the rational expression $\dfrac{1}{x-2}$ is undefined when x is 2. A rational expression is **undefined** when its denominator has the value 0, regardless of the value of its numerator.

Undefined Rational Expression

If P and Q are polynomials, the rational expression

$$\frac{P}{Q}$$

is *undefined* for all numbers that make Q have a value of zero.

► ### EXAMPLE 1

Determine the numbers for which each rational expression is undefined.

(a) $\dfrac{x}{x+3}$ (b) $\dfrac{x+2}{(x+1)(x-2)}$ (c) $\dfrac{5}{x^2+1}$

Solutions

(a) If x is -3, the rational expression $\dfrac{x}{x+3}$ becomes $\dfrac{-3}{0}$. So $\dfrac{x}{x+3}$ is undefined when x is -3.

(b) If x is -1, the rational expression $\dfrac{x+2}{(x+1)(x-2)}$ becomes $\dfrac{1}{0}$, and if x is 2, it becomes $\dfrac{4}{0}$. So $\dfrac{x+2}{(x+1)(x-2)}$ is undefined when x is -1 or 2.

(c) The denominator x^2+1 is never 0 for any value of x. (Try some numbers!) The rational expression $\dfrac{5}{x^2+1}$ is defined for all real numbers.   ◄

From the rules of signed number arithmetic, we know that if p and q are real numbers, with $q \neq 0$, then

$$\frac{-p}{q} = \frac{p}{-q} = -\frac{p}{q}$$

This idea is also true for rational expressions and is very important.

Sign Property of Rational Expressions

If A and B are polynomials and $\dfrac{A}{B}$ is defined, then

$$\frac{-A}{B} = \frac{A}{-B} = -\frac{A}{B}$$

► **EXAMPLE 2**

Use the Sign Property of Rational Expressions to write each rational expression two different ways.

(a) $\dfrac{-5}{x}$ (b) $-\dfrac{x}{x^2 + 3}$ (c) $\dfrac{-x + 2}{x^2}$

Solutions

Be Careful!

(a) $\dfrac{-5}{x} = \dfrac{5}{-x} = -\dfrac{5}{x}$

(b) $-\dfrac{x}{x^2 + 3} = \dfrac{-x}{x^2 + 3} = \dfrac{x}{-(x^2 + 3)}$ The parentheses in the denominator are necessary, because $-(x^2 + 3)$ is not the same as $-x^2 + 3$.

(c) Be very careful with this one! The leading negative in $-x + 2$ **does not** indicate the opposite of $(x + 2)$. However, because $-x + 2 = -(x - 2)$, we can write

$$\dfrac{-x + 2}{x^2} = \dfrac{-(x - 2)}{x^2}$$

so

$$\dfrac{-x + 2}{x^2} = \dfrac{-(x - 2)}{x^2} = \dfrac{x - 2}{-x^2} = -\dfrac{x - 2}{x^2}$$ ◄

There are many ways of writing the same fraction. For example, $\dfrac{1}{2}$ can be written as $\dfrac{3}{6}, \dfrac{4}{8}$, or $\dfrac{6}{12}$. The form we use depends on what we are doing with the fraction. The same idea applies to rational expressions, as stated in the following principle.

Fundamental Principle of Rational Expressions

If A, B, and C are polynomials, then

$$\dfrac{AC}{BC} = \dfrac{A}{B}$$

where each rational expression is defined.

This principle states that we can divide the numerator and denominator of a rational expression by the same nonzero *factor*. When we write a rational expression so that the numerator and denominator have no common factor, we say that the rational expression is in **lowest terms** or that we have **reduced** to lowest terms.

Consider the rational expression $\dfrac{2(x + 1)}{a(x + 1)}$. By the Fundamental Principle of Rational Expressions,

$$\dfrac{2\,(x + 1)}{a\,(x + 1)} = \dfrac{2}{a}$$

Notice that $x + 1$ is a factor in both the numerator and the denominator. By dividing both numerator and denominator by the only common factor, we have written $\dfrac{2(x + 1)}{a(x + 1)}$ in lowest terms.

Each rational expression used in the Fundamental Principle of Rational Expressions must be defined. This means that, when we divide by $x + 1$, we must assume that $x + 1 \neq 0$ or $x \neq -1$. So it would be proper to write

$$\frac{2(x + 1)}{a(x + 1)} = \frac{2}{a}, \qquad x \neq -1$$

Be Careful!

However, we usually do not write the $x \neq -1$. Instead, we understand that when we divide by a common factor its value must not be zero.

When using the Fundamental Principle of Rational Expressions to reduce, **factoring** is the key. We divide by common *factors*. So make sure that both the numerator and the denominator are **factored** before reducing.

> ### To Reduce a Rational Expression to Lowest Terms
>
> **1.** Factor the numerator and denominator.
>
> **2.** Apply the Fundamental Principle of Rational Expressions.

► **EXAMPLE 3**

Reduce each rational expression to lowest terms.

(a) $\dfrac{12x^2}{-18x^3}$ (b) $\dfrac{5x + 5y}{x^2 - y^2}$ (c) $\dfrac{x + 3}{x^2 + 6x + 9}$

(d) $\dfrac{x^2 + xy - xz - yz}{ax - az + bx - bz}$

Solutions

(a) First factor the numerator and the denominator.

$$\frac{12x^2}{-18x^3} = \frac{(6)\,(2)\,x^2}{-(6)\,(3)\,x^2\,x}$$

Then divide by any common factors. Both 6 and x^2 are factors in both the numerator and denominator. So we divide both the numerator and the denominator by $6x^2$.

$$\frac{12x^2}{-18x^3} = \frac{(6)\,(2)\,x^2}{-(6)\,(3)\,x^2\,x} = \frac{2}{-3x}$$

This answer could be written in various forms:

$$\frac{2}{-3x} = \frac{-2}{3x} = -\frac{2}{3x}$$

(b) $\dfrac{5x + 5y}{x^2 - y^2} = \dfrac{5(x + y)}{(x + y)(x - y)}$ Factor

$= \dfrac{5}{x - y}$ Reduce

(c) $\dfrac{x + 3}{x^2 + 6x + 9} = \dfrac{x + 3}{(x + 3)^2}$ Factor

$= \dfrac{1}{x + 3}$ Reduce

(d) $\dfrac{x^2 + xy - xz - yz}{ax - az + bx - bz} = \dfrac{x(x + y) - z(x + y)}{a(x - z) + b(x - z)}$

Be Careful!

Although we have factored some *terms* in order to factor by grouping, at this point neither the numerator nor the denominator is factored. So we **cannot** divide out any common factors here. Continuing to factor by grouping gives us

$$= \dfrac{(x + y)(x - z)}{(x - z)(a + b)} = \dfrac{x + y}{a + b} \qquad \blacktriangleleft$$

► **EXAMPLE 4**

Reduce each rational expression to lowest terms.

(a) $\dfrac{a - b}{b - a}$ (b) $\dfrac{x^2 - y^2}{y^3 - x^3}$

Solutions

(a) The numerator and the denominator are opposites. Remember that $a - b = -(b - a)$. So we write

$$\dfrac{a - b}{b - a} = \dfrac{-(b - a)}{b - a}$$

Now $b - a$ is a factor in both the numerator and the denominator. So

$$\dfrac{a - b}{b - a} = \dfrac{-(b - a)}{b - a} = -\dfrac{b - a}{b - a} = -1$$

(b) $\dfrac{x^2 - y^2}{y^3 - x^3} = \dfrac{(x - y)(x + y)}{(y - x)(y^2 + yx + x^2)}$ Factor

We replace $x - y$ with $-(y - x)$ in the numerator.

$$= \dfrac{-(y - x)(x + y)}{(y - x)(y^2 + yx + x^2)}$$

Now $(y - x)$ is a factor of both the numerator and denominator.

$$= \dfrac{-(x + y)}{x^2 + xy + y^2} = -\dfrac{x + y}{x^2 + xy + y^2}$$

Notice in the last step we wrote the denominator in descending powers of x.

The most common mistake made in reducing fractions is to divide by a *term* and not a *factor. Make a habit of checking carefully to see that anything being divided out of a fraction is a* **factor.**

There are two ways to look at the Fundamental Principle of Rational Expressions. If we write $\dfrac{AC}{BC} = \dfrac{A}{B}$, we are reducing. If we write it the other way, $\dfrac{A}{B} = \dfrac{AC}{BC}$, we are multiplying the numerator and denominator by C. Later we will need to do this when adding and subtracting rational expressions. This is sometimes called *building up a denominator.*

Let's write the rational expression $\dfrac{10m}{7n^2}$ with a denominator of $21n^3$. First notice that $21n^3 = 7n^2(3n)$. So use the Fundamental Principle of Rational Expressions and multiply both numerator and denominator by $3n$.

$$\frac{10m}{7n^2} = \frac{10m\,(3n)}{7n^2\,(3n)}$$

$$= \frac{30mn}{21n^3}$$

► **EXAMPLE 5**

Write each rational expression with the denominator indicated.

(a) $\dfrac{-xy}{a}; a^3$ (b) $\dfrac{x+2}{x-1}; x^2 - 1$

(c) $\dfrac{x}{x+1}; x^2 + 2x + 1$ (d) $\dfrac{x-2}{x-3}; x^2 - x - 6$

Solutions

(a) First note that $a^3 = a(a^2)$.

$$\frac{-xy}{a} = \frac{-xy\,(a^2)}{a\,(a^2)} = \frac{-a^2xy}{a^3}$$

(b) Notice that $x^2 - 1 = (x - 1)(x + 1)$. So we multiply numerator and denominator by $x + 1$.

$$\frac{x+2}{x-1} = \frac{(x+2)\,(x+1)}{(x-1)\,(x+1)}$$

$$= \frac{x^2 + 3x + 2}{x^2 - 1}$$

(c) Because $x^2 + 2x + 1 = (x + 1)^2$, multiply numerator and denominator by $x + 1$.

$$\frac{x}{x+1} = \frac{x\,(x+1)}{(x+1)\,(x+1)}$$

$$= \frac{x^2 + x}{x^2 + 2x + 1}$$

(d) Notice that $x^2 - x - 6 = (x - 3)(x + 2)$. So we multiply the numerator and denominator by $x + 2$.

$$\frac{x - 2}{x - 3} = \frac{(x - 2)\,(x + 2)}{(x - 3)\,(x + 2)}$$

$$= \frac{x^2 - 4}{x^2 - x - 6} \qquad \blacktriangleleft$$

► **EXAMPLE 6**

Write each rational expression with the denominator indicated.

(a) $\dfrac{5}{7 - x}$; $x - 7$ (b) $\dfrac{3}{2 - x}$; $x^2 - 4$

Solutions

(a) Since $7 - x$ and $x - 7$ are opposites, we replace $7 - x$ with $-(x - 7)$.

$$\frac{5}{7 - x} = \frac{5}{-(x - 7)}$$

$$= \frac{-5}{x - 7}$$

(b) Because $x^2 - 4 = (x - 2)(x + 2)$, and $2 - x$ and $x - 2$ are opposites, replace $2 - x$ with $-(x - 2)$, and then multiply the numerator and denominator by $x + 2$.

$$\frac{3}{2 - x} = \frac{3}{-(x - 2)}$$

$$= \frac{-3}{x - 2}$$

$$= \frac{-3\,(x + 2)}{(x - 2)\,(x + 2)}$$

$$= \frac{-3x - 6}{x^2 - 4} \qquad \blacktriangleleft$$

Problem Set 3.1

Warm-Ups

In Problems 1 through 16, determine the numbers, if any, for which each rational expression is undefined. See Example 1.

1. $\dfrac{5}{x - 3}$ **2.** $\dfrac{1}{x + 1}$ **3.** $\dfrac{-6}{x - 5}$ **4.** $\dfrac{3}{x}$

5. $\dfrac{x}{6}$

6. $\dfrac{4}{x^2}$

7. $\dfrac{7}{x^2 + 1}$

8. $\dfrac{x}{x - 9}$

9. $\dfrac{x - 5}{x - 2}$

10. $\dfrac{x + 3}{x + 8}$

11. $\dfrac{x}{(x + 3)(x - 1)}$

12. $\dfrac{x^2 + 2}{(x - 6)(x - 4)}$

13. $\dfrac{5}{x(x - 2)}$

14. $\dfrac{x - 7}{x^2 - x - 6}$

15. $\dfrac{x^2 - 2x - 3}{x^2 - 4}$

16. $\dfrac{x}{x^2 + 6x + 9}$

In Problems 17 through 24, use the Sign Property of Rational Expressions to write each rational expression in two different ways. See Example 2.

17. $\dfrac{-3}{x}$

18. $\dfrac{x^2}{-8}$

19. $-\dfrac{7}{x}$

20. $\dfrac{-(x - 1)}{x}$

21. $\dfrac{2x}{-(x + 5)}$

22. $-\dfrac{x}{x + 3}$

23. $\dfrac{x - 1}{-x^2}$

24. $\dfrac{-x - 6}{x + 3}$

In Problems 25 through 36, reduce each rational expression to lowest terms. See Example 3.

25. $\dfrac{x^7}{x^2}$

26. $\dfrac{y^2}{y^3}$

27. $\dfrac{-2^3 xy}{2^2 x^2 y^2}$

28. $\dfrac{(-3)^2 a^2}{3a}$

29. $\dfrac{5x^2(x + y)^2}{10x(x + y)^3}$

30. $\dfrac{(x + 1)(3 + x)}{(x - 1)(x + 3)^2}$

31. $\dfrac{12x^2 - 12x}{18x^2}$

32. $\dfrac{x^2 - 16}{x - 4}$

33. $\dfrac{x + 1}{x^3 + 1}$

34. $\dfrac{x^2 - 3x - 10}{x^2 - 2x - 15}$

35. $\dfrac{8x^3 - 8}{2x^2 - 6x + 4}$

36. $\dfrac{x^3 + xz^2 + x^2 z + z^3}{x^3 + xz^2 - x^2 z - z^3}$

In Problems 37 through 42, reduce each rational expression to lowest terms. See Example 4.

37. $\dfrac{s - 1}{1 - s}$

38. $\dfrac{(x - 2)(x + 3)}{(2 - x)(1 + x)}$

39. $\dfrac{(2x - 1)^2}{(1 - 2x)(x + 2)}$

40. $\dfrac{5x^2 - 10x}{10 - 5x}$

41. $\dfrac{x^2 - 3x + 2}{3 - 2x - x^2}$

42. $\dfrac{m^3 - n^3}{n^2 - m^2}$

In Problems 43 through 46, write each rational expression with the denominator indicated. See Example 5.

43. $\dfrac{5}{x}$; $10x$

44. $\dfrac{15b}{2c}$; $10bc^2$

45. $\dfrac{r^2}{6}$; $6m + 6n$

46. $\dfrac{-3x}{x + 1}$; $x^2 - 1$

For Problems 47 through 50, write each rational expression with the denominator indicated. See Example 6.

47. $\dfrac{x + 1}{x - 1}$; $x - x^2$

48. $\dfrac{2z}{3 - z}$; $z^2 - 6z + 9$

49. $\dfrac{x}{3 - x}$; $x^2 - 9$

50. $\dfrac{3z}{y - 2z}$; $4z^2 - y^2$

Practice Exercises

In Problems 51 through 60, determine the numbers, if any, for which each rational expression is undefined.

51. $\dfrac{5}{x}$

52. $\dfrac{x}{7}$

53. $\dfrac{6}{x^2}$

54. $\dfrac{1}{x^2 + 4}$

55. $\dfrac{x}{x - 7}$

56. $\dfrac{x - 3}{x - 4}$

57. $\dfrac{x^2 + 3}{(x - 5)(x - 2)}$

58. $\dfrac{5}{x(x - 1)}$

59. $\dfrac{x^2 - 2x - 8}{x^2 - 1}$

60. $\dfrac{x}{x^2 - 6x + 9}$

In Problems 61 through 108, reduce each rational expression to lowest terms.

61. $\dfrac{x^8}{x^3}$

62. $\dfrac{y^3}{y^7}$

63. $\dfrac{48t^3}{16t}$

64. $\dfrac{-15x^3 y}{5xy^2}$

65. $\dfrac{-2x^2 y}{-8xy}$

66. $\dfrac{18x^3 yz^2}{-27xy^3 z^3}$

67. $\dfrac{-3^3 xy}{3^2 x^3 y^3}$

68. $\dfrac{(-5)^2 a^2}{5a}$

69. $\dfrac{7^2 5^3 x^2 y}{7^3 5^2 xy^3}$

70. $\dfrac{5x^2 (x - y)^2}{15x(x - y)^3}$

71. $\dfrac{-n^5 (n - m)}{n^4 (n - m)^4}$

72. $\dfrac{(x + y)^2}{(x + y)^7}$

73. $\dfrac{(x + 2)(5 + x)}{(x - 1)(x + 5)^2}$

74. $\dfrac{t - 1}{1 - t}$

75. $\dfrac{(a - 3)(a + 3)}{(3 - a)(1 + a)}$

76. $\dfrac{(3x - 1)^2}{(1 - 3x)(x + 1)}$

77. $\dfrac{4a^2 - 8a}{8 - 4a}$

78. $\dfrac{18x^3 - 12x}{6x - 9x^3}$

79. $\dfrac{5x}{25x - 15x^2}$

80. $\dfrac{xy^3}{xy + y}$

81. $\dfrac{3y + 12y^4}{15y^3 - 25y}$

82. $\dfrac{4x^3 - 24x}{x^3 - 6x}$

83. $\dfrac{x + 3}{x^2 + 5x + 6}$

84. $\dfrac{x^2 - 25}{x + 5}$

85. $\dfrac{x - 1}{x^3 - 1}$

86. $\dfrac{p^3 - 8}{p^2 - 5p + 6}$

87. $\dfrac{x^2 - 8x + 16}{x - 4}$

88. $\dfrac{b^2 - 1}{b^2 + b - 2}$

89. $\dfrac{x^2 - x - 12}{x^2 - 3x - 4}$

90. $\dfrac{x^2 + x - 2}{x^2 + 4x + 4}$

91. $\dfrac{p^3 + 27}{p^2 + 5p + 6}$

92. $\dfrac{x^2 - 4x}{x^3 - 64}$

93. $\dfrac{x^2 + 3x - 10}{8 - 2x - x^2}$

94. $\dfrac{6x^2 + x - 2}{3x^2 - x - 2}$

95. $\dfrac{s^2 - 1}{1 - s^3}$

96. $\dfrac{m^2 - n^2}{n^3 - m^3}$

97. $\dfrac{3x^4 - 3x}{6x^3 - 6x}$

98. $\dfrac{x^4 - 16}{x^2 - 4}$

99. $\dfrac{x^4 - x^2 - 2}{x^4 + x^2 - 6}$

100. $\dfrac{x^4 - x^2 - 6}{x^4 - 3x^2 - 10}$

101. $\dfrac{9 - 4x^2}{2x^2 - x - 3}$

102. $\dfrac{25 - 4x^2}{2x^2 + x - 15}$

103. $\dfrac{am + 3m - 2a - 6}{cm - 2c + 6m - 12}$

104. $\dfrac{a^3 + ab^2 + a^2 c + b^2 c}{a^3 + ab^2 - a^2 c - b^2 c}$

105. $\dfrac{ps - pt - qs + qt}{pt + qt - ps - qs}$

106. $\dfrac{2x^2 + 6ax + 4a^2}{x^2 + bx + ax + ab}$

107. $\dfrac{x^3 - xy + x^2 y - y^2}{x^3 - x^2 y - xy + y^2}$

108. $\dfrac{y^2 + yc + yd + dc}{3y^2 - 3yc + 3yd - 3dc}$

In Problems 109 through 130, write each rational expression with the denominator indicated.

109. $\dfrac{5}{x}$; $12x^2$

110. $\dfrac{3}{a}$; $4a^2$

111. $\dfrac{15b}{2c^2}$; $10bc^2$

112. $\dfrac{-10a}{3xy}$; $30x^2y^3$

113. $\dfrac{p}{p-5}$; $2p-10$

114. $\dfrac{a^2}{a+b}$; $4a+4b$

115. $\dfrac{-3x}{x-1}$; x^2-1

116. $\dfrac{a}{a+2}$; a^2+a-2

117. $\dfrac{2z}{z+3}$; z^2+6z+9

118. $\dfrac{x-1}{x}$; x^2+2x

119. $\dfrac{3}{x-y}$; $y-x$

120. $\dfrac{5}{m-n}$; $n-m$

121. $\dfrac{x}{2-x}$; x^2-4

122. $\dfrac{x}{3-x}$; x^2-9

123. $\dfrac{2x}{x+y}$; x^3+y^3

124. $\dfrac{x-1}{x^2+2x+4}$; x^3-8

125. $\dfrac{2x}{x-y}$; x^3-y^3

126. $\dfrac{x-1}{x^2-2x+4}$; x^3+8

127. $a-5$; 3

128. $x+2$; 7

129. $x-7$; x

130. $x-1$; $x+1$

In Problems 131–136, solve each problem.

131. A Canadian Pacific passenger train leaves Vancouver for Banff and travels $160x$ miles in $x+12$ hours. Write a rational expression that gives the average rate of this train for that part of the trip. (HINT: Rate is distance divided by time.)

132. The acceleration of a certain rocket is given by the following relationship: twice the distance divided by the elapsed time squared. Find a rational expression that gives the acceleration of this rocket if it travels $14t^3 - 2t + 77$ meters in $t+5$ seconds.

133. Sharon runs N times around a 2-mile track at an average rate of $8-N$ miles per hour. Write a rational expression that gives the time it takes Sharon to run these laps.

134. Power is the time rate of doing work. It is given by the relationship

$$\text{power} = \frac{\text{work}}{\text{time}}$$

and work is calculated as force times distance. Write a rational expression that gives the power required for a force of measure $x+5$ to move through a distance of $2x-3$ units in x^2+1 hours.

135. The cost, in millions of dollars, of removing $x\%$ of a pollutant from a scenic river is predicted by the formula

$$C = \frac{88x}{500-5x}$$

Predict the approximate cost of removing 10%, 30%, 50%, 75%, and 95% of the pollutant from the river. Estimate, if possible, the cost of removing 100% of the pollutant. Write a sentence describing the relationship between cost and the percentage of the pollutant removed.

136. The *Sun Daily News* purchased printing presses with a life expectancy of 16 years. The net earnings for the company from these printing presses after t years can be predicted by the formula

$$E = 2752 - 300t + \frac{1}{2}t^3$$

where E is in thousands of dollars. Suppose that the

salvage value of the printing presses at time t years is given by the formula

$$S = \frac{7200}{7 + t}$$

where S is in thousands of dollars. Calculate E and S for t equal to 2, 3, 5, 8, 9, and 10 years. Estimate when

$E = S$ during those years. When should the company sell the printing press?

Challenge Problems

In Problems 137 and 138, reduce each rational expression to lowest terms (n is a natural number).

137. $\dfrac{(x + 2)^{n+1}}{(x + 2)^{n}}$

138. $\dfrac{x^{2n+1}y^{2n-1}}{x^{2n+3}y^{2n}}$

In Problems 139 through 141, determine the values of x, if any, for which each rational expression is undefined.

139. $\dfrac{x + 1}{x^2 + 4x + 3}$

140. $\dfrac{x - 2}{x^2 - 4x + 4}$

141. $\dfrac{x + 2}{x^4 - 16}$

In Your Own Words

142. Explain what the Fundamental Principle of Rational Expressions allows us to do with a fraction.

143. The rational expression $\dfrac{x^2 - 4}{x^2 + x - 6}$ can be reduced to $\dfrac{x + 2}{x + 3}$. Is this reduction true for any value of x? Explain. Why can't we divide the x^2 terms out of the original expression and reduce it to $\dfrac{-4}{x - 6}$?

144. Explain what we mean when we say that a rational expression is undefined.

3.2 Multiplication and Division

TAPE 6

The operations of multiplication and division are performed with rational expressions just as they are with rational numbers.

> ### Multiplication of Rational Expressions
>
> If A, B, C, and D are polynomials,
>
> $$\frac{A}{B} \cdot \frac{C}{D} = \frac{AC}{BD}$$
>
> where each rational expression is defined.

To multiply rational expressions, multiply the numerators and multiply the denominators. This gives a single rational expression that should be reduced, if possible.

> **Procedure to Multiply Rational Expressions**
>
> 1. Factor the numerators and denominators.
> 2. Multiply numerators and multiply denominators.
> 3. Reduce to lowest terms.

▶ **EXAMPLE 1**

Perform the operation indicated.

(a) $\dfrac{abc^3}{d^4} \cdot \dfrac{d^2}{a^2 b^4 c}$ (b) $(x + 1)^2 \cdot \dfrac{x}{x + 1}$

(c) $\dfrac{a + 2}{a + 1} \cdot \dfrac{a^2 - 1}{a^2 - 4}$ (d) $\dfrac{t^2 - t - 2}{2t - 6} \cdot \dfrac{t - 3}{t^3 - 8}$

Solutions

(a) As the fraction is already factored, we multiply numerators and multiply denominators.

$$\frac{abc^3}{d^4} \cdot \frac{d^2}{a^2 b^4 c} = \frac{(abc^3)d^2}{d^4(a^2 b^4 c)} \qquad \text{Multiply}$$

$$= \frac{abc^3 d^2}{a^2 b^4 c d^4}$$

$$= \frac{c^2}{ab^3 d^2} \qquad \text{Reduce}$$

(b) Again we begin by multiplying numerators and multiplying denominators.

$$(x + 1)^2 \cdot \frac{x}{x + 1} = \frac{(x + 1)^2 \cdot x}{x + 1} \qquad \text{Multiply}$$

We do not want to multiply further in the numerator because our next step is to reduce if possible. Look for common factors to reduce.

$$= x(x + 1) \qquad \text{Reduce}$$

(c) First we factor in each fraction.

$$\frac{a + 2}{a + 1} \cdot \frac{a^2 - 1}{a^2 - 4} = \frac{a + 2}{a + 1} \cdot \frac{(a + 1)(a - 1)}{(a + 2)(a - 2)} \qquad \text{Factor}$$

Next, we multiply numerators and denominators.

$$= \frac{(a + 2)\ (a + 1)\ (a - 1)}{(a + 1)\ (a + 2)\ (a - 2)} \qquad \text{Multiply}$$

$$= \frac{a - 1}{a - 2} \qquad \text{Reduce}$$

(d) Since multiplication is just a copying step, we can factor at the same time.

$$\frac{t^2 - t - 2}{2t - 6} \cdot \frac{t - 3}{t^3 - 8} = \frac{(t - 2)(t + 1)(t - 3)}{2(t - 3)(t - 2)(t^2 + 2t + 4)}$$

$$= \frac{t + 1}{2(t^2 + 2t + 4)} \qquad \text{Reduce}$$

It is not necessary to multiply further in the denominator. However, it would not be wrong to do so. ◄

► EXAMPLE 2

Perform the operation indicated.

$$\frac{m^2 - n^2}{m^5} \cdot \frac{m^2}{n - m}$$

Solution

$$\frac{m^2 - n^2}{m^5} \cdot \frac{m^2}{n - m} = \frac{(m - n)(m + n)m^2}{m^5(n - m)}$$

Notice that $m - n$ and $n - m$ are opposites. We replace $m - n$ with $-(n - m)$ and then reduce.

$$= \frac{-(n - m)(m + n)m^2}{m^5(n - m)}$$

$$= \frac{-(m + n)}{m^3}$$

$$= -\frac{m + n}{m^3} \qquad ◄$$

Next, notice that division is just a form of multiplication.

Division of Rational Expressions

If A, B, C, D are polynomials,

$$\frac{A}{B} \div \frac{C}{D} = \frac{A}{B} \cdot \frac{D}{C}$$

$$= \frac{AD}{BC}$$

where each rational expression is defined.

To divide two polynomials, multiply by the reciprocal of the divisor.

Procedure to Divide Two Rational Expressions

1. Change the division to multiplication by multiplying by the reciprocal of the divisor.
2. Factor numerators and denominators.
3. Multiply numerators and multiply denominators.
4. Reduce to lowest terms.

► **EXAMPLE 3**

Perform the operation indicated.

(a) $\dfrac{6a^3b}{5c^4} \div \dfrac{2ab^2}{c^2}$ (b) $\dfrac{x^2 + 3x + 2}{x^4} \div \dfrac{x^2 + 2x + 1}{x^2}$

Solutions

(a) We change the division to multiplication.

$$\dfrac{6a^3b}{5c^4} \div \dfrac{2ab^2}{c^2} = \dfrac{6a^3b}{5c^4} \cdot \dfrac{c^2}{2ab^2} \qquad \text{Change to multiplication}$$

$$= \dfrac{(6a^3b)(c^2)}{(5c^4)(2ab^2)} \qquad \text{Multiply}$$

$$= \dfrac{3a^2}{5bc^2} \qquad \text{Reduce}$$

(b) $\dfrac{x^2 + 3x + 2}{x^4} \div \dfrac{x^2 + 2x + 1}{x^2} = \dfrac{x^2 + 3x + 2}{x^4} \cdot \dfrac{x^2}{x^2 + 2x + 1}$

$$= \dfrac{(x + 2)(x + 1)x^2}{x^4(x + 1)^2}$$

$$= \dfrac{x + 2}{x^2(x + 1)}$$

◄

► **EXAMPLE 4**

Perform the operation indicated.

$$\dfrac{x^3 + 1}{3 - x} \div \dfrac{2(x + 1)}{x^2 - 9}$$

Solution

We convert the division to multiplication, multiply, and factor.

$$\dfrac{x^3 + 1}{3 - x} \div \dfrac{2(x + 1)}{x^2 - 9} = \dfrac{x^3 + 1}{3 - x} \cdot \dfrac{x^2 - 9}{2(x + 1)}$$

$$= \dfrac{(x + 1)(x^2 - x + 1)(x - 3)(x + 3)}{2(3 - x)(x + 1)}$$

The factors of the numerator include $(x - 3)$, and the factors of the denominator include $(3 - x)$. They are opposites. We replace $(3 - x)$ with $-(x - 3)$.

$$= \frac{(x + 1)(x^2 - x + 1)(x - 3)(x + 3)}{-2(x - 3)(x + 1)}$$

$$= \frac{(x^2 - x + 1)(x + 3)}{-2}$$

$$= -\frac{(x^2 - x + 1)(x + 3)}{2} \quad \blacktriangleleft$$

Be Careful!   The key to multiplying or dividing rational expressions is factoring. Make sure that the numerator and denominator are **factored** before reducing.

Problem Set 3.2

Warm-Ups

In Problems 1 through 8, perform the operation indicated. Write answers in reduced form. See Example 1.

1. $\dfrac{xy^2}{z} \cdot \dfrac{z^2}{x^4 y}$

2. $\dfrac{-8ab^4}{c^4} \cdot \dfrac{c^5}{4a}$

3. $\dfrac{17a + 17b}{3} \cdot \dfrac{-39}{51a + 51b}$

4. $\dfrac{p^2 q - p^2 q^2}{p + p^2} \cdot \dfrac{pq + q}{p - pq}$

5. $\dfrac{-4}{a^4 - 5a^2 + 6} \cdot (a^4 - 4)$

6. $\dfrac{s - 2t}{s^2} \cdot \dfrac{s^2 + s^6}{s^3 - 8t^3}$

7. $\dfrac{2x^2 + 2x - 12}{x^2 - x - 12} \cdot \dfrac{x^2 - 3x - 4}{4x^2 - 4x - 8}$

8. $\dfrac{m^2 - n^2}{m^3 + n^3} \cdot \dfrac{m^2 - mn + n^2}{m - n}$

In Problems 9 through 12, perform the indicated multiplication. Reduce answers. See Example 2.

9. $\dfrac{s - t}{t^2} \cdot \dfrac{t}{t - s}$

10. $\dfrac{y^2 - 4}{y} \cdot \dfrac{y^3}{2y - y^2}$

11. $(x^2 - x) \cdot \dfrac{y^2}{x^2 - x^3}$

12. $\dfrac{16 - x^2}{x^2 + 3x + 2} \cdot \dfrac{x^2 + x - 2}{x^2 - 5x + 4}$

In Problems 13 through 20, perform the indicated division. Reduce answers. See Example 3.

13. $\dfrac{abc^2}{xyz^2} \div \dfrac{ab^2 c}{xy^4}$

14. $\dfrac{-2^2 x}{ab} \div \dfrac{2^3 x^4 y}{a^2}$

15. $\dfrac{(x + a)^2}{4a^2} \div (x + a)$

16. $\dfrac{50 - 2x^2}{x^2 + 2x} \div \dfrac{4x + 20}{4 - x^2}$

17. $\dfrac{x^2 + 3x - 4}{x^2 + 4x + 4} \div \dfrac{x^2 + 2x - 3}{x^2 + 3x + 2}$

18. $\dfrac{2x^2 + 3x - 2}{x^2 - 2x - 3} \div \dfrac{x^2 - 3x - 10}{x^2 - 7x + 12}$

19. $\dfrac{b^3 + 8c^3}{x^4 - 1} \div \dfrac{ab + 2ac}{x^2 + 1}$

20. $\dfrac{ax - ay - bx + by}{cx + cy + dx + dy} \div \dfrac{ax + ay - bx - by}{cx - cy - dx + dy}$

In Problems 21 and 22, perform the indicated division. Reduce answers. See Example 4.

21. $(x - y) \div \dfrac{y - x}{y}$

22. $\dfrac{s^2 - r^2}{r^2 + 2rs + s^2} \div \dfrac{r^3 - s^3}{r + s}$

Practice Exercises

In Problems 23 through 90, perform the operation indicated. Write answers in reduced form.

23. $\dfrac{xy^3}{z} \cdot \dfrac{z^4}{x^2 y}$

24. $\dfrac{-6ab^3}{c^5} \cdot \dfrac{c^7}{3a}$

25. $\dfrac{-3^3 m}{n} \cdot \dfrac{n}{(-3)^2 m^4}$

26. $\dfrac{a^5 (bc)^3}{d} \cdot (-cd)^2$

27. $\dfrac{z}{x^4 y} \cdot \left(\dfrac{-xy}{z^2}\right)^3$

28. $\dfrac{120 s^3}{(-t)^2} \cdot \dfrac{st^3}{320 s}$

29. $\dfrac{-2^3 m}{n} \cdot \dfrac{n}{(-2)^2 m^3}$

30. $\dfrac{a^4 (bc)^2}{d} \cdot \dfrac{(-cd)^2}{a^2}$

31. $\dfrac{z}{x^2 y} \cdot \left(\dfrac{-xy}{z}\right)^3$

32. $\dfrac{360 s^2}{(-t)^2} \cdot \dfrac{st^2}{270 s}$

33. $\dfrac{(ab^4)^3}{c(dx)^2} \cdot \dfrac{(c^3 d)^2}{(-a)^3 b^3}$

34. $\dfrac{(2x)^3}{5y} \cdot \dfrac{15 y^4}{4x}$

35. $\dfrac{x - 1}{x} \cdot \dfrac{x^2}{x - 1}$

36. $\dfrac{a - b}{t^3} \cdot \dfrac{t}{b - a}$

37. $\dfrac{13a + 13b}{5} \cdot \dfrac{-55}{39a + 39b}$

38. $\dfrac{pq + pq^2}{p - p^2} \cdot \dfrac{q - qp}{p^2 q^2 + p^2 q}$

39. $\dfrac{a^4 + a^2}{b^8} \cdot \dfrac{b^2}{a^3 + a}$

40. $\dfrac{r^4 s^5 - r^3 s^3}{t^2 u^2 + t^4 u^4} \cdot \dfrac{t^3 u^2}{r^2 s^5}$

41. $\dfrac{(z + 2)^2}{z^2 - 1} \cdot \dfrac{z - 1}{z + 2}$

42. $\dfrac{s + t}{s - t} \cdot \dfrac{t^2 - s^2}{t + s}$

43. $\dfrac{c^2 - 16}{2c - 8} \cdot \dfrac{-8}{c^2 + 3c - 4}$

44. $\dfrac{6a^2 - 6b^2}{a^2 + 2ab + b^2} \cdot \dfrac{a^2 - ab - 2b^2}{3a^2 + 3ab - 6b^2}$

45. $(x^3 - 125) \cdot \dfrac{x^3}{5 - x}$

46. $\dfrac{r^3 + s^3}{r^2 - s^2} \cdot \dfrac{r + s}{r^2 - rs + s^2}$

47. $\dfrac{2x^2 + 6x + 4}{x^2 - 4x + 3} \cdot \dfrac{x^2 - x - 6}{4x^2 - 4x - 8}$

48. $\dfrac{16 - x^2}{x^2 + 7x + 10} \cdot \dfrac{x^2 + 3x - 10}{x^2 - 6x + 8}$

49. $\dfrac{ax - az + bx - bz}{ax - 2bx + az - 2bz} \cdot \dfrac{ax + az + bx + bz}{ax - az - bx + bz}$

50. $\dfrac{as + st + a + t}{ax^2 - bx^2 + 2a - 2b} \cdot \dfrac{bx^2 - ax^2 + 2b - 2a}{as - 2a + st - 2t}$

51. $\dfrac{6x^2 + 7x - 3}{10x^2 - x - 2} \cdot \dfrac{5x^2 - 3x - 2}{4x^2 + 4x - 3} \cdot \dfrac{6x^2 - 5x + 1}{3x^2 - 4x + 1}$

52. $\dfrac{a^2 + 2ab + b^2}{2a^2 - 3ab + b^2} \cdot \dfrac{a^2 - b^2}{a^3 + 3a^2 b + 3ab^2 + b^3} \cdot \dfrac{6a^2 - ab - b^2}{3a^2 - 5ab - 2b^2}$

53. $\dfrac{abc^3}{xyz^4} \div \dfrac{ab^2 c}{xy^3}$

54. $\dfrac{(2x)^3 y}{25} \div \dfrac{(2x)^2}{85}$

55. $\dfrac{m^5 (n^3 p)^2}{rst^3} \div \dfrac{(mnp)^3}{r^3 s^2 t^2}$

56. $\dfrac{-16a^3}{(bc)^5} \div \dfrac{24a}{b}$

57. $\dfrac{-5^2x}{ab} \div \dfrac{5^3x^2y}{a^3}$

58. $\dfrac{(-2d)^3}{x^4} \div \dfrac{(-2d)^2}{x^3}$

59. $\dfrac{(2x)^2y}{15} \div \dfrac{(2x)^3}{75}$

60. $\dfrac{(-2a)^3}{x^5} \div \dfrac{(2a)^2}{x^2}$

61. $\dfrac{(-ab)^3}{r^2s} \div \dfrac{a^2b}{(rs^2)^2}$

62. $\dfrac{(-3)^2r}{(2st)^2} \div \dfrac{3r^2}{8t}$

63. $\dfrac{(-ab)^3}{r^3s} \div \dfrac{a^5b}{(rs^3)^4}$

64. $\dfrac{(-4)^2r}{(2st)^3} \div \dfrac{4r^2}{4t}$

65. $\dfrac{p}{p-q} \div \dfrac{q}{q-p}$

66. $\dfrac{(x-r)^3}{4a^3} \div (x-r)$

67. $\dfrac{ax-bx}{ay+by} \div \dfrac{a^2-ab}{2a+2b}$

68. $\dfrac{5u^3+10u^2}{2u-2} \div \dfrac{15u^2+30u}{au-a}$

69. $\dfrac{p^2+pt}{p-p^2} \div \dfrac{pq+qt}{pq-q}$

70. $\dfrac{a^3b+2a^2b^2}{c^2b+cbd} \div \dfrac{a^2+2ab}{c^2d+c^3}$

71. $\dfrac{16x^2-1}{x^2+8x+16} \div \dfrac{64x^3-1}{x+4}$

72. $\dfrac{8-2a^2}{a^2+3a} \div \dfrac{5a+10}{9-a^2}$

73. $\dfrac{x^2-4x-12}{x^2-4x-5} \div \dfrac{x^2-3x-18}{x^2-7x+10}$

74. $\dfrac{2x^2+7x+3}{3x^2-x-10} \div \dfrac{2x^2-x-1}{x^2-4}$

75. $\dfrac{x^2-3x-10}{2x^2+5x+3} \div \dfrac{5+4x-x^2}{2x^2+7x+6}$

76. $\dfrac{8t^3-s^3}{32t^2-2} \div \dfrac{4t^2+2st+s^2}{16t^2-4t}$

77. $\dfrac{x^2+10x+21}{x^2-4x+3} \div \dfrac{x^3+7x^2}{x^2-2x+1}$

78. $\dfrac{2x^2+16x-18}{x^3+3x^2} \div \dfrac{x^2+7x-18}{x^2-4x-21}$

79. $\dfrac{2x^2+5x-3}{2x^2+5x+3} \div \dfrac{2x^3-x^2}{2x^2+x-3}$

80. $\dfrac{2x^2-8x-42}{x^2+2x+1} \div \dfrac{4x^2+8x-12}{x+x^2}$

81. $\dfrac{b^3-2b^2+3b-6}{b^2-b-6} \div \dfrac{b^4-9}{b^2-9}$

82. $\dfrac{x^3+3x^2+4x+12}{x^2+x-2} \div \dfrac{x^4-16}{1-x^2}$

83. $\dfrac{ac+bc+2ad+2bd}{a^2-d^2} \div \dfrac{ac-bc+2ad-2bd}{a^2+2ab-ad-2bd}$

84. $\dfrac{rt-ru+st-su}{2rt+ru+4st+2su} \div \dfrac{rt-ru-st+su}{r^2+3rs+2s^2}$

85. $\dfrac{2x^2+5x-3}{3x^2+2x-5} \cdot \dfrac{3x^2-x-10}{x^2+2x-3} \cdot \dfrac{x^2-2x+1}{2x^2+x-1}$

86. $\dfrac{a^2+ab-2b^2}{2a^2+5ab-3b^2} \cdot \dfrac{2a^2+ab-b^2}{2a^2+ab-3b^2} \cdot \dfrac{2a^2+9ab+9b^2}{a^2+3ab+2b^2}$

87. $\dfrac{p^2+pt-2t^2}{2p^2-5pt-3t^2} \cdot \dfrac{p^2-2pt-3t^2}{p^2-3pt+2t^2} \cdot \dfrac{4p^2+4pt+t^2}{2p^2+3pt-2t^2}$

88. $\dfrac{2x^2+x-10}{3x^2+7x+4} \cdot \dfrac{3x^2+x-4}{6x^2+13x-5} \div \dfrac{x^2+5x-6}{3x^2+2x-1}$

89. $\dfrac{x^2-y^2}{2x^2+3xy-2y^2} \cdot \dfrac{x^3+2x^2y}{x^2+2xy+y^2} \div \dfrac{x^2+xy-2y^2}{x^3+x^2y+x+y}$

90. $\dfrac{x^2+x-6}{2x^2-x-1} \cdot \dfrac{x^2+2x+4}{2x^2+9x+9} \div \dfrac{x^3-8}{2x^2+3x+1}$

Challenge Problems

In Problems 91 and 92, perform the operations indicated. Write answers in lowest terms. Assume that n, s, and p are natural numbers.

91. $\dfrac{x^{2n}-x^n-6}{x^{2p}+2x^p-3} \cdot \dfrac{x^{2p}-1}{x^{2n}+4x^n+4}$

92. $\dfrac{x^{3s}-8}{x^{2s}+2x^s-3} \div \dfrac{x^{2s}-4}{x^{2s}+3x^s}$

In Your Own Words

93. Explain how to multiply two rational expressions.

94. Explain how to divide two rational expressions.

3.3 Addition and Subtraction

TAPE 7

Addition and subtraction of rational expressions are performed using the same procedures that we use for rational numbers.

> **Addition and Subtraction of Rational Expressions**
>
> If A, B, C are polynomials and if each rational expression is defined,
>
> $$\frac{A}{C} + \frac{B}{C} = \frac{A + B}{C} \quad \text{and} \quad \frac{A}{C} - \frac{B}{C} = \frac{A - B}{C}$$

To add or subtract rational expressions, we first write each rational expression with the same denominator and then write the sum or difference of the numerators divided by the denominator. This gives us a single rational expression that we reduce to lowest terms.

► **EXAMPLE 1**

Perform the operation indicated.

(a) $\dfrac{x + y}{xy} + \dfrac{1}{xy}$ (b) $\dfrac{a}{a + b} - \dfrac{a - b}{a + b}$

Solutions

(a) $\dfrac{x + y}{xy} + \dfrac{1}{xy} = \dfrac{x + y + 1}{xy}$

Be Careful!

Note the parentheses in the next example. Be careful to subtract the entire numerator of the second fraction.

(b) $\dfrac{a}{a + b} - \dfrac{a - b}{a + b} = \dfrac{a - (a - b)}{a + b}$

$$= \dfrac{a - a + b}{a + b}$$

$$= \dfrac{b}{a + b}$$

◄

► **EXAMPLE 2**

Perform the operation indicated.

$$\frac{p}{p - q} + \frac{q}{q - p}$$

Solution

Notice that the denominators are opposites. By substituting $-(p - q)$ for $q - p$, we make the denominators the same.

$$\frac{p}{p - q} + \frac{q}{q - p} = \frac{p}{p - q} + \frac{q}{-(p - q)}$$

$$= \frac{p}{p - q} - \frac{q}{p - q}$$

$$= \frac{p - q}{p - q}$$

$$= 1 \qquad \blacktriangleleft$$

Often the denominators are not the same. We must figure out an appropriate denominator and write each rational expression with this denominator. It must be a polynomial that is divisible by each denominator. We call this denominator the **least common denominator (LCD).**

Procedure to Find the Least Common Denominator (LCD)

1. Factor each denominator completely, using exponents.
2. List all different prime factors from all denominators.
3. Write the LCD. The LCD is the product of the factors in step 2 each raised to the highest power of that factor in any single denominator.

► **EXAMPLE 3**

Assume that each expression is a denominator of a fraction. Find the LCD for each pair of denominators.

(a) $x + y$ and $x - y$ (b) $x^2 + 2x - 15$ and $x + 5$

(c) $x^2 + 6x + 9$ and $x^2 + 3x$ (d) $x^7 y^3$ and xy^4

Solutions

(a) $x + y$ and $x - y$

STEP 1   Factor each denominator.
$x + y$ and $x - y$ are prime.

STEP 2   The different prime factors are $x + y$ and $x - y$.
Each factor occurs one time in each denominator.

STEP 3   The LCD is $(x + y)(x - y)$.

(b) $x^2 + 2x - 15 = (x + 5)(x - 3)$
Factor each denominator.
$x + 5$ is prime.

List the different prime factors.

Prime factors are $x + 5$ and $x - 3$.

Count the number of times prime factors occur.

$x + 5$ occurs one time in $(x + 5)(x - 3)$ and one time in $x + 5$.

$x - 3$ occurs one time in $(x + 5)(x - 3)$ and no times in $x + 5$.

So we must use each factor one time in the LCD.

The LCD is $(x + 5)(x - 3)$.

(c) $x^2 + 6x + 9$ and $x^2 + 3x$.

$$x^2 + 6x + 9 = (x + 3)^2 \quad \text{and} \quad x^2 + 3x = x(x + 3)$$

The different prime factors are $x + 3$ and x.

$x + 3$ occurs two times in $(x + 3)^2$ and one time in $x(x + 3)$.

x occurs no times in $(x + 3)^2$ and one time in $x(x + 3)$.

We must use $x + 3$ two times and x one time.

The LCD is $x(x + 3)^2$.

(d) x^7y^3 and xy^4.

Both are factored.

The different prime factors are x and y.

x occurs seven times in x^7y^3 and one time in xy^4.

y occurs three times in x^7y^3 and four times in xy^4.

The LCD is x^7y^4.   ◀

Procedure to Add or Subtract Rational Expressions

1. Find the least common denominator (LCD).
2. Rewrite each rational expression with the LCD as its denominator.
3. Write the sum or difference of the numerators divided by the LCD.
4. Reduce to lowest terms.

► EXAMPLE 4

Perform the operations indicated and reduce answers to lowest terms.

(a) $\dfrac{5}{a^2b} - \dfrac{3}{ab^4}$

(b) $\dfrac{8}{m + n} - \dfrac{1}{m - n}$

(c) $\dfrac{-28}{x^2 - 2x - 3} + \dfrac{7}{x - 3}$

(d) $\dfrac{t}{t-2} - 1$

(e) $\dfrac{1}{x^2 + 4x + 4} + \dfrac{2}{x^2 + 5x + 6}$

Solutions

(a) The LCD is $a^2 b^4$.

$$\frac{5}{a^2 b} - \frac{3}{ab^4} = \frac{5(b^3)}{a^2 b(b^3)} - \frac{3(a)}{ab^4(a)}$$

$$= \frac{5b^3}{a^2 b^4} - \frac{3a}{a^2 b^4}$$

$$= \frac{5b^3 - 3a}{a^2 b^4}$$

(b) The LCD is $(m+n)(m-n)$.

$$\frac{8}{m+n} - \frac{1}{m-n} = \frac{8(m-n)}{(m+n)(m-n)} - \frac{1(m+n)}{(m-n)(m+n)}$$

$$= \frac{8(m-n) - 1(m+n)}{(m+n)(m-n)}$$

*Be
Careful!*

Notice that the numerator is not factored at this point. So **do not** try to reduce!

$$= \frac{8m - 8n - m - n}{(m+n)(m-n)}$$

$$= \frac{7m - 9n}{(m+n)(m-n)}$$

We could multiply in the denominator, but it is not necessary.

(c) $\dfrac{-28}{x^2 - 2x - 3} + \dfrac{7}{x-3} = \dfrac{-28}{(x-3)(x+1)} + \dfrac{7}{x-3}$ Factor

The LCD is $(x-3)(x+1)$.

$$\frac{-28}{(x-3)(x+1)} + \frac{7(x+1)}{(x-3)(x+1)} = \frac{-28 + 7(x+1)}{(x-3)(x+1)}$$

$$= \frac{-28 + 7x + 7}{(x-3)(x+1)}$$

$$= \frac{7x - 21}{(x-3)(x+1)}$$

$$= \frac{7(x-3)}{(x-3)(x+1)}$$

$$= \frac{7}{x+1}$$

(d) $\dfrac{t}{t-2} - 1 = \dfrac{t}{t-2} - \dfrac{t-2}{t-2}$ The LCD is $t-2$

When subtracting, *use parentheses.*

$$= \frac{t - (t-2)}{t-2}$$

Be careful when removing parentheses!

$$= \frac{t - t + 2}{t-2}$$

$$= \frac{2}{t-2}$$

(e) $\dfrac{1}{x^2 + 4x + 4} + \dfrac{2}{x^2 + 5x + 6} = \dfrac{1}{(x+2)^2} + \dfrac{2}{(x+2)(x+3)}$ Factor

The LCD is $(x+2)^2(x+3)$.

$$\frac{x+3}{(x+2)^2(x+3)} + \frac{2(x+2)}{(x+2)^2(x+3)} = \frac{x+3+2(x+2)}{(x+2)^2(x+3)}$$

$$= \frac{x+3+2x+4}{(x+2)^2(x+3)}$$

$$= \frac{3x+7}{(x+2)^2(x+3)} \quad \blacktriangleleft$$

▶ **EXAMPLE 5**

Perform the operation indicated.

$$\frac{3}{x^2 - 4} + \frac{1}{2 - x}$$

Solution

$$\frac{3}{x^2 - 4} + \frac{1}{2 - x} = \frac{3}{(x-2)(x+2)} + \frac{1}{2 - x}$$

We replace $2 - x$ with $-(x-2)$, which will make two of our factors the same.

$$= \frac{3}{(x-2)(x+2)} + \frac{1}{-(x-2)}$$

$$= \frac{3}{(x-2)(x+2)} + \frac{-1}{x-2}$$

Now the LCD is $(x-2)(x+2)$.

$$= \frac{3}{(x-2)(x+2)} + \frac{-1(x+2)}{(x-2)(x+2)}$$

$$= \frac{3 - 1(x+2)}{(x-2)(x+2)}$$

$$= \frac{1 - x}{(x-2)(x+2)} \quad \blacktriangleleft$$

Problem Set 3.3

Warm-Ups

In Problems 1 through 8, perform the operation indicated. Write answers in lowest terms. See Example 1.

1. $\dfrac{7}{x^2} - \dfrac{x+2}{x^2}$

2. $\dfrac{2a}{t} - \dfrac{a+b}{t}$

3. $\dfrac{x+y}{xy} + \dfrac{x-y}{xy}$

4. $\dfrac{x-y}{x+y} + \dfrac{y-x}{y+x}$

5. $\dfrac{x}{x+y} + \dfrac{x+2y}{x+y}$

6. $\dfrac{2}{r-2} - \dfrac{1}{r-2}$

7. $\dfrac{3}{m-2} - \dfrac{2-m}{m-2}$

8. $\dfrac{6}{p+1} - \dfrac{5-p}{p+1}$

For Problems 9 and 10, perform the indicated operation. See Example 2.

9. $\dfrac{3y}{x-5} - \dfrac{2y}{5-x}$

10. $\dfrac{x}{z-t} + \dfrac{3}{t-z}$

For Problems 11 through 22, see Examples 3 and 4.

11. $\dfrac{p}{pq^2} + \dfrac{q}{p^2q}$

12. $\dfrac{1}{x^2y^3} - \dfrac{3}{x^4y}$

13. $\dfrac{1}{r+2} + \dfrac{2}{r-3}$

14. $\dfrac{a}{a+2} - \dfrac{1}{a-3}$

15. $\dfrac{2}{5a+10} + \dfrac{7}{3a+6}$

16. $\dfrac{x}{x^2-4x+4} + \dfrac{2}{x^2-4}$

17. $\dfrac{x+5}{x^2-2x-15} - \dfrac{x}{x^2-6x+5}$

18. $\dfrac{3y+6}{8-y^3} + \dfrac{4}{4-y^2}$

19. $x + \dfrac{2}{x}$

20. $a + \dfrac{3}{a+b}$

21. $\dfrac{t}{t-1} - 1$

22. $\dfrac{1}{a} - \dfrac{1}{a+1} + \dfrac{1}{a^2+a}$

For Problems 23 through 26, see Example 5.

23. $\dfrac{10}{4-2a} - \dfrac{12}{3a-6}$

24. $\dfrac{3}{p^2-p} + \dfrac{7}{1-p}$

25. $\dfrac{x}{x^2-6x+8} - \dfrac{2}{2+x-x^2}$

26. $\dfrac{2}{x-2} + \dfrac{3}{x+2} - \dfrac{5}{4-x^2}$

Practice Exercises

In Problems 27 through 76, perform the operations indicated. Write answers in lowest terms.

27. $\dfrac{8}{y^3} - \dfrac{x+2}{y^3}$

28. $\dfrac{x+2}{x+1} - \dfrac{x+4}{x+1}$

29. $\dfrac{a}{b^3 c} + \dfrac{b}{bc^2}$

30. $\dfrac{2}{r^2 t} - \dfrac{1}{rt^4}$

31. $\dfrac{2}{x^2 y} - \dfrac{4}{xyz} + \dfrac{1}{x}$

32. $\dfrac{1}{x} + \dfrac{1}{y} - \dfrac{1}{z}$

33. $\dfrac{2}{y-1} + \dfrac{1}{y}$

34. $\dfrac{u}{u-v} + \dfrac{2v}{u+v}$

35. $\dfrac{3}{s+1} - \dfrac{1}{s-1}$

36. $\dfrac{5}{a+3} - \dfrac{1}{a+1}$

37. $\dfrac{q}{q-4} + \dfrac{q}{q+6}$

38. $\dfrac{v}{v-3} - \dfrac{1}{v+7}$

39. $\dfrac{a}{x-1} + \dfrac{4a}{1-x}$

40. $\dfrac{4}{x-y} - \dfrac{2}{y-x}$

41. $\dfrac{2}{3r+6} + \dfrac{5}{4r+8}$

42. $\dfrac{x}{x^2-2x} + \dfrac{4}{2x^2-x^3}$

43. $\dfrac{3}{5m-5n} + \dfrac{4}{2n-2m}$

44. $\dfrac{3r+1}{1-r^2} + \dfrac{2}{r-1}$

45. $\dfrac{2t}{t-1} - \dfrac{t^2-4t+3}{t^2-2t+1}$

46. $\dfrac{a}{b+a} - \dfrac{2b^2}{a^2-b^2}$

47. $\dfrac{d}{d+3} + \dfrac{5d+9}{d^2+4d+3}$

48. $\dfrac{2x}{x-5} - \dfrac{14x+20}{x^2-x-20}$

49. $\dfrac{x^2-x}{x^2-9} + \dfrac{1}{3-x}$

50. $\dfrac{y}{y-5} + \dfrac{y^2}{5+4y-y^2}$

51. $\dfrac{1}{x^2-7x+12} + \dfrac{x}{x^2-12x+32}$

52. $\dfrac{2t}{t^2-2t-3} - \dfrac{t}{t^2-8t+15}$

53. $\dfrac{z+2}{2z^2-21z+10} + \dfrac{1}{2z^2-7z+3}$

54. $\dfrac{3b-23}{b^2+8b-9} + \dfrac{35}{b^2+11b+18}$

55. $\dfrac{-3y+9}{y^3+27} + \dfrac{4}{y^2-9}$

56. $\dfrac{14x+63}{x^2+9x+14} + \dfrac{8x}{7-6x-x^2}$

57. $\dfrac{3z+17}{2z^2+z-3} + \dfrac{3z+7}{2z^2+7z+6}$

58. $\dfrac{3t+19}{t^2+t-2} + \dfrac{3t-46}{t^2+4t-5}$

59. $\dfrac{y}{y^2-10y+25} - \dfrac{y+1}{y^2-25}$

60. $\dfrac{w}{w^2+w-12} - \dfrac{w}{w^2-2w-3}$

61. $\dfrac{s}{s^2-5s-24} - \dfrac{s-1}{s^2-10s+16}$

62. $\dfrac{2y}{y^2-9y+14} - \dfrac{y-1}{y^2-8y+7}$

63. $\dfrac{u^2}{u-1} - u$

64. $\dfrac{5}{p+1} + p$

65. $\dfrac{a}{a+b} - 2$

66. $r + \dfrac{rt}{r+t}$

67. $\dfrac{2}{x} - \dfrac{3}{x+2} + \dfrac{4}{x^2+2x}$

68. $\dfrac{3}{a+b} + \dfrac{4}{a-b} + \dfrac{6a}{b^2-a^2}$

69. $\dfrac{1}{x+4} - \dfrac{1}{x+3} - \dfrac{1}{x+2}$

70. $\dfrac{2}{y-1} - \dfrac{1}{y+1} - \dfrac{1}{y-2}$

71. $\dfrac{qt}{q^2-qt} + \dfrac{1}{q-t} + 1$

72. $\dfrac{n^3}{m^3+m^2 n} + \dfrac{n}{m} - 1$

73. $\dfrac{3}{z+1} - \dfrac{1}{z-1} + \dfrac{1}{z+2}$

74. $\dfrac{p}{p+2} + \dfrac{p+1}{p+3} + \dfrac{2}{p^2+5p+6}$

75. $\dfrac{7}{x^2+5x+6} - \dfrac{49}{x^2-x-12} + \dfrac{x}{x^2-2x-8}$

76. $\dfrac{3(r+1)}{r^2-9} - \dfrac{40}{r^2-4r-21} + \dfrac{16}{r^2-10r+21}$

77. If f is the focal length of a lens, then

$$\frac{1}{f} = \frac{1}{\text{object distance}} + \frac{1}{\text{image distance}}$$

Find a rational expression that gives $\dfrac{1}{f}$ for a lens if the object distance is x centimeters and the image distance is $x+5$ cm.

78. Find a rational expression that gives $\dfrac{1}{f}$ for a lens if the object distance is x^2+6x+9 meters and the image distance is x^2-9 meters. (See Problem 77.)

79. When lights, toasters, and other appliances are plugged into wall outlets at home, they are connected in *parallel*. If R_1 and R_2 are two resistances connected in parallel, the total resistance, R_t, can be found from the relationship

$$\frac{1}{R_t} = \frac{1}{R_1} + \frac{1}{R_2}$$

If a coffee maker with a resistance of $x+1$ ohms and a toaster with a resistance of $2x-3$ ohms are plugged into a kitchen circuit, find a rational expression for $\dfrac{1}{R_t}$.

80. If a lamp with a resistance of k ohms and a steam iron with a resistance of k^3-k^2 ohms are plugged into the same household circuit, find a rational expression for $\dfrac{1}{R_t}$. (See Problem 79.)

Challenge Problems

In Problems 81 through 83, perform the operations indicated. Write answers in lowest terms. Consider whether or not to work inside the parentheses first.

81. $(x+2)\left(\dfrac{3}{x+2} + \dfrac{1}{x-1}\right)$

82. $\left(\dfrac{1}{x} + \dfrac{1}{y}\right) \div \left(\dfrac{1}{x} - \dfrac{1}{y}\right)$

83. $\left(\dfrac{1}{x-3} - \dfrac{1}{x+3}\right)\left(\dfrac{1}{x+3} + \dfrac{1}{x-3}\right)$

In Your Own Words

84. Explain how to find a least common denominator.

3.4   Rational Equations and Inequalities

TAPE 7

In Section 2.1, linear equations in one variable were solved. Some of the coefficients were fractions, but none of the equations contained variables in denominators. An equation that contains variables in a denominator is called a **fractional equation.** In this section, we learn how to solve some fractional equations.

When studying rational expressions in Section 3.1, we noted that rational expressions are undefined when a denominator has a value of zero. The same situation exists with rational equations. For example, in the equation

$$\frac{1}{x - 3} = 3$$

Be Careful!

the fraction $\dfrac{1}{x - 3}$ is not defined when x is 3. Thus 3 *cannot* be in the solution set of the equation, because it cannot make the equation true. Remember this as we solve rational equations.

To solve a rational equation, use the two tools from Section 2.1. The general strategy is to clear the equation of fractions, solve the resulting equation, and then check the proposed solution(s) to see if they make any denominator have a value of zero.

To clear the preceding equation of fractions, multiply both sides by the LCD, $(x - 3)$.

$$\frac{1}{x - 3} = 3$$

$$(x - 3)\frac{1}{x - 3} = (x - 3)3$$

$$1 = 3x - 9$$

$$10 = 3x$$

$$\frac{10}{3} = x$$

Since $\dfrac{10}{3}$ does not make a denominator in the original equation have the value zero, the solution set is

$$\left\{ \frac{10}{3} \right\}$$

► **EXAMPLE 1**

Solve $\dfrac{1}{x} + \dfrac{1}{2x} = \dfrac{3}{2}$.

Solution

To clear the fractions, multiply both sides of the equation by $2x$, the LCD of all the denominators in the equation.

$$2x \left(\frac{1}{x} + \frac{1}{2x} \right) = 2x \left(\frac{3}{2} \right)$$

$$2x \cdot \frac{1}{x} + 2x \cdot \frac{1}{2x} = 2x \cdot \frac{3}{2} \qquad \text{Distributive Property}$$

$$2 + 1 = 3x$$

$$3 = 3x$$

$$1 = x$$

Since 1 does not make any denominator in the original equation have the value zero, write the solution set,

$$\{1\}$$

◄

> **A Procedure for Solving Rational Equations**
>
> 1. Multiply both sides by the least common denominator of all denominators in the equation.
> 2. Solve the resulting equation for possible solutions.
> 3. Check to see if any of the possible solutions make a denominator have the value zero. If so, **do not** include them in the solution set.
> 4. Write the solution set.

► **EXAMPLE 2**

Solve $\dfrac{7}{x - 5} + 2 = \dfrac{x + 3}{x - 5}$.

Solution

We multiply both sides by the LCD, which is $(x - 5)$.

$$(x - 5)\left(\frac{7}{x - 5} + 2\right) = (x - 5) \cdot \frac{x + 3}{x - 5}$$

Now apply the Distributive Property.

$$(x - 5)\frac{7}{x - 5} + (x - 5)\,2 = (x - 5)\frac{x + 3}{x - 5}$$

$$7 + 2(x - 5) = x + 3$$

$$7 + 2x - 10 = x + 3$$

$$2x - 3 = x + 3$$

$$x = 6$$

Since 6 does not make any denominator have the value 0, we write the solution set,

$$\{6\}$$ ◄

When multiplying both sides of an equation by a number, the Distributive Property allows us to multiply *each* term in the equation by the number. This is a convenient shortcut when clearing an equation of fractions.

► **EXAMPLE 3**

Solve $\dfrac{2}{x - 2} - \dfrac{1}{x + 1} = \dfrac{5}{x^2 - x - 2}$.

Solution

We must factor so that we can find the LCD.

$$\frac{2}{x - 2} - \frac{1}{x + 1} = \frac{5}{(x - 2)(x + 1)} \qquad \text{Factor}$$

Multiply each term by the LCD, $(x - 2)(x + 1)$.

$$(x - 2)(x + 1) \cdot \frac{2}{x - 2} - (x - 2)(x + 1) \cdot \frac{1}{x + 1}$$

$$= (x - 2)(x + 1) \cdot \frac{5}{(x - 2)(x + 1)}$$

$$2(x + 1) - (x - 2) = 5$$

Carefully remove the parentheses!

$$2x + 2 - x + 2 = 5 \qquad \text{Distributive Property}$$

$$x + 4 = 5$$

$$x = 1$$

Be Careful!

Since 1 does not make a denominator in the original equation have a value of 0, 1 is the solution.

$$\{1\}$$
◄

► EXAMPLE **4**

Solve $\dfrac{2}{x + 2} - \dfrac{1}{x} = \dfrac{-4}{x(x + 2)}$.

Solution

The LCD is $x(x + 2)$. Multiply each term by $x(x + 2)$.

$$x(x + 2) \cdot \frac{2}{x + 2} - x(x + 2) \cdot \frac{1}{x} = x(x + 2) \cdot \frac{-4}{x(x + 2)}$$

$$2x - (x + 2) = -4 \qquad \text{Multiplication}$$

$$x - 2 = -4$$

$$x = -2$$

Since -2 makes a denominator in the original equation have the value zero, it cannot be in the solution set. Thus the solution set is

$$\varnothing$$
◄

► EXAMPLE **5**

Solve $\dfrac{3}{x - 2} - \dfrac{2}{4 - x^2} = \dfrac{1}{x + 2}$.

Solution

To compute the LCD, we need to factor.

$$\frac{3}{x - 2} - \frac{2}{(2 - x)(2 + x)} = \frac{1}{x + 2}$$

We replace $(2 - x)$ by $-(x - 2)$.

$$\frac{3}{x - 2} - \frac{2}{-(x - 2)\,(2 + x)} = \frac{1}{x + 2}$$

$$\frac{3}{x - 2} + \frac{2}{(x - 2)(2 + x)} = \frac{1}{x + 2}$$

Now we see that the LCD is $(x - 2)(x + 2)$. We multiply each term by the LCD.

$$3(x + 2) + 2 = 1(x - 2) \qquad \text{Distributive Property}$$

$$3x + 6 + 2 = x - 2$$

$$2x = -10$$

$$x = -5$$

Since -5 does not make a denominator in the original equation have the value zero, we write the solution set,

$$\{-5\} \qquad \blacktriangleleft$$

Rational Inequalities

A rational inequality has boundary numbers at all solutions of the associated equation, just like other inequalities. In addition, a rational inequality has a boundary number where any denominator has a value of zero. Furthermore, boundary numbers where a denominator has a value of zero are **never** in the solution set of the inequality. We call these boundary numbers **free boundary numbers.** *A free boundary number is never in a solution set.*

Procedure to Solve Rational Inequalities

1. Find the *free boundary numbers;* that is, find the numbers that make any denominator in the inequality have a value of 0.
2. Solve the associated *equation* to find other boundary numbers.
3. Locate *all* the boundary numbers on a number line.
4. Determine which regions formed by the boundary numbers make the *original inequality* true by testing with one number inside the region.
5. Shade only the regions that test true.
6. Check the boundary numbers themselves.
7. Write the solution set.

▶ **EXAMPLE 6**

Solve $\dfrac{x + 2}{x - 1} < 2$. Write the solution set in interval notation.

Solution

The number 1 is a *free* boundary number because it would make the denominator have a value of zero. For other boundary numbers, we solve the equation

$$\frac{x + 2}{x - 1} = 2$$

$$x + 2 = 2(x - 1) \qquad \text{Multiply by } x - 1$$

$$x + 2 = 2x - 2$$

$$4 = x$$

and get a second boundary number, 4. The boundary numbers are 1 and 4. We summarize testing each region in a table.

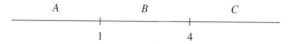

Region	Number in Region	Statement $\dfrac{x + 2}{x - 1} < 2$	Truth of Statement	Region in Solution Set?
A	0	$\dfrac{0 + 2}{0 - 1} < 2$	True	Yes
B	2	$\dfrac{2 + 2}{2 - 1} < 2$	False	No
C	5	$\dfrac{5 + 2}{5 - 1} < 2$	True	Yes

Since the original problem does not allow equality, the boundary number, 4, is not in the solution set. The other boundary number, 1, is a number that makes a denominator in the inequality have a value of zero. Such boundary numbers are *never* in the solution set.

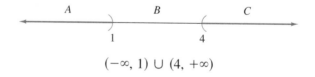

$$(-\infty, 1) \cup (4, +\infty)$$

◀

► ## EXAMPLE 7

Solve $\dfrac{3x - 5}{2x + 1} \leq 1$.

Solution

The first step is to find the boundary numbers. We solve the associated equation

$$\frac{3x - 5}{2x + 1} = 1$$

$$3x - 5 = 2x + 1$$

$$x = 6$$

So we have a boundary number 6, the solution of the associated equation, and a free boundary number $-\frac{1}{2}$, where the denominator has a value of zero. We locate the boundary numbers on a number line.

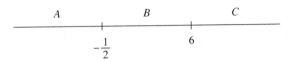

Test region A with -1.

$$\frac{3(-1) - 5}{2(-1) + 1} \leq 1$$

$$\frac{-8}{-1} \leq 1$$

$$8 \leq 1 \qquad \text{False. Region } A \text{ is not included}$$

Test region B with 0.

$$\frac{3(0) - 5}{2(0) + 1} \leq 1$$

$$-5 \leq 1 \qquad \text{True. Region } B \text{ is included}$$

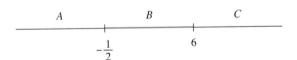

Test region C with 7.

$$\frac{3(7) - 5}{2(7) + 1} \leq 1$$

$$\frac{16}{15} \leq 1 \qquad \text{False. Region } C \text{ is not included}$$

The boundary number 6 is in the solution set because equality is allowed in the original statement. However, the boundary number $-\frac{1}{2}$ is *not* in the solution set because it is a free boundary number (makes a denominator have a value of zero).

$$\left\{ x \mid -\frac{1}{2} < x \leq 6 \right\} \quad \text{or} \quad \left(-\frac{1}{2}, 6 \right]$$

Rational Equations and Inequalities as Models

Rational equations and inequalities can be mathematical models for word problems.

► **EXAMPLE 8**

Mary Kay drives 1 mile in the same amount of time that Lettie drives 1.5 miles. If Lettie's rate is 10 mph more than Mary Kay's rate, how fast does Mary Kay drive?

Solution

Let r be Mary Kay's rate in mph. Then $r + 10$ is Lettie's rate in mph.
We make a chart, showing both distances and both rates. Using the distance–rate–time formula, $t = \dfrac{d}{r}$, we complete the chart by filling in time.

	d	r	t
Mary Kay	1	r	
Lettie	1.5	$r + 10$	

	d	r	t
Mary Kay	1	r	$1/r$
Lettie	1.5	$r + 10$	$1.5/(r + 10)$

Mary Kay's time is the same as Lettie's time. We form an equation and solve.

$$\frac{1}{r} = \frac{1.5}{r + 10}$$

Multiply by $r(r + 10)$.

$$r(r + 10) \cdot \frac{1}{r} = r(r + 10) \cdot \frac{1.5}{r + 10}$$

$$r + 10 = 1.5r$$

$$10 = 0.5r$$

$$20 = r \qquad \text{Divide by 0.5}$$

20 does not make a denominator have a value of 0, and 20 makes sense and will answer the question.

Mary Kay drives at a rate of 20 mph.   ◄

► **EXAMPLE 9**

The ratio of wins to losses for the Texas Rangers in one season was 7 to 4. How many games did they lose if the played 143 games that season?

Solution

Let x be the number of games lost. Since a total of 143 games were played, the number of games won is $143 - x$.

We represent a ratio of 7 to 4 as $\dfrac{7}{4}$.

$$\frac{7}{4} = \frac{\text{wins}}{\text{losses}}$$

$$\frac{7}{4} = \frac{143 - x}{x}$$

$$4x \cdot \frac{7}{4} = 4x \cdot \frac{143 - x}{x} \qquad \text{Multiply by } 4x$$

$$7x = 4(143 - x)$$

$$7x = 572 - 4x$$

$$11x = 572$$

$$x = 52$$

The Rangers lost 52 games. ◄

► EXAMPLE 10

The reciprocal of twice a number is not more than the reciprocal of one more than twice the number. Find all possible numbers.

Solution

Let n be such a number. The given information translates into an inequality.

$$\frac{1}{2n} \leq \frac{1}{2n + 1}$$

To solve the inequality, we use the method of boundary numbers.

There are two free boundary numbers, 0 and $-\dfrac{1}{2}$.

Next, we solve the associated equation.

$$\frac{1}{2n} = \frac{1}{2n + 1}$$

$$2n + 1 = 2n \qquad\qquad \text{Multiply both sides by } 2n(2n + 1)$$
$$1 = 0 \qquad\qquad\qquad\qquad \text{to clear fractions}$$

There are no solutions to this equation. So we have two boundary numbers, $-\dfrac{1}{2}$ and 0. We locate them on a number line.
We test the three regions.

Region	Number in Region	Statement $\dfrac{1}{2n} \leq \dfrac{1}{2n+1}$	Truth of Statement	Region in Solution Set?
A	-1	$\dfrac{1}{-2} \leq \dfrac{1}{-1}$	False	No
B	$-\dfrac{1}{4}$	$-2 \leq 2$	True	Yes
C	1	$\dfrac{1}{2} \leq \dfrac{1}{3}$	False	No

The boundary numbers are not included since they were both free boundary numbers. Thus the solution set is region B.

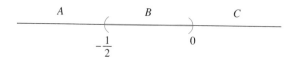

All numbers between $-\dfrac{1}{2}$ and 0 meet the conditions of the problem.   ◀

Problem Set 3.4

Warm-Ups

In Problems 1 through 8, find the solution set. See Example 1.

1. $\dfrac{1}{x} + 2 = 3$

2. $\dfrac{2}{x} + 3 = 5$

3. $\dfrac{3}{x} - 1 = \dfrac{1}{2}$

4. $\dfrac{7}{s} - 2 = \dfrac{1}{3}$

5. $\dfrac{1}{2x} + \dfrac{1}{x} = \dfrac{1}{2}$

6. $\dfrac{1}{3x} - \dfrac{1}{x} = \dfrac{1}{2}$

7. $\dfrac{2}{3t} + \dfrac{1}{4} = \dfrac{3}{4t}$

8. $\dfrac{1}{5} + \dfrac{7}{10x} = \dfrac{3}{5x}$

In Problems 9 through 14, solve each equation. See Example 2.

9. $\dfrac{3}{x-1} + 2 = \dfrac{5}{x-1}$

10. $\dfrac{4}{x+1} - 3 = \dfrac{-2}{x+1}$

11. $\dfrac{5}{2x+3} = \dfrac{1}{2x+3} + 1$

12. $\dfrac{4}{3x-2} = \dfrac{-4}{3x-2} - 1$

13. $\dfrac{1.4}{x} + \dfrac{3.2}{2x} = -1.2$

14. $\dfrac{0.3}{x} + 1.2 = \dfrac{1}{0.2x}$

For Problems 15 through 19, find the solution set. See Example 3.

15. $\dfrac{1}{x+2} + \dfrac{1}{x} = \dfrac{12}{x^2 + 2x}$

16. $\dfrac{2}{x} - \dfrac{1}{x+1} = \dfrac{3}{x^2 + x}$

17. $\dfrac{2}{x-3} = \dfrac{4}{x+4}$

18. $\dfrac{3}{x-1} - \dfrac{1}{x+3} = \dfrac{8}{x^2 + 2x - 3}$

19. $\dfrac{6}{x-3} - \dfrac{3}{x+2} = \dfrac{12}{x^2 - x - 6}$

For Problems 20 through 22, find the solution set. See Example 4.

20. $\dfrac{1}{z+1} + \dfrac{1}{z-1} = \dfrac{2}{z^2-1}$

21. $\dfrac{-4}{5(x+2)} = \dfrac{3}{x+2}$

22. $\dfrac{3}{x-2} + \dfrac{1}{x-5} = \dfrac{-9}{x^2-7x+10}$

For Problems 23 through 25, find the solution set. See Example 5.

23. $\dfrac{4}{1-x} + \dfrac{1}{x} = \dfrac{5}{x^2-x}$

24. $\dfrac{1}{x} + \dfrac{2}{1-x} + \dfrac{1}{x+1} = 0$

25. $\dfrac{3}{2+x} + \dfrac{2}{2-x} = \dfrac{2x}{x^2-4}$

For Problems 26 through 31, find the solution set and express it in interval notation. See Examples 6 and 7.

26. $\dfrac{x-5}{x+5} \geq 0$

27. $\dfrac{x-4}{1-x} \leq 0$

28. $\dfrac{x+5}{x-5} \geq 2$

29. $\dfrac{3-x}{x+6} \geq 2$

30. $\dfrac{3}{x+1} > \dfrac{2}{x-1}$

31. $\dfrac{1}{x-1} < \dfrac{2}{x+7}$

For Problems 32 through 34, solve each problem. See Example 8.

32. Jeri sails 10 miles in the same amount of time as Douglas sails 12 miles. If Douglas's rate is 5 mph faster than Jeri's rate, find Jeri's rate.

33. John Phillip averages 2 mph faster than Kenneth when riding a bicycle. John Phillip gives Kenneth a 7-mile headstart and then catches him after going 56 miles. What rate does John Phillip average on his bike?

34. A United jet flies 400 miles in the same amount of time that a Delta jet flies 320 miles. If the rate of the Delta plane is 50 mph slower than the United plane, find the speed of the United jet.

In Problems 35 through 38, solve each problem. See Example 9.

35. The ratio of boys to girls in the algebra class is 4 to 5. If there are 32 boys in the class, how many girls are in the class?

36. The ratio of bluebirds to woodpeckers in a certain area is 3 to 7. If 600 of both kinds of birds have been counted, how many of them are bluebirds?

37. If 1 inch is 2.54 centimeters, how many centimeters are in 1 foot?

38. If 1 kilogram is 2.2 pounds, how many pounds are in 5 kilograms?

In Problems 39 and 40, solve each problem. See Example 10.

39. One number is twice another number. If the sum of their reciprocals is more than $\dfrac{1}{2}$, find all such numbers.

40. The reciprocal of 1 more than a number is less than twice the reciprocal of the number. Find all such numbers.

Practice Exercises

In Problems 41 through 80, find the solution set of each equation or inequality. Perform the indicated operations on the expressions.

41. $\dfrac{1}{x} + \dfrac{1}{3} = \dfrac{1}{4}$

42. $\dfrac{1}{x} - \dfrac{2}{3} = \dfrac{1}{2}$

43. $\dfrac{2}{y} - \dfrac{1}{5}$

44. $\dfrac{1}{x} + \dfrac{1}{3}$

45. $\dfrac{2}{x} - \dfrac{3}{4} = \dfrac{1}{x}$

46. $\dfrac{5}{x} - \dfrac{2}{7} = \dfrac{1}{x}$

47. $\dfrac{1}{x+1} - \dfrac{2}{x} = 0$

48. $\dfrac{2}{x-1} + \dfrac{1}{x} = 0$

49. $\dfrac{4}{2-x} - 1 = \dfrac{3}{2-x}$

50. $\dfrac{7}{2x-1} - 1 = \dfrac{9}{2x-1}$

51. $\dfrac{3}{x} = \dfrac{5}{x+4}$

52. $\dfrac{7}{x} = \dfrac{8}{2x-6}$

53. $\dfrac{14.3}{w} - \dfrac{3.1}{2w} = 1.1$

54. $\dfrac{0.5}{x} + 1.2 = \dfrac{1}{1.2x}$

55. $\dfrac{1}{x-5} + \dfrac{x}{25-x^2} = 0$

56. $\dfrac{4}{x} + \dfrac{3}{5-x} = \dfrac{-20}{x^2-5x}$

57. $\dfrac{1}{x+3} = \dfrac{x}{x+1} - 1$

58. $\dfrac{w}{w+2} = 1 - \dfrac{1}{w-3}$

59. $\dfrac{x}{x-3} = \dfrac{x+1}{x-1}$

60. $\dfrac{u-2}{u+3} = \dfrac{u-1}{u+2}$

61. $\dfrac{x}{x+4} - \dfrac{2}{x-3} = 1$

62. $\dfrac{3}{x-1} + x = \dfrac{x^2}{x-1}$

63. $\dfrac{3}{x-1} - 1 + \dfrac{x^2-5}{x^2+2x-3} = 0$

64. $\dfrac{2}{z-2} + \dfrac{3}{z} = \dfrac{1}{2-z}$

65. $\dfrac{2}{x+1} + \dfrac{3}{x+2} = \dfrac{-3}{x^2+3x+2}$

66. $\dfrac{x-2}{x+1} = \dfrac{x+2}{x-1}$

67. $\dfrac{3}{x-1} - \dfrac{1}{x+2} = \dfrac{9}{x^2+x-2}$

68. $\dfrac{2}{x-2} + \dfrac{1}{x-5} = \dfrac{-6}{x^2-7x+10}$

69. $\dfrac{x+1}{x-1} > 0$

70. $\dfrac{x-2}{x+3} < 0$

71. $\dfrac{x-4}{x-5} \le 0$

72. $\dfrac{x-5}{x+1} \le 0$

73. $\dfrac{x+2}{x} < 0$

74. $\dfrac{x}{x-1} > 0$

In Problems 75 through 80, write the solution set in interval notation.

75. $\dfrac{r-5}{r+2} \le 2$

76. $\dfrac{x+7}{x+3} \ge 3$

77. $\dfrac{2x+3}{x+4} \ge 1$

78. $\dfrac{2}{w-3} < \dfrac{1}{w+2}$

79. $\dfrac{1}{x-1} \le \dfrac{1}{x+1}$

80. $\dfrac{1}{2x-1} \le \dfrac{1}{2x+1}$

In Problems 81 through 94, solve each problem.

81. Jessye jogs 6 miles in the same amount of time that Lila jogs 8 miles. If Jessye's rate is 2 mph slower than Lila's rate, find Lila's rate.

82. Manuel lives 10 miles from Riverfront Stadium and Maria lives 16 miles away from the stadium. They plan to meet at the stadium before the game. If each travels at the same rate, Manuel will arrive $\dfrac{1}{2}$ hour earlier than Maria. How long will it take Maria to get to the stadium?

83. Bill Anderson drives from home to work on Memorial Drive every morning. This is a distance of 15 miles. He uses I-285 to come home. Although this is 10 miles farther, he can average 20 mph faster. If he drives to work in the same amount of time that he drives home from work, find his speed going to work.

84. Toby and Bart both work at a lighthouse. Toby lives 3 miles from the lighthouse and Bart, 5 miles. They ride their bikes to work in the same amount of time. If Bart's average speed is 4 mph faster than Toby's, what is Toby's speed? How long does it take them to get to work?

85. The ratio of Democrats to Republicans in the state legislature is 3 to 8. If 45 of them are Democrats, how many Republicans are there?

86. The ratio of pine trees to oak trees in the park is 5 to 9. If 42,000 of both kinds of trees have been counted, how many of them are pine trees?

87. If 1 gallon is 3.8 liters, how many liters are in 5 gallons?

88. If 1 foot is 30.5 centimeters, how many centimeters are in 10 yards?

89. One number is three times another number. If the sum of their reciprocals is $\frac{1}{4}$, find the two numbers.

90. The reciprocal of one less than a number is twice the reciprocal of the number. Find the number.

91. The reciprocal of three times a number is equal to the reciprocal of 2 more than the number. Find the number.

92. The reciprocal of one more than a number is negative. Find all such numbers.

93. The reciprocal of one less than a number is positive. Find all such numbers.

94. If one more than a number is divided by one less than the number, the quotient is nonnegative. Find all possible numbers.

Challenge Problems

In Problems 95 and 96, find the solution set; a, b, and c are constants.

95. $\frac{a}{x} + \frac{b}{2x} = c$

96. $\frac{a}{x - 1} - \frac{2a}{x + 1} = 0$

97. $\frac{x - a}{x - b} \leq 0; a > 0, b < 0$

98. $\frac{x - a}{x - b} \leq 0; a < 0, b > 0$

In Your Own Words

99. What must be checked when solving fractional equations, and why?

100. What are free boundary numbers and how do they differ from other boundary numbers?

3.5 COMPLEX FRACTIONS

TAPE 7

A fraction that contains a fraction is called a **complex fraction.** Two methods are used to simplify complex fractions. The first is to think of the fraction as a division problem, and the second is to use the Fundamental Principle of Rational Expressions.

▶ **EXAMPLE 1**

Simplify $\dfrac{\dfrac{1}{x}}{\dfrac{2}{x^2}}$.

Solution

We will work this by the two methods.

Method 1—Division method:

This complex fraction is $\dfrac{1}{x}$ divided by $\dfrac{2}{x^2}$.

$$\frac{\dfrac{1}{x}}{\dfrac{2}{x^2}} = \frac{1}{x} \div \frac{2}{x^2}$$

$$= \frac{1}{x} \cdot \frac{x^2}{2}$$

$$= \frac{x^2}{2x}$$

$$= \frac{x}{2}$$

Method 2—Multiplication method:

The LCD for x and x^2 is x^2. We use the Fundamental Principle of Rational Expressions and multiply the numerator and denominator by x^2.

$$\frac{\dfrac{1}{x}}{\dfrac{2}{x^2}} = \frac{\dfrac{1}{x} \cdot x^2}{\dfrac{2}{x^2} \cdot x^2}$$

$$= \frac{\dfrac{x^2}{x}}{\dfrac{2x^2}{x^2}}$$

$$= \frac{x}{2}$$

PROCEDURES TO SIMPLIFY COMPLEX FRACTIONS

Division Method:

1. Write the fraction as a division problem.
2. Perform the division.
3. Reduce to lowest terms.

Multiplication Method:

1. Find the LCD of all denominators.
2. Multiply numerator and denominator by the LCD.
3. Reduce to lowest terms.

► **EXAMPLE 2**

Simplify $\dfrac{\dfrac{a}{x+1}+2}{\dfrac{b}{x+1}-1}$.

Solution

Again we show both methods.

Division Method

Before writing the fraction as a division problem, we make a single fraction in the numerator and the denominator. The numerator becomes

$$\frac{a}{x+1}+2=\frac{a}{x+1}+\frac{2(x+1)}{x+1}$$

$$=\frac{a+2x+2}{x+1}$$

The denominator becomes

$$\frac{b}{x+1}-1=\frac{b}{x+1}-\frac{x+1}{x+1}$$

$$=\frac{b-(x+1)}{x+1}$$

$$=\frac{b-x-1}{x+1}$$

Now we begin the simplification of the complex fraction.

$$\frac{\dfrac{a}{x+1}+2}{\dfrac{b}{x+1}-1}=\frac{\dfrac{a+2x+2}{x+1}}{\dfrac{b-x-1}{x+1}}$$

Notice that there is now a single fraction (or one term) in the numerator and in the denominator. Now we write the division problem.

$$=\frac{a+2x+2}{x+1}\div\frac{b-x-1}{x+1}$$

$$=\frac{a+2x+2}{x+1}\cdot\frac{x+1}{b-x-1}$$

$$=\frac{(a+2x+2)(x+1)}{(x+1)(b-x-1)}$$

$$=\frac{a+2x+2}{b-x-1}$$

Multiplication Method

The LCD is $x + 1$. So we multiply the numerator and denominator by $x + 1$.

$$\frac{\left(\dfrac{a}{x+1} + 2\right)(x+1)}{\left(\dfrac{b}{x+1} - 1\right)(x+1)} =$$

Notice the distributive property.

$$\frac{\dfrac{a}{x+1}(x+1) + 2(x+1)}{\dfrac{b}{x+1}(x+1) - 1(x+1)} = \frac{a + 2x + 2}{b - x - 1}$$

◄

A complex fraction may be simplified by either method. However, as Example 2 indicates, one method might be better than the other. In general, it is usually better to use the division method when the complex fraction contains a single term in its numerator and one term in its denominator. If the numerator *or* denominator of a complex fraction contains more than one term, it is usually better to use the multiplication method.

► **EXAMPLE 3**

Simplify each complex fraction.

(a) $\dfrac{\dfrac{1}{a} - \dfrac{1}{b}}{\dfrac{1}{b} + \dfrac{1}{a}}$ (b) $\dfrac{\dfrac{x}{x^2 - x - 2}}{\dfrac{1}{x - 2} + \dfrac{1}{x + 1}}$

Solutions

(a) Since both the numerator and the denominator have two terms, we will use the multiplication method. The least common denominator is ab.

$$\frac{\dfrac{1}{a} - \dfrac{1}{b}}{\dfrac{1}{b} + \dfrac{1}{a}} = \frac{\left(\dfrac{1}{a} - \dfrac{1}{b}\right)ab}{\left(\dfrac{1}{b} + \dfrac{1}{a}\right)ab}$$ Multiply the numerator and denominator by ab

$$= \frac{\dfrac{1}{a} \cdot ab - \dfrac{1}{b} \cdot ab}{\dfrac{1}{b} \cdot ab + \dfrac{1}{a} \cdot ab}$$ Distributive Property

$$= \frac{b - a}{a + b}$$

(b) Since the denominator has two terms, we will use the multiplication method. We first factor $x^2 - x - 2$ to determine the LCD.

$$\frac{\dfrac{x}{x^2 - x - 2}}{\dfrac{1}{x - 2} + \dfrac{1}{x + 1}} = \frac{\dfrac{x}{(x - 2)(x + 1)}}{\dfrac{1}{x - 2} + \dfrac{1}{x + 1}} \qquad \text{Factor}$$

$$= \frac{\dfrac{x}{(x - 2)(x + 1)} \cdot (x - 2)(x + 1)}{\left(\dfrac{1}{x - 2} + \dfrac{1}{x + 1}\right)(x - 2)(x + 1)} \qquad \text{LCD is } (x - 2)(x + 1).$$

Now we must use the distributive property in the denominator.

$$= \frac{\dfrac{x}{(x - 2)(x + 1)} \cdot (x - 2)(x + 1)}{\dfrac{1}{x - 2}(x - 2)(x + 1) + \dfrac{1}{x + 1}(x - 2)(x + 1)}$$

$$= \frac{x}{x + 1 + x - 2}$$

$$= \frac{x}{2x - 1} \qquad \blacktriangleleft$$

► **EXAMPLE 4**

Simplify $\dfrac{\dfrac{p}{p - 2} + p}{\dfrac{p}{2 - p} - p}$

Solution

We choose the multiplication method because there are two terms in both the numerator and denominator. Notice that $p - 2$ and $2 - p$ are opposites.

$$\frac{\dfrac{p}{p - 2} + p}{\dfrac{p}{2 - p} - p} = \frac{\dfrac{p}{p - 2} + p}{\dfrac{p}{-(p - 2)} - p}$$

$$= \frac{\dfrac{p}{p - 2} + p}{\dfrac{-p}{p - 2} - p}$$

The LCD is $p - 2$.

$$= \frac{\left(\dfrac{p}{p - 2} + p\right)(p - 2)}{\left(\dfrac{-p}{p - 2} - p\right)(p - 2)}$$

$$= \frac{\dfrac{p}{p - 2}(p - 2) + p(p - 2)}{\dfrac{-p}{p - 2}(p - 2) - p(p - 2)} \qquad \text{Distributive Property}$$

$$= \frac{p + p^2 - 2p}{-p - p^2 + 2p}$$

$$= \frac{p^2 - p}{p - p^2}$$

$$= \frac{p(p - 1)}{p(1 - p)}$$

$$= \frac{p - 1}{1 - p}$$

$$= \frac{p - 1}{-(p - 1)}$$

$$= -1 \qquad \blacktriangleleft$$

Negative exponents often appear in complex fractions.

▶ **EXAMPLE 5**

Simplify $\dfrac{x^{-1} + y^{-1}}{y^{-1}}$

Solution

Use the definition of negative exponent to rewrite the problem.

$$\frac{x^{-1} + y^{-1}}{y^{-1}} = \frac{\dfrac{1}{x} + \dfrac{1}{y}}{\dfrac{1}{y}}$$

Since the numerator has two terms, we will use the multiplication method to simplify.

$$= \frac{\left(\dfrac{1}{x} + \dfrac{1}{y}\right)xy}{\dfrac{1}{y} \cdot xy} \qquad \text{Multiply the numerator and denominator by } xy.$$

$$= \frac{y + x}{x} \qquad \text{Distributive Property} \qquad \blacktriangleleft$$

Problem Set 3.5

Warm-Ups

In Problems 1 through 8, simplify each complex fraction. Write answers in lowest terms. See Example 1.

1. $\dfrac{\dfrac{2}{a}}{\dfrac{4}{a^2}}$

2. $\dfrac{\dfrac{xyz^2}{t^4}}{\dfrac{x^3y}{t^2}}$

3. $\dfrac{\dfrac{-3r^3}{s^4}}{\dfrac{18r^4}{s^6}}$

4. $\dfrac{\dfrac{32a^3b}{m^2n}}{\dfrac{-48ab^2}{mn^3}}$

5. $\dfrac{\dfrac{4s^2}{p-2}}{\dfrac{12s}{p-2}}$

6. $\dfrac{\dfrac{4}{s-1}}{\dfrac{8}{1-s}}$

7. $\dfrac{\dfrac{z-2}{24}}{\dfrac{z-2}{36}}$

8. $\dfrac{\dfrac{-12}{5r+5s}}{\dfrac{18}{r+s}}$

In Problems 9 through 14, simplify each complex fraction. Write answers in lowest terms. See Examples 2, 3, and 4.

9. $\dfrac{\dfrac{1}{y}-1}{\dfrac{1}{y}+1}$

10. $\dfrac{\dfrac{1}{a+2}-1}{\dfrac{1}{a+2}+1}$

11. $\dfrac{\dfrac{1}{m}+\dfrac{1}{n}}{\dfrac{1}{m}-\dfrac{1}{n}}$

12. $\dfrac{\dfrac{1}{a}-\dfrac{1}{b}}{\dfrac{a^2-b^2}{ab}}$

13. $\dfrac{\dfrac{1}{b-1}+2}{3-\dfrac{1}{1-b}}$

14. $\dfrac{\dfrac{2x-1}{x^2-x}}{\dfrac{2x}{x-1}+\dfrac{1}{x}}$

In Problems 15 and 16, simplify each complex fraction. See Example 5.

15. $\dfrac{a^{-2}+b^{-1}}{(ab)^{-1}}$

16. $\dfrac{2^{-1}}{x^{-1}-3^{-1}}$

Practice Exercises

In Problems 17 through 50, simplify each complex fraction. Write answers in lowest terms.

17. $\dfrac{\dfrac{3}{a}}{\dfrac{9}{a^3}}$

18. $\dfrac{\dfrac{xyz^3}{t^5}}{\dfrac{x^2y}{t^7}}$

19. $\dfrac{\dfrac{xy}{x-2}}{\dfrac{ax}{x-2}}$

20. $\dfrac{\dfrac{7s^3}{p-3}}{\dfrac{28s}{p-3}}$

21. $\dfrac{\dfrac{z-2}{64}}{\dfrac{z-2}{48}}$

22. $\dfrac{\dfrac{15}{t-1}}{\dfrac{35}{1-t}}$

23. $\dfrac{\dfrac{-14}{3r+3s}}{\dfrac{21}{r+s}}$

24. $\dfrac{\dfrac{m-1}{m^2-1}}{\dfrac{3}{m+1}}$

25. $\dfrac{\dfrac{m+1}{m^2-1}}{\dfrac{3}{m-1}}$

26. $\dfrac{\dfrac{x}{x^2-x-6}}{\dfrac{x}{x-3}}$

27. $\dfrac{\dfrac{24}{x-5}}{\dfrac{128}{125-x^3}}$

28. $\dfrac{\dfrac{11}{3-x}}{\dfrac{33}{x^3-27}}$

29. $\dfrac{1 - \dfrac{1}{r-1}}{1 + \dfrac{1}{r-1}}$

30. $\dfrac{\dfrac{1}{a-2} + 1}{\dfrac{1}{a-2} - 1}$

31. $\dfrac{3y^{-1} - 2}{y^{-1} + 4}$

32. $\dfrac{m^{-1} - n^{-1}}{m^{-1} + n^{-1}}$

33. $\dfrac{\dfrac{1}{2-b} + 1}{1 - \dfrac{1}{b-2}}$

34. $\dfrac{\dfrac{x}{x-y} - 1}{2 - \dfrac{2y}{y-x}}$

35. $\dfrac{\dfrac{3t^2 + 2st - s^2}{st}}{\dfrac{3}{s} - \dfrac{1}{t}}$

36. $\dfrac{\dfrac{2t^2 - st - s^2}{st}}{\dfrac{2}{s} + \dfrac{1}{t}}$

37. $\dfrac{2a^{-1} + 2b^{-1}}{\dfrac{a^3 + b^3}{ab}}$

38. $\dfrac{\dfrac{x^3 - 8}{2x}}{2^{-1} - x^{-1}}$

39. $\dfrac{\dfrac{t+5}{t^2 - 16}}{1 + \dfrac{1}{t+4}}$

40. $\dfrac{\dfrac{2}{x-2} + \dfrac{1}{x+4}}{\dfrac{x+2}{x^2 + 2x - 8}}$

41. $\dfrac{\dfrac{x}{x^2 - x - 12}}{\dfrac{x}{x-4}}$

42. $\dfrac{\dfrac{b}{a-2}}{\dfrac{b}{a^2 - a - 2}}$

43. $\dfrac{\dfrac{2x+1}{x^2 + x}}{\dfrac{2x}{x+1} - \dfrac{1}{x}}$

44. $\dfrac{\dfrac{w}{w-3} - \dfrac{108}{w^2 - 9}}{\dfrac{2}{w+3} - \dfrac{1}{w-3}}$

45. $\dfrac{\dfrac{m+3}{m-3} - \dfrac{m+3}{m-3}}{\dfrac{m+3}{m-3} + \dfrac{m+3}{m-3}}$

46. $\dfrac{\dfrac{a+b}{a-b} + \dfrac{a+b}{a-2b}}{\dfrac{a-b}{a-2b} - \dfrac{a+b}{a-b}}$

47. $\dfrac{1 + \dfrac{1}{x} - \dfrac{1}{x+1}}{\dfrac{x^2+1}{x+1} - \dfrac{1}{x}}$

48. $\dfrac{\dfrac{3}{w-2} + \dfrac{1}{w} + 2}{\dfrac{w}{w-2} + \dfrac{1}{w}}$

49. $\dfrac{\dfrac{3}{x} - \dfrac{2}{y} - \dfrac{4}{z}}{\dfrac{1}{x} - \dfrac{1}{y} - \dfrac{1}{z}}$

50. $\dfrac{1 - x + \dfrac{12}{x+3}}{1 + x - \dfrac{8}{x+3}}$

In Problems 51 and 53, solve each problem.

51. For the lens shown below, the formula for finding the image distance i, in feet, given the focal length, in feet, of the lens, f, is

$$i = \dfrac{f}{1 - \dfrac{f}{7}}$$

Find the image distance when the focal length is

(a) $\dfrac{7}{2}$ ft   **(b)** $\dfrac{1}{2}$ ft

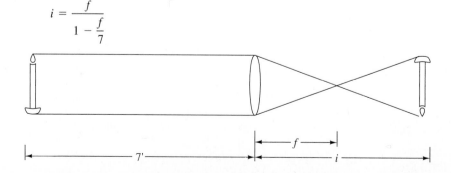

52. An electric water heater with a resistance of 12 ohms is connected in parallel to an electric outlet in a kitchen. The formula for the total resistance in the circuit, R_t, is

$$R_t = \frac{R}{1 + \dfrac{R}{12}}$$

where R is the resistance of any other appliance plugged into the outlet. What is the total resistance when:

(a) An electric iron with a resistance of 20 ohms is plugged into the outlet?

(b) A small lamp with a resistance of $\dfrac{5}{6}$ ohm is plugged into the outlet?

Challenge Problems

In Problems 53 through 55, simplify each complex fraction.

53. $1 + \dfrac{1}{1 + \dfrac{1}{1 + 1}}$

54. $x + \dfrac{x}{x + \dfrac{x}{x + x}}$

55. $\dfrac{1 + \dfrac{1}{1 - \dfrac{1}{x}}}{1 - \dfrac{1}{1 + \dfrac{1}{x}}}$

In Your Own Words

56. Explain the two methods that we use to simplify a complex fraction. When do we use each method?

► Chapter Summary

GLOSSARY

Rational expression The quotient of two polynomitals. A rational expression is **undefined** when its denominator has a value of zero.

To **reduce** a rational expression to lowest terms: To write the expression so that there are no common factors in the numerator and the denominator.

Multiplication of Rational Expressions	$\dfrac{A}{B} \cdot \dfrac{C}{D} = \dfrac{AC}{BD}$
Division of Rational Expressions	$\dfrac{A}{B} \div \dfrac{C}{D} = \dfrac{A}{B} \cdot \dfrac{D}{C} = \dfrac{AD}{BC}$
Addition and Subtraction of Rational Expressions	$\dfrac{A}{B} + \dfrac{C}{B} = \dfrac{A + C}{B}, \qquad \dfrac{A}{B} - \dfrac{C}{B} = \dfrac{A - C}{B}$

Procedure to Find the Least Common Denominator

1. Factor each denominator completely, using exponents.

2. List all different prime factors from all denominators.

3. Write the LCD. The LCD is the product of the factors in step 2 each raised to the largest power of that factor in any single denominator.

Rational Equations

1. Clear the equation of fractions (multiply both sides by the least common denominator).

2. Solve the resulting equation for possible solutions.

3. Possible solutions that make any denominator have a value of zero *cannot* be included in the solution set.

Rational Inequalities

1. Find the *free boundary numbers;* that is, find the numbers that make any denominator in the inequality have a value of 0.
2. Solve the associated *equation* to find other boundary numbers.
3. Locate *all* the boundary numbers on a number line.
4. Determine which regions formed by the boundary numbers make the *original inequality* true by testing with one number inside each region.
5. Shade only the regions that test true.
6. Check the boundary numbers themselves.
7. Write the solution set.

There are two techniques to use in simplifying a complex fraction.

Complex Fractions

1. Treat the complex fraction as a division problem.
2. Multiply the numerator and denominator by the LCD of all the denominators.

CHECK UPS

1. Determine the values of x for which $\dfrac{x + 2}{(x + 1)(x - 2)}$ is undefined.

Section 3.1; Example 1b

2. Reduce $\dfrac{5x + 5y}{x^2 - y^2}$ to lowest terms.

Section 3.1; Example 3b

3. Write the rational expression $\dfrac{x - 2}{x - 3}$ with a denominator of $x^2 - x - 6$.

Section 3.1; Example 5d

4. Perform the operation indicated.
$$\frac{t^2 - t - 2}{2t - 6} \cdot \frac{t - 3}{t^3 - 8}$$

Section 3.2; Example 1d

5. Perform the operation indicated.
$$\frac{x^2 + 3x + 2}{x^4} \div \frac{x^2 + 2x + 1}{x^2}$$

Section 3.2; Example 3b

6. Perform the operation indicated.
$$\frac{3}{x^2 - 4} + \frac{1}{2 - x}$$

Section 3.3; Example 5

7. Solve $\dfrac{2}{x - 2} - \dfrac{1}{x + 1} = \dfrac{5}{x^2 - x - 2}$.

Section 3.4; Example 3

8. Solve $\dfrac{x + 2}{x - 1} < 2$ and write the solution set in interval notation.

Section 3.4; Example 6

9. Simplify $\dfrac{\dfrac{a}{x + 1} + 2}{\dfrac{b}{x + 1} - 1}$.

Section 3.5; Example 2

► Review Problems

In Problems 1 through 5, determine the values of x, if any, for which each rational expression is undefined.

1. $\dfrac{14}{x}$

2. $\dfrac{x}{x - 8}$

3. $\dfrac{x^2}{67}$

4. $\dfrac{x + 4}{(x + 2)(x - 8)}$

5. $\dfrac{x - 1}{x^2 + 4x + 3}$

In Problems 6 through 10, express each rational expression with the denominator indicated.

6. $\dfrac{3}{x^4 y^2}$; $x^8 y^3$

7. $\dfrac{4}{3p}$; $6p^2(p - 2)$

8. $\dfrac{x}{x - 1}$; $x^2 - 7x + 6$

9. $\dfrac{x + 3}{x - 4}$; $x^2 - 8x + 16$

10. $\dfrac{4}{x + 2}$; $x^2 + 2x$

In Problems 11 through 15, reduce each rational expression to lowest terms.

11. $\dfrac{m^5 n}{m^7 n^3}$

12. $\dfrac{4 - x}{x^2 - 16}$

13. $\dfrac{x^2 - 4x - 5}{x^3 + 1}$

14. $\dfrac{r^2 + 6r + 9}{r^2 - 9}$

15. $\dfrac{s^3 t - st}{s^2 + s}$

In Problems 16 through 35, perform the operations indicated. Write answers in lowest terms.

16. $\dfrac{3}{x^2 y} - \dfrac{x + y}{x}$

17. $\dfrac{12 - 3x^2}{2x^2 + x - 15} \cdot \dfrac{2x^2 - 3x - 5}{x^2 - 4x + 4}$

18. $\dfrac{4}{t - 7} + \dfrac{3}{t + 7}$

19. $\dfrac{a^2 - 3a + 9}{a^3 - ab^2} \div \dfrac{a^3 + 27}{a^4 + 2a^3 b + a^2 b^2}$

20. $\dfrac{2x - 2}{x^2 - 2x - 8} - \dfrac{1}{4 - x}$

21. $\dfrac{r^2 st^3}{u^4 v} \cdot \dfrac{-t^5}{u} \cdot \dfrac{u^2 v^3}{rst}$

22. $\dfrac{2}{x + 1} + \dfrac{3}{x^2 + 2x + 1}$

23. $\dfrac{u^2 - uv + 2uw - 2vw}{u^3 + 8w^3} \div \dfrac{u^2 + 4uw + 4w^2}{u^2 - 2uw + 4w^2}$

24. $\dfrac{4}{x + 3} - \dfrac{x + 1}{x + 7}$

25. $\dfrac{1 - 2y + y^2}{6y^2 - y - 1} \cdot \dfrac{y^2 + 7y + 12}{3y^2 + y - 4} \cdot \dfrac{6y^2 + 5y - 4}{y^2 + 2y - 3}$

26. $\dfrac{15x}{x^2 + x - 6} - \dfrac{11x - 3}{x^2 + 2x - 3}$

27. $\dfrac{12x^2 - 5x - 2}{x^2 + 2xy + y^2} \div \dfrac{12x^2 + x - 6}{x + y}$

28. $\dfrac{1}{xy} + \dfrac{3}{y} - \dfrac{4}{xyz}$

29. $\dfrac{12t^3 - 27t}{t^8 - 16} \cdot \dfrac{t^4 + 2t^2}{2t^2 - 3t - 9}$

30. $\dfrac{25x}{x^2 - 3x - 4} - \dfrac{21x - 64}{12 - 7x + x^2}$

31. $\dfrac{96t^7}{375s^2} \div \dfrac{-72t^4}{125s^7}$

32. $\dfrac{1}{x + 2} - \dfrac{3}{x - 1} + \dfrac{5}{x - 3}$

33. $\dfrac{a^3 + a^2 b + ab^2 + b^3}{32a^7 b^3} \cdot \dfrac{-768a^2 b}{a^4 - b^4}$

34. $\dfrac{7}{11 - 2x} + \dfrac{3}{2x - 11}$

35. $\dfrac{m^2 - 16n^2}{6m^4 + 3m^2} \div \dfrac{64n^3 - m^3}{12m^2}$

In Problems 36 through 45, simplify each complex fraction.

36. $\dfrac{\dfrac{2x^3 z^2}{y}}{\dfrac{-48x}{y^6}}$

37. $\dfrac{\dfrac{x^2 y}{x + 1}}{\dfrac{x}{x + 1}}$

38. $\dfrac{\dfrac{5x}{p - 1}}{\dfrac{30x}{p - 1}}$

39. $\dfrac{\dfrac{1}{x} - \dfrac{2}{xy}}{\dfrac{2}{x} + \dfrac{1}{xy}}$

40. $\dfrac{\dfrac{2}{a+b} - \dfrac{3}{a-b}}{\dfrac{3}{a-b} - \dfrac{2}{a+b}}$

41. $\dfrac{\dfrac{3t^2+5t}{t^2-25}}{\dfrac{2}{t-5} + \dfrac{1}{t+5}}$

42. $\dfrac{\dfrac{t^2-4z^2}{zt}}{\dfrac{1}{z} - \dfrac{2}{t}}$

43. $\dfrac{\dfrac{1}{x-5} - \dfrac{1}{x+3}}{\dfrac{16x^2+16}{x^2-2x-15}}$

44. $\dfrac{\dfrac{1}{x+2} + x + 2}{\dfrac{1}{x+2} - x - 2}$

45. $\dfrac{\dfrac{1}{s-5} + \dfrac{s+5}{s^2+5s+25}}{\dfrac{2s+5}{s^3-125}}$

In Problems 46 through 57, find the solution set.

46. $\dfrac{x-7}{x+5} > 0$

47. $\dfrac{x}{x+7} \le 2$

48. $\dfrac{1}{x-3} + \dfrac{1}{x+3} = \dfrac{4}{x^2-9}$

49. $\dfrac{1}{x} - \dfrac{1}{3x} = \dfrac{1}{2}$

50. $\dfrac{x-2}{x} \ge 4$

51. $\dfrac{x-3}{x+2} \ge 6$

52. $\dfrac{1}{4-x} + \dfrac{8}{16-x^2} = \dfrac{1}{4+x}$

53. $\dfrac{1}{x-2} = \dfrac{2}{x-2}$

54. $\dfrac{3}{x-1} < \dfrac{2}{x+1}$

55. $\dfrac{1}{x+2} < \dfrac{2}{x-1}$

56. $\dfrac{3x}{x+2} = \dfrac{3}{2}$

57. $\dfrac{5x}{2x-3} = \dfrac{3}{2}$

58. Lief-Ann runs 10 km in the same time that Hillary runs 8 km. If Lief's rate is 2 kilometers per hour faster than Hillary's, what are their rates?

59. A truck can travel 675 miles in the same time that a car can travel 750 miles. If the truck's average speed is 6 mph less than the car's speed, find the speed of the truck.

60. If a piece of gold jewelry is 14-karat gold, then the ratio of gold to a mixture of metals (called an alloy) is 14 to 10. Ryan made a 14-karat gold locket that had 2 more ounces of gold than alloy in it. How many ounces of alloy were in the locket? How much did the locket weigh?

61. According to the ancient Greeks, a rectangle whose dimensions are in the ratio of approximately 1 to 1.6 is the most pleasing. Hershel wants to represent this ratio (the Golden Ratio) in a model of the Parthenon. One wall of the model will be 6 inches longer than its height. What should the wall's dimensions be?

62. A baseball batting average is the ratio of the number of hits to the number of times at bat. If Cal Ripken, Jr., of the Baltimore Orioles had a 1991 season batting average of 0.323 with 210 hits, how many times did he bat?

63. In 1912, President Taft officially set the ratio of the width to the length of an American flag as $\dfrac{1}{1.9}$. If James makes a flag with a length that is 6 in. longer than its width, find the dimensions he must use to keep the official ratio.

64. The reciprocal of four times a number is one less than the reciprocal of the number. Find the number.

65. One number is twice another number. If the sum of their reciprocals is 1, find the number.

66. The quotient of a number and 2 less than the number is negative. Find all such numbers.

67. The quotient of a number and 1 more than the number is negative. Find all such numbers.

Let's Not Forget . . .

Identify the expressions that are in factored form. Factor those that are not factored.

68. $(3x-1)^3$

69. $27x^3 - 1$

70. $x(a-b) + y(a-b)$

71. $3s^2t(st+1)$

72. $3x^4y$

How many terms are in each expression? Which expressions have x − 1 as a factor?

73. $x^3 + 1$

74. $(x - 1)^3$

75. $x^2 - x - 2$

76. $x^2 - 2x + 1$

77. $3(x - 1) - (x - 1)^2$

78. $a(x - 1) + b(1 - x)$

Find each product.

79. $(x + 2y)^2$

80. $(c - 2)^3$

Simplify each expression.

81. -2^2 **82.** $(-2)^2$ **83.** $(-2)^3$ **84.** $(-a)^4$ **85.** $(-a)^5$

Reduce, if possible.

86. $\dfrac{3(p + q) - 2(p - q)}{(p + q)(p - q)}$

87. $\dfrac{a(y + z) - b(y + z)}{(y + z)(y - z)}$

The following problems can be worked by using a least common denominator. Follow the directions in each and notice how the LCD is used.

88. Perform the operation indicated: $\dfrac{1}{x - 2} + \dfrac{4}{x + 2}$.

89. Simplify $\dfrac{\dfrac{4}{x - 2} + 1}{\dfrac{2x}{x + 2} - 1}$.

90. Solve $\dfrac{1}{2x} - \dfrac{1}{3} = \dfrac{1}{x}$

Label each as an equation, inequality, or expression. Solve the equations and inequalities, and perform the indicated operations on the expressions.

91. $\dfrac{1}{x - 1} \leq \dfrac{2}{x + 1}$

92. $\dfrac{x + y}{x - y} - \dfrac{(x - y)^2}{x^2 - y^2}$

93. $2x - 5 = 3(1 - x)$

94. $|x + 1| < 1.5$

95. $-3^2 + 3^{-2}$

96. $\dfrac{x^{-1} + 1}{x^{-1} - 1}$

► You Decide

Due to circumstances too bizarre to relate, you find yourself in a deserted cabin, 96 miles from Nome, Alaska. It's early September and the weather has taken a turn for the worst. Two snowmobiles are available. The new snowmobile can travel 24 miles in the same time it takes the old snowmobile to travel 16 miles. The new snowmobile is 6 mph faster than the old snowmobile, but it only gets 12 miles to the gallon while the old snowmobile gets 16 miles to the gallon.

You have only 6 gallons of gasoline and your only source of heat is a gasoline stove that burns a gallon of gas every 12 hours. It is 5 pm and already quite dark. What time do you decide to leave for Nome, and what time do you arrive? Why did you decide on your course of action?

▶ Chapter 3 Test

In Problems 1 through 5, choose the correct answer.

1. The rational expression $\dfrac{x-2}{x(x+3)}$ is undefined if x has the value(s) of (?)

A. 2   **C.** 0 and -3

B. -2 **D.** 3

2. If the rational expression $\dfrac{3x}{x-7}$ is written with a denominator of $x^2 - 14x + 49$, it becomes (?)

A. $\dfrac{3x^2 - 7}{x^2 - 14x + 49}$ **C.** $\dfrac{3x}{x^2 - 14x + 49}$

B. $\dfrac{3x^2 - 21x}{x^2 - 14x + 49}$ **D.** $\dfrac{3x^2 + 21x}{x^2 - 14x + 49}$

3. When reduced to lowest terms, the rational expression $\dfrac{x^3 - 8y^3}{x^2 - 4y^2}$ becomes (?)

A. $\dfrac{x^2 + 2xy + 4y^2}{x + 2y}$ **C.** $x - 2y$

B. $\dfrac{(x - 2y)^2}{x + 2y}$ **D.** $x + 2y$

4. $\dfrac{x}{x+4} - \dfrac{x-3}{x-2} = $ (?)

A. $-\dfrac{x + 12}{(x+4)(x-2)}$ **C.** $\dfrac{-3(x-4)}{(x+4)(x-2)}$

B. $\{4\}$ **D.** $-3(x+4)$

5. When simplified, the complex fraction $\dfrac{\dfrac{3}{x} + \dfrac{1}{y}}{\dfrac{9y}{x} - \dfrac{x}{y}}$ becomes (?)

A. $\dfrac{4}{9y - x}$ **C.** $\dfrac{1}{3y + x}$

B. $\dfrac{1}{3y - x}$ **D.** $\dfrac{3x + y}{9y - x}$

In Problems 6 through 9, perform the operation indicated. Write answers in lowest terms.

6. $\dfrac{a}{ab - b^2} + \dfrac{b}{a^2 - ab}$

7. $\dfrac{x^3 + 3x^2}{x^3 - 6x^2 + 9x} \div \dfrac{x^2 + 2x - 3}{x^2 - 9}$

8. $\dfrac{y + 1}{y - 2} - \dfrac{y^2 + 5y + 1}{y^2 + y - 6}$

9. $\dfrac{6x^2 + 13x + 6}{9x^2 - 4} \cdot \dfrac{9x^2 - 12x + 4}{2x^2 + x - 3}$

10. Simplify the complex fraction.

$$\dfrac{\dfrac{2t}{s + t} + 1}{\dfrac{4t}{s + t} - 1}$$

In Problems 11 through 13, find the solution set.

11. $\dfrac{x - 3}{x + 1} \geq 2$

12. $\dfrac{1}{3x} - \dfrac{3}{2x} = \dfrac{1}{3}$

13. $\dfrac{2}{x - 1} - \dfrac{1}{x + 2} = \dfrac{8}{x^2 + x - 2}$

14. The reciprocal of one less then a number is less than or equal to the reciprocal of one more than the number. Find all such numbers.

15. Jane can bike 10 miles in the same time that Bob can bike 6 miles. If Jane's speed is 3 mph faster than Bob's speed, find their speeds.

4 Radicals and Exponents

See Problem Set 4.5, Exercise 69.

► Connections

The way we write algebra today is mainly due to the French mathematician René Descartes (1596–1650). Descartes popularized the use of superscripts, which we call exponents, to indicate a number multiplied by itself. For example, we write x^3 to mean $x \cdot x \cdot x$. Soon the properties of exponents were discovered, and this concise notation was expanded to include zero, negative integers, and rational number exponents.

An important consideration in the design of an auditorium is the time difference in sound traveling to a seat along different paths. If the time difference is too great, poor acoustics will result. For this study, the formula $t = \dfrac{d}{\sqrt{grT}}$ gives the time for sound to travel a distance, d. Formulas that contain square roots, such as this one, are common in all engineering fields.

In this chapter we will see that exponents are related to square roots and other radicals. We will investigate this relationship. This study and the complex numbers also developed in this chapter are necessary to solve quadratic equations in Chapter 5.

4.1   Roots and Radicals

TAPE 7

There are two real numbers whose squares are 4. They are 2 and -2. There are two real numbers whose squares are 9. They are 3 and -3. How about 7? Are there two real numbers whose squares are 7? The answer to this question is yes, there are two such numbers. However, they are not integers or even rational numbers. They are irrational numbers. How do we express them? The following notation has been chosen for such numbers. We let $\sqrt{7}$ be the *positive* number whose square is 7, and $-\sqrt{7}$ is the *negative* number whose square is 7.

$$(\sqrt{7})^2 = 7 \qquad (-\sqrt{7})^2 = 7$$

The number $\sqrt{7}$ is called the **principal square root** of 7. Because $\sqrt{4}$ is the positive number whose square is 4, we have $\sqrt{4} = 2$. Is there a real number whose square is -4? The answer is no. Such a number would violate our rules for signed-number arithmetic. So expressions such as $\sqrt{-4}$ are undefined in the system of real numbers.

Square Roots of Negative Numbers

If $k > 0$, then $\sqrt{-k}$ is undefined in the system of real numbers.

There is a real number whose cube is 8. It is 2. Also, there is a real number whose cube is -8. It is -2. Is there a real number whose cube is 11? Yes, it is denoted by $\sqrt[3]{11}$. Furthermore, there is a real number whose cube is -11; it is $\sqrt[3]{-11}$. In fact, $\sqrt[3]{-11} = -\sqrt[3]{11}$. We call $\sqrt[3]{11}$ the **cube root** of 11. Note that $\sqrt[3]{8} = 2$ and $\sqrt[3]{-8} = -2$. Every real number has a cube root.

In general, there are roots for every natural number, n, denoted by the expression

$$\sqrt[n]{q}$$

where $(\sqrt[n]{q})^n = q$. We call $\sqrt[n]{q}$ a **radical.** $\sqrt{}$ is called a **radical sign.** The real number q is called the **radicand,** and the natural number n is called the **index.** If the index is omitted, the radical is a square root.

Definition of $\sqrt[n]{q}$

If n is an *even* natural number,

1. $\sqrt[n]{q}$ $\begin{cases} \text{is a real number when } q \geq 0. \\ \text{is } not \text{ a real number when } q < 0. \end{cases}$
2. $\sqrt[n]{q}$ is the *nonnegative* number such that $(\sqrt[n]{q})^n = q$.

If n is an *odd* natural number,

1. $\sqrt[n]{q}$ exists for all real numbers q.
2. $\sqrt[n]{q}$ is the real number such that $(\sqrt[n]{q})^n = q$.
3. $\sqrt[n]{-q} = -\sqrt[n]{q}$.

► **EXAMPLE 1**

Find each root.

(a) $\sqrt{121}$ (b) $\sqrt[3]{125}$ (c) $\sqrt[4]{81}$ (d) $\sqrt[3]{-27}$ (e) $\sqrt{\dfrac{9}{16}}$ (f) $\sqrt[3]{\dfrac{8}{125}}$

Solutions

(a) Since $11^2 = 121$ and 11 is *positive,*
$$\sqrt{121} = 11$$

(b) Since $5^3 = 125$,
$$\sqrt[3]{125} = 5$$

(c) Since $3^4 = 81$ and 3 is *positive,*
$$\sqrt[4]{81} = 3$$

(d) Since $(-3)^3 = -27$,
$$\sqrt[3]{-27} = -3$$

(e) Since $\left(\dfrac{3}{4}\right)^2 = \dfrac{9}{16}$ and $\dfrac{3}{4}$ is *positive,*
$$\sqrt{\dfrac{9}{16}} = \dfrac{3}{4}$$

(f) Since $\left(\dfrac{2}{5}\right)^3 = \dfrac{8}{125}$,
$$\sqrt[3]{\dfrac{8}{125}} = \dfrac{2}{5}$$

◄

We must be *very* careful when simplifying radicals with *even* indexes if there are variables in the radicand. If we *know* that x is *not negative*, we can write $\sqrt{x^2} = x$. However, if x has the value -3, then to say that $\sqrt{x^2} = x$ would be to say that $\sqrt{(-3)^2} = -3$ or $\sqrt{9} = -3$, which is incorrect! The Challenge Problems at the end of this section investigate this situation further. Unless otherwise stated, we will assume that letters appearing in the radicand represent nonnegative real numbers.

▶ **EXAMPLE 2**

Find each of the following roots. Assume that all variables represent **nonnegative** real numbers.

(a) $\sqrt{25x^2}$　　(b) $\sqrt{16y^4}$　　(c) $\sqrt[3]{-8x^3}$　　(d) $\sqrt{576x^4y^8}$

Solutions

(a) $\sqrt{25x^2} = 5x$
   [since $5x$ is nonnegative and $(5x)^2 = 25x^2$]

(b) $\sqrt{16y^4} = 4y^2$
   [$4y^2$ is nonnegative and $(4y^2)^2 = 16y^4$]

(c) $\sqrt[3]{-8x^3} = -2x$
   [since $(-2x)^3 = -8x^3$]
   Sometimes it is difficult to recognize a perfect square. In the next example, notice how factoring helps.

(d) $\sqrt{576x^4y^8} = \sqrt{2^6 \cdot 3^2 x^4 y^8}$

$$= 2^3 \cdot 3x^2 y^4$$

$$= 24x^2 y^4 \qquad ◀$$

Because radicals are real numbers, we utilize them in real-number algebra. The following properties are useful.

Properties of Radicals

If $\sqrt[n]{p}$ and $\sqrt[n]{q}$ represent real numbers, then

1. $\sqrt[n]{p \cdot q} = \sqrt[n]{p} \cdot \sqrt[n]{q}$　　Root of a product

2. $\sqrt[n]{\dfrac{p}{q}} = \dfrac{\sqrt[n]{p}}{\sqrt[n]{q}}, \quad q \neq 0$　　Root of a quotient

---

These properties are used for simplifying roots of products and quotients. Consider $\sqrt{75}$. Since 75 is not a perfect square, $\sqrt{75}$ is not an integer. However, 75 has a factor that is a perfect square. Notice how the *root of a product* property works to simplify a product.

$$\sqrt{75} = \sqrt{5^2 \cdot 3} \qquad \text{Factor}$$

$$= \sqrt{5^2} \cdot \sqrt{3} \qquad \text{Root of a product}$$

$$= 5\sqrt{3}$$

The *root of a product* property allows us to take out perfect square **factors** of the radicand. The same idea applies to cube roots. Perfect cube factors can be taken out of the radicand. The idea works for *n*th roots in general. First, we factor the radicand to determine what can be taken out.

► **EXAMPLE 3**

Simplify each radical expression. Assume that all variables represent nonnegative real numbers.

(a) $\sqrt{8}$ (b) $\sqrt{x^3 y^5}$ (c) $2\sqrt{72x^{12}}$

Solutions

(a) The idea is to find the largest perfect square factor of 8.

$$\sqrt{8} = \sqrt{2^2 \cdot 2} \qquad \text{Factor}$$
$$= \sqrt{2^2} \cdot \sqrt{2} \qquad \text{Root of a product}$$
$$= 2\sqrt{2}$$

(b) Again the idea is to find largest perfect square factors of the radicand.

$$\sqrt{x^3 y^5} = \sqrt{x^2 \cdot xy^4 \cdot y} \qquad \text{Factor}$$
$$= \sqrt{x^2 y^4 xy} \qquad \text{Place the perfect square factors first}$$
$$= \sqrt{x^2 y^4} \cdot \sqrt{xy} \qquad \text{Root of a product}$$
$$= xy^2 \sqrt{xy}$$

(c) $$2\sqrt{72x^{12}} = 2\sqrt{6^2 \cdot 2 \cdot x^{10} \cdot x} \qquad \text{Factor}$$
$$= 2\sqrt{6^2 x^{10} \cdot 2x} \qquad \text{Place perfect square factors first}$$
$$= 2\sqrt{6^2 x^{10}} \cdot \sqrt{2x} \qquad \text{Root of a product}$$
$$= 2 \cdot 6x^5 \sqrt{2x}$$
$$= 12x^5 \sqrt{2x} \qquad \blacktriangleleft$$

► **EXAMPLE 4**

Simplify each radical. Assume that all variables are nonnegative real numbers.

(a) $\sqrt[3]{-54x^3}$ (b) $\sqrt[3]{\dfrac{-16}{27}}$ (c) $\sqrt[5]{96x^{12}y^{10}}$

Solutions

(a) $$\sqrt[3]{-54x^3} = \sqrt[3]{(-3)^3 \cdot 2x^3} \qquad \text{Factor}$$
We place the cubes first.
$$= \sqrt[3]{(-3)^3 x^3 \cdot 2}$$
$$= \sqrt[3]{(-3)^3 x^3} \cdot \sqrt[3]{2} \qquad \text{Root of a product}$$
$$= -3x\sqrt[3]{2}$$

(b) $\sqrt[3]{\dfrac{-16}{27}} = \dfrac{\sqrt[3]{-16}}{\sqrt[3]{27}}$ Root of a quotient

$= \dfrac{\sqrt[3]{(-2)^3 \cdot 2}}{\sqrt[3]{27}}$

$= \dfrac{\sqrt[3]{(-2)^3} \cdot \sqrt[3]{2}}{\sqrt[3]{27}}$ Root of a product

$= \dfrac{-2\sqrt[3]{2}}{3}$

(c) $\sqrt[5]{96x^{12}y^{10}} = \sqrt[5]{2^5 \cdot 3x^{10}x^2y^{10}}$

$= \sqrt[5]{2^5x^{10}y^{10} \cdot 3x^2}$

$= \sqrt[5]{2^5x^{10}y^{10}} \cdot \sqrt[5]{3x^2}$ Root of a product

$= 2x^2y^2\sqrt[5]{3x^2}$ ◀

Example 5 illustrates that the *root of a product* property will split up **factors,** but not **terms.**

► EXAMPLE 5

Simplify each of the following expressions. Assume that all variables represent *nonnegative* real numbers.

(a) $\sqrt{x^2y^2}$ (b) $\sqrt{x^2 + y^2}$ (c) $\sqrt{(x + y)^2}$

Solutions

(a) Because x^2 and y^2 are *factors,*

$\sqrt{x^2y^2} = \sqrt{x^2} \cdot \sqrt{y^2}.$ Root of a product
$= xy$

because x and y are both nonnegative.

(b) $\sqrt{x^2 + y^2}$ does not simplify, because x^2 and y^2 are not *factors* of the radicand. They are *terms.*

(c) $\sqrt{(x + y)^2} = x + y$
because x and y are both nonnegative.   ◀

Be Careful!   Be sure to note the difference between parts (b) and (c) in Example 5. We must be careful in part (b). If x has the value 3 and y has the value 4, we can see that $\sqrt{x^2 + y^2}$ and $\sqrt{x^2} + \sqrt{y^2}$ represent *different* numbers.

$$\sqrt{3^2 + 4^2} = \sqrt{9 + 16} = \sqrt{25} = 5$$

$$\sqrt{3^2} + \sqrt{4^2} = 3 + 4 = 7$$

A **radical expression** is an expression containing one or more radicals. We say that a radical expression is simplified if it satisfies the following conditions.

> **Simplified Radical Expression**
>
> 1. The power of any factor in a radicand is *less* than the index of the radical.
> 2. No radicand contains fractions or negative numbers, and there are no radicals in a denominator.

Problem Set 4.1

Warm-Ups

In this problem set, assume that all variables represent nonnegative real numbers.
In Problems 1 through 8, simplify each radical. See Example 1.

1. $2\sqrt{49}$

2. $-\sqrt{64}$

3. $\sqrt[3]{125}$

4. $3\sqrt[3]{-64}$

5. $2\sqrt{36} - \sqrt{16}$

6. $\sqrt{\dfrac{25}{49}}$

7. $\sqrt{\dfrac{256}{625}}$

8. $\sqrt[3]{\dfrac{27}{64}}$

In Problems 9 through 14, simplify each expression. See Example 2.

9. $-\sqrt{4x^2}$

10. $\sqrt[3]{-8x^6y^3}$

11. $\sqrt{100z^6}$

12. $-\sqrt{144\,j^6 l^4}$

13. $\sqrt{\dfrac{121y^4}{x^2z^2}}$

14. $\sqrt{\dfrac{36x^{10}}{y^8}}$

In Problems 15 through 18, simplify each radical. See Example 3.

15. $\sqrt{50}$

16. $-2\sqrt{45}$

17. $\sqrt{252}$

18. $\sqrt{240}$

In Problems 19 through 22, simplify each radical. See Example 4.

19. $\sqrt[3]{40}$

20. $-3\sqrt[3]{-108}$

21. $\sqrt[3]{27x^7y^3}$

22. $\sqrt[5]{\dfrac{x^6y^{14}}{32}}$

For Problems 23 and 24, simplify each radical. See Example 5.

23. (a) $\sqrt{4y^2}$ (b) $\sqrt{4 + y^2}$
      (c) $\sqrt{(2 + y)^2}$

24. (a) $\sqrt[3]{64k^6}$ (b) $\sqrt[3]{64 + k^6}$
      (c) $\sqrt[3]{(2 + k)^6}$

Practice Exercises

In Problems 25 through 69, simplify each expression, if possible.

25. $3\sqrt{36}$

26. $\sqrt{81}$

27. $-\sqrt{100}$

28. $\sqrt{169} - 2\sqrt{49}$

29. $\sqrt[3]{216}$

30. $2\sqrt[3]{-125}$

31. $-\sqrt{16x^4y^2}$

32. $\sqrt[3]{27x^9y^3}$

33. $\sqrt{72}$

34. $-3\sqrt{20}$

35. $\sqrt{288}$

36. $\sqrt{275}$

37. $\sqrt[3]{54}$

38. $-5\sqrt[3]{-192}$

39. $\sqrt[5]{-486}$

40. $\sqrt{-256}$

41. $\sqrt{\dfrac{16}{81}}$

42. $\sqrt{\dfrac{144}{169}}$

43. $\sqrt{\dfrac{27}{49}}$

44. $\sqrt{\dfrac{40}{81}}$

45. $-3\sqrt{\dfrac{8}{9}}$

46. $\sqrt[3]{\dfrac{-81}{125}}$

47. $\sqrt{9x^3y^2}$

48. $2\sqrt{5xy^3z^2}$

49. $\sqrt[3]{\dfrac{64}{27}}$

50. $\sqrt{\dfrac{8}{25}}$

51. $\sqrt{\dfrac{75}{49}}$

52. $-2\sqrt{\dfrac{7}{25}}$

53. $\sqrt[3]{\dfrac{56}{27}}$

54. $\sqrt{4x^2y^3}$

55. $3\sqrt{7x^2y^3z}$

56. $\sqrt{169z^8}$

57. $-\sqrt{450x^3y^9}$

58. $-\sqrt{100j^4k^2l^3}$

59. $\sqrt{\dfrac{196}{p^4q^2}}$

60. $\sqrt{\dfrac{135x^7}{y^{12}}}$

61. $\sqrt[3]{16z^3}$

62. $\sqrt[3]{-128x^9y^4}$

63. $\sqrt[3]{\dfrac{216a^6b^{12}c^4}{d^{27}}}$

64. $\sqrt[4]{243xy^6z^{10}}$

65. $\sqrt[5]{-96x^{15}z^{26}}$

66. $\sqrt[5]{486x^6k^{12}}$

67. $\sqrt{(13+2x)^2}$

68. $\sqrt{x^4+16}$

69. $\sqrt{36+9v^2}$

70. A first-century Greek mathematician whose name was Hero or Heron expressed the area of a triangle in terms of the lengths of its sides. If the three sides of a triangle have lengths of a, b, and c units, then the area of the triangle is

$$\sqrt{s(s-a)(s-b)(s-c)} \text{ square units}$$

where s is the semiperimeter of the triangle; that is, $s = \dfrac{a+b+c}{2}$.

Find the area of a triangle with sides of

(a) 3, 4, and 5 feet

(b) 16 inches each

(c) 6, 9, and 12 centimeters

(d) 6, 18, and 18 feet

Problem 70

Challenge Problems

71. If x is positive (or zero), $\sqrt{x^2}$ is x. Is $\sqrt{x^2} = x$ always true? Try -2 for x and work very carefully.

72. If x is *negative*, what is the value of $\sqrt{x^2}$?

73. Write a formula for $\sqrt{x^2}$ that is valid for all real numbers. (HINT: Remember the definition of $|x|$.)

In Problems 74 through 79, simplify each expression. Assume that a, b, and c are real numbers with a positive and b negative.

74. $\sqrt{a^2}$

75. $\sqrt{b^2}$

76. $\sqrt{c^2}$

77. $\sqrt{(5+a)^2}$

78. $\sqrt{(b-1)^2}$

79. $\sqrt{(a+b)^2}$

In Your Own Words

80. Explain what is meant by the "square root of a positive number."

4.2   Addition and Subtraction of Radical Expressions

TAPE 8

We continue our study of radicals with a look at addition. We know that

$$3x + 4x = 7x$$

is always true because we added like terms, whereas

$$3x + 4y$$

cannot be simplified because x and y are **not** like terms. Compare the following two expressions.

$$\boxed{1} \quad 3\sqrt{5} + 4\sqrt{5}$$
$$\boxed{2} \quad 3\sqrt{5} + 4\sqrt{7}$$

The first can be simplified,

$$3\sqrt{5} + 4\sqrt{5} = 7\sqrt{5}$$

whereas the second cannot, because $\sqrt{5}$ and $\sqrt{7}$ are not like radicals. Only radicals with the same index and the same radicand can be combined.

► **EXAMPLE 1**

Perform the indicated operation, if possible.

(a) $6\sqrt{7} + 4\sqrt{7}$ (b) $9\sqrt{2} - 2\sqrt{2}$

(c) $3 + 2\sqrt{3}$ (d) $\sqrt{6} - 8\sqrt{6}$

Solutions

(a) $6\sqrt{7} + 4\sqrt{7}$
Notice that in the two radical expressions being added only the numerical coefficients are different.
$$6\sqrt{7} + 4\sqrt{7} = 10\sqrt{7} \qquad \text{Add like terms}$$

(b) $9\sqrt{2} - 2\sqrt{2} = 7\sqrt{2} \qquad \text{Subtract like terms}$

(c) $3 + 2\sqrt{3}$
The order of operations requires that *multiplication* be done *before addition*. Since 3 and $2\sqrt{3}$ are *not like terms*, no more simplification can be done to this example.

(d) $\sqrt{6} - 8\sqrt{6} = -7\sqrt{6} \qquad \text{Subtract like terms}$ ◄

Consider the addition problem

$$2\sqrt{3} + \sqrt{27}$$

Since $\sqrt{3}$ and $\sqrt{27}$ are not like radicals, it seems that the expression will not simplify. However, the radical $\sqrt{27}$ will simplify. Sometimes simplifying the individual radicals will help simplify an expression.

$$2\sqrt{3} + \sqrt{27} = 2\sqrt{3} + \sqrt{9 \cdot 3}$$
$$= 2\sqrt{3} + 3\sqrt{3} \qquad \text{Simplify } \sqrt{27}$$
$$= 5\sqrt{3} \qquad \text{Add like terms}$$

► **EXAMPLE 2**

Perform the indicated operation, if possible.

(a) $\sqrt{5} + \sqrt{45}$ (b) $\sqrt{32} - \sqrt{8}$

(c) $4\sqrt[3]{16} - 3\sqrt[3]{54} + 5\sqrt[3]{2}$ (d) $\sqrt[4]{16} + \sqrt[4]{32}$

Solutions

(a) $\sqrt{5} + \sqrt{45} = \sqrt{5} + \sqrt{9 \cdot 5}$
$$= \sqrt{5} + 3\sqrt{5} \qquad \text{Simplify } \sqrt{45}$$
$$= 4\sqrt{5} \qquad \text{Add like terms}$$

(b) $\sqrt{32} - \sqrt{8} = \sqrt{16 \cdot 2} - \sqrt{4 \cdot 2}$
$$= 4\sqrt{2} - 2\sqrt{2} \qquad \text{Simplify radicals}$$
$$= 2\sqrt{4} \qquad \text{Subtract like terms}$$

(c) $4\sqrt[3]{16} - 3\sqrt[3]{54} + 5\sqrt[3]{2} = 4\sqrt[3]{8 \cdot 2} - 3\sqrt[3]{27 \cdot 2} + 5\sqrt[3]{2}$
$$= 4 \cdot 2 \sqrt[3]{2} - 3 \cdot 3 \sqrt[3]{2} + 5 \sqrt[3]{2} \quad \text{Simplify radicals}$$
$$= 8 \sqrt[3]{2} - 9 \sqrt[3]{2} + 5 \sqrt[3]{2}$$
$$= 4 \sqrt[3]{2} \qquad \text{Combine like terms}$$

(d) $\sqrt[4]{16} + \sqrt[4]{32} = 2 + \sqrt[4]{16 \cdot 2}$
$$= 2 + 2\sqrt[4]{2} \qquad \text{Simplify } \sqrt[4]{32}$$

The order of operations will not let us simplify this expression further.

◄

Radical expressions may contain radicals that contain letters that represent variables or unknown constants. Remember, we can gather like terms only when everything except the numerical coefficients is the same.

► **EXAMPLE 3**

Perform the indicated operation, if possible. All letters represent nonnegative real numbers.

(a) $2x\sqrt{3x} + \sqrt{12x^3}$ (b) $2\sqrt{5x^3} - x\sqrt{80x}$

(c) $2\sqrt[4]{x^5y^3} - 3x\sqrt[4]{16xy^3}$ (d) $3\sqrt{8x^2y} + x\sqrt{2y^3} - x\sqrt{50y}$

Solutions

(a) $2x\sqrt{3x} + \sqrt{12x^3} = 2x\sqrt{3x} + \sqrt{4 \cdot 3x^2 \cdot x}$

$\qquad\qquad\qquad\quad = 2x\sqrt{3x} + 2x\sqrt{3x}$ \qquad Simplify a radical

$\qquad\qquad\qquad\quad = 4x\sqrt{3x}$ \qquad\qquad\qquad Combine like terms

(b) $2\sqrt{5x^3} - x\sqrt{80x} = 2\sqrt{x^2 \cdot 5x} - x\sqrt{16 \cdot 5x}$

$\qquad\qquad\qquad\quad\; = 2x\sqrt{5x} - 4x\sqrt{5x}$ \qquad Simplify radicals

$\qquad\qquad\qquad\quad\; = -2x\sqrt{5x}$ \qquad\qquad\quad Combine like terms

(c) $2\sqrt[4]{x^5 y^3} - 3x\sqrt[4]{16xy^3} = 2\sqrt[4]{x^4 xy^3} - 3x\sqrt[4]{16xy^3}$

$\qquad\qquad\qquad\qquad\quad\;\; = 2x\sqrt[4]{xy^3} - 6x\sqrt[4]{xy^3}$

$\qquad\qquad\qquad\qquad\quad\;\; = -4x\sqrt[4]{xy^3}$

(d) $3\sqrt{8x^2 y} + x\sqrt{2y^3} - x\sqrt{50y} = 6x\sqrt{2y} + xy\sqrt{2y} - 5x\sqrt{2y}$

$\qquad\qquad\qquad\qquad\qquad\qquad\;\; = x\sqrt{2y} + xy\sqrt{2y}$

This expression will not combine further because the terms with x and xy are not like terms. However, we could factor the final expression to $x\sqrt{2y}(1 + y)$ if we wished. ◀

Problem Set 4.2

Warm-Ups

In Problems 1 through 4, perform the indicated operation, if possible. See Example 1.

1. $3\sqrt{13} + 5\sqrt{13}$

2. $4\sqrt{12} - 3\sqrt{12}$

3. $11\sqrt{3} - 19\sqrt{3}$

4. $7 - 2\sqrt{5}$

In Problems 5 through 8, perform the indicated operation, if possible. See Example 2.

5. $\sqrt{54} + \sqrt{6}$

6. $\sqrt{27} - \sqrt{18}$

7. $\sqrt[3]{-8} + \sqrt[3]{81}$

8. $3\sqrt[4]{32} - 2\sqrt[4]{162} + \sqrt[4]{625}$

In Problems 9 through 12, perform the indicated operation, if possible. All letters represent nonnegative real numbers. See Example 3.

9. $\sqrt{8t^3} + 3t\sqrt{2t}$

10. $5\sqrt{2x^5} + 2x\sqrt{18x^3}$

11. $3\sqrt[3]{8x^4 y^2} - 5x\sqrt[3]{-xy^2}$

12. $5x\sqrt{27y^5} - xy\sqrt{3y^3} + 2y^2\sqrt{12x^2 y}$

Practice Exercises

In Problems 13 through 36, perform the indicated operation, if possible. All letters represent nonnegative real numbers.

13. $3\sqrt{13} + 8\sqrt{13}$

14. $2\sqrt{11} + 11\sqrt{11}$

15. $4\sqrt{18} - 5\sqrt{18}$

16. $3\sqrt{8} - 7\sqrt{8}$

17. $\sqrt{75} + 6\sqrt{12}$

18. $\sqrt{45} + 3\sqrt{80}$

19. $\sqrt[3]{128} - \sqrt[3]{16}$

20. $\sqrt[4]{243} - \sqrt[4]{48}$

21. $9\sqrt{3x} + \sqrt{27x}$

22. $2\sqrt{8t} - \sqrt{2t}$

23. $2\sqrt{24} + 3\sqrt{54} - \sqrt{6}$

24. $3\sqrt{80} - 5\sqrt{5} + \sqrt{20}$

25. $t^2\sqrt{8t} + 2t\sqrt{2t^3} - \sqrt{2t^5}$

26. $x^5\sqrt{12x} - 5x^3\sqrt{27x^5} + 17x\sqrt{3x^9}$

27. $\dfrac{3}{8}\sqrt{48x^3} + \dfrac{1}{2}x\sqrt{3x}$

28. $\dfrac{2r}{9}\sqrt{18r} - \dfrac{1}{3}\sqrt{2r^3}$

29. $2\sqrt[3]{54} - \sqrt[3]{250} + 7\sqrt[3]{2}$

30. $8\sqrt[3]{16} + \sqrt[3]{128} - 3\sqrt[3]{54}$

31. $\sqrt[3]{-192} - 2\sqrt[3]{-375}$

32. $\sqrt[3]{-320} - \sqrt[3]{135}$

33. $\dfrac{2}{9}\sqrt{18x} + \dfrac{1}{3}\sqrt{8x}$

34. $\dfrac{5t}{6}\sqrt{8s^3t^3} - \dfrac{t^2}{9}\sqrt{18s^3t}$

35. $6\sqrt{\dfrac{8x^3y}{81}} - 5x\sqrt{\dfrac{32xy}{225}}$

36. $14k\sqrt{\dfrac{27kp}{16}} - \dfrac{15}{4}\sqrt{\dfrac{12k^3p}{25}}$

Challenge Problems

In Problems 37 through 40, perform the indicated operation, if possible. All letters represent nonnegative real numbers, with m and n integers.

37. $x^m\sqrt{x^{2n}} + x^n\sqrt{x^{2m}}$

38. $\sqrt[n]{4^n x} + \sqrt[n]{2^{2n}x}$

39. $3 x^n\sqrt[n]{x^{mn}} - x^m\sqrt[m]{x^{mn}}$

40. $\sqrt[3]{5^{3n}} - \sqrt{5^{2n}}$

In Your Own Words

41. Explain "like terms" and "unlike terms" when dealing with radical expressions.

4.3 Multiplication of Radical Expressions

TAPE 8

Next, we look at multiplication. The *root of a product* property in Section 4.1 shows how to multiply radicals with the same index.

> ### Multiplication of Radicals
>
> If $\sqrt[n]{p}$ and $\sqrt[n]{q}$ represent real numbers, then
>
> $$\sqrt[n]{p} \cdot \sqrt[n]{q} = \sqrt[n]{pq}$$

▶ EXAMPLE 1

Perform the operations indicated. Assume that all letters represent nonnegative real numbers.

(a) $\sqrt{x} \cdot \sqrt{x}$ (b) $\sqrt{2xy^3} \cdot \sqrt{8xy}$ (c) $\sqrt[3]{A} \cdot \sqrt[3]{A^2}$

(d) $\sqrt[4]{10x^3y} \cdot \sqrt[4]{1000xy^2}$ (e) $(5\sqrt{3})^2$

Solutions

(a) $\sqrt{x} \cdot \sqrt{x} = \sqrt{x^2}$ Multiplication of radicals

       $= x$ Definition of $\sqrt{}$

(b) $\sqrt{2xy^3} \cdot \sqrt{8xy} = \sqrt{16x^2y^4}$ Multiplication of radicals

         $= 4xy^2$

(c) $\sqrt[3]{A} \cdot \sqrt[3]{A^2} = \sqrt[3]{A^3}$ Multiplication of radicals

        $= A$

(d) $\sqrt[4]{10x^3y} \cdot \sqrt[4]{1000xy^2} = \sqrt[4]{10000x^4y^3}$

           $= 10x\sqrt[4]{y^3}$

(e) $(5\sqrt{3})^2 = 5^2 \cdot (\sqrt{3})^2$ Product to a power

        $= 25(3) = 75$ ◄

      Since radicals represent numbers, the properties of real numbers apply. The next example shows the distributive property at work.

► EXAMPLE 2

Perform the operations indicated. Assume that all letters represent nonnegative real numbers.

(a) $\sqrt{2}(\sqrt{3} + \sqrt{5})$ (b) $(\sqrt[3]{3} - \sqrt[3]{9x^2y})\sqrt[3]{9}$

Solutions

(a) $\sqrt{2}(\sqrt{3} + \sqrt{5}) = \sqrt{2} \cdot \sqrt{3} + \sqrt{2} \cdot \sqrt{5}$

            $= \sqrt{6} + \sqrt{10}$

(b) $(\sqrt[3]{3} - \sqrt[3]{9x^2y})\sqrt[3]{9} = \sqrt[3]{27} - \sqrt[3]{81x^2y}$

            $= 3 - 3\sqrt[3]{3x^2y}$ ◄

      Multiplication of radical expressions involving several terms follows the same pattern as the multiplication of polynomials. Likewise, special products also work.

► EXAMPLE 3

Perform the operations indicated. Assume that all letters represent nonnegative real numbers.

(a) $(\sqrt{2} + 2\sqrt{3})(5\sqrt{2} - \sqrt{3})$ (b) $(\sqrt{3} + 2\sqrt{7})^2$

(c) $(2 - \sqrt{3})(2 + \sqrt{3})$

Solutions

(a) The same pattern is used as when multiplying two binomials.

$$(\sqrt{2} + 2\sqrt{3})(5\sqrt{2} - \sqrt{3}) = \sqrt{2} \cdot 5\sqrt{2} - \sqrt{2} \cdot \sqrt{3} + 2\sqrt{3} \cdot 5\sqrt{2}$$
$$- 2\sqrt{3} \cdot \sqrt{3}$$
$$= 5\sqrt{4} - \sqrt{6} + 10\sqrt{6} - 2\sqrt{9}$$
$$= 10 + 9\sqrt{6} - 6$$
$$= 4 + 9\sqrt{6}$$

(b) $(\sqrt{3} + 2\sqrt{7})^2$

We must be careful here. The same pattern holds as when squaring a binomial.

$$(A + B)^2 = A^2 + 2 \cdot A \cdot B + B^2$$
$$(\sqrt{3} + 2\sqrt{7})^2 = (\sqrt{3})^2 + 2 \cdot \sqrt{3} \cdot 2\sqrt{7} + (3\sqrt{7})^2$$
$$= 3 + 4\sqrt{21} + 9 \cdot 7$$
$$= 66 + 4\sqrt{21}$$

(c) $(2 - \sqrt{3})(2 + \sqrt{3})$

Look carefully; this is the difference in two squares.

$$(2 - \sqrt{3})(2 + \sqrt{3}) = 2^2 - (\sqrt{3})^2$$
$$= 4 - 3$$
$$= 1$$

◄

The pair of numbers $2 - \sqrt{3}$ and $2 + \sqrt{3}$ in Example 3c is called **conjugates**.

Problem Set 4.3

In this Problem Set, assume that all letters represent nonnegative real numbers.

Warm-Ups

In Problems 1 through 6, perform the operations indicated. See Example 1.

1. $2\sqrt{z} \cdot \sqrt{z}$

2. $\sqrt{6xy} \cdot \sqrt{3yz}$

3. $\sqrt[3]{24xy^2} \cdot \sqrt[3]{9xy}$

4. $\sqrt[3]{-16x} \cdot \sqrt[3]{4x^2}$

5. $(3\sqrt{7})^2$

6. $(2\sqrt{2})^2$

In Problems 7 through 10, perform the operations indicated. See Example 2.

7. $\sqrt{3}(\sqrt{2} - \sqrt{3})$

8. $(2\sqrt{3x} + 3\sqrt{2x})\sqrt{6x}$

9. $2\sqrt{3}(3\sqrt{3} + \sqrt{6})$

10. $\sqrt[3]{3}(\sqrt[3]{9} - \sqrt[3]{18x})$

In Problems 11 through 14, perform the operations indicated. See Example 3.

11. $(2 + \sqrt{3})(3 - \sqrt{3})$

12. $(\sqrt{2} - 2\sqrt{3})(3\sqrt{2} + \sqrt{3})$

13. $(2\sqrt{3} - 3\sqrt{2})^2$

14. $(\sqrt{5} + 2\sqrt{2})(\sqrt{5} - 2\sqrt{2})$

Practice Exercises

In Problems 15 through 54, perform the operations indicated.

15. $\sqrt{x} \cdot \sqrt{x} \cdot \sqrt{x}$

16. $\sqrt[3]{y} \cdot \sqrt[3]{y} \cdot \sqrt[3]{y}$

17. $\sqrt{7xy^3} \cdot 2\sqrt{14xy}$

18. $3\sqrt{15st} \cdot \sqrt{5s^3t}$

19. $\sqrt{7}(2\sqrt{2} - \sqrt{3})$

20. $\sqrt{5}(3\sqrt{3} + \sqrt{2})$

21. $\sqrt[3]{3s^2t^4} \cdot \sqrt[3]{18s}$

22. $\sqrt[4]{8x^2y^3} \cdot \sqrt[4]{8x^2y^5}$

23. $(3\sqrt{3})^2$

24. $(2\sqrt{5})^2$

25. $(1 - \sqrt{2})(1 - 3\sqrt{2})$

26. $(\sqrt{3} + 1)(2\sqrt{3} - 1)$

27. $(s - \sqrt{2t})^2$

28. $(\sqrt{3x} + y)^2$

29. $\sqrt{6}(2\sqrt{2} - \sqrt{3})$

30. $\sqrt{3}(3\sqrt{6} + \sqrt{3})$

31. $2\sqrt{3}(2\sqrt{5} + 4\sqrt{3})$

32. $3\sqrt{7}(2\sqrt{7} + 7\sqrt{5})$

33. $\sqrt[3]{2x^2y^2}(\sqrt[3]{4xy^4} + 2y\sqrt[3]{12xy})$

34. $\sqrt[3]{3x^2y}(\sqrt[3]{9xy^2} + 2y\sqrt[3]{54xy})$

35. $\sqrt[5]{4k^9}\left(2\sqrt[5]{8k} - \sqrt[5]{\dfrac{1}{4}k}\right)$

36. $\sqrt[3]{5z^4}\left(\sqrt[3]{25z^3} + \sqrt[3]{\dfrac{2}{5}z^2}\right)$

37. $(2\sqrt{3} + 3)(2\sqrt{3} - 2)$

38. $(1 + 3\sqrt{5})(1 - 2\sqrt{5})$

39. $(x\sqrt{2} - y\sqrt{3})(x\sqrt{2} + y\sqrt{3})$

40. $(x\sqrt{5} - y\sqrt{2})(x\sqrt{5} + y\sqrt{2})$

41. $(2\sqrt{2x} + \sqrt{3})^2$

42. $(\sqrt{3x} - 3\sqrt{2})^2$

43. $(\sqrt{3} + 2\sqrt{6})(2\sqrt{3} - 3\sqrt{6})$

44. $(5\sqrt{2} - 2\sqrt{5})(2\sqrt{2} + 5\sqrt{5})$

45. $(\sqrt{2} + 1)(\sqrt{2} - 2)(\sqrt{2} - 1)$

46. $(3 + \sqrt{3})(1 - \sqrt{3})(3 - \sqrt{3})$

47. $\left(\dfrac{1}{2} - \sqrt{2}\right)\left(2 - \dfrac{1}{2}\sqrt{2}\right)$

48. $\left(\dfrac{2}{3} + \sqrt{3}\right)\left(1 - \dfrac{1}{3}\sqrt{3}\right)$

49. $(\sqrt[3]{2} + 1)^3$

50. $(1 + \sqrt[3]{3})^3$

51. $(\sqrt[3]{5} + 1)^2(\sqrt[3]{5} + 2)$

52. $(1 - \sqrt[3]{2})^2(2 + \sqrt[3]{2})$

53. $(\sqrt[3]{2} - \sqrt[3]{3})^3$

54. $(\sqrt[3]{2} + \sqrt[3]{4})^3$

In Problems 55 through 60, factor each expression.

55. $\sqrt{10} + \sqrt{6}$

56. $\sqrt{14} + \sqrt{6}$

57. $\sqrt{6} - \sqrt{3}$

58. $\sqrt{10} - \sqrt{5}$

59. $\sqrt{12} + \sqrt{20}$

60. $\sqrt{18} + \sqrt{27}$

Challenge Problems

In Problems 61 and 62, perform the indicated operation.

61. $(\sqrt[3]{3} - \sqrt[3]{2})(\sqrt[3]{9} + \sqrt[3]{6} + \sqrt[3]{4})$

62. $(\sqrt[3]{3} + \sqrt[3]{2})(\sqrt[3]{9} - \sqrt[3]{6} + \sqrt[3]{4})$

In Problems 63 and 64, multiply each expression, and simplify. Each problem involves a factoring rule. Identify it.

63. $(\sqrt[3]{a} - \sqrt[3]{b})(\sqrt[3]{a^2} + \sqrt[3]{ab} + \sqrt[3]{b^2})$

64. $(\sqrt[3]{a} + \sqrt[3]{b})(\sqrt[3]{a^2} - \sqrt[3]{ab} + \sqrt[3]{b^2})$

In Your Own Words

65. Compare the two expressions $(\sqrt{x} + \sqrt{y})^2$ and $(\sqrt{x})^2 + (\sqrt{y})^2$. What lesson can be drawn from comparison?

4.4   Division of Radical Expressions and Rationalizing

TAPE 8

The next operation with rational expressions that we will discuss is division. Since division is a form of multiplication, similar rules apply. To divide two rational expressions with the **same** index, we use the *root of a quotient* property of Section 4.1.

> **Division of Radicals**
>
> If $\sqrt[n]{p}$ and $\sqrt[n]{q}$ represent real numbers and $q \neq 0$, then
>
> $$\frac{\sqrt[n]{p}}{\sqrt[n]{q}} = \sqrt[n]{\frac{p}{q}}$$

► **EXAMPLE 1**

Perform each division and simplify.

(a) $\dfrac{\sqrt{24}}{\sqrt{2}}$ (b) $\dfrac{\sqrt[3]{54}}{\sqrt[3]{2}}$

Solutions

(a) $\dfrac{\sqrt{24}}{\sqrt{2}} = \sqrt{\dfrac{24}{2}}$ Division of radicals

$\qquad = \sqrt{12}$

$\qquad = 2\sqrt{3}$

(b) $\dfrac{\sqrt[3]{54}}{\sqrt[3]{2}} = \sqrt[3]{\dfrac{54}{2}}$ Division of radicals

$\qquad = \sqrt[3]{27}$

$\qquad = 3$

◄

► **EXAMPLE 2**

Perform each operation and simplify, if possible.

(a) $\dfrac{\sqrt{6} + \sqrt{10}}{\sqrt{2}}$ (b) $\dfrac{2\sqrt{3} - 3\sqrt{6}}{\sqrt{3}}$ (c) $\dfrac{\sqrt{5}}{\sqrt{10} + 3\sqrt{5}}$

Solutions

(a) Quotients of this form may be rewritten by using the addition property in reverse. That is, $\dfrac{A + B}{C} = \dfrac{A}{C} + \dfrac{B}{C}$.

$$\frac{\sqrt{6} + \sqrt{10}}{\sqrt{2}} = \frac{\sqrt{6}}{\sqrt{2}} + \frac{\sqrt{10}}{\sqrt{2}}$$

$$= \sqrt{\frac{6}{2}} + \sqrt{\frac{10}{2}} \qquad \text{Division of radicals}$$

$$= \sqrt{3} + \sqrt{5}$$

(b) $$\frac{2\sqrt{3} - 3\sqrt{6}}{\sqrt{3}} = \frac{2\sqrt{3}}{\sqrt{3}} - \frac{3\sqrt{6}}{\sqrt{3}}$$

$$= 2 - 3\sqrt{2}$$

The order of operations will not allow further simplification.

(c) The approach used in parts (a) and (b) will not work here because the sum is in the *denominator*. However, we can factor the denominator.

$$\frac{\sqrt{5}}{\sqrt{10} + 3\sqrt{5}} = \frac{\sqrt{5}}{\sqrt{5} \cdot \sqrt{2} + 3\sqrt{5}} \qquad \text{Factor } \sqrt{10}$$

$$= \frac{\sqrt{5}}{\sqrt{5}(\sqrt{2} + 3)} \qquad \text{Common factor in denominator}$$

$$= \frac{1}{\sqrt{2} + 3} \qquad \text{Fundamental Principle of Fractions} \blacktriangleleft$$

Fractions with radicals in the denominator, such as the answer to part (c) of Example 2, are sometimes considered not simplified. The process of removing a radical from the denominator of a fraction is called **rationalizing the denominator.**

The first technique we will learn for rationalizing the denominator applies if a **square root** is a **factor** of the denominator.

► **EXAMPLE 3**

Rationalize the denominator.

(a) $\dfrac{5}{\sqrt{6}}$ (b) $\dfrac{2\sqrt{3}}{5\sqrt{2}}$ (c) $\sqrt{\dfrac{7}{12}}$

Solutions

(a) $\sqrt{6}$ is a factor of the denominator. Multiply the numerator and denominator by this factor.

$$\frac{5}{\sqrt{6}} = \frac{5 \cdot \sqrt{6}}{\sqrt{6} \cdot \sqrt{6}} = \frac{5\sqrt{6}}{6}$$

(b) $\sqrt{2}$ is a factor of the denominator. Multiply the numerator and denominator by this factor.

$$\frac{2\sqrt{3}}{5\sqrt{2}} = \frac{2\sqrt{3} \cdot \sqrt{2}}{5\sqrt{2} \cdot \sqrt{2}}$$

$$= \frac{2\sqrt{6}}{5 \cdot 2} = \frac{\sqrt{6}}{5}$$

(c) First, use the division property of radicals and simplify the denominator.

$$\sqrt{\frac{7}{12}} = \frac{\sqrt{7}}{\sqrt{12}} = \frac{\sqrt{7}}{2\sqrt{3}}$$

$\sqrt{3}$ is a factor of the denominator.

$$= \frac{\sqrt{7} \cdot \sqrt{3}}{2\sqrt{3} \cdot \sqrt{3}} = \frac{\sqrt{21}}{2 \cdot 3} = \frac{\sqrt{21}}{6} \qquad \blacktriangleleft$$

If a radical of index higher than 2 is a **factor** of the denominator, we modify the procedure as shown in the next example.

▶ **EXAMPLE 4**

Rationalize each denominator.

(a) $\dfrac{1}{\sqrt[3]{4}}$ (b) $\dfrac{3\sqrt[4]{7}}{2\sqrt[4]{3}}$

Solutions

(a) Since $4 = 2^2$, we can get $\sqrt[3]{2^3}$ in the denominator by multiplying the numerator and denominator by $\sqrt[3]{2}$. Then we will have a perfect third power in the denominator radicand.

$$\frac{1}{\sqrt[3]{4}} = \frac{1 \cdot \sqrt[3]{2}}{\sqrt[3]{4} \cdot \sqrt[3]{2}}$$

$$= \frac{\sqrt[3]{2}}{\sqrt[3]{8}}$$

$$= \frac{\sqrt[3]{2}}{2}$$

(b) Multiply by $\sqrt[4]{3^3}$ to make a perfect fourth power in the denominator radicand.

$$\frac{3\sqrt[4]{7}}{2\sqrt[4]{3}} = \frac{3\sqrt[4]{7} \cdot \sqrt[4]{3^3}}{2\sqrt[4]{3} \cdot \sqrt[4]{3^3}}$$

$$= \frac{3\sqrt[4]{7 \cdot 3^3}}{2\sqrt[4]{3 \cdot 3^3}}$$

$$= \frac{3\sqrt[4]{7 \cdot 27}}{2\sqrt[4]{3^4}}$$

$$= \frac{3\sqrt[4]{189}}{2 \cdot 3}$$

$$= \frac{\sqrt[4]{189}}{2} \qquad \blacktriangleleft$$

The third technique for rationalizing denominators applies if the denominator is a sum or difference of two terms containing **square roots.** We take advantage of the special product the *difference of two squares.* That is,

$$(a - b)(a + b) = a^2 - b^2$$

► **EXAMPLE 5**

Rationalize the denominator.

(a) $\dfrac{1}{5 - \sqrt{2}}$ (b) $\dfrac{2\sqrt{3} - \sqrt{2}}{2\sqrt{3} + \sqrt{2}}$

Solutions

(a) Multiply both the numerator and denominator by the number $5 + \sqrt{2}$, which is the conjugate of the denominator.

$$\frac{1}{5 - \sqrt{2}} = \frac{1(5 + \sqrt{2})}{(5 - \sqrt{2})(5 + \sqrt{2})}$$

Notice that the denominator is now the difference of two squares.

$$= \frac{5 + \sqrt{2}}{5^2 - (\sqrt{2})^2} = \frac{5 + \sqrt{2}}{25 - 2}$$

$$= \frac{5 + \sqrt{2}}{23}$$

(b) Again, multiply the numerator and denominator by the conjugate of the denominator.

$$\frac{2\sqrt{3} - \sqrt{2}}{2\sqrt{3} + \sqrt{2}} = \frac{(2\sqrt{3} - \sqrt{2})(2\sqrt{3} - \sqrt{2})}{(2\sqrt{3} + \sqrt{2})(2\sqrt{3} - \sqrt{2})}$$

$$= \frac{(2\sqrt{3})^2 - 2 \cdot 2\sqrt{3} \cdot \sqrt{2} + (\sqrt{2})^2}{(2\sqrt{3})^2 - (\sqrt{2})^2}$$

$$= \frac{12 - 4\sqrt{6} + 2}{12 - 2}$$

$$= \frac{14 - 4\sqrt{6}}{12 - 2}$$

$$= \frac{14 - 4\sqrt{6}}{10}$$

Be Careful!

Because 2 is a factor of both the numerator and denominator, this fraction can be simplified. However, *be careful* either to factor the numerator first or to write the fraction as two fractions.

$$= \frac{14}{10} - \frac{4\sqrt{6}}{10}$$

$$= \frac{7}{5} - \frac{2\sqrt{6}}{5}$$

◄

Before the advent of the hand-held calculator, rationalizing denominators was an important aid in calculation. It is much easier to approximate $\frac{1}{2}\sqrt{10}$ with a table of square roots than it is to approximate $\frac{\sqrt{5}}{\sqrt{2}}$. Although today most of us have calculators, it is still important to develop these skills in manipulating radicals. In further studies, it will be necessary to rationalize the *numerator* as well as the denominator.

► **EXAMPLE 6**

Rationalize each *numerator*. Simplify, if possible.

(a) $\dfrac{\sqrt{5}}{x}$ (b) $\sqrt{\dfrac{7}{12}}$ (c) $\dfrac{\sqrt{7}+2}{\sqrt{7}-2}$

Solutions

In each part we use the same techniques as when rationalizing denominators, except that we rationalize the *numerator*.

(a) $\dfrac{\sqrt{5}}{x} = \dfrac{\sqrt{5}\cdot\sqrt{5}}{x\cdot\sqrt{5}}$

$\qquad = \dfrac{5}{\sqrt{5}\,x}$

(b) $\sqrt{\dfrac{7}{12}} = \dfrac{\sqrt{7}}{\sqrt{12}}$

$\qquad = \dfrac{\sqrt{7}\cdot\sqrt{7}}{\sqrt{12}\cdot\sqrt{7}}$

$\qquad = \dfrac{7}{\sqrt{84}}$

$\qquad = \dfrac{7}{2\sqrt{21}}$

(c) $\dfrac{\sqrt{7}+2}{\sqrt{7}-2} = \dfrac{(\sqrt{7}+2)(\sqrt{7}-2)}{(\sqrt{7}-2)(\sqrt{7}-2)}$

$\qquad = \dfrac{(\sqrt{7})^{2}-2^{2}}{(\sqrt{7})^{2}-2\cdot2\cdot\sqrt{7}+2^{2}}$

$\qquad = \dfrac{7-4}{7-4\sqrt{7}+4}$

$\qquad = \dfrac{3}{11-4\sqrt{7}}$ ◄

Problem Set 4.4

Warm-Ups

In Problems 1 through 4, perform the indicated operation. See Example 1.

1. $\dfrac{\sqrt{48}}{\sqrt{6}}$

2. $\dfrac{\sqrt[3]{24}}{\sqrt[3]{-3}}$

3. $\dfrac{\sqrt{90}}{\sqrt{5}}$

4. $\dfrac{\sqrt[3]{108}}{\sqrt[3]{4}}$

In Problems 5 and 6, perform the indicated operation. See Example 2.

5. $\dfrac{\sqrt{6} - \sqrt{15}}{\sqrt{3}}$

6. $\dfrac{\sqrt{2} + 4\sqrt{14}}{\sqrt{2}}$

In Problems 7 through 10, rationalize each denominator. See Example 3.

7. $\dfrac{1}{\sqrt{5}}$

8. $\dfrac{2}{3\sqrt{3}}$

9. $\dfrac{6}{2\sqrt{6}}$

10. $\sqrt{\dfrac{7}{8}}$

In Problems 11 through 14, rationalize each denominator. See Example 4.

11. $\dfrac{1}{\sqrt[3]{7}}$

12. $\dfrac{1}{\sqrt[3]{36}}$

13. $\dfrac{1}{\sqrt[4]{4}}$

14. $\dfrac{x^2}{\sqrt[5]{8}}$

In Problems 15 through 18, rationalize each denominator. See Example 5.

15. $\dfrac{5}{\sqrt{5} + 2}$

16. $\dfrac{2\sqrt{2}}{1 - \sqrt{5}}$

17. $\dfrac{2\sqrt{3}}{\sqrt{2} - \sqrt{3}}$

18. $\dfrac{2 - \sqrt{6}}{\sqrt{3} - \sqrt{2}}$

In Problems 19 through 20, rationalize each numerator. See Example 6.

19. $\dfrac{\sqrt{3}}{\sqrt{3} + 3\sqrt{2}}$

20. $\dfrac{\sqrt{5} - \sqrt{2}}{\sqrt{5} + \sqrt{2}}$

Practice Exercises

In Problems 21 through 40, perform the indicated operation. Assume that all letters represent nonnegative real numbers.

21. $\dfrac{\sqrt{72}}{\sqrt{24}}$

22. $\dfrac{\sqrt{30}}{\sqrt{15}}$

23. $\dfrac{\sqrt[3]{32}}{\sqrt[3]{2}}$

24. $\dfrac{\sqrt[4]{243}}{\sqrt[4]{3}}$

25. $\dfrac{\sqrt{120}}{\sqrt{10}}$

26. $\dfrac{\sqrt{90}}{\sqrt{5}}$

27. $\dfrac{\sqrt{6} - \sqrt{3}}{\sqrt{3}}$

28. $\dfrac{\sqrt{8} + \sqrt{2}}{\sqrt{2}}$

29. $\dfrac{\sqrt[3]{-56}}{\sqrt[3]{-7}}$

30. $\dfrac{\sqrt[3]{135}}{\sqrt[3]{-5}}$

31. $\dfrac{\sqrt{15x}}{\sqrt{3x}}$

32. $\dfrac{\sqrt{21rt}}{\sqrt{7r}}$

33. $\dfrac{6\sqrt{7} + 9\sqrt{21}}{3\sqrt{7}}$

34. $\dfrac{8\sqrt{5} - 6\sqrt{15}}{2\sqrt{5}}$

35. $\dfrac{x\sqrt{6xy^3}}{\sqrt{3x^3}}$

36. $\dfrac{w\sqrt{12w^3z}}{\sqrt{4w^5}}$

37. $\dfrac{3\sqrt{12}}{2\sqrt{3}}$

38. $\dfrac{5\sqrt{18}}{3\sqrt{2}}$

39. $\dfrac{10\sqrt{3} + 4\sqrt{15}}{2\sqrt{3}}$

40. $\dfrac{12\sqrt{11} - 24\sqrt{22}}{12\sqrt{11}}$

In Problems 41 through 60, rationalize each denominator.

41. $\dfrac{2}{\sqrt{11}}$

42. $\dfrac{3}{\sqrt{6}}$

43. $\dfrac{1}{2\sqrt{5}}$

44. $\dfrac{s}{3\sqrt{2}}$

45. $\dfrac{\sqrt{3}}{\sqrt{5}}$

46. $\dfrac{\sqrt{2}}{\sqrt{7}}$

47. $\dfrac{3}{3+\sqrt{2}}$

48. $\dfrac{2}{\sqrt{5}-3}$

49. $\dfrac{2\sqrt{5}}{5\sqrt{2}}$

50. $\dfrac{5\sqrt{11}}{2\sqrt{5}}$

51. $\dfrac{\sqrt{3}}{2+\sqrt{3}}$

52. $\dfrac{\sqrt{2}}{3-\sqrt{2}}$

53. $\dfrac{3\sqrt{2}}{\sqrt{6}-2\sqrt{2}}$

54. $\dfrac{2\sqrt{3}}{2\sqrt{3}+\sqrt{6}}$

55. $\dfrac{1-\sqrt{5}}{1+\sqrt{5}}$

56. $\dfrac{\sqrt{3}+2}{\sqrt{3}-2}$

57. $\dfrac{1}{\sqrt[3]{2}}$

58. $\dfrac{1}{\sqrt[3]{3}}$

59. $\dfrac{4}{\sqrt[4]{8}}$

60. $\dfrac{9}{\sqrt[4]{27}}$

In Problems 61 through 70, rationalize each numerator.

61. $\dfrac{\sqrt{6}}{5}$

62. $\dfrac{\sqrt{3}}{17}$

63. $\dfrac{\sqrt{2}}{1-\sqrt{2}}$

64. $\dfrac{\sqrt{5}}{\sqrt{5}+\sqrt{2}}$

65. $\dfrac{-5\sqrt{3}}{3+\sqrt{6}}$

66. $\dfrac{-7\sqrt{2}}{2\sqrt{2}-\sqrt{6}}$

67. $\dfrac{3+\sqrt{5}}{5-\sqrt{5}}$

68. $\dfrac{2-\sqrt{3}}{2+\sqrt{3}}$

69. $\dfrac{2\sqrt{2}+\sqrt{3}}{\sqrt{2}-3\sqrt{3}}$

70. $\dfrac{\sqrt{5}-\sqrt{3}}{2\sqrt{3}+\sqrt{5}}$

Challenge Problems

Problems 71 and 72 illustrate an important manipulation in calculus. In each problem, rationalize the numerator and simplify.

71. $\dfrac{\sqrt{x+h}-\sqrt{x}}{h}, \quad h\neq 0$

72. $\dfrac{\sqrt{x}-\sqrt{x_0}}{x-x_0}, \quad x\neq x_0$

*In Problem 73 rationalize the denominator, and in Problem 74, rationalize the numerator. (*HINT: *Use the results of Problems 63 and 64 of Section 4.3)*

73. $\dfrac{1}{\sqrt[3]{3}-\sqrt[3]{2}}$

74. $\dfrac{\sqrt[3]{2}+1}{\sqrt[3]{2}}$

In Your Own Words

75. What does it mean to "rationalize" a denominator or a numerator?

4.5 **Equations Containing Square Roots**

TAPE 8

The equation $\sqrt{x + 1} = 2$ cannot be solved by adding a number to both sides or by multiplying both sides by the same nonzero number. Somehow we must undo the square root.

First we have to ask: how do the solutions of the equation $P = Q$ compare to the solutions of the equation $P^2 = Q^2$? What happens if both sides of an equation are squared?

Squaring Property

If the equation $P = Q$ has solutions, these solutions are found in the solutions of the equation $P^2 = Q^2$.

Consider the equation

$$x = 1$$

Its solution set is

$$\{1\}$$

Now if both sides of the equation are squared, it becomes

$$x^2 = 1$$

and we see by inspection that the solution set of the "squared" equation is

$$\{-1, 1\}$$

Note that 1 is a solution to $x = 1$, but -1 is not.

Notice that the solutions of the "squared" equation $P^2 = Q^2$ are *possible* solutions of the equation $P = Q$, but some solutions of the squared equation *may not* be solutions of the original equation.

The general strategy to solve an equation with variables under a square root is to solve the squared equation to find all possible solutions and then *check each* possible solution in the original equation.

In Section 2.1 we discussed checking possible solutions. Be sure to use a correct procedure for checking. Be careful to evaluate the left side and the right side separately.

► **EXAMPLE 1**

Solve $\sqrt{z + 1} = 2$.

Solution

$\boxed{A}$ $\qquad\qquad\qquad\qquad \sqrt{z + 1} = 2$

Square both sides.

$\boxed{B}$ $\qquad\qquad\qquad\qquad (\sqrt{z + 1})^2 = 2^2$

$$z + 1 = 4$$

$$z = 3$$

3 is a solution to equation $\boxed{B}$. It may or may not be a solution to equation $\boxed{A}$. We **must** check to see if it is.

Check 3 in equation $\boxed{A}$. Evaluate each side when z is 3.

- LS: $\sqrt{z+1} = \sqrt{3+1} = \sqrt{4} = \boxed{2}$
- RS: $\boxed{2}$

Since the values of the LS and the RS are the same, 3 is a solution to equation $\boxed{A}$ and the solution set is

$$\{3\}$$

◄

► **EXAMPLE 2**

Solve $\sqrt{x+1} + 2 = 0$.

Solution

The first step is to isolate the radical. Subtract 2 from both sides.

$\boxed{*}$ $\qquad\qquad\qquad\qquad \sqrt{x+1} = -2$

Now square both sides and solve.

$$(\sqrt{x+1})^2 = (-2)^2$$

$$x + 1 = 4$$

$$x = 3$$

We check 3 in the original equation.

- LS: $\sqrt{x+1} + 2 = \sqrt{3+1} + 2 = 2 + 2 = \boxed{4}$
- RS: $\boxed{0}$

Since we get different values in the left side and the right side, 3 is not in the solution set of $\sqrt{x+1} + 2 = 0$. Thus the solution set is the empty set,

$$\varnothing$$

Notice that it could be seen that the solution set would be empty at the step marked $\boxed{*}$, because a square root cannot be negative. ◄

Possible solutions that do not check are often called **extraneous** solutions.

A Procedure for Solving Equations Containing Square Roots

1. Isolate one radical.

2. Square both sides.

3. Solve the resulting equation for all possible solutions.

4. Check all possible solutions in the original equation.

5. Write the solution set.

▶ **EXAMPLE 3**

Solve each equation.

(a) $\sqrt{3x + 1} - \sqrt{2x + 6} = 0$ (b) $\sqrt{x + 1} = \sqrt{x + 4}$

Solutions

Be Careful!

(a) $\sqrt{3x + 1} - \sqrt{2x + 6} = 0$

The first step (isolate one radical) is very important. Notice what happens if both sides of the original equation are squared. The left side becomes

$$(3x + 1) - 2\sqrt{3x + 1} \cdot \sqrt{2x + 6} + (2x + 6)$$

[Remember, $(p - q)^2 = p^2 - 2pq + q^2$.] This is certainly no improvement over the original! So the first thing we do is isolate one radical.

$$\sqrt{3x + 1} = \sqrt{2x + 6}$$

Square both sides and solve.

$$(\sqrt{3x + 1})^2 = (\sqrt{2x + 6})^2$$
$$3x + 1 = 2x + 6$$
$$x = 5$$

Check 5 in the original equation.

- LS: $\sqrt{3x + 1} - \sqrt{2x + 6} = \sqrt{15 + 1} - \sqrt{10 + 6} = 4 - 4 = \boxed{0}$
- RS: $\boxed{0}$

Since we get the same value in the LS and the RS, the solution set is

$$\{5\}$$

(b) $\sqrt{x + 1} = \sqrt{x + 4}$

$$(\sqrt{x + 1})^2 = (\sqrt{x + 4})^2 \qquad \text{Square both sides}$$
$$x + 1 = x + 4 \qquad \text{Solve}$$
$$1 = 4$$

Since this statement is false no matter what the value of x, the solution set is the empty set,   Ø   ◀

▶ **EXAMPLE 4**

Solve $5\sqrt{x - 3} = 3\sqrt{2x + 1}$.

Solution

$$5\sqrt{x - 3} = 3\sqrt{2x + 1}$$
$$(5\sqrt{x - 3})^2 = (3\sqrt{2x + 1})^2 \qquad \text{Square both sides}$$
$$25(x - 3) = 9(2x + 1)$$
$$25x - 75 = 18x + 9 \qquad \text{Solve}$$
$$7x = 84$$
$$x = 12$$

Check 12 in the original equation.

- LS: $5\sqrt{x-3} = 5\sqrt{12-3} = 5\sqrt{9} = 5 \cdot 3 = \boxed{15}$
- RS: $3\sqrt{2x+1} = 3\sqrt{24+1} = 3\sqrt{25} = 3 \cdot 5 = \boxed{15}$

$\{12\}$ Write the solution set   ◄

► **EXAMPLE 5**

Solve $\sqrt{x^2 + 8} - 4 = x$.

Solution

$$\sqrt{x^2 + 8} - 4 = x$$
$$\sqrt{x^2 + 8} = x + 4 \qquad \text{Isolate radical}$$
$$(\sqrt{x^2 + 8})^2 = (x + 4)^2 \qquad \text{Square both sides}$$
$$x^2 + 8 = x^2 + 8x + 16$$

Don't forget the middle term on the right side!

$$8 = 8x + 16 \qquad \text{Solve}$$
$$-8 = 8x$$
$$-1 = x$$

Check -1 in the original equation.

- LS: $\sqrt{x^2 + 8} - 4 = \sqrt{(-1)^2 + 8} - 4 = \sqrt{9} - 4 = 3 - 4$
$= \boxed{-1}$
- RS: $x = \boxed{-1}$

$\{-1\}$ Write solution set   ◄

Problem Set 4.5

Warm-Ups

In Problems 1 through 31, find the solution set. For problems 1 through 10, see Examples 1 and 2.

1. $\sqrt{x+1} = 2$ **2.** $\sqrt{x-3} = -4$ **3.** $\sqrt{2t+3} = 5$

4. $\sqrt{3x-1} = 5$ **5.** $\sqrt{2-x} = 3$ **6.** $\sqrt{4-x} = 2$

7. $\sqrt{x} = 0$ **8.** $\sqrt{x+1} = 0$ **9.** $\sqrt{x-3} = -8$

10. $\sqrt{2x+1} = 4$

For Problems 11 through 17, see Example 3.

11. $\sqrt{w+1} = \sqrt{3w-1}$ **12.** $\sqrt{2x-3} = \sqrt{6-x}$ **13.** $\sqrt{3x+2} + \sqrt{2x+7} = 0$

14. $\sqrt{3x+3} - \sqrt{4x+2} = 0$ **15.** $\sqrt{x+1} = \sqrt{2x-2}$ **16.** $\sqrt{2x-1} - \sqrt{x} = 0$

17. $\sqrt{x-1} = \sqrt{x+1}$

For Problems 18 through 21, see Example 4.

18. $2\sqrt{x-3} = \sqrt{x}$ **19.** $12 = 3\sqrt{x-1}$ **20.** $2\sqrt{2x+5} = 5\sqrt{x-6}$

21. $\sqrt{2x+3} - 3\sqrt{x-2} = 0$

For Problems 22 through 31, see Example 5.

22. $\sqrt{x^2+2} = x$ **23.** $\sqrt{x^2+3} - x = 1$ **24.** $\sqrt{x^2-2} = x+2$

25. $\sqrt{y^2-2y} = -y$ **26.** $\sqrt{x^2+24} - x = -4$ **27.** $\sqrt{x^2+x} = x$

28. $\sqrt{x^2-9} = 3-x$ **29.** $\sqrt{x^2-4} = x+2$ **30.** $\sqrt{3+x^2} + x = 3$

31. $\sqrt{4x^2+9} + 1 = 2x$

Practice Exercises

In Problems 32 through 61, find the solution set.

32. $\sqrt{x+3} = 4$ **33.** $\sqrt{z-5} = -7$ **34.** $\sqrt{2x-5} = 15$

35. $\sqrt{3x+4} = 4$ **36.** $\sqrt{3-x} = 3$ **37.** $\sqrt{1-x} = 5$

38. $\sqrt{2-3x} = 3$ **39.** $\sqrt{3-5x} = -2$ **40.** $\sqrt{x-1} = 0$

41. $\sqrt{2-t} = 0$ **42.** $\sqrt{x+4} = -5$ **43.** $\sqrt{2x-1} = 3$

44. $\sqrt{x-3} = \sqrt{2x-7}$ **45.** $\sqrt{2-x} = \sqrt{x+3}$ **46.** $\sqrt{x-7} - \sqrt{2x-18} = 0$

47. $\sqrt{1-3s} + \sqrt{2s+11} = 0$ **48.** $\sqrt{x+3} = \sqrt{2x-1}$ **49.** $\sqrt{1+3x} = \sqrt{3x}$

50. $2\sqrt{1-3x} - \sqrt{x+17} = 0$

51. $3\sqrt{6-x} = 2\sqrt{x+46}$ **52.** $\sqrt{x^2+2x} = x$ **53.** $2\sqrt{x} = \sqrt{3x+1}$

54. $2\sqrt{2x+1} - \sqrt{7x+8} = 0$ **55.** $\sqrt{x^2-1} = 1-x$ **56.** $\sqrt{x^2+3x} = x+3$

57. $\sqrt{x^2-3x} = x-3$ **58.** $\sqrt{x^2+4} = x-2$ **59.** $\sqrt{x^2+11} = x+1$

60. $\sqrt{x^2+12} - x = 6$ **61.** $\sqrt{x^2-1} = x+1$

In Problems 62 through 67, approximate x to the nearest thousandth in each equation.

62. $55.321 = \sqrt{14.007x}$ **63.** $22.822 = \sqrt{x-61.190}$ **64.** $0.00456 = 3\sqrt{\dfrac{x}{123456}}$

65. $8.631 = 15.889\sqrt{6.331-x}$ **66.** $\sqrt{\dfrac{x+1}{x-1}} = \pi$ **67.** $\sqrt{2x+1.133} - \sqrt{x+4} = 0$

68. The period of a simple pendulum is given by $T = 2\pi\sqrt{\dfrac{L}{32}}$, where T is the period in seconds and L is the length of the pendulum in feet. Find the length of a pendulum whose period is 2 seconds.

69. The velocity of a wave in shallow water is given by the formula $v = \sqrt{32h}$, where v is the velocity of the wave and h is the wave height. Find the wave height in feet if the velocity is 4 ft/sec.

70. A formula to describe the number of facts, represented by y, learned in x hours is $y = 8\sqrt{x}$. Find how many facts were learned after 1 hour. How many hours are required to learn 16 facts?

71. The Feline Company sells cat food in standard-size cans. The radius of a can is related to its height by $r = \sqrt{1.9h}$. What would be the height of the can if the radius is 2 inches? (Round the answer to the nearest tenth.)

Problems 68 and 72

72. The period, in seconds, of a compound pendulum of mass 12 units is given by the formula $T = 2\pi\sqrt{\dfrac{I}{384d}}$, where I is the moment of inertia and d is the distance from the center of mass to the axis of rotation in feet. Estimate d for such a pendulum of period 4 seconds whose moment of inertia is 340 units.

In Your Own Words

73. Why do we isolate a radical before we square both sides of an equation containing square roots?

74. Why is checking part of the procedure for solving an equation containing square roots?

4.6 Rational Exponents

TAPE 9

Let's see if we can define exponents as fractions in a manner consistent with our earlier definitions and the five properties of exponents. If we have such a definition, then

$$(x^{1/2})^2 = x^{(1/2)2} \qquad \text{Power of a power}$$
$$= x^1 = x$$

That is, $x^{1/2}$ is a number that, when squared, is x. We know such a number from Section 4.1; it is $\sqrt{x}$. So $x^{1/2} = \sqrt{x}$ is a definition of the one-half power that is consistent with the exponent property *power of a power*. Notice that, because x is a number squared, it **cannot** be negative.

By a similar argument,

$$(x^{1/3})^3 = x^{(1/3)3} = x^1 = x$$

and we see that $x^{1/3}$ is a number that, when cubed, is x. $\sqrt[3]{x}$ is such a number. Notice that x **can** be negative in this case.

Since this can be generalized, we state the following definition, which is consistent with the five properties of exponents.

Exponent of the Form $\dfrac{1}{n}$

If n is a natural number,

$$x^{1/n} = \sqrt[n]{x}, \qquad x \geq 0 \quad \text{if } n \text{ is even}$$

▶ **EXAMPLE 1**

Simplify each expression.

(a) $4^{1/2}$ (b) $27^{1/3}$ (c) $-256^{1/4}$ (d) $(-32)^{1/5}$

Solutions

(a) $4^{1/2} = \sqrt{4}$
$$= 2$$

(b) $27^{1/3} = \sqrt[3]{27}$
$$= 3$$

(c) $-256^{1/4} = -\sqrt[4]{256}$ The base is 256, not -256

$\qquad = -4$

(d) $(-32)^{1/5} = \sqrt[5]{-32}$ The base is -32

$\qquad = -2$ ◄

Exponent definitions and properties are summarized as follows.

Definitions

For n a natural number:

1. $x^n = \underset{n\ x\text{'s}}{x \cdot x \cdot x \ldots x}$ Natural number power

2. $x^0 = 1, \quad x \neq 0$ Zero power

3. $x^{-n} = \dfrac{1}{x^n}, \quad x \neq 0$ Negative power

4. $x^{1/n} = \sqrt[n]{x}, \quad x \geq 0$ if n even   Rational power

Properties

For x and y real numbers, and s and t reduced rational numbers:

1. $x^s \cdot x^t = x^{s+t}$ Product with same base

2. $(x^s)^t = x^{s \cdot t}$ Power of a power

3. $(xy)^s = x^s y^s$ Product to a power

4. $\left(\dfrac{x}{y}\right)^s = \dfrac{x^s}{y^s}, \quad y \neq 0$ Quotient to a power

5. $\dfrac{x^s}{x^t} = x^{s-t}, \quad x \neq 0$ Quotient with same base

---

With these definitions and properties, any rational number can be used as an exponent. Notice how to simplify numbers such as $4^{3/2}$.

► **EXAMPLE 2**

Simplify each expression.

(a) $4^{3/2}$ (b) $81^{-3/4}$

Solutions

(a) $4^{3/2} = 4^{(1/2)3}$

$\qquad = (4^{1/2})^3$ Power of a power

$\qquad = (\sqrt{4})^3 = 2^3$

$\qquad = 8$

(b) $81^{-3/4} = 81^{(1/4)(-3)}$

$\qquad = (81^{1/4})^{-3}$ $\qquad\qquad$ Power of a power

$\qquad = (\sqrt[4]{81})^{-3}$

$\qquad = 3^{-3} = \dfrac{1}{3^3} = \dfrac{1}{27}$ $\qquad\qquad\qquad\qquad\qquad$ ◀

In general, note the following result.

Exponent Form ↔ Radical Form

If $\sqrt[n]{x}$ represents a real number, then

$$x^{m/n} = (\sqrt[n]{x})^m \quad \text{and} \quad x^{m/n} = \sqrt[n]{x^m}$$

In computations, generally $(\sqrt[n]{x})^m$ is the most convenient. Consider evaluating $81^{3/4}$. The choice is between

$\boxed{A}$ $81^{3/4} = (\sqrt[4]{81})^3 = (3)^3 = 27$ or $\boxed{B}$ $81^{3/4} = \sqrt[4]{81^3} = \sqrt[4]{531441} = 27.$

Notice that in $\boxed{B}$ we had to cube 81 and then find the fourth root of 531,441.

▶ **EXAMPLE 3**

Simplify.

(a) $(-8)^{2/3}$ $\qquad$ (b) $-100^{3/2}$ $\qquad$ (c) $\left(\dfrac{256}{625}\right)^{-3/2}$ $\qquad$ (d) $\left(\dfrac{27}{1000}\right)^{2/3}$

Solutions

(a) $(-8)^{2/3} = (\sqrt[3]{-8})^2$ $\qquad\qquad$ The base is -8

$\qquad\qquad = (-2)^2 = 4$

(b) $-100^{3/2} = -(\sqrt{100})^3$ $\qquad\qquad$ The base is 100

$\qquad\qquad = -(10)^3 = -1000$

(c) $\left(\dfrac{256}{625}\right)^{-3/2} = \left(\sqrt{\dfrac{256}{625}}\right)^{-3}$

$\qquad\qquad = \left(\dfrac{\sqrt{256}}{\sqrt{625}}\right)^{-3}$

$\qquad\qquad = \left(\dfrac{16}{25}\right)^{-3} = \dfrac{16^{-3}}{25^{-3}} = \dfrac{25^3}{16^3}$

$\qquad\qquad = \dfrac{15{,}625}{4096}$

(d) $\left(\dfrac{27}{1000}\right)^{2/3} = \left(\sqrt[3]{\dfrac{27}{1000}}\right)^2$

$= \left(\dfrac{\sqrt[3]{27}}{\sqrt[3]{1000}}\right)^2$

$= \left(\dfrac{3}{10}\right)^2 = \dfrac{3^2}{10^2} = \dfrac{9}{100}$ ◄

► **EXAMPLE 4**

Use the properties of exponents to simplify each expression.

(a) $3^{2/3} \cdot 3^{4/3}$ (b) $(16^{5/8})^{6/5}$ (c) $(9x^2)^{3/2}, \quad x > 0$ (d) $\dfrac{32^{3/10}}{32^{1/10}}$

Solutions

(a) $3^{2/3} \cdot 3^{4/3} = 3^{2/3+4/3}$ Product with same base

$= 3^{6/3} = 3^2$

$= 9$

(b) $(16^{5/8})^{6/5} = 16^{(5/8) \cdot (6/5)}$ Power of a power

$= 16^{3/4}$ *Note:* $\dfrac{5}{8} \cdot \dfrac{6}{5} = \dfrac{3}{4}$

$= (\sqrt[4]{16})^3 = 2^3$

$= 8$

(c) $(9x^2)^{3/2} = (\sqrt{9x^2})^3$

$= (3x)^3$ Because $x > 0$

$= 27x^3$ Product to a power

(d) $\dfrac{32^{3/10}}{32^{1/10}} = 32^{3/10-1/10}$ Quotient with same base

$= 32^{2/10} = 32^{1/5} = \sqrt[5]{32} = 2$ ◄

► **EXAMPLE 5**

Simplify. Assume that all variables represent nonnegative real numbers. Express answers in a form containing nonnegative exponents.

(a) $(49m^3n^2)^{1/2}$ (b) $(-125x^{-3}y^6)^{-2/3}$

Solutions

(a) $(49m^3n^2)^{1/2} = 49^{1/2}(m^3)^{1/2}(n^2)^{1/2}$ Product to a power

$= \sqrt{49}m^{3/2}n^{2/2}$

$= 7m^{3/2}n$

(b) $(-125x^{-3}y^6)^{-2/3} = (-125)^{-2/3}(x^{-3})^{-2/3}(y^6)^{-2/3}$

$$= (\sqrt[3]{-125})^{-2}x^{-3\cdot(-2/3)}y^{6\cdot(-2/3)}$$

$$= (-5)^{-2}x^2y^{-4}$$

$$= \frac{x^2}{(-5)^2y^4} = \frac{x^2}{25y^4}$$

◀

► **EXAMPLE 6**

Perform the operations indicated. Assume that all variables represent nonnegative real numbers.

(a) $x^{1/2}(x^{1/2} + x^{-1/2})$ (b) $5x^3(2x^{1/3} - 3x^{-1/6})$

Solutions

(a) $x^{1/2}(x^{1/2} + x^{-1/2})$

By the distributive property,

$$x^{1/2}(x^{1/2} + x^{-1/2}) = x^{1/2}\cdot x^{1/2} + x^{1/2}\cdot x^{-1/2}$$

Next, we use the *product with same base* property.

$$= x^{1/2+1/2} + x^{1/2-1/2}$$

$$= x^1 + x^0 = x + 1$$

(b) $5x^3(2x^{1/3} - 3x^{-1/6}) = 5x^3\cdot 2x^{1/3} - 5x^3\cdot 3x^{-1/6}$ Distributive
$$= 10x^{3+1/3} - 15x^{3-1/6}$$ Property
Product with
$$= 10x^{10/3} - 15x^{17/6}$$ same base

◀

Calculator Box

Roots and Fractional Exponents with a Calculator

With an nth Root Key

A few calculators have an nth root key. Let's approximate $\sqrt[5]{22.5 + \pi}$ with such a calculator.

Scientific calculators:

Press $\boxed{(}$ $\boxed{22.5}$ $\boxed{+}$ $\boxed{\pi}$ $\boxed{)}$ $\boxed{\sqrt[n]{}}$ $\boxed{5}$ $\boxed{=}$

Graphing calculators:

Press $\boxed{5}$ $\boxed{\sqrt[n]{}}$ $\boxed{(}$ $\boxed{22.5}$ $\boxed{+}$ $\boxed{\pi}$ $\boxed{)}$ $\boxed{\text{EXE}}$

In either case, we read $\boxed{1.91332612}$ on the display. On some calculators the nth root key looks like $\boxed{\sqrt[x]{}}$

Without an *n*th Root Key

If there is no *n*th root key, we resort to fractional exponents. Using the fact that the *n*th root is the $\frac{1}{n}$ th power, we find $\sqrt[5]{22.5 + \pi}$ by calculating $(22.5 + \pi)^{1/5}$ with the exponent key.

Press [( ] [22.5] [+] [π] [ )] [y^x] [( ] [1] [÷] [5] [ )] [=]   (or [ENTER] or [EXE])

These keystrokes work on both the *Scientific* and *Graphing* calculators.

The same strategy is used to raise a number to a fractional power. Let's approximate $977^{3/7}$ to the nearest ten-thousandth.

Press [977] [y^x] [( ] [3] [÷] [7] [ )] [=]   (or [ENTER] or [EXE])

and read **19.115399743** on the display. Therefore, $977^{3/7} \approx 19.1154$.

Calculator Exercises

Use a calculator to approximate each expression to the nearest thousandth.

1. $923^{3/5}$ **2.** $\sqrt[6]{741}$ **3.** $(0.975)^{5/9}$

4. $177^{4/7}$ **5.** $\sqrt[8]{421}$ **6.** $(1.175)^{7/9}$

Answers

1. 60.134   **2.** 3.008   **3.** 0.986

4. 19.256   **5.** 2.128   **6.** 1.134

Problem Set 4.6

Warm-Ups

In Problems 1 through 9, simplify if the expression represents a real number. See Example 1.

1. $49^{1/2}$ **2.** $121^{1/2}$ **3.** $169^{1/2}$ **4.** $-100^{1/2}$

5. $(-100)^{1/2}$ **6.** $(-1000)^{1/3}$ **7.** $81^{1/4}$ **8.** $32^{1/5}$

9. $1024^{1/10}$

In Problems 10 through 15, simplify. See Example 2.

10. $27^{2/3}$ **11.** $64^{5/6}$ **12.** $144^{3/2}$ **13.** $4^{7/2}$

14. $9^{-1/2}$ **15.** $27^{-2/3}$

In Problems 16 through 24, simplify if possible. See Example 3.

16. $-49^{-3/2}$ **17.** $(-49)^{-3/2}$ **18.** $2048^{-3/11}$

19. $\left(\dfrac{8}{125}\right)^{2/3}$

20. $\left(\dfrac{81}{16}\right)^{3/4}$

21. $\left(\dfrac{196}{169}\right)^{3/2}$

22. $\left(\dfrac{1}{64}\right)^{-5/6}$

23. $\left(\dfrac{100}{121}\right)^{-3/2}$

24. $\left(-\dfrac{343}{216}\right)^{-2/3}$

In Problems 25 through 39, simplify, assuming that all variables represent nonnegative real numbers. Express answers in a form containing nonnegative exponents. For Problems 25 through 30, see Example 4.

25. $2^{1/2} \cdot 2^{5/2}$

26. $7^{8/3} \cdot 7^{-2/3}$

27. $(25^{3/8})^{4/3}$

28. $\left(\dfrac{4^{3/12}}{s^2}\right)^6$

29. $\dfrac{2^{7/3}}{2^{1/3}}$

30. $\dfrac{27^{11/6}}{27^{7/6}}$

For Problems 31 through 39, see Example 5.

31. $(36x^2)^{1/2}$

32. $(144x^2y^4)^{1/2}$

33. $(49x^2y^{-2})^{1/2}$

34. $(8a^3b)^{1/3}$

35. $(-64x^{-6}z)^{-1/3}$

36. $-(81x^4y^{-4})^{3/4}$

37. $(8x^{-2}y)^{1/2}$

38. $-11(x^{1/2}y^{-1/2})^2$

39. $(3a^{2/3}x^{-1/3})^3$

In Problems 40 through 44, perform the operation indicated. Assume that all variables represent nonnegative real numbers. Express answers in a form containing nonnegative exponents. See Example 6.

40. $x(x^{1/2} + x^{-1/2})$

41. $2x^2(3x^{1/6} - 5x^{-5/6})$

42. $x^{2/3}(2x^{2/3} + 3x^{-2/3})$

43. $(x^{1/2} + x^{-1/2})^2$

44. $(x^{1/2} + x^{-1/2})(x^{1/2} - x^{-1/2})$

Practice Exercises

In Problems 45 through 74, simplify if the expression represents a real number.

45. $36^{1/2}$

46. $196^{1/2}$

47. $225^{1/2}$

48. $(-144)^{1/2}$

49. $-(100)^{1/2}$

50. $(-125)^{1/3}$

51. $81^{1/2}$

52. $81^{1/4}$

53. $-64^{1/6}$

54. $27^{1/3}$

55. $216^{1/3}$

56. $512^{1/3}$

57. $(-81)^{1/4}$

58. $(-32)^{1/5}$

59. $1024^{3/10}$

60. $8^{2/3}$

61. $32^{7/5}$

62. $100^{3/2}$

63. $9^{5/2}$

64. $25^{-1/2}$

65. $64^{-2/3}$

66. $-64^{-3/2}$

67. $(-27)^{-2/3}$

68. $4096^{-3/12}$

69. $\left(\dfrac{64}{343}\right)^{2/3}$

70. $\left(\dfrac{625}{256}\right)^{3/4}$

71. $\left(\dfrac{256}{225}\right)^{3/2}$

72. $\left(\dfrac{1}{32}\right)^{-4/5}$

73. $\left(\dfrac{169}{324}\right)^{-3/2}$

74. $\left(-\dfrac{8}{729}\right)^{-2/3}$

In Problems 75 through 89, simplify, assuming that all variables represent nonnegative real numbers. Express answers in a form containing nonnegative exponents.

75. $(25y^2)^{1/2}$

76. $(100x^6y^2)^{1/2}$

77. $(16x^4y^{-6})^{1/2}$

78. $(-27a^9b^6)^{1/3}$

79. $(64x^{-3}z)^{-1/3}$

80. $-(16x^8y^{-8})^{3/4}$

81. $(18x^2y^{-2})^{1/2}$

82. $(x^{1/3}y^{-1/3})^3$

83. $-3(3a^{3/2}x^{-1/2})^2$

84. $\dfrac{x^{3/2}}{x^{1/2}y^{-1/2}}$

85. $\dfrac{5x^{-1/3}y^{1/3}}{y^{2/3}}$

86. $\dfrac{a^{3/5}b^{-1/4}}{5a^{-2/5}b}$

87. $\left(\dfrac{2x^{1/3}y}{3z^{2/3}}\right)^{-3}$

88. $\left(\dfrac{16x^4z^{-6}}{49y^{-8}}\right)^{1/2}$

89. $\left(\dfrac{25x^{-2/3}}{y^{2/3}}\right)^{-3/2}$

In Problems 90 through 95, perform the operation indicated. Assume that all variables represent nonnegative real numbers. Express answers in a form containing nonnegative exponents.

90. $x(x^{-1/2} + x^{-3/2})$

91. $3x^3(2x^{1/3} - 5x^{-5/3})$

92. $x^{3/2}(3x^{3/2} + 2x^{-3/2})$

93. $2a^{-1/2}b^{3/2}(3a^{1/2}b^{-3/2} + 1)$

94. $(x^{3/2} - x^{-3/2})^2$

95. $(x^{3/2} + x^{-3/2})(x^{3/2} - x^{-3/2})$

Challenge Problems

In Problems 96 through 99, perform the operation indicated. Assume that all variables represent nonnegative real numbers. Express answers in a form containing nonnegative exponents.

96. $(2x^{1/2} + x^{-1/2})(x^{3/2} - x^{1/2})$

97. $(a^{1/3} + b^{1/3})^3$

98. $(a^{1/3} + b^{1/3})(a^{2/3} - a^{1/3}b^{1/3} + b^{2/3})$

99. $(x^{1/3} - x^{-1/3})^3$

100. Factor $x^{1/3}$ from the expression $x^{2/3} + x^{1/3}$.

101. Factor $x^{2/3} - x^{1/3} - 6$.

102. Factor $z^{2/3} - 4$.

103. What is wrong with the following "proof" that $-1 = 1$?
$$-1 = (-1)^1 = (-1)^{2/2} = \sqrt{(-1)^2} = \sqrt{1} = 1$$

The property $\sqrt[n]{x^m} = x^{m/n}$ (provided that $\sqrt[n]{x}$ exists) allows us to simplify radicals when there are common factors between an exponent of the radicand and the index of the radical. Consider $\sqrt[6]{x^4}$. We can write this radical with a smaller index.

$$\sqrt[6]{x^4} = x^{4/6} = x^{2/3}$$
$$= \sqrt[3]{x^2}$$

In Problems 104 through 106, write each radical with a smaller index.

104. $\sqrt[4]{x^6}$

105. $\sqrt[8]{x^4y^4}$

106. $\sqrt[6]{x^2}$

In Your Own Words

107. Explain the relationship between radicals and fractional exponents.

4.7 Complex Numbers

TAPE 9

In Section 4.1 it was noted that expressions such as $\sqrt{-4}$ have no meaning as real numbers because there is no real number that equals -4 when squared. In Chapter 5 we will solve equations like

$$x^2 + 4 = 0$$

Any solution to this equation must be a number whose square is -4. Because there is no real number with this property, we will have to *enlarge* our set of numbers to include solutions to such equations. This new, larger set of numbers is called the set of **complex numbers.**

The set of complex numbers is constructed by introducing a new "number" to the set of real numbers. The new number is the number that when squared is -1. We call it i. That is, i is a number such that

$$i^2 = -1$$

Other numbers must be included in this set so that the sum or product of two complex numbers will be a complex number. This results in the following set.

The Set of Complex Numbers

The set of complex numbers is the set of all numbers that can be written as

$$a + bi$$

where a and b are real numbers and i has the property

$$i^2 = -1$$

Some examples of complex numbers are

$$5 + 3i, \qquad -\frac{6}{7} + 13i, \qquad 1 - 2i$$

$$1 + \sqrt{22}i, \qquad 6i, \qquad -17$$

Notice that the set of real numbers is a subset of the set of complex numbers. (The real number -17 can be written as $-17 + 0i$; thus it belongs to the set of complex numbers.)

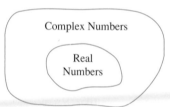

Next, let's look at $\sqrt{-5}$. If this were defined, it would be the number that when squared is -5. Of course, it cannot be a real number, but is it a complex number? To answer this question, we note that

$$(\sqrt{5}i)^2 = (\sqrt{5})^2 \cdot i^2 = 5 \cdot (-1) = -5$$

Thus a sensible definition is $\sqrt{-5} = \sqrt{5}i$. Or, in general:

Definition of Negative Radicand

If k is any positive real number,

$$\sqrt{-k} = \sqrt{k}i$$

Be Careful! Note that the i is **not** under the radical. To emphasize this, $\sqrt{k}i$ may be written $i\sqrt{k}$.

► **EXAMPLE 1**

Write each complex number in the form $a + bi$.

(a) $\sqrt{-7}$ (b) $3 + \sqrt{-4}$ (c) $1 - \sqrt{-12}$ (d) 6

Solutions

(a) $\sqrt{-7} = \sqrt{7}i$ Definition of negative radicand

To answer the question, we put this complex number in the form $a + bi$.

$$= 0 + \sqrt{7}i$$

(b) $3 + \sqrt{-4} = 3 + \sqrt{4}i$ Definition of negative radicand

$$= 3 + 2i$$

(c) $1 - \sqrt{-12} = 1 - \sqrt{12}i$ Definition of negative radicand

$$= 1 - 2\sqrt{3}i$$

In the strictest sense, this is not quite in the form $a + bi$.

$$= 1 + (-2\sqrt{3})i$$

(d) Remember, real numbers *are* complex numbers.

$$6 = 6 + 0i$$ ◄

Be Careful!   When dealing with symbols such as $\sqrt{-5}$ or $\sqrt{-k}$, with k positive, it is **important** to rewrite them as $\sqrt{5}i$ or $\sqrt{k}i$ **before** doing any arithmetic.

$$\sqrt{-2} \cdot \sqrt{-3} \quad does \; not \text{ equal} \quad \sqrt{(-2)(-3)} = \sqrt{6}$$

$$\sqrt{-2}\sqrt{-3} \quad \text{equals} \quad \sqrt{2}i \cdot \sqrt{3}i = \sqrt{6}i^2 = \sqrt{6}(-1) = -\sqrt{6}$$

A complex number written in the form $a + bi$, where a and b are real numbers, is said to be in **standard form.** The real number a in the standard form is called the **real part** of the complex number, and the real number b is called the **imaginary part.**

► **EXAMPLE 2**

Identify the real part and the imaginary part of each complex number.

(a) $5 + 17i$ (b) $3 - \sqrt{5}i$ (c) $\sqrt{-24} - 2$ (d) $\sqrt{18}$

Solutions

(a) $5 + 17i$
The real part is 5 and the imaginary part is 17.

(b) Because $3 - \sqrt{5}i = 3 + (-\sqrt{5})i$:
The real part is 3 and the imaginary part is $-\sqrt{5}$.

(c) First, write the number in standard form.

$$\sqrt{-24} - 2 = \sqrt{24}i - 2$$

$$= 2\sqrt{6}i - 2$$

$$= -2 + 2\sqrt{6}i$$

The real part is -2 and the imaginary part is $2\sqrt{6}$.

(d) Because $\sqrt{18} = 3\sqrt{2} + 0i$:
The real part is $3\sqrt{2}$ and the imaginary part is 0.   ◄

Let's look at some powers of i. Using the usual definitions,

$$i^0 = 1$$
$$i^1 = i$$
$$i^2 = -1$$
$$i^3 = i^2 \cdot i = (-1) \cdot i = -i$$
$$i^4 = (i^2)^2 = (-1)^2 = 1$$
$$i^5 = i^4 \cdot i = 1 \cdot i = i$$

and the pattern repeats $(1, i, -1, -i, 1, i,$ and so on). Can any power of i be found using this pattern? Yes, for example i^{23}:

$$i^{23} = i^{20+3} = i^{20} \cdot i^3 = i^{4 \cdot 5} \cdot i^3 = (i^4)^5 \cdot i^3 = 1^5 \cdot i^3 = i^3 = -i$$

Notice that 3 is the remainder when you divide 23 by 4, *and*

$$i^{23} = i^3$$

The Principal Powers of i

$$i^0 = 1$$
$$i^1 = i$$
$$i^2 = -1$$
$$i^3 = -i$$

► **EXAMPLE 3**

Rewrite each expression as $1, -1, i,$ or $-i$.

(a) i^{13} (b) i^{92} (c) i^{462}

Solutions

(a) Four divides into 13 three times, with a remainder of 1, so

$$i^{13} = i^1$$
$$= i$$

(b) Division of 92 by 4 gives a remainder of 0. Thus

$$i^{92} = i^0$$
$$= 1$$

(c) If we divide 462 by 4, we get a remainder of 2.

$$i^{462} = i^2$$
$$= -1$$

◄

> **Sum, Difference, and Product of Complex Numbers**
>
> **1.** Sum and difference:
> $$(a + bi) + (c + di) = (a + c) + (b + d)i$$
> $$(a + bi) - (c + di) = (a - c) + (b - d)i$$
>
> **2.** Product:
> $$(a + bi) \cdot (c + di) = (ac - bd) + (ad + bc)i$$

Rather than memorize the definitions above, we usually do arithmetic with complex numbers the same way that we do arithmetic with polynomials, remembering to replace powers of i by 1, -1, i, or $-i$.

► **EXAMPLE 4**

Perform the operation indicated.

(a) $(3 + 5i) + (2 - 3i)$ (b) $(3 + 5i) - (2 - 3i)$
(c) $(3 + 5i)(2 - 3i)$ (d) $\sqrt{-2}(1 - \sqrt{-2})$

Solutions

(a) $(3 + 5i) + (2 - 3i) = (3 + 2) + (5i - 3i)$
$= 5 + 2i$

(b) $(3 + 5i) - (2 - 3i) = (3 + 5i) + (-2 + 3i)$
$= (3 - 2) + (5i + 3i)$
$= 1 + 8i$

(c) $(3 + 5i)(2 - 3i) = 6 - 9i + 10i - 15i^2$
$= 6 + i - 15(-1)$
$= 6 + i + 15$
$= 21 + i$

(d) It is important to rewrite the square roots first.
$$\sqrt{-2}(1 - \sqrt{-2}) = \sqrt{2}i(1 - \sqrt{2}i)$$
$$= \sqrt{2}i - 2i^2$$
$$= \sqrt{2}i - 2(-1)$$
$$= 2 + \sqrt{2}i \qquad ◄$$

Special products work with complex numbers as well.

► **EXAMPLE 5**

Perform the operation indicated.

(a) $(2 + 3i)(2 - 3i)$ (b) $(2 + 3i)^2$

Solutions

(a) The complex numbers $a + bi$ and $a - bi$ are called **complex conjugates.**

The product of conjugates is an important special product.

$$(2 + 3i)(2 - 3i) = 2^2 - (3i)^2$$
$$= 4 - 9i^2 = 4 - 9(-1)$$
$$= 4 + 9 = 13$$

(b) $(2 + 3i)^2 = 4 + 12i + 9i^2$
$$= 4 + 12i + 9(-1)$$
$$= 4 + 12i - 9$$
$$= -5 + 12i$$ ◄

To do a division problem in complex numbers, write it as a fraction; then rationalize the denominator. That is, we multiply the numerator and denominator by the conjugate of the denominator. Since we multiply conjugates every time, the following result saves some steps.

The Product of Complex Conjugates
$$(a + bi)(a - bi) = a^2 + b^2$$

► **EXAMPLE 6**

Perform the operations indicated.

(a) $(2 + 3i) \div (1 + i)$ (b) $\dfrac{i}{1 - 2i}$

Solutions

(a) First rewrite the problem as a fraction.

$$(2 + 3i) \div (1 + i) = \frac{2 + 3i}{1 + i}$$

Next, rationalize the denominator by multiplying the numerator and denominator by the **conjugate** of the **denominator.**

$$= \frac{(2 + 3i)(1 - i)}{(1 + i)(1 - i)}$$

$$= \frac{2 - 2i + 3i - 3i^2}{1^2 - i^2}$$

$$= \frac{2 - 2i + 3i + 3}{1 - (-1)}$$

$$= \frac{5 + i}{2}$$

$$= \frac{5}{2} + \frac{1}{2}i$$

(b) Again, rationalize the denominator.

$$\frac{i}{1 - 2i} = \frac{i(1 + 2i)}{(1 - 2i)(1 + 2i)}$$

$$= \frac{i + 2i^2}{1^2 + 2^2} \qquad \text{Product of conjugates}$$

$$= \frac{i + 2(-1)}{1 + 4}$$

$$= \frac{-2 + i}{5} = -\frac{2}{5} + \frac{1}{5}i \qquad \blacktriangleleft$$

▶ **EXAMPLE 7**

Find the reciprocal of $4 - 7i$.

Solution

$$\frac{1}{4 - 7i} = \frac{1(4 + 7i)}{(4 - 7i)(4 + 7i)}$$

$$= \frac{4 + 7i}{4^2 + 7^2} \qquad \text{Product of conjugates}$$

$$= \frac{4 + 7i}{16 + 49}$$

$$= \frac{4 + 7i}{65} = \frac{4}{65} + \frac{7}{65}i \qquad \blacktriangleleft$$

▶ **EXAMPLE 8**

Evaluate i^{-43}.

Solution

By the definition of negative exponents,

$$i^{-43} = \frac{1}{i^{43}}$$

Because 43 divided by 4 gives a remainder of 3, $i^{43} = i^3$.

$$= \frac{1}{i^3} = \frac{1}{-i}, \qquad \text{which we rationalize}$$

$$= \frac{1(i)}{-i(i)} = \frac{i}{-i^2} = \frac{i}{-(-1)} = \frac{i}{1} = i \qquad \blacktriangleleft$$

Problem Set 4.7

Warm-Ups

In Problems 1 through 9, write each number as a simplified complex number in standard form. See Example 1.

1. $\sqrt{-9}$

2. $\sqrt{-49}$

3. $-\sqrt{-121}$

4. $1 + \sqrt{-3}$

5. $3 - \sqrt{-4}$

6. $\sqrt{-5}$

7. $\sqrt{50}$

8. $3\sqrt{-3}$

9. $(\sqrt{-2})^2$

In Problems 10 through 15, simplify the given powers of i. See Examples 3 and 8.

10. i^6

11. i^{11}

12. i^{29}

13. i^{48}

14. i^{-1}

15. i^{-235}

In Problems 16 through 39, perform the operations indicated.
For Problems 16 through 25, see Example 4.

16. $(1 + i) + (2 + 3i)$

17. $(3 + 5i) - (2 + 3i)$

18. $(5 - 2i) + (3 - 5i)$

19. $(-4 - i) + (1 + 5i)$

20. $i(-7i)$

21. $i(14i)$

22. $i(3 + 2i)$

23. $3i(-2 - 5i)$

24. $(1 + i)(2 + 3i)$

25. $(1 - i)(5 + 2i)$

For Problems 26 through 33, see Example 5.

26. $(1 + i)^2$

27. $(2 - 3i)^2$

28. $(\sqrt{2} - \sqrt{3}i)^2$

29. $(\sqrt{8} - \sqrt{-12})^2$

30. $(2 - i)(2 + i)$

31. $(5 + 2i)(5 - 2i)$

32. $(\sqrt{3} - \sqrt{2}i)(\sqrt{3} + \sqrt{2}i)$

33. $(\sqrt{7} + \sqrt{-5})(\sqrt{7} - \sqrt{-5})$

For Problems 34 through 39, see Example 6.

34. $\dfrac{1}{1 + i}$

35. $\dfrac{1}{2 - i}$

36. $\dfrac{2}{1 - 3i}$

37. $\dfrac{1 + 2i}{1 + 3i}$

38. $\dfrac{2 - i}{3 + 2i}$

39. $(2 - i) \div (2 + i)$

Practice Exercises

In Problems 40 through 48, write each number as a simplified complex number in standard form.

40. $\sqrt{-4}$

41. $\sqrt{-25}$

42. $-\sqrt{-169}$

43. $2 - \sqrt{-5}$

44. $3 + \sqrt{-9}$

45. $\sqrt{-7}$

46. $\sqrt{27}$

47. $2\sqrt{-2}$

48. $(\sqrt{-5})^2$

In Problems 49 through 54, simplify the given powers of i.

49. i^7

50. i^{12}

51. i^{30}

52. i^{49}

53. i^{-3}

54. i^{-237}

In Problems 55 through 98, perform the operations indicated.

55. $(1 - i) + (2 - 3i)$

56. $(3 - 5i) - (2 - 3i)$

57. $(5 + 2i) + (3 + 5i)$

58. $(-4 + i) - (1 - 5i)$

59. $(3 + 3i) - (5 + 5i)$

60. $(6 - 7i) - (4 + 5i)$

61. $i(9i)$

62. $i(-11i)$

63. $i(2 - 3i)$

64. $2i(-3 + 4i)$

65. $(1 - i)(3 + 2i)$

66. $(1 + i)(2 + 4i)$

67. $(1 - 3i)(2 - 3i)$

68. $(3 + 2i)(2 - 5i)$

69. $(-2 - 3i)(1 - 2i)$

70. $(-4 - 5i)(-3 + 3i)$

71. $(1 - i)^2$

72. $(2 + 3i)^2$

73. $(3 + 4i)^2$

74. $(3 - i)^2$

75. $(\sqrt{3} + 3i)^2$

76. $(3 - \sqrt{3}i)^2$

77. $(\sqrt{3} + \sqrt{2}i)^2$

78. $(\sqrt{8} + \sqrt{-18})^2$

79. $(3 - i)(3 + i)$

80. $(4 + 3i)(4 - 3i)$

81. $(7 - 10i)(7 + 10i)$

82. $(3 + 2i)(2i - 3)$

83. $(\sqrt{3} - 2i)(\sqrt{3} + 2i)$

84. $(2 + \sqrt{3}i)(2 - \sqrt{3}i)$

85. $(\sqrt{2} - \sqrt{3}i)(\sqrt{2} + \sqrt{3}i)$

86. $(\sqrt{8} + \sqrt{-3})(\sqrt{8} - \sqrt{-3})$

87. $\dfrac{1}{1 - i}$

88. $\dfrac{1}{3 + i}$

89. $\dfrac{3}{1 - 2i}$

90. $\dfrac{2}{3 + 2i}$

91. $\dfrac{i}{4 + 4i}$

92. $3i \div (5 + i)$

93. $\dfrac{2 + i}{1 - 3i}$

94. $\dfrac{3 - i}{2 + 3i}$

95. $(5 - i) \div (5 + i)$

96. $(1 - i)^3$

97. $(2 + 3i)^3$

98. $(1 - 5i)^3$

Challenge Problems

99. Show that i is a solution of $x^2 + 1 = 0$. Can you find another solution?

100. Find two solutions of $x^2 + 4 = 0$.

101. Factor $x^2 + 9$. [HINT: $9 = -(3i)^2$.]

102. Show that $(a + bi)(a - bi) = a^2 + b^2$.

In Your Own Words

103. Using the language of sets, describe the relationship between the set of real numbers and the set of complex numbers.

104. Why was it necessary to introduce the number i?

► Chapter Summary

GLOSSARY

The set of **complex numbers**   The set of all numbers that can be expressed as $a + bi$, where a and b are real numbers and i has the property $i^2 = -1$.

Principal square root of a positive number x $\sqrt{x}$ is the nonnegative number such that $(\sqrt{x})^2 = x$.

Radical expression   An expression containing one or more radicals.

Rationalizing the denominator (numerator)   The process of removing a radical from the denominator (numerator) of a fraction.

Definition of nth Roots of Real Numbers

If n is an *even* natural number:

1. $\sqrt[n]{x}$ is a real number *only if* $x \geq 0$.

2. $\sqrt[n]{x}$ is the *positive* number such that $(\sqrt[n]{x})^n = x$.

If n is an *odd* natural number.

1. $\sqrt[n]{x}$ is a real number for all real numbers x.

2. $\sqrt[n]{x}$ is the real number such that $(\sqrt[n]{x})^n = x$.

3. $\sqrt[n]{-x} = -\sqrt[n]{x}$.

Exponent of the form $\frac{1}{n}$

If n is a natural number, $x^{1/n} = \sqrt[n]{x}$; $x \geq 0$ if n is even.

Equations Containing Square Roots

1. Isolate one radical.

2. Square both sides.

3. Solve the resulting equation.

4. *Check* all possible solutions in the original equation.

5. Write the solution set.

Rational Exponents and Radicals

$$x^{m/n} = (\sqrt[n]{x})^m \qquad \text{provided that } \sqrt[n]{x} \text{ is a real number}$$

$$(\sqrt[n]{x})^m = \sqrt[n]{x^m} \qquad \text{provided that } \sqrt[n]{x} \text{ is a real number}$$

Square Roots of Negative Numbers

If $k > 0$, then $\sqrt{-k}$ exists in the set of complex numbers and, furthermore, $\sqrt{-k} = \sqrt{k}i$.

CHECKUPS

1. Simplify $\sqrt[3]{-54x^3}$.
Section 4.1; Example 4a

2. Simplify $3\sqrt{8x^2y} + x\sqrt{2y^3} - x\sqrt{50y}$, x and y positive.
Section 4.2; Example 3d

3. Perform the operation indicated.

$$(\sqrt{2} + 2\sqrt{3})(5\sqrt{2} - \sqrt{3})$$

Section 4.3; Example 3a

4. Rationalize the denominator.

$$\frac{2\sqrt{3} - \sqrt{2}}{2\sqrt{3} + \sqrt{2}}$$

Section 4.4; Example 5b

5. Solve $\sqrt{x^2 + 8} - 4 = x$.
Section 4.5; Example 5

6. Simplify $(-125x^{-3}y^6)^{-2/3}$.
Section 4.6; Example 5b

7. Perform the operation indicated.

$$(3 + 5i)(2 - 3i)$$

Section 4.7; Example 4c

► Review Problems

In Problems 1 through 15, simplify, if possible. Assume that all variables represent positive real numbers. The answers should not contain negative exponents.

1. $\sqrt{75}$

2. $\sqrt[3]{-128}$

3. $\sqrt{8x^3yz^4}$

4. $\sqrt{\dfrac{24x^5}{54x}}$

5. $\sqrt[3]{\dfrac{-8x^{-2}}{xy^3}}$

6. $\sqrt{4 + x^2}$

7. $-49^{1/2}$

8. $32^{4/5}$

9. $8^{-2/3}$

10. $\left(\dfrac{49}{121}\right)^{-3/2}$

11. $\left(\dfrac{2x^{1/2}y^{-1/3}}{z^{1/3}}\right)^6$

12. $\left(\dfrac{4x^6y^{-4}}{9z^{-2}}\right)^{-1/2}$

13. $\dfrac{x^{1/3}y^{-2/3}}{x^{-2/3}y^{1/3}}$

14. $\left(\dfrac{-8x^{3/2}y^{-1/2}}{z^{-3}}\right)^{-2/3}$

15. $\dfrac{4^{-1/2} + 1}{4^{1/2}}$

In Problems 16 through 25, perform the operations indicated and simplify. Assume that all variables represent positive real numbers.

16. $\sqrt{50} - 3\sqrt{8}$

17. $\sqrt{432} - 2\sqrt{147} + 3\sqrt{3}$

18. $\sqrt[3]{16x^4} + x\sqrt[3]{2x} - \sqrt[3]{54x^4}$

19. $\sqrt{2}(\sqrt{6} - 1)$

20. $(\sqrt{3} - 1)(2 + \sqrt{2})$

21. $(\sqrt{2} - \sqrt{3})^2$

22. $(2\sqrt{5} + \sqrt{7})(2\sqrt{5} - \sqrt{7})$

23. $x^{1/2}(x^{1/2}y - x^{-1/2})$

24. $(x^{1/2} + y^{-1/2})^2$

25. $(2x^{1/2} - 3y^{-1/2})(2x^{1/2} + 3y^{-1/2})$

In Problems 26 through 34, rationalize the denominator.

26. $\dfrac{2}{\sqrt{6}}$

27. $\dfrac{11}{7\sqrt{11}}$

28. $\dfrac{\sqrt{3}}{\sqrt{5}}$

29. $\dfrac{1}{\sqrt[3]{6}}$

30. $\dfrac{4}{\sqrt[5]{16}}$

31. $\dfrac{\sqrt[3]{4} + \sqrt[3]{2}}{\sqrt[3]{2}}$

32. $\dfrac{6}{\sqrt{5} - 1}$

33. $\dfrac{\sqrt{2}}{\sqrt{2} + \sqrt{6}}$

34. $\dfrac{\sqrt{3} - 2}{\sqrt{3} + 2}$

In Problems 35 through 40, solve each equation.

35. $\sqrt{3x + 2} = 2$

36. $\sqrt{x^2 + 3} = x + 3$

37. $\sqrt{6s - 2} = \sqrt{3s + 1}$

38. $\sqrt{t + 3} = \sqrt{2 - t}$

39. $\sqrt{x + 2} = \sqrt{x - 5}$

40. $\sqrt{x^2 - 15} - x = 5$

In Problems 41 through 43, write each complex number in standard form.

41. $\sqrt{-225}$

42. $\sqrt{18} + \sqrt{-18}$

43. $\sqrt{-2}(\sqrt{2} - \sqrt{-2})$

In Problems 44 through 51, perform the operation indicated.

44. $(2 - 3i) + (7 - 11i)$

45. $(5 - 4i) - (2 - 6i)$

46. $4i(3 - 2i)$

47. $(1 + 2i)(5 - 3i)$

48. $(3 - 5i)^2$

49. $(6 - 7i)(6 + 7i)$

50. $\dfrac{3 - 7i}{5 + 5i}$

51. $(2 - 3i)^3$

Let's Not Forget . . .

Identify the expressions that are in factored form. Factor those that are not factored.

52. $(x - 3)^2$

53. $r^2(x - y) + t^2(x - y)$

54. $(2x + 3)(4x^2 - 6x + 9)$

55. $(2x + 3)^3$

56. $8x^3 - 27$

57. $4x^2 - 12xy + 9y^2$

How many terms are in each expression? Which expressions have $1 - x$ as a factor?

58. $1 - x^2$

59. $1 - x^3$

60. $2 - x - x^2$

61. $(1 - x)^2$

62. $1 - x + x^2$

Simplify each expression, if possible, leaving only nonnegative exponents in your answer.

63. $\dfrac{8x^3 - 1}{1 - 4x^2}$

64. $\dfrac{-5x^{-1}}{y^2 z^{-3}}$

65. $-4^{-1/2}$

66. $\left(\dfrac{27a^{-6}}{8b^{-9}}\right)^{-1/3}$

67. $\dfrac{2^{-1} + 3^{-1}}{2^{-2}}$

68. $\dfrac{\sqrt{12} - \sqrt{-4}}{2}$

69. $\sqrt{(x^2 + y^2)^2}$

Find each product.

70. $\sqrt{6}(\sqrt{2} + \sqrt{6})$

71. $(x + \sqrt{2})^2$

72. $(2 - x)^3$

Reduce, if possible.

73. $\dfrac{r^2(x - a) - s^2(x - a)}{(x - a)(r + s)}$

74. $\dfrac{4 + \sqrt{8}}{4}$

The following problems can be worked by using a least common denominator. Follow the directions in each and notice how the LCD is used.

75. Solve $\dfrac{1}{x + 1} + \dfrac{1}{x - 2} = \dfrac{1}{x^2 - x - 2}$.

76. Perform the subtraction indicated: $\dfrac{4}{x + 1} - \dfrac{1}{x - 2}$.

77. Simplify $\dfrac{1 - \dfrac{1}{x + 1}}{\dfrac{1}{x - 2} + 1}$.

Label each as an equation or an expression. Solve the equations and perform the operations indicated on the expressions.

78. $\left|\dfrac{2x - 5}{3}\right| = 3$

79. $\sqrt{x + 3} = 2$

80. $\dfrac{x}{x^2 + x - 6} - \dfrac{1}{x + 3}$

81. $\dfrac{3(x + 1)}{2} - 1 = \dfrac{x + 2}{5}$

82. $(4 - 3x)^2$

83. $(3 - 2i)(1 + i)$

▶ You Decide

While building a backyard shed, you need to hold a frame temporarily so that the corners are "square". You have a scrap 2 by 4 that is 13 ft long to use for the bracing. However, you have misplaced your carpenter's square. So you decide to brace as shown.

Of course, you may saw the board into smaller lengths for more than one brace, if you so choose. What is your plan? Exactly where do you drive your nails?

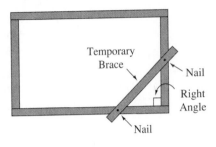

▶ Chapter 4 Test

In Problems 1 through 9, choose the correct answer.

1. $(6^{-1/2})^{-2} = (?)$

   A. 6   **B.** 9   **C.** $\dfrac{1}{6}$ **D.** $-\dfrac{1}{9}$

2. $\dfrac{(b^{1/2})^6}{(b^{-1})^2} = (?)$

   A. $-b^2$ **B.** b^5 **C.** $\dfrac{1}{b^2}$ **D.** $\dfrac{1}{b}$

3. $\sqrt[3]{-8x^4y^2} = (?)$

   A. $-2x^3\sqrt[3]{xy^2}$ **B.** $-4x^2y$

   C. $-2x\sqrt[3]{xy^2}$ **D.** Not defined

4. $\sqrt{32} + \sqrt{18} - \sqrt{50} = (?)$

   A. 2   **B.** $2\sqrt{2}$ **C.** 0   **D.** 4

5. $i^{117} = (?)$

   A. 1   **B.** i **C.** -1 **D.** $-i$

6. $(\sqrt{5} - 2)^2 = (?)$

   A. 1   **B.** 21   **C.** $9 - 2\sqrt{5}$ **D.** $9 - 4\sqrt{5}$

7. $\left(-\dfrac{729}{64}\right)^{2/3} = (?)$

   A. $\dfrac{81}{16}$ **B.** $-\dfrac{243}{32}$ **C.** $-\dfrac{81}{16}$ **D.** $\dfrac{9}{4}$

8. $\sqrt{-3}(2 + \sqrt{-3}) = (?)$

   A. $3 + 2\sqrt{3}i$ **B.** $-3 - 6i$

   C. $-3 + 2\sqrt{3}i$ **D.** $3 - 2\sqrt{3}i$

9. $(5 - 2i)^2 = (?)$

   A. 29   **B.** 21   **C.** $29 - 20i$ **D.** $21 - 20i$

In Problems 10 and 11, solve each equation.

10. $\sqrt{x + 1} = \sqrt{2x - 2}$

11. $\sqrt{x + 3} + \sqrt{2x - 3} = 0$

In Problems 12 and 13, perform the operations indicated. Assume that all variables represent nonnegative real numbers. Simplify each answer and express it in a form containing no negative exponents.

12. $\left(\dfrac{36a^4b^{-12}}{4c^{-6}}\right)^{-1/2}$

13. $\dfrac{i}{1 + i}$

In Problems 14 and 15, rationalize the denominator.

14. $\dfrac{3}{2\sqrt{6}}$

15. $\dfrac{7 - \sqrt{2}}{1 + \sqrt{2}}$

5 Nonlinear Equations and Inequalities in One Variable

See Problem Set 5.5, Exercise 66.

► Connections

During the fifteenth and sixteenth centuries, mathematicians tried to solve more complicated equations. They were hampered by their inability to accept negative numbers and complex numbers (Section 4.7).

Girolamo Cardano (1501–1576), an Italian, worked on solving the equation $x(10 - x) = 40$. He found the two solutions $5 + \sqrt{-15}$ and $5 - \sqrt{-15}$, but these numbers were most puzzling to him. Today, equations such as this arise as mathematical models in electrical engineering as well as in other scientific fields.

In this chapter, we examine equations and inequalities in one variable that are not linear. We call such equations and inequalities nonlinear. In particular, we look at the second-degree equation in detail. With the addition of an important property of our number system, we will have the tools at hand to solve all members of this important class of equations. We look at three useful methods fo solving second-degree equations: factoring, completing the square, and the quadratic formula. Also, we examine higher–degree equations and inequalities, and we solve rational and radical equations that lead to second-degree equations.

5.1   Solving Quadratic Equations and Inequalities by Factoring

TAPE 9

In this section, second-degree equations and inequalities are solved by factoring.

Definition

A **quadratic equation** in one variable is an equation of the form

$$ax^2 + bx + c = 0$$

where a, b, and c are real numbers with $a \neq 0$.
If $a > 0$, this form is called **standard form.**

For example,

$$2x^2 - 5x + 3 = 0$$

is a quadratic equation where $a = 2$, $b = -5$, and $c = 3$.

$$3x^2 - x = 0$$

is an example where $a = 3$, $b = -1$, and $c = 0$, and

$$\frac{1}{2}x^2 - 5 = 0$$

is another example, where $a = \frac{1}{2}$, $b = 0$, and $c = -5$. The equation

$$2x^2 = -5x + 2$$

is also a quadratic equation. Rewrite it as

$$2x^2 + 5x - 2 = 0$$

to see that $a = 2$, $b = 5$, and $c = -2$.

If $a = 0$, the equation becomes linear. The two tools used to solve linear equations, adding the same number to both sides and multiplying both sides by the same nonzero number, are not enough to solve higher-degree equations. However, our number system has the following important property, which will provide the other tool necessary to solve higher-degree equations.

Property of Zero Products

The statement $pq = 0$ is *true* if either $p = 0$ or $q = 0$ and *false* if neither p nor q is 0.

Consider the quadratic equation

$$x^2 - 5x + 6 = 0$$

The left side of the equation will factor:

$$(x - 3)(x - 2) = 0$$

This equation is certainly equivalent to the first. Notice that this is a statement that the product of two real numbers equals zero. The Property of Zero Products says that this statement will be true if *either* $(x - 3)$ *or* $(x - 2)$ is zero. That is,

$$x - 3 = 0 \quad \text{or} \quad x - 2 = 0$$

$$x = 3 \quad \text{or} \quad x = 2$$

Thus 3 and 2 are each solutions of $x^2 - 5x + 6 = 0$, so the solution set is

$$\{2, 3\}$$

▶ **EXAMPLE 1**

Solve $x^2 + 3x + 2 = 0$.

Solution

$$x^2 + 3x + 2 = 0$$

$$(x + 2)(x + 1) = 0 \qquad \text{Factor}$$

$$x + 2 = 0 \quad \text{or} \quad x + 1 = 0 \qquad \text{Property of Zero Products}$$

$$x = -2 \quad \text{or} \quad x = -1$$

$$\{-2, -1\}$$

◀

> **Solving a Quadratic Equation by Factoring**
>
> 1. Write the equation in standard form.
> 2. Factor.
> 3. Apply the Property of Zero Products.
> 4. Check the solutions in the original equation if required.
> 5. Write the solution set.

▶ **EXAMPLE 2**

Solve $2x^2 + 5x = 3$.

Solution

$$2x^2 + 5x - 3 = 0$$

$$(2x - 1)(x + 3) = 0 \qquad \text{Factor}$$

$$2x - 1 = 0 \quad \text{or} \quad x + 3 = 0 \qquad \text{Property of Zero Products}$$

$$2x = 1 \quad \text{or} \qquad x = -3$$

$$x = \frac{1}{2} \quad \text{or} \qquad x = -3$$

$$\left\{ \frac{1}{2}, -3 \right\} \qquad\qquad\qquad ◀$$

Notice that we *first* put the equation in standard form and *then* factored it.

▶ **EXAMPLE 3**

Solve $x^2 + 2x = 0$.

Solution

$$x^2 + 2x = 0$$

$$x(x + 2) = 0 \qquad \text{Factor}$$

$$x = 0 \quad \text{or} \quad x + 2 = 0$$

$$\{0, -2\} \qquad\qquad ◀$$

Be Careful!   A common mistake is to lose the zero solution in quadratics like the example above. Note that with only a little mental effort the solution set can be written directly after the equation has been factored.

► **EXAMPLE 4**

Solve $(3y + 5)(2y + 1) = -1$.

Be Careful!

Solution

Be very careful to write the equation in standard form first. It is very tempting to set each factor equal to -1 and solve. *That would be wrong!*

$$(3y + 5)(2y + 1) = -1$$
$$6y^2 + 13y + 5 = -1$$
$$6y^2 + 13y + 6 = 0 \qquad \text{Standard form}$$
$$(3y + 2)(2y + 3) = 0 \qquad \text{Factor}$$
$$\left\{ -\frac{2}{3}, -\frac{3}{2} \right\}$$
◄

► **EXAMPLE 5**

Solve $x^2 + \frac{2}{3}x - \frac{8}{3} = 0$.

Solution

Multiply both sides by 3 to clear fractions.

$$3x^2 + 2x - 8 = 0$$
$$(x + 2)(3x - 4) = 0$$
$$\left\{ -2, \frac{4}{3} \right\}$$
◄

► **EXAMPLE 6**

Solve $x^2 - 4 = 0$.

Solution

$$x^2 - 4 = 0$$
$$(x + 2)(x - 2) = 0$$
$$\{-2, 2\}$$
◄

A common abbreviation for $\{-2, 2\}$ is $\{\pm 2\}$. Also, we often write

$$x = \pm k \quad \text{which means} \quad x = -k \quad \text{or} \quad x = k$$

► **EXAMPLE 7**

Solve $2t^2 = 18$.

Solution

$$2t^2 = 18$$

$$2t^2 - 18 = 0$$

Be sure to factor out the common factor first!

$$2(t^2 - 9) = 0$$

$$t^2 - 9 = 0 \qquad \text{Divide by 2}$$

$$(t - 3)(t + 3) = 0 \qquad \text{Factor}$$

$$\{-3, 3\} \quad \text{or} \quad \{\pm 3\} \qquad \blacktriangleleft$$

Notice how the factor of 2 in Example 7 did not become part of the solution set.

▶ **EXAMPLE 8**

Solve $-2x^2 + 9x + 35 = 0$.

Solution

$$-2x^2 + 9x + 35 = 0$$

Factoring is much easier if the leading coefficient is positive. So multiply both sides of the equation by -1 to write the equation in standard form.

$$2x^2 - 9x - 35 = 0$$

$$(2x + 5)(x - 7) = 0$$

$$\left\{ -\frac{5}{2}, 7 \right\} \qquad \blacktriangleleft$$

▶ **EXAMPLE 9**

Solve $z^2 - 6z + 9 = 0$.

Solution

$$z^2 - 6z + 9 = 0$$

$$(z - 3)(z - 3) = 0$$

$$\{3\} \qquad \blacktriangleleft$$

The solution set of Example 9 has only one number because both factors of the quadratic were the same. Such a single solution is called a **repeated root** or a root of **multiplicity 2.**

Quadratic Inequalities

The method of boundary numbers reduces the problem of solving a quadratic inequality to the problem of solving a quadratic equation.

► **EXAMPLE 10**

Solve the inequality $(x - 1)(x + 2) \leq 0$.

Solution

First, we find the boundary numbers

$$(x - 1)(x + 2) = 0$$

The boundary numbers are -2 and 1.

We locate the boundary numbers on a number line.

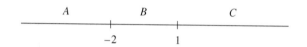

We can make a table to summarize testing the regions.

Region	Number in Region	Statement $(x - 1)(x + 2) \leq 0$	Truth of Statement	Region in Solution Set?
A	-3	$(-4)(-1) \leq 0$	False	No
B	0	$(-1)(2) \leq 0$	True	Yes
C	2	$(1)(4) \leq 0$	False	No

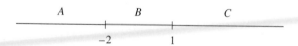

Since $(x - 1)(x + 2) \leq 0$ allows equality, both -2 and 1 are in the solution set.

Finally, we write the solution set.

$$\{x \mid -2 \leq x \leq 1\}$$

◄

► **EXAMPLE 11**

Solve $-x^2 + x + 2 < 0$. Write the solution set in interval notation.

Solution

First, the boundary numbers.

$$-x^2 + x + 2 = 0$$
$$x^2 - x - 2 = 0 \qquad \text{Multiply by } -1$$
$$(x + 1)(x - 2) = 0 \qquad \text{Factor}$$

The boundary numbers are -1 and 2.

We locate the boundary numbers on a number line.

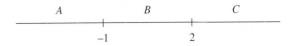

We check region A with any number in it, say -5.

$$-(-5)^2 + (-5) + 2 < 0$$
$$-25 - 5 + 2 < 0$$
$$-28 < 0$$

A true statement. Region A is included.

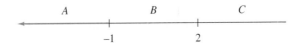

For region B, we will test 0.

$$-0^2 + 0 + 2 < 0$$
$$2 < 0$$

A false statement. Region B is not included.

For region C, we will test 3.

$$-3^2 + 3 + 2 < 0$$
$$-9 + 5 < 0$$
$$-4 < 0$$

So region C is included.

Since $-x^2 + x + 2 < 0$ does not include the equality, the boundary numbers are not included in the solution set.

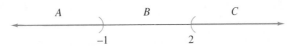

The solution set is

$$(-\infty, -1) \cup (2, +\infty) \qquad \blacktriangleleft$$

Problem Set 5.1

Warm-Ups

In Problems 1 through 14, write the solution set.

1. $(x + 7)(x + 3) = 0$ **2.** $(x - 5)(x + 11) = 0$ **3.** $(x + 5)(x - 5) = 0$

4. $x(x - 3) = 0$

5. $2x(x + 5) = 0$

6. $(2x - 1)(x + 3) = 0$

7. $-2x(3x - 1) = 0$

8. $(3x + 7)(2x - 5) = 0$

9. $(x + 4)(x + 4) = 0$

10. $(4x - 9)^2 = 0$

11. $-2(t - 3)(2t - 9) = 0$

12. $(y - \sqrt{3})(y + \sqrt{3}) = 0$

13. $(z - 2i)(z + 2i) = 0$

14. $(z + \sqrt{3}i)(z - \sqrt{3}i) = 0$

In Problems 15 through 20, find the solution set by factoring. See Examples 1 and 2.

15. $x^2 - x - 2 = 0$

16. $x^2 + 6x = -5$

17. $x^2 - 4x + 3 = 0$

18. $3x^2 - 5x + 2 = 0$

19. $4v^2 + 5v + 1 = 0$

20. $2x^2 + 5x - 3 = 0$

In Problems 21 and 22, find the solution set. See Example 3.

21. $x^2 - 2x = 0$

22. $x^2 = -7x$

In Problems 23 and 24, solve each equation. See Example 4.

23. $(x + 6)(x - 1) = -10$

24. $(x + 4)^2 = 9$

In Problems 25 through 28, solve each equation. See Example 5.

25. $\dfrac{2}{5}x^2 - \dfrac{3}{5}x - 1 = 0$

26. $x^2 - \dfrac{x}{6} - \dfrac{1}{3} = 0$

27. $\dfrac{3}{2}t^2 + \dfrac{5}{2}t = 1$

28. $\dfrac{1}{2}x^2 + \dfrac{7}{6}x - 1 = 0$

In Problems 29 and 30, find the solution set. See Example 6.

29. $x^2 - 4 = 0$

30. $x^2 = 16$

In Problems 31 and 32, find the solution set. See Example 7.

31. $2x^2 - 6x = 0$

32. $6x^2 - 16x = 0$

In Problems 33 and 34, find the solution set. See Example 8.

33. $-6x^2 + 5x + 1 = 0$

34. $-7x^2 - 4x + 3 = 0$

In Problems 35 and 36, solve each equation. See Example 9.

35. $x^2 - 16x + 64 = 0$

36. $4x^2 - 4x = -1$

For Problems 37 through 40, solve each inequality. See Example 10.

37. $(x + 1)(x - 2) < 0$

38. $(x + 4)(x - 7) > 0$

39. $(x - 3)(x + 1) \leq 0$

40. $(2 - x)(x + 3) \geq 0$

In Problems 41 through 46, solve each inequality and write the solution set in interval notation. See Example 11.

41. $x^2 - 4x + 3 < 0$

42. $x^2 - x - 6 > 0$

43. $x^2 + 6x > -5$

44. $x^2 - 2x \geq 3$

45. $-x^2 - x + 2 > 0$

46. $x^2 - 4 \leq 0$

Practice Exercises

In Problems 47 through 100, find the solution set by factoring.

47. $x^2 - 5x - 6 = 0$

48. $x^2 + 5x - 6 = 0$

49. $x^2 + 7x + 12 = 0$

50. $x^2 - 12x - 13 = 0$

51. $x^2 - 6x - 7 = 0$

52. $x^2 - 8x + 16 = 0$

53. $x^2 - 6x - 16 = 0$

54. $x^2 + 10x = -24$

55. $(x - 7)(x - 8) = 6$

56. $(x - 6)(x + 5) = 12$

57. $x^2 - 24x + 144 = 0$

58. $y^2 - 10y + 25 = 0$

59. $5x^2 - 11x + 2 = 0$

60. $15w^2 + 12 = 28w$

61. $3x^2 + 5x + 2 = 0$

62. $5x^2 + 8x + 3 = 0$

63. $2x^2 - 7x + 6 = 0$

64. $3x^2 - 108x = 0$

65. $5x^2 - 6x - 8 = 0$

66. $2x^2 - 5x - 12 = 0$

67. $(x - 5)(x - 4) = 12$

68. $7x^2 + 39x - 18 = 0$

69. $(x - 6)(x - 2) = -3$

70. $7x^2 + 45x - 28 = 0$

71. $6x^2 + x = 12$

72. $9x^2 = 1$

73. $4x^2 + 4x - 3 = 0$

74. $6x^2 - 29x - 22 = 0$

75. $2s^2 - 2s - 4 = 0$

76. $6x^2 + 15x - 9 = 0$

77. $-x^2 + x + 12 = 0$

78. $-y^2 + 5y - 6 = 0$

79. $-2t^2 + 24t + 26 = 0$

80. $-3x^2 - 9x - 6 = 0$

81. $5x^2 + \dfrac{1}{3}x = 2$

82. $\dfrac{9}{5}t^2 - \dfrac{9}{5}t - 2 = 0$

83. $5x^2 + \dfrac{7}{4}x - \dfrac{3}{2} = 0$

84. $\dfrac{4}{3}w^2 - \dfrac{1}{3}w - \dfrac{1}{2} = 0$

85. $(3x + 1)(x - 4) < 0$

86. $(x + 3)(5 - 2x) > 0$

87. $(x + 1)(x + 1) < 0$

88. $(x + 3)(x - 3) > 0$

89. $(5 + x)(5 + x) \leq 0$

90. $(x - 6)(x - 6) \geq 0$

91. $x^2 - 7x + 6 > 0$

92. $x^2 - 2x - 8 < 0$

93. $x^2 - 7x > 8$

94. $x^2 + 6x < -8$

95. $x^2 > 16$

96. $x^2 + 2x < -1$

97. $-t^2 - 5t \geq 0$

98. $2y^2 - y < 0$

99. $s^2 + 3 < -4s$

100. $x^2 > 0$

101. $5x + 14 \geq x^2$

102. $2x^2 + 11x - 21 < 0$

103. $10x^2 + x - 3 > 0$

104. $2z^2 + 5 > 11z$

In Problems 105 through 110, write the solution set in interval notation.

105. $21x^2 + 19x - 12 \leq 0$

106. $-7x^2 + 5x + 2 \leq 0$

107. $12x^2 + 5x - 2 < 0$

108. $6y^2 + 5y > 6$

109. $10x^2 - 21x - 49 \geq 0$

110. $3t^2 + 17t > 6$

Challenge Problems

In Problems 111 through 116, find the solution set by factoring.

111. $x^2 - 5 = 0$ [HINT: Find the product $(x - \sqrt{5})(x + \sqrt{5})$.]

112. $x^2 + 4 = 0$ [HINT: Find the product $(x - 2i)(x + 2i)$.]

113. $w^2 + 5 = 0$ **114.** $x^2 + 8 = 0$

115. $4x^2 + 25 = 0$ **116.** $2v^2 + 11 = 0$

In Problems 117 through 122, find a quadratic equation with the solution set given.

117. $\{1, 2\}$ **118.** $\{-1, 3\}$ **119.** $\{\pm 2\}$

120. $\{\pm\sqrt{5}\}$ **121.** $\{\pm 2i\}$ **122.** $\left\{\dfrac{1}{2}, \dfrac{3}{2}\right\}$

In Your Own Words

123. Explain how factoring can be used to solve a quadratic equation.

5.2   Quadratic Equations and Inequalities as Models for Applications

TAPE 9

Many word problems lead to quadratic equations and inequalities. In this section, such applications are examined and the techniques developed in Section 5.1 are applied to solve them.

Often, there will be more than one solution to a quadratic equation. Each solution of the equation must be tested in the statement of the word problem to answer the original question.

Recall the procedures developed in Chapter 2 for using mathematical models to solve word problems.

> ### A Procedure to Model Word Problems
>
> **1.** Read the problem to determine what quantities are to be found.
>
> **2.** Assign a variable, such as x, to represent one of the quantities to be found.
>
> **3.** Express all other quantities to be found in terms of x (or the variable chosen).
>
> **4.** Draw a figure or picture to illustrate the problem, if possible. Label it.
>
> **5.** Read the problem and form a mathematical model (an equation or inequality).
>
> **6.** Solve the equation (inequality) and find the value of all quantities to be found.
>
> **7.** Check the solution(s) of the equation (inequality) in the original word problem. It should make sense and answer the question.
>
> **8.** Write an answer to the original question.

► **EXAMPLE 1**

Find two consecutive positive even integers whose product is 224.

Solution

We follow the general procedure for solving word problems and choose a variable to represent one of the numbers.

Let n be the smaller of the two integers. Then $n + 2$ is the larger of the two.

The product of the two numbers is 224.

$$n(n + 2) = 224$$

Solve the equation.

$$n^2 + 2n - 224 = 0$$
$$(n + 16)(n - 14) = 0$$
$$n = -16 \quad \text{or} \quad n = 14$$

However, the question asked for the *positive* integers, so discard -16. Checking the other solution in the problem, note that 14 and 16 are consecutive positive even integers and $14(16) = 224$. So we write the answer:

The integers 14 and 16.   ◄

► **EXAMPLE 2**

A missile is shot into the air. The height h in feet above the ground is given by the formula $h = 128t - 16t^2$, where t is the time in seconds since launch. Find the time interval during which the missile is at least 112 feet above the ground.

Solution

The mathematical model for this problem is an inequality. We want h to be greater than or equal to 112.

$$h \geq 112$$

$$128t - 16t^2 \geq 112$$

We will solve the quadratic inequality using the method of boundary numbers. We find the boundary numbers by solving the equation.

$$128t - 16t^2 = 112$$
$$-16t^2 + 128t - 112 = 0 \qquad \text{Standard form}$$
$$t^2 - 8t + 7 = 0 \qquad \text{Divide both sides by } -16$$
$$(t - 7)(t - 1) = 0 \qquad \text{Factor}$$

The boundary numbers are 1 and 7.

Testing the three regions and the boundary numbers shows that the solution set is [1, 7]. Now we can answer the question.

The missile is at least 112 ft above ground from 1 sec to 7 sec after launch.   ◄

► **EXAMPLE 3**

A 3 ft by 4 ft rectangular flag is to be made with a blue rectangular center and a uniformly wide white border. If the blue part and the white part have the same area, how wide should the white border be?

Solution

Let x be the width of the uniform border, in feet.

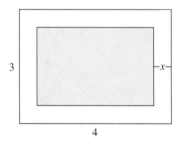

The figure shows that the area of the center rectangle (shown in blue) is $(3 - 2x)(4 - 2x)$, and from the problem it must be the same as the area of the white border. Thus the area of the blue rectangle is one-half the total area of the flag (3 times 4). So form the equation

$$(3 - 2x)(4 - 2x) = \frac{1}{2}(3)(4)$$

$$12 - 14x + 4x^2 = 6$$

$$4x^2 - 14x + 6 = 0$$

$$2x^2 - 7x + 3 = 0$$

$$(2x - 1)(x - 3) = 0$$

$$x = \frac{1}{2} \quad \text{or} \quad x = 3$$

A 3-ft border does not make sense since the width of the flag is 3 ft.

The border should be $\frac{1}{2}$ ft wide.   ◄

► **EXAMPLE 4**

If the area of a triangle is 14 ft² and its base is 3 ft longer than its height, find the length of the base and the height.

Solution

Let h be the height of the triangle in feet. Then $h + 3$ is the length of the base in feet. Remembering that the area of a triangle is one-half the base times the

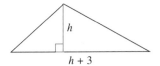

height (altitude), form the equation

$$14 = \frac{1}{2}(h + 3)h$$

Multiply both sides by 2; then apply the Distributive Property.

$$28 = (h + 3)h = h^2 + 3h$$

Then write in standard form and solve.

$$0 = h^2 + 3h - 28$$
$$0 = (h + 7)(h - 4)$$
$$h = -7 \quad \text{or} \quad h = 4$$

Since -7 does not make sense in this problem, h must be 4.

The height is 4 ft and the base is 7 ft.   ◄

► ## EXAMPLE 5

The lengths of the sides of a right triangle are three consecutive even integers. What are they?

Solution

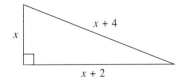

Let x be the smallest of the three consecutive even integers. Then $x + 2$ is the next one, and $x + 4$ is the largest. Since the hypotenuse is the longest side of a right triangle, its length must be given by $x + 4$. By the Pythagorean Theorem, we have the equation

$$x^2 + (x + 2)^2 = (x + 4)^2$$

Square the two binomials and simplify.

$$x^2 + x^2 + 4x + 4 = x^2 + 8x + 16$$
$$x^2 - 4x - 12 = 0$$
$$(x - 6)(x + 2) = 0$$
$$x = 6 \quad \text{or} \quad x = -2$$

Since a length cannot be negative, the smallest side must be 6.

The sides of the triangle are 6, 8, and 10 units.   ◄

Problem Set 5.2

Warm-Ups

For Problems 1 through 6, see Example 1.

1. The two Taylor girls were born in consecutive, even years. If the product of their ages is 288, how old is each girl?

2. Find two consecutive negative odd numbers whose product is 143.

3. Two bike trails have a total length of 20 miles. If the product of the lengths of the trails is 91, find the length of each trail.

4. Find two numbers whose sum is 21 and product is 68.

For Problems 7 and 8, see Example 2.

7. The height in feet of a stone catapulted upward is given by $h = 80t - 16t^2$, where t is the time in seconds after launch. During what time interval is the stone higher than 96 feet?

For Problems 9 and 10, see Example 3.

9. An artist wishes to put a frame of uniform width around a painting whose dimensions are 4 ft by 6 ft. How wide should she make the frame if the total area is to be 48 ft²?

For Problems 11 and 12, see Example 4.

11. The area of a triangle is 56 ft². If the base is 9 ft more than the altitude, find the length of the base and the altitude.

For Problems 13 and 14, see Example 5.

13. The hypotenuse of a right triangle is 13 ft. The longer leg is two more than twice the length of the shorter leg. Find the length of each leg.

5. Find three consecutive positive integers such that the sum of their squares is 77.

6. Find three consecutive negative integers such that the sum of their squares is 110.

8. A new tractor costs $96,000. Suppose the yearly maintenance cost in dollars is given by $C = t^2 + 1180t$, where t is the age of the tractor in years. When will the maintenance cost be greater than 25% of the purchase price?

10. A rectangular garden is to be made with a uniform path completely around the outside. If the garden is 8 ft by 12 ft and the total area of the garden and path is 221 ft², how wide should the path be?

12. The area of a triangle is 25 yd². If the base is twice the height, find the length of the base and height.

14. The lengths of the sides of a right triangle are three consecutive integers. Find the lengths of the sides.

Practice Exercises

15. Find two consecutive negative integers whose product is 132.

16. Jack and Bob played in their first varsity basketball game with Jack scoring two more points than Bob. If the product of their points scored is 168, how many points did each score?

17. Find two numbers whose sum is 9 and whose product is 20.

18. Find two numbers whose product is 51 and whose sum is 20.

19. Find three consecutive odd positive integers such that the sum of their squares is 35.

20. Find three consecutive even positive integers such that the sum of their squares is 200.

21. Find the dimensions of a rectangle whose length is 3 m less than twice its width and whose area is 65 m².

22. A helicopter is searching a rectangular zone of the Talledega National Forest for wreckage from a plane crash. The perimeter of the search zone is 52 miles and its area is 165 sq mi. Find the dimensions of the search zone.

Problem 22

23. When Rose fills her backyard reflecting pool, the volume of water in cubic feet in the pool t minutes after the water is turned on is given by the formula $V = t(3t - 14)$. During what time period will the pool contain less than 160 ft^3 of water?

24. In Jefferson, Georgia, the population T years after 1985 is given by the formula $P = T^2 - 80T + 2000$. When will the population of Jefferson be at most 4000?

25. A sidewalk of uniform width is to be built outside a square parking lot, 40 ft on each side. How wide should the sidewalk be if the total area of parking lot and sidewalk is to be 2025 ft^2?

26. A rectangular field is fenced in on three sides and bordered by a river on the fourth. If the area of the field is 1250 ft^2 and the length of the fence is 100 ft, what are the dimensions of the field?

27. A metal box with no top was made by cutting a 3-in. square from each corner of a square sheet of cooper and then bending up the sides. What size sheet of copper was used so that the box contains 192 in.^{3}? (HINT: $V = lwh$.)

28. An open box with no top was made from a square piece of cardboard by cutting a 2-in. square from each corner and turning up the sides. If the volume of the box is 98 in.^{3}, what size was the original piece of cardboard? (HINT: $V = lwh$.)

29. Find the dimensions of a rectangular poster of length 4 in. less than twice its width and with an area of 70 in.^{2}

30. Find the dimensions of a rectangular picture whose length is 2 in. greater than three times its width and whose area is 56 in.^{2}

31. The perimeter of a rectangular billboard is 44 m and its area is 120 m^2. What are its dimensions?

32. What are the dimensions of a rectangular placement of perimeter 56 cm and area 187 cm^2?

33. The profit per month of Mary's Dress Shop is given by the formula $P = x^2 + 20x - 200$, where x is the number of dresses sold. How many dresses must Mary sell this month to make a profit in excess of $600?

34. The area of a square playground must be at most 10,000 yd^2. How long could a side be?

35. Plastic soft-drink bottles can be recycled for use by carpet manufacturers. One square inch of carpet requires 32 of the recycled 2-liter bottles. Suppose the area of an office is 600 ft^2 and its length is 10 feet less than twice the width.
    (a) What are the dimensions of the office?
    (b) How many 2-liter bottles are required to manufacture the office carpet?

36. Yetunde framed a picture that was 2 inches longer than it was wide. The frame had a uniform width of 2 inches. If the area of the framed picture was 168 in.^{2}, find the dimensions of the unframed picture.

37. The area of a triangle is 49 yd^2. If the base is twice the height, find the length of the base and the height.

38. The area of a triangle is 8 ft^2. If the base is four times the height, find the length of the base.

39. The hypotenuse of a right triangle is 26 m. If the length of the longer leg is 4 more than twice the shorter leg, find the length of each leg.

40. A ball is thrown straight upward from the ground with an initial velocity of 48 ft/sec. t seconds later it will be at a height of $h = 48t - 16t^2$. How many seconds will it take the ball to reach a height of 32 ft? What is the meaning of two answers?

Challenge Problems

41. Two trains leave the station at the same time. One travels south at a rate 70 mph faster than the other train, which is traveling east. After 2 hours they are 260 miles apart. Find the speed of each.

42. Winfield throws a baseball from home plate toward third base, and at the same time, Puckett throws a baseball from home plate toward first base. Winfield's ball travels 20 ft/sec faster than Puckett's. If the balls are 100 ft apart after 1 second, how fast is each ball traveling?

In Your Own Words

43. Make up a word problem about a right triangle that uses a quadratic equation as its mathematical model.

5.3   Completing the Square

TAPE 10

Notice the similarity in solving the following four quadratic equations.

$$\boxed{1} \qquad x^2 = 4 \qquad\qquad \boxed{2} \qquad x^2 = 5$$
$$x^2 = 2^2 \qquad\qquad\qquad x^2 = (\sqrt{5})^2$$
$$x^2 - 2^2 = 0 \qquad\qquad x^2 - (\sqrt{5})^2 = 0$$
$$(x + 2)(x - 2) = 0 \qquad (x + \sqrt{5})(x - \sqrt{5}) = 0$$
$$\{\pm 2\} \qquad\qquad\qquad \{\pm\sqrt{5}\}$$

$$\boxed{3} \qquad x^2 = -4 \qquad\qquad \boxed{4} \qquad x^2 = -5$$
$$x^2 = 4(-1) \qquad\qquad\qquad x^2 = 5(-1)$$
$$x^2 = (2i)^2 \qquad\qquad\qquad x^2 = (\sqrt{5}i)^2$$
$$x^2 - (2i)^2 = 0 \qquad\qquad x^2 - (\sqrt{5}i)^2 = 0$$
$$(x + 2i)(x - 2i) = 0 \qquad (x + \sqrt{5}i)(x - \sqrt{5}i) = 0$$
$$\{\pm 2i\} \qquad\qquad\qquad \{\pm\sqrt{5}i\}$$

In each case note that the equation

$$x^2 = A$$

led to the solution set $\{\pm\sqrt{A}\}$. (Remember immediately to replace such expressions as $\sqrt{-17}$ with $\sqrt{17}i$.) These examples suggest the following useful result, which allows us to omit the factoring steps.

Square Root Property

$$X^2 = A$$

is equivalent to the two equations

$$X = \pm\sqrt{A}$$

► **EXAMPLE 1**

Solve using the Square Root Property.

(a) $x^2 = 27$ (b) $2x^2 + 16 = 0$

Solutions

(a) $x^2 = 27$

$$x = \pm\sqrt{27} \qquad\qquad \text{Square Root Property}$$
$$x = \pm 3\sqrt{3}$$
$$\{\pm 3\sqrt{3}\}$$

(b) $2x^2 + 16 = 0$

$$2x^2 = -16$$

$$x^2 = -8 \qquad \text{Divide by 2}$$

$$x = \pm\sqrt{-8} \qquad \text{Square Root Property}$$

$$x = \pm\sqrt{8}i$$

$$x = \pm2\sqrt{2}i$$

$$\{\pm2\sqrt{2}i\} \qquad \blacktriangleleft$$

The Square Root Property says that the equation $X^2 = A$ is equivalent to the pair of equations $X = \pm\sqrt{A}$. Note that the X in this theorem need not be just a single variable, but can be an expression such as $x + 1$.

▶ **EXAMPLE 2**

Solve $(x + 1)^2 = 9$.

Solution

$$(x + 1)^2 = 9$$

$$x + 1 = \pm3 \qquad \text{Square Root Property}$$

$$x = -1 \pm 3$$

This is equivalent to the *two* equations

$$x = -1 + 3 \quad \text{or} \quad x = -1 - 3$$

$$x = 2 \qquad \text{or} \quad x = -4$$

$$\{2, -4\} \qquad \blacktriangleleft$$

▶ **EXAMPLE 3**

Solve $x^2 - 6x + 9 = 5$.

Solution

$$x^2 - 6x + 9 = 5$$

$$(x - 3)^2 = 5 \qquad \text{Factoring}$$

$$x - 3 = \pm\sqrt{5} \qquad \text{Square Root Property}$$

$$x = 3 \pm \sqrt{5}$$

$$\{3 + \sqrt{5}, 3 - \sqrt{5}\} \quad \text{or} \quad \{3 \pm \sqrt{5}\} \qquad \blacktriangleleft$$

The preceding examples suggest a general technique that will apply to *all* quadratic equations. The idea is to make one side of the equation a perfect square trinomial and then apply the Square Root Property.

► **EXAMPLE 4**

Solve $2x^2 + 30 = 16x$.

Solution

First write the equation in standard form.

$$2x^2 - 16x + 30 = 0$$

Divide both sides by 2, the leading coefficient.

$$x^2 - 8x + 15 = 0$$

Subtract 15 from both sides.

$$x^2 - 8x = -15$$

Before we can continue, we must learn how to *complete the square*. What must we add to $x^2 - 8x$ to make a perfect square? Notice that $x^2 - 8x + 16$ is a perfect square and 16 is $\frac{1}{2}(-8)$ squared.

Square one-half of -8, which gives 16, and add it to both sides.

$$x^2 - 8x \boxed{+ 16} = -15 \boxed{+ 16}$$

Factor the left side and simplify the right.

$$(x - 4)^2 = 1$$

Apply the Square Root Property and simplify.

$$x - 4 = \pm\sqrt{1}$$
$$x - 4 = \pm 1$$
$$x = 4 \pm 1$$
$$\{3, 5\}$$ ◄

► **EXAMPLE 5**

Calculate the term to add to each binomial to make each a perfect square.

(a) $x^2 + 6x$ (b) $x^2 - 3x$

Solutions

(a) $x^2 + 6x$

The coefficient of x is 6. Divide 6 by 2 and square this result.

$$\left(\frac{6}{2}\right)^2 = (3)^2 = 9$$

$x^2 + 6x + 9$ is a perfect square.

(b) $x^2 - 3x$

The coefficient of x is -3.

$$\left(\frac{-3}{2}\right)^2 = \frac{9}{4}$$

$x^2 - 3x + \frac{9}{4}$ is a perfect square. ◄

> ### Solving a Quadratic Equation by Completing the Square
>
> **1.** Write the equation in standard form.
> **2.** Divide both sides by the coefficient of x^2 if the coefficient is not 1.
> **3.** Subtract the constant term from both sides.
> **4.** Divide the coefficient of x by 2, square the result, and then add this number to both sides.
> **5.** Factor the left side (it will be a perfect square) and simplify the right side.
> **6.** Apply the Square Root Property and simplify.
> **7.** Write the solution set.

Note the foregoing steps in the following worked examples.

► **EXAMPLE 6**

Solve $x^2 - 5x + 3 = 0$ by completing the square.

Solution

$$x^2 - 5x + 3 = 0 \qquad \text{Standard form}$$

$$x^2 - 5x = -3 \qquad \text{Subtract 3 from both sides}$$

$$x^2 - 5x + \left(-\frac{5}{2}\right)^2 = -3 + \left(-\frac{5}{2}\right)^2 \qquad \text{Complete the square}$$

$$\left(x - \frac{5}{2}\right)^2 = -3 + \frac{25}{4}$$

$$\left(x - \frac{5}{2}\right)^2 = \frac{-12}{4} + \frac{25}{4} = \frac{13}{4} \qquad \text{Common denominator}$$

$$x - \frac{5}{2} = \pm\sqrt{\frac{13}{4}} \qquad \text{Square Root Property}$$

$$x = \frac{5}{2} \pm \frac{\sqrt{13}}{2}$$

$$\left\{\frac{5}{2} \pm \frac{\sqrt{13}}{2}\right\} \quad \text{or} \quad \left\{\frac{5 \pm \sqrt{13}}{2}\right\} \qquad ◄$$

► **EXAMPLE 7**

Solve $2x^2 - x + 4 = 0$ by completing the square.

Solution

$$2x^2 - x + 4 = 0$$

$$x^2 - \frac{1}{2}x + 2 = 0 \qquad \text{Divide both sides by 2}$$

$$x^2 - \frac{1}{2}x = -2 \qquad \text{Subtract 2 from both sides}$$

$$x^2 - \frac{1}{2}x + \left(-\frac{1}{4}\right)^2 = -2 + \left(-\frac{1}{4}\right)^2 \qquad \text{Complete the square}$$

$$x^2 - \frac{1}{2}x + \frac{1}{16} = -2 + \frac{1}{16}$$

$$\left(x - \frac{1}{4}\right)^2 = \frac{-31}{16}$$

$$x - \frac{1}{4} = \pm\sqrt{\frac{-31}{16}} \qquad \text{Square Root Property}$$

$$x - \frac{1}{4} = \pm\sqrt{\frac{31}{16}}\, i = \pm\frac{\sqrt{31}}{4}\, i$$

$$x = \frac{1}{4} \pm \frac{\sqrt{31}}{4}\, i$$

$$\left\{\frac{1}{4} \pm \frac{\sqrt{31}}{4}\, i\right\} \qquad \blacktriangleleft$$

Calculator Box

Storing and Recalling Numbers

Storing a Number in Memory

First, we learn how to store a number for later use. Let's put $\sqrt{3^2 + 16}$ in memory. Look for a key like $\boxed{\text{STO}}$ $\boxed{\text{M}_{\text{IN}}}$ or $\boxed{\rightarrow}$

Scientific calculator:

Press $\boxed{3}$ $\boxed{x^2}$ $\boxed{+}$ $\boxed{16}$ $\boxed{=}$ $\boxed{\sqrt{}}$ $\boxed{\text{STO}}$

to store the display number.

Graphing calculator:

Graphing calculators usually have at least 26 memories indexed by letters of the alphabet and the $\boxed{\text{ALPHA}}$ key.

Press $\boxed{\sqrt{}}$ $\boxed{(}$ $\boxed{3}$ $\boxed{x^2}$ $\boxed{+}$ $\boxed{16}$ $\boxed{)}$ $\boxed{\text{STO}}$ $\boxed{\text{ALPHA}}$ $\boxed{\text{J}}$ $\boxed{\text{EXE}}$

to store the result of $\sqrt{3^2 + 16}$ in memory J.

Some graphing calculators do not require the $\boxed{\text{ALPHA}}$ key when *storing* a number. With the TI–81, press

$\boxed{\sqrt{}}$ $\boxed{(}$ $\boxed{3}$ $\boxed{x^2}$ $\boxed{+}$ $\boxed{16}$ $\boxed{)}$ $\boxed{\text{STO}}$ $\boxed{\text{J}}$ $\boxed{\text{ENTER}}$.

Recalling a Number from Memory

Now let's recall our stored number from memory and use it to calculate $\dfrac{-3 + \sqrt{3^2 + 16}}{2}$.

Scientific calculator:

Clear the display and press

| (| | 3 | | +/− | | + | | RCL | |) | | ÷ | | 2 | | = | and read ▮**1** on

the display. Some calculators use | MR | instead of | RCL | to recall a stored number.

Graphing calculator:

Press | (| | (−) | | 3 | | + | |ALPHA| | J | |) | | ÷ | | 2 | | ENTER |

to compute $\dfrac{-3 + \sqrt{3^2 + 16}}{2}$.

Calculator Exercises

Write all results to five decimal places.

1. Store 1234 in memory; then recall it to approximate $\sqrt{1234}$.
2. Store $2 - \sqrt{2}$ in memory; then recall it to approximate $(2 - \sqrt{2})^3$.

Store $2.345^{7/3} + \pi$ in memory for Problems 3 and 4.

3. Approximate $5 + \sqrt{\text{stored number}}$.
4. Approxmiate $5 - \sqrt[3]{\text{stored number}}$.

Answers

1. ≈ 35.12834 **2.** ≈ 0.20101 **3.** ≈ 8.23224 **4.** ≈ 2.81390

Problem Set 5.3

Warm-Ups

In Problems 1 through 8, use the Square Root Property to solve each quadratic equation. See Example 1.

1. $x^2 - 25 = 0$ 　　　　　　　　　　　**2.** $x^2 + 9 = 0$
3. $x^2 - 5 = 0$ 　　　　　　　　　　　**4.** $x^2 + 5 = 0$
5. $x^2 - 48 = 0$ 　　　　　　　　　　**6.** $x^2 + 18 = 0$
7. $2t^2 + 16 = 0$ 　　　　　　　　　　**8.** $-6s^2 - 18 = 0$

In Problems 9 through 12, solve using the Square Root Property. See Example 2.

9. $(x + 3)^2 = 4$ 　　　　　　　　　　**10.** $(y + 3)^2 = 12$
11. $(x - 5)^2 = -7$ 　　　　　　　　　**12.** $(x + 8)^2 = -18$

In Problems 13 through 18, calculate the term that must be added to each binomial to make it a perfect square. See Example 5.

13. $x^2 + 4x$ 　　　　　　　　　　　　**14.** $x^2 - 6x$

15. $x^2 + 7x$

16. $x^2 - 5x$

17. $x^2 + x$

18. $k^2 - 13k$

In Problems 19 through 28, solve each equation by completing the square. For Problems 19 through 24, see Examples 3 and 6.

19. $x^2 - 4x - 5 = 0$

20. $x^2 - 2x + 2 = 0$

21. $x^2 + x - 6 = 0$

22. $x^2 + 2x = 4$

23. $x^2 - 6x + 1 = 0$

24. $t^2 + 3t + 4 = 0$

For Problems 25 through 28, see Examples 4 and 7.

25. $3w^2 - 6w - 9 = 0$

26. $2x^2 - 4x + 4 = 0$

27. $4x^2 + 12x + 6 = 0$

28. $2x^2 - 5x + 1 = 0$

Practice Exercises

In Problems 29 through 48, solve each equation by completing the square.

29. $x^2 - 2x - 8 = 0$

30. $x^2 - 4x - 12 = 0$

31. $x^2 - 2x + 1 = 0$

32. $2x^2 - 4x - 6 = 0$

33. $3x^2 - 12x = -9$

34. $2x^2 - 6x + 1 = 0$

35. $x^2 - 3x - 4 = 0$

36. $x^2 - x = -2$

37. $4x^2 - 20x + 16 = 0$

38. $5z^2 - 5z + 15 = 0$

39. $2u^2 - 2u = 2$

40. $3x^2 + 2x - 5 = 0$

41. $2x^2 + 5x - 4 = 0$

42. $5y^2 - 2y + 7 = 0$

43. $3v^2 - 4v + 10 = 0$

44. $2w^2 - 3w + 1 = 0$

45. $2x^2 - 7x + 5 = 0$

46. $3x^2 - 7x + 2 = 0$

47. $4x^2 - 6x + 10 = 0$

48. $6x^2 - 8x - 2 = 0$

In Problems 49 through 54, approximate solutions to the nearest thousandth.

49. $x^2 = 444$

50. $x^2 - 3.4 = 0$

51. $3t^2 = -18$

52. $2z^2 = 11$

53. $4x^2 - 3.613 = 0$

54. $5x^2 - 4 = 2\pi$

In Problems 55 through 64, solve each problem.

55. Sam's Steak House is buying an advertisement in the newspaper. The ad is in the shape of a rectangle with the length three times the width.
(a) If the ad area is 4.6875 in.², what are the dimensions of the ad?
(b) On weekdays, the newspaper charges $27.90 for each square inch. What is the cost of Sam's ad?
(c) In Sunday editions of the newspaper, the rate for advertising is $42.63 for each square inch. What does it cost Sam to run a Sunday ad?

(d) If Sam's advertising budget is $600/month, can Sam run at least one ad per week, with two of those weeks being in the Sunday newspaper?

56. The second grades at Main Street Elementary School are painting a big red right triangle on the gym wall. If the hypotenuse is 8 m and extends from the floor to a point 4 m up the wall, how far along the wall does the painting extend?

57. What is the distance between the opposite corners of a square video game screen whose width is 12.5 inches? Answer to the nearest tenth of an inch.

58. The area of the YWCA baseball field is 1600 square feet. How far will a home-run hitter run when she runs around the bases?

59. An acre was deemed by royal decree in the fourteenth century to be 43,560 ft^2. What would be the measure of one side of a square-shaped acre of farmland? (Answer to the nearest foot.)

60. The hypotenuse of a right triangle is $3\sqrt{5}$ feet long and one leg is 6 ft long. Find the length of the other leg.

61. The hypotenuse of a right triangle is 10 m long and one leg is $5\sqrt{2}$ meters long. Find the length of the other leg.

62. What radius is necessary so that a traffic circle has an area of at least 1000 m^2? Answer to the nearest tenth.

63. Wenjian is going to fertilize the ground to plant a garden. The bag of fertilizer he purchased will adequately fertilize 20 square feet of soil. If he wants to till a rectangular plot in which the length is twice the width, what dimension might he choose for the width in order to have enough fertilizer? (Answer to the nearest tenth of a foot.)

64. Elija has enough grass seed to cover 150 ft^2. He wishes to start a lawn area in the shape of a square. What is the largest integer value possible for the length of the side of the square?

Challenge Problems

In Problems 65 through 70, use the Square Root Property to solve each equation. Assume that a and b are real numbers.

65. $x^2 = a + b$

66. $x^2 = a^2 + b^2$

67. $(x + 1)^2 = a$

68. $(x - a)^2 = -16$

69. $(x + b)^2 = 8$

70. $(x - b)^2 = -12$

In Problems 71 through 75, solve each equation by completing the square. Assume that a, b, and c are real numbers.

71. $4x^2 + 4ax - a = 0$

72. $\frac{1}{2}x^2 - \frac{1}{3}x + \frac{4}{9} = 0$

73. $0.8x^2 - 0.48x + 0.01 = 0$

74. $ax^2 + 3ax = a^2, \quad a \neq 0$

75. $ax^2 + bx + c = 0, \quad a \neq 0$

In Your Own Words

76. Explain how to solve the equation $x^2 - 6x + 4 = 0$ by completing the square.

5.4 The Quadratic Formula

TAPE 10

Completing the square is often a cumbersome method to use in solving a quadratic equation. If the quadratic can be factored, that method is easier and faster. However, not all quadratics will factor, so a general method is needed. Since completing the square works for all quadratics, we can solve the general quadratic equation by completing the square and find a formula to use in solving all such equations.

We follow the procedure given in Section 5.3 for completing the square:

$$ax^2 + bx + c = 0, \quad a > 0 \qquad \text{Standard form}$$

$$x^2 + \frac{b}{a}x + \frac{c}{a} = 0 \qquad \text{Divide by } a$$

$$x^2 + \frac{b}{a}x = -\frac{c}{a} \qquad \text{Subtract } \frac{c}{a}$$

$$x^2 + \frac{b}{a}x + \left(\frac{b}{2a}\right)^2 = -\frac{c}{a} + \left(\frac{b}{2a}\right)^2 \qquad \text{Complete the square}$$

$$\left(x + \frac{b}{2a}\right)^2 = -\frac{c}{a} + \left(\frac{b}{2a}\right)^2 \qquad \text{Factor left side}$$

$$\left(x + \frac{b}{2a}\right)^2 = \frac{-4ac + b^2}{4a^2} \qquad \text{Simplify right side}$$

$$x + \frac{b}{2a} = \pm\sqrt{\frac{b^2 - 4ac}{4a^2}} \qquad \text{Square Root Property}$$

$$x = -\frac{b}{2a} \pm \frac{\sqrt{b^2 - 4ac}}{2a} \qquad \text{Subtract } \frac{b}{2a}$$

$$x = \frac{-b \pm \sqrt{b^2 - 4ac}}{2a}$$

This is an important formula and should be memorized. It is called the Quadratic Formula. Although we derived the formula for $a > 0$, it works just as well if $a < 0$.

The Quadratic Formula

The quadratic equation $ax^2 + bx + c = 0$, $a \neq 0$, is equivalent to

$$x = \frac{-b \pm \sqrt{b^2 - 4ac}}{2a}$$

► EXAMPLE 1

Solve $x^2 + 6 = -5x$ by the Quadratic Formula.

Solution

Write the equation in standard form.

$$x^2 + 5x + 6 = 0$$

Thus $a = 1$, $b = 5$, and $c = 6$. Substitute into the Quadratic Formula.

$$x = \frac{-5 \pm \sqrt{5^2 - 4(1)(6)}}{2(1)}$$

$$= \frac{-5 \pm \sqrt{25 - 24}}{2}$$

$$= \frac{-5 \pm 1}{2}$$

$$x = \frac{-5 + 1}{2} \quad \text{or} \quad x = \frac{-5 - 1}{2}$$

$$x = -2 \quad \text{or} \quad x = -3$$

$$\{-2, -3\}$$

◄

► **EXAMPLE 2**

Solve $2x^2 + 3x - 4 = 0$.

Solution

Here $a = 2$, $b = 3$, and $c = -4$.

$$x = \frac{-3 \pm \sqrt{9 - 4(2)(-4)}}{2(2)}$$

$$= \frac{-3 \pm \sqrt{9 + 32}}{4}$$

$$= \frac{-3 \pm \sqrt{41}}{4} = -\frac{3}{4} \pm \frac{\sqrt{41}}{4}$$

There are several acceptable ways to write this solution set.

$$\left\{ \frac{-3 \pm \sqrt{41}}{4} \right\} \quad \text{or} \quad \left\{ \frac{-3 + \sqrt{41}}{4}, \frac{-3 - \sqrt{41}}{4} \right\} \quad \text{or} \quad \left\{ -\frac{3}{4} \pm \frac{\sqrt{41}}{4} \right\}$$

$$\text{or} \quad \left\{ -\frac{3}{4} + \frac{\sqrt{41}}{4}, -\frac{3}{4} - \frac{\sqrt{41}}{4} \right\}$$

◄

► **EXAMPLE 3**

Solve $t^2 - 3t + 3 = 0$.

Solution

For this equation, $a = 1$, $b = -3$, and $c = 3$.

Be Careful!

Be careful substituting -3 for b in the formula; $-b$ will become $-(-3)$.

$$t = \frac{-(-3) \pm \sqrt{(-3)^2 - 4(1)(3)}}{2(1)}$$

$$= \frac{3 \pm \sqrt{9 - 12}}{2}$$

$$= \frac{3 \pm \sqrt{-3}}{2}$$

$$= \frac{3 \pm \sqrt{3}i}{2} \qquad\qquad \text{Simplify } \sqrt{-3}$$

$$= \frac{3}{2} \pm \frac{\sqrt{3}}{2}i \qquad\qquad \text{Standard form of complex numbers}$$

$$\left\{\frac{3}{2} \pm \frac{\sqrt{3}}{2}i\right\} \quad \text{or} \quad \left\{\frac{3}{2} + \frac{\sqrt{3}}{2}i, \frac{3}{2} - \frac{\sqrt{3}}{2}i\right\}$$

The solutions are complex conjugates.   ◄

The Quadratic Formula often leads to an expression such as the following. Notice how it is simplified.

$$\frac{8 \pm \sqrt{20}}{4} = \frac{8 \pm \sqrt{4(5)}}{4}$$

$$= \frac{8 \pm 2\sqrt{5}}{4}$$

$$= \frac{8}{4} \pm \frac{2\sqrt{5}}{4}$$

$$= 2 \pm \frac{\sqrt{5}}{2}$$

This may also be written with a common denominator as $\dfrac{4 \pm \sqrt{5}}{2}$.

► **EXAMPLE 4**

Solve $3x^2 - 2x + 5 = 0$.

Solution

With $a = 3$, $b = -2$, and $c = 5$, we have

$$x = \frac{-(-2) \pm \sqrt{4 - 4(3)(5)}}{2(3)}$$

$$= \frac{2 \pm \sqrt{4 - 60}}{6}$$

$$= \frac{2 \pm \sqrt{-56}}{6} = \frac{2 \pm \sqrt{56}i}{6}$$

Immediately rewrite $\sqrt{-56}$ as $\sqrt{56}i$ and note that there are two complex solutions. Notice that $\sqrt{56} = \sqrt{4(14)} = 2\sqrt{14}$, so

$$x = \frac{2 \pm 2\sqrt{14}i}{6}$$

$$= \frac{2}{6} \pm \frac{2\sqrt{14}i}{6}$$

$$= \frac{1}{3} \pm \frac{\sqrt{14}i}{3}$$

$$= \frac{1}{3} \pm \frac{\sqrt{14}}{3}i \qquad \text{Standard form of complex}$$
$$\text{numbers}$$

$$\left\{ \frac{1}{3} \pm \frac{\sqrt{14}}{3}i \right\} \quad \text{or} \quad \left\{ \frac{1}{3} + \frac{\sqrt{14}}{3}i, \frac{1}{3} - \frac{\sqrt{14}}{3}i \right\} \qquad \blacktriangleleft$$

Notice that the solutions are complex conjugates.

► **EXAMPLE 5**

Solve $2x^2 + 2x - 1 = 0$.

Solution

$$x = \frac{-2 \pm \sqrt{2^2 - 4(2)(-1)}}{2(2)} \qquad \text{Substitute into formula}$$

$$= \frac{-2 \pm \sqrt{4 + 8}}{4} = \frac{-2 \pm \sqrt{12}}{4}$$

$$= \frac{-2 \pm 2\sqrt{3}}{4} \qquad \text{Simplify } \sqrt{12}$$

$$= \frac{-2}{4} \pm \frac{2\sqrt{3}}{4}$$

$$= -\frac{1}{2} \pm \frac{\sqrt{3}}{2} \qquad \text{Reduce}$$

$$\left\{ -\frac{1}{2} \pm \frac{\sqrt{3}}{2} \right\} \qquad \blacktriangleleft$$

► **EXAMPLE 6**

Solve $\frac{1}{6}x^2 + \frac{1}{3}x - \frac{1}{6} = 0$.

Solution

It is a good idea to clear fractions first and then use the Quadratic Formula. Multiply by 6 to get

$$x^2 + 2x - 1 = 0$$

So $a = 1$, $b = 2$, and $c = -1$, and with the formula

$$x = \frac{-2 \pm \sqrt{4 - 4(1)(-1)}}{2(1)}$$

$$= \frac{-2 \pm \sqrt{8}}{2}$$

$$= \frac{-2 \pm 2\sqrt{2}}{2}$$

$$= -\frac{2}{2} \pm \frac{2\sqrt{2}}{2}$$

$$= -1 \pm \sqrt{2}$$

$$\{-1 \pm \sqrt{2}\}$$

Solving a Quadratic Equation Using the Quadratic Formula

1. Write the equation in standard form.
2. Identify the values of a, b, and c.
3. Substitute these values into the formula and simplify.
4. Check the solutions if required.
5. Write the solution set.

A quadratic equation must be in standard form before the Quadratic Formula can be used.

▶ **EXAMPLE 7**

Solve $(2r + 1)(r - 1) = 5$ using the Quadratic Formula.

Solution

Be Careful!

Be careful to write the equation in standard form first.

$$(2r + 1)(r - 1) = 5$$

$$2r^2 - r - 1 = 5$$

$$2r^2 - r - 6 = 0 \qquad \text{Standard form}$$

Thus $a = 2$, $b = -1$, and $c = -6$.

$$r = \frac{-(-1) \pm \sqrt{(-1)^2 - 4(2)(-6)}}{2(2)} \qquad \begin{array}{l}\text{Substitute into}\\\text{Quadratic Formula}\end{array}$$

$$= \frac{1 \pm \sqrt{1 + 48}}{4}$$

$$= \frac{1 \pm \sqrt{49}}{4}$$

$$= \frac{1 \pm 7}{4}$$

$$\left\{-\frac{3}{2}, 2\right\}$$

Notice the important role played by the expression that appears under the radical, $b^2 - 4ac$. The expression, called the **discriminant,** gives the following information about the nature of the solutions of the quadratic equation.

The Discriminant

$b^2 - 4ac$	**Nature of solutions**
Positive	Two real solutions
Zero	One real solution of multiplicity 2
Negative	Two complex, nonreal solutions

► **EXAMPLE 8**

Find the discriminant and determine the nature of the solutions for each equation.

(a) $3x^2 + 2x + 4 = 0$

(b) $x^2 + 2x + 1 = 0$

(c) $3x^2 + 5x - 2 = 0$

Solutions

(a) $b^2 - 4ac = 2^2 - 4(3)(4)$

$$= 4 - 48$$

$$= -44$$

The value of the discriminant is -44 and the equation has two complex, nonreal solutions.

(b) $b^2 - 4ac = 2^2 - 4(1)(1)$

$$= 4 - 4$$

$$= 0$$

Hence the equation has one real solution of multiplicity 2.

(c) $b^2 - 4ac = 5^2 - 4(3)(-2)$

$$= 25 + 24$$

$$= 49$$

The equation has two real solutions.   ◄

We have learned how to solve a quadratic equation by factoring, the Square Root Property, completing the square, and the Quadratic Formula. Which method should be used in solving a quadratic equation?

Any quadratic equation can be solved by using the Quadratic Formula or by completing the square. However, factoring is much easier if it can be used. The Square Root Property saves some steps if it can be used.

> **Choosing a Method to Use in Solving a Quadratic Equation**
>
> **1** Use the Square Root Property if the equation is of the form $X^2 = A$.
> **2.** Next, try factoring.
> **3.** Use the Quadratic Formula if factoring won't work.
> **4.** Use completing the square when directed to do so.

Quadratic Inequalities

▶ **EXAMPLE 9**

Solve each inequality and write the solution set in interval notation.

(a) $x^2 - 4x + 2 < 0$ (b) $x^2 - 4x + 5 > 0$

Solutions

(a) The inequality can be solved by the method of boundary numbers. We first find the boundary numbers by solving the associated equation.

$$x^2 - 4x + 2 = 0$$

$$x = \frac{4 \pm \sqrt{16 - 4(1)(2)}}{2} \qquad \text{Substitute into the Quadratic Formula}$$

$$= \frac{4 \pm \sqrt{8}}{2} = 2 \pm \sqrt{2}$$

We locate the boundary numbers on a number line.

At first, it might seem difficult to select test numbers in the regions. However, we notice that 2 is exactly halfway between $2 + \sqrt{2}$ and $2 - \sqrt{2}$ and thus in region *B*. To find test numbers in the other regions, estimate the value of $2 - \sqrt{2}$ and $2 + \sqrt{2}$. Since $2 - \sqrt{2} \approx 0.6$ and $2 + \sqrt{2} \approx 3.4$, zero is in region *A* and 5 is in region *C*. Testing shows that region *B* is in the solution set. The original inequality does not include equality. Thus the solution set is $(2 - \sqrt{2}, 2 + \sqrt{2})$.

(b) We use the method of boundary numbers and first find the boundary numbers.

$$x^2 - 4x + 5 = 0 \qquad \text{The associated equation}$$

$$x = \frac{4 \pm \sqrt{16 - 4(1)(5)}}{2} \qquad \text{Substitute into the Quadratic Formula}$$

$$= \frac{4 \pm \sqrt{-4}}{2} = \frac{4 \pm 2i}{2} = 2 \pm i$$

Calculator Box

Quadratic Formula with a Calculator

Quadratics with Real Solutions

Let's approximate the solutions for $2x^2 + 6x + 3 = 0$. Substituting in the quadratic formula yields

$$\frac{-6 \pm \sqrt{6^2 - 4 \cdot 2 \cdot 3}}{2 \cdot 2}.$$

It is convenient to calculate the solution using the $+$ of the $\pm$ first. We will store the square root in memory so that we won't have to calculate it twice.

Scientific calculator: Press

$\boxed{36}$ $\boxed{-}$ $\boxed{24}$ $\boxed{=}$ $\boxed{\sqrt{}}$ $\boxed{\text{STO}}$ (store the square root) $\boxed{-}$ $\boxed{6}$ $\boxed{=}$ $\boxed{\div}$ $\boxed{4}$ $\boxed{=}$

and we read $\boxed{-0.633974596}$ on the display. (Notice that we did simple arithmetic in our head.) Now, for the second solution, press

$\boxed{6}$ $\boxed{+/-}$ $\boxed{-}$ $\boxed{\text{RCL}}$ $\boxed{=}$ $\boxed{\div}$ $\boxed{4}$ $\boxed{=}$ and read $\boxed{-2.3660254}$.

Graphing calculator:

Press $\boxed{\sqrt{}}$ $\boxed{(}$ $\boxed{36}$ $\boxed{-}$ $\boxed{24}$ $\boxed{)}$ $\boxed{\text{STO}}$ $\boxed{\text{H}}$ $\boxed{\text{ENTER}}$

(to store in H)

$\boxed{-}$ $\boxed{6}$ $\boxed{\text{ENTER}}$ $\boxed{\div}$ $\boxed{4}$ $\boxed{\text{ENTER}}$

for the first solution, and

$\boxed{(}$ $\boxed{(-)}$ $\boxed{6}$ $\boxed{-}$ $\boxed{\text{ALPHA}}$ $\boxed{\text{H}}$ $\boxed{)}$ $\boxed{\div}$ $\boxed{4}$ $\boxed{\text{ENTER}}$

for the second.

Our two solutions are approximately -0.6340 and -2.3360.

Quadratics with Complex Nonreal Solutions

Next, let's approximate the solutions for $3x^2 - x + 2 = 0$. Substituting in the quadratic formula gives $\dfrac{1 \pm \sqrt{1^2 - 4 \cdot 3 \cdot 2}}{3 \cdot 2}$. Here, when we calculate the expression under the square root, we get -23, a negative number. We will have two complex solutions. The value of $\sqrt{-23}$ is $\sqrt{23}\, i$. Therefore, our solutions are $\dfrac{1}{6} \pm \dfrac{\sqrt{23}}{6} i$. Or, with either type of calculator, they are approximately $1.6667 \pm 0.7993i$.

Calculator Exercises

Approximate the solutions to each quadratic equation to the nearest thousandth.

1. $2x^2 + 3x - 4 = 0$ **2.** $23x^2 - 77x + 29 = 0$

3. $0.1234x^2 + 9x - 1.0987 = 0$ **4.** $3x^2 - 2x + 5 = 0$

Answers

1. ≈ -2.351 and 0.851 **2.** ≈ 0.432 and 2.915 **3.** ≈ -73.055 and 0.123 **4.** $\approx 0.333 \pm 1.247i$

The next step calls for us to locate these boundary numbers on a number line. The numbers $2 \pm i$ are complex numbers and therefore we cannot locate them on a number line of real numbers. So there are no *real* boundary numbers. If we locate no numbers on the number line, we have one region, the entire number line.

If we test any real number in the original inequality, we will know the solution set. Testing 0 shows that the statement $x^2 - 4x + 5 > 0$ is true. Thus the entire number line is in the solution set. The solution set written in interval notation is $(-\infty, +\infty)$.   ◀

Problem Set 5.4

Warm-Ups

In Problems 1 through 6, solve each equation using the Quadratic Formula. See Examples 1 and 2.

1. $x^2 + 2x - 8 = 0$ **2.** $x^2 = 4$ **3.** $x^2 + 2x = 0$

4. $x^2 - x - 1 = 0$ **5.** $x^2 + x - 1 = 0$ **6.** $x^2 + 4x + 4 = 0$

In Problems 7 through 12, solve each equation using the Quadratic Formula. See Example 3.

7. $x^2 - x + 1 = 0$ **8.** $x^2 - 3x + 12 = 0$

9. $x^2 + 1 = 0$ **10.** $x^2 + 2 = 0$

11. $x^2 + 5x + 9 = 0$ **12.** $x^2 - x + 3 = 0$

In Problems 13 through 18, solve each equation using the Quadratic Formula. See Examples 4 and 5.

13. $x^2 - 2x = 1$ **14.** $x^2 + 2x - 1 = 0$

15. $x^2 + 4x - 2 = 0$ **16.** $x^2 + 6x + 10 = 0$

17. $x^2 + 2x + 2 = 0$ **18.** $x^2 + 4x = -10$

In Problems 19 through 22, solve each equation using the Quadratic Formula. See Example 6.

19. $\dfrac{1}{2}x - 1 = \dfrac{1}{4}x^2$ **20.** $\dfrac{1}{3}x = \dfrac{1}{6}x^2 + 1$

21. $\dfrac{2}{3}x^2 + \dfrac{11}{3}x - 7 = 0$ **22.** $5v^2 - \dfrac{11}{2}v - 3 = 0$

In Problems 23 through 26, solve each equation using the Quadratic Formula. See Example 7.

23. $r(r - 6) - (2r - 1) = 0$ **24.** $(x - 1)(x + 1) = x$

25. $(2y + 3)(y - 1) = 3$

26. $t(t + 3) + 2(t + 1) = 1$

In Problems 27 through 30, find the discriminant for each equation and determine the nature of the solutions. See Example 8.

27. $x^2 - x + 4 = 0$

28. $x^2 + 2x - 1 = 0$

29. $2x^2 + 3x = 4$

30. $x^2 - 4x + 4 = 0$

In Problems 31 through 34, solve each inequality and write the solution set in interval notation. See Example 9.

31. $x^2 - 6x + 4 \leq 0$

32. $x^2 - 6x + 7 > 0$

33. $x^2 - 6x + 10 < 0$

34. $x^2 - 2x + 10 > 0$

Practice Exercises

In Problems 35 through 74, solve each equation using any method.

35. $x^2 + 6x + 9 = 0$

36. $x(x + 6) + 6(x + 6) = 0$

37. $2x^2 - x - 1 = 0$

38. $2x^2 + 8 = 0$

39. $x^2 - 2(x + 4) = 0$

40. $x^2 = 4(x - 3)$

41. $5x^2 = 13x + 6$

42. $\left(x - \dfrac{2}{3}\right)^2 = \dfrac{2}{9}$

43. $6x^2 + x - 2 = 0$

44. $x^2 = -x$

45. $x(x - 1) = 0$

46. $5x^2 = 2x$

47. $x^2 + 25 = 0$

48. $\left(x + \dfrac{1}{2}\right)^2 = \dfrac{13}{4}$

49. $x^2 + x - 4 = 0$

50. $x^2 - x + 4 = 0$

51. $\left(y - \dfrac{3}{2}\right)^2 = -\dfrac{11}{4}$

52. $x^2 - 10x + 5 = 0$

53. $x^2 - 2x + 2 = 0$

54. $\dfrac{1}{4}z^2 + z + \dfrac{7}{2} = 0$

55. $x^2 - 6x + 10 = 0$

56. $(x - 1)(2x + 1) = 1$

57. $(2x + 1)(x + 3) = 2$

58. $x^2 - 1.8x - 1.44 = 0$

59. $y^2 + 0.1y - 0.06 = 0$

60. $\dfrac{5}{2}x^2 - 7x - \dfrac{3}{2} = 0$

61. $x^2 + 8x + 7 = 0$

62. $\left(x - \dfrac{5}{2}\right)^2 = \dfrac{81}{4}$

63. $2x^2 - 6 = 0$

64. $2x^2 - 7x - 4 = 0$

65. $3x^2 - 2x - 1 = 0$

66. $(x - 5)^2 + 5 = 0$

67. $2x^2 - x = 0$

68. $y^2 + 8 = 0$

69. $z^2 = 4$

70. $(x - 1)(x + 2) = 0$

71. $2x^2 - 2x + 1 = 0$

72. $2x^2 + 2x + 1 = 0$

73. $9x^2 + 12x + 4 = 0$

74. $9y^2 - 6y + 1 = 0$

In Problems 75 through 82, write the solution set in interval notation.

75. $x^2 - 6x + 6 \leq 0$

76. $x^2 - 2x - 1 \leq 0$

77. $x^2 - 3x + 3 > 0$

78. $x^2 - 2x + 2 > 0$

79. $x^2 + 4x - 2 > 0$

80. $x^2 - 6x - 2 > 0$

81. $x^2 + 3x + 4 < 0$

82. $x^2 - 3x + 5 < 0$

83. What is the discriminant for the equation $ax^2 + bx = 0$? Will the solutions be real or complex?

84. What is the discriminant for the equation $ax^2 + c = 0$? Will the solutions be real or complex?

In Problems 85 through 88, find all real values of k that make each condition true.

85. $x^2 - 2x + k = 0$ has two complex solutions.

86. $x^2 - kx - 1 = 0$ has two distinct real solutions.

87. $x^2 - 2x + k = 0$ has one solution of multiplicity 2.

88. $x^2 + kx + 4 = 0$ has one solution of multiplicity 2.

89. A flare is shot upward from the ground. The height in feet is given by the formula $h = 144t - 16t^2$, where t is the number of seconds after firing. Approximate the number of seconds it takes the flare to first reach a height of 300 ft.

90. A homemade rocket is shot skyward such that its height in feet is given by the formula $h = 160t - 16t^2$,

where t is the number of seconds since lift-off. Approximate the length of time that the rocket is above 250 feet.

91. The output power of an old electric generator, measured in volts, in the basement of the Student Union building is given by $P = 73I - 5I^2$, where I is the current in amperes. What current I is necessary for an output of 120 volts? (Approximate to the nearest hundredth of an ampere.)

92. The length of the cover on a technical manual is 7.4 centimeters less than twice the width. If the area of the cover is 277 cm², find the width to the nearest tenth of a centimeter.

Challenge Problems

In Problems 93 through 99, solve each equation for the variables indicated using the Quadratic Formula or the Square Root Property.

93. $ax^2 + 3x - 2 = 0$; solve for x.

94. $x^2 + 2mx - n = 0$; solve for x ($m > 0$ and $n > 0$).

95. $x^2 + (n - m)x - mn = 0$; solve for x ($m > 0$ and $n > 0$).

96. $\frac{1}{2}gx^2 - kx + L = 0$; solve for x.

97. $d = \frac{D}{L}(L^2 - a^2)$; solve for L.

98. $V = \frac{2}{3}pR^2h$; solve for R.

99. $L = \frac{2}{R^2} - d^2$; $R > 0$; solve for d.

100. Suppose that $b^2 - 4ac > 0$. What can be said about the solutions to $ax^2 + bx + c = 0$ if $b^2 - 4ac$ is a perfect square? If it is not a perfect square?

In Problems 101 and 102, suppose that R_1 and R_2 are two solutions of the general quadratic equation $ax^2 + bx + c = 0$.

101. Find the "sum of the roots formula"; that is, find a formula for $R_1 + R_2$.

102. Find the "product of the roots formula"; that is, find a formula for $R_1 R_2$.

In Problems 103 through 106, find the sum of the roots and the product of the roots without solving the equation.

103. $2x^2 + 3x + 4 = 0$

104. $x^2 - 5x + 10 = 0$

105. $2x - 17 = 5x^2$

106. $3x^2 - 11 = 0$

72. The design of a certain rectangular building calls for a base in the form of a square. Each side of the square base is to be 100 feet less than the height. The local code calls for at least 4000 cubic feet of volume for each occupant. What height will meet the code if the building is designed to house 500 people?

Challenge Problems

In Problems 73 through 80, find the solution set.

73. $x^6 - 64 = 0$

74. $x^6 - 1 = 0$

75. $x^3 - 2x^2 + 4x - 8 = 0$

76. $x^3 + x^2 - x - 1 = 0$

77. $(x + 1)^4 - 5(x + 1)^2 + 4 = 0$

78. $(x - 3)^4 + 2(x - 3)^2 - 15 = 0$

79. Find an equation whose solution set is $\{4, 2, -1\}$.

80. Find an equation whose solution set is $\{\pm 3, \pm i\}$.

In Problems 81 through 84, find the solution set.

81. $x^3 + 8 > 0$

82. $x^2 + x + 1 \geq 0$

83. $x^2(x - 2) \geq 0$

84. $x(x - 1)^2 < 0$

In Your Own Words

85. How can factoring be used to solve some equations?

5.6   More Rational Equations and Inequalities

TAPE 10

In Section 3.4 we solved rational equations that led to linear equations. In this section we solve rational equations that lead to quadratic equations. The procedure we used was to multiply both sides of the equation by the least common denominator of all the fractions in the equation. If any denominator contains a variable, there is a possibility of multiplication by zero. (Remember, we can multiply both sides of an equation by any nonzero number.)

The best strategy is to clear fractions and solve the resulting equation for possible solutions. Then check to see if any of the possible solutions would make any denominator of the original equation have a value of zero. If so, that number cannot be included in the solution set.

> ### Solving Rational Equations
>
> **1.** Multiply both sides of the equation by the least common denominator to clear fractions.
>
> **2.** Find possible solutions by solving the resulting equation.
>
> **3.** Discard possible solutions that make any denominator in the original equation have a value of zero.
>
> **4.** Write the solution set.

▶ **EXAMPLE 1**

Solve $\dfrac{x}{x-2} - \dfrac{7}{x+5} = \dfrac{17}{x^2+3x-10}$.

Solution

$$\frac{x}{x-2} - \frac{7}{x+5} = \frac{17}{x^2+3x-10}$$

Factor all denominators.

$$\frac{x}{x-2} - \frac{7}{x+5} = \frac{17}{(x-2)(x+5)}$$

Multiply both sides by $(x-2)(x+5)$, the LCD.

$$x(x+5) - 7(x-2) = 17$$
$$x^2 + 5x - 7x + 14 = 17$$
$$x^2 - 2x - 3 = 0$$
$$(x-3)(x+1) = 0$$
$$x = 3 \quad \text{or} \quad x = -1$$

Because neither -1 nor 3 makes a denominator in the original equation have a value of zero, they are both in the solution set.

$$\{-1, 3\}$$

◀

▶ **EXAMPLE 2**

Solve $\dfrac{1}{x-6} + \dfrac{6}{x^2-11x+30} - \dfrac{x+1}{5-x} = 0$.

Solution

$$\frac{1}{x-6} + \frac{6}{x^2-11x+30} - \frac{x+1}{5-x} = 0$$

Factor all denominators.

$$\frac{1}{x-6} + \frac{6}{(x-5)(x-6)} - \frac{x+1}{5-x} = 0$$

Since $x-5$ and $5-x$ are opposites, replace $5-x$ with $-(x-5)$.

$$\frac{1}{x-6} + \frac{6}{(x-5)(x-6)} - \frac{x+1}{-(x-5)} = 0$$

$$\frac{1}{x-6} + \frac{6}{(x-5)(x-6)} + \frac{x+1}{x-5} = 0$$

Multiply both sides by $(x-5)(x-6)$ to clear fractions.

$$(x - 5) + 6 + (x + 1)(x - 6) = 0$$
$$x - 5 + 6 + x^2 - 5x - 6 = 0$$
$$x^2 - 4x - 5 = 0$$
$$(x - 5)(x + 1) = 0$$
$$x = 5 \quad \text{or} \quad x = -1$$

In this example, note that replacing x by 5 in the original equation makes two of the denominators have a value of zero, while -1 does not make any denominator have a value of zero. Thus -1 is in the solution set and 5 is not.

$$\{-1\} \qquad \blacktriangleleft$$

► **EXAMPLE 3**

Solve $\dfrac{x}{x - 7} = \dfrac{35}{x^2 - 9x + 14} + \dfrac{7}{x - 2}$.

Solution

$$\frac{x}{x - 7} = \frac{35}{x^2 - 9x + 14} + \frac{7}{x - 2}$$

$$\frac{x}{x - 7} = \frac{35}{(x - 7)(x - 2)} + \frac{7}{x - 2} \qquad \text{Factor}$$

$$x(x - 2) = 35 + 7(x - 7) \qquad \text{Multiply by } (x - 7)(x - 2)$$

$$x^2 - 2x = 35 + 7x - 49 \qquad \text{Distributive Property}$$

$$x^2 - 9x + 14 = 0 \qquad \text{Standard form}$$

$$(x - 7)(x - 2) = 0 \qquad \text{Factor}$$

$$x = 7 \quad \text{or} \quad x = 2 \qquad \text{Property of Zero Products}$$

Neither 7 nor 2 is in the solution set, because each makes a denominator of the original equation have a value of zero. So, as they are the only possibilities, the solution set must be empty.

$$\varnothing \qquad \blacktriangleleft$$

► **EXAMPLE 4**

Solve $\dfrac{x}{6} - \dfrac{1}{x} \le \dfrac{1}{6}$. Write the solution set in interval notation.

Solution

We use the method of boundary numbers. First we find all the boundary numbers. There is a *free* boundary number. It is 0. We solve the associated equation to find other boundary numbers.

$$\frac{x}{6} - \frac{1}{x} = \frac{1}{6} \qquad \text{The associated equation}$$

$$x^2 - 6 = x \qquad \text{Multiply both sides by } 6x$$

$$x^2 - x - 6 = 0$$

$$(x - 3)(x + 2) = 0$$

Thus, we see that 3 and -2 are also boundary numbers. We locate *all* the boundary numbers on a number line.

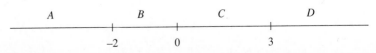

We test the four regions.

Region	Number in Region	Statement $\dfrac{x}{6} - \dfrac{1}{x} \leq \dfrac{1}{6}$	Truth of Statement	Region in Solution Set?
A	-3	$\dfrac{-3}{6} - \dfrac{1}{-3} \leq \dfrac{1}{6}$	True	Yes
B	-1	$\dfrac{-1}{6} - \dfrac{1}{-1} \leq \dfrac{1}{6}$	False	No
C	1	$\dfrac{1}{6} - \dfrac{1}{1} \leq \dfrac{1}{6}$	True	Yes
D	4	$\dfrac{4}{6} - \dfrac{1}{4} \leq \dfrac{1}{6}$	False	No

The boundary numbers -2 and 3 are included but 0 is *not* in the solution set because it is a free boundary number. We now have a graph of the solution set.

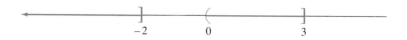

In interval notation this set is $(-\infty, -2] \cup (0, 3]$. ◄

Recalling the definition of negative exponents, we see that equations containing them are often just equations containing fractions.

► **EXAMPLE 5**

Solve $x^{-4} - 10x^{-2} + 9 = 0$.

Solution

METHOD I

$$x^{-4} - 10x^{-2} + 9 = 0$$

$$\frac{1}{x^4} - \frac{10}{x^2} + 9 = 0$$

Multiply both sides by x^4.

$$1 - 10x^2 + 9x^4 = 0$$

$$(1 - 9x^2)(1 - x^2) = 0$$

$$1 - 9x^2 = 0 \quad \text{or} \quad 1 - x^2 = 0$$

$$x^2 = \frac{1}{9} \quad \text{or} \quad x^2 = 1$$

$$x = \pm\frac{1}{3} \quad \text{or} \quad x = \pm 1$$

As none of these values makes any of the denominators in the original equation have a value of zero, all four possibilities are in the solution set.

$$\left\{ \pm\frac{1}{3},\ \pm1 \right\}$$

There is another approach to this problem. The original equation will factor.

METHOD 2

$$x^{-4} - 10x^{-2} + 9 = 0$$

$$(x^{-2} - 1)(x^{-2} - 9) = 0$$

$x^{-2} = 1$ or $x^{-2} = 9$	Factor
$\dfrac{1}{x^2} = 1$ or $\dfrac{1}{x^2} = 9$	Definition of x^{-2}
$1 = x^2$ or $\dfrac{1}{9} = x^2$	Multiply by x^2
$x = \pm1$ or $x = \pm\dfrac{1}{3}$	Square Root Property

$$\left\{ \pm\frac{1}{3},\ \pm1 \right\}$$ ◄

► ## EXAMPLE 6

Tony entered a 17-mile race. He ran for 15 miles and hurt his leg and was forced to walk the rest of the race. His running rate was 3 mph faster than his walking rate. If the race took 2 hours to complete, find his walking speed.

Solution

Let x be his walking speed in mph.

We make a chart and fill in D, R, and T for the two parts, running and walking. The let statement gives us the rates. The problem tells us the distances.

	Distance	Rate	Time
Walking		x	
Running		$x + 3$	

	Distance	Rate	Time
Walking	2	x	
Running	15	$x + 3$	

Use the formula $T = \dfrac{D}{R}$ to find the time for each part.

	Distance	Rate	Time
Walking	2	x	$\dfrac{2}{x}$
Running	15	$x + 3$	$\dfrac{15}{x + 3}$

We form an equation by noting that the total time is 2 hours.

$$\frac{2}{x} + \frac{15}{x + 3} = 2$$

Multiply both sides by the LCD, $x(x + 3)$.

$$2(x + 3) + 15x = 2x(x + 3)$$
$$2x + 6 + 15x = 2x^2 + 6x$$
$$0 = 2x^2 - 11x - 6$$
$$0 = (2x + 1)(x - 6)$$
$$x = -\frac{1}{2} \quad \text{or} \quad x = 6$$

We see that $-\dfrac{1}{2}$ cannot be a speed.

His walking speed was 6 mph.   ◄

► EXAMPLE 7

Jim and Roy can both clear their camping site in 3 hours if they work together. If each worked alone, it would take Roy 8 hours longer than Jim to clear the same site. How long would it take Jim to clear the site if he were working alone?

Solution

Let x be the number of hours it would take Jim, working alone. Then $x + 8$ is the number of hours it would take Roy alone. So in 1 hour Jim would clear $\dfrac{1}{x}$ of the site, and in 1 hour Roy would clear $\dfrac{1}{x + 8}$ of the site, while Jim and Roy together would clear $\dfrac{1}{3}$ of the site. Thus in 1 hour we have

$$\frac{1}{x} + \frac{1}{x + 8} = \frac{1}{3}$$

Multiply both sides by $3x(x + 8)$, the LCD.

$$3(x + 8) + 3x = x(x + 8)$$

$$3x + 24 + 3x = x^2 + 8x$$

$$0 = x^2 + 2x - 24$$

$$= (x - 4)(x + 6)$$

$$x = 4 \quad \text{or} \quad x = -6$$

Since -6 cannot be an answer to this problem, we may write the solution. It would take Jim 4 hours working alone. ◀

Problem Set 5.6

Warm-Ups

In Problems 1 through 5, solve each equation. See Example 1.

1. $\dfrac{6}{x - 2} - \dfrac{3}{x} = 1$

2. $\dfrac{2}{(x - 3)^2} - \dfrac{x}{x - 3} = 1$

3. $\dfrac{2}{x} - 1 = \dfrac{4}{x + 3}$

4. $\dfrac{x}{x + 4} = \dfrac{3}{x - 1}$

5. $\dfrac{4}{x + 5} - \dfrac{5}{x - 2} = 3$

In Problems 6 through 9, solve each equation. See Example 2.

6. $\dfrac{18}{9 - y^2} + \dfrac{y}{y - 3} = \dfrac{1}{y + 3}$

7. $\dfrac{z}{z - 2} + \dfrac{2}{z + 1} = \dfrac{7z + 1}{z^2 - z - 2}$

8. $\dfrac{2}{w + 5} - \dfrac{7}{w - 4} = \dfrac{w^2 + 6w - 13}{w^2 + w - 20}$

9. $\dfrac{x}{x + 1} - \dfrac{2}{1 - x} = \dfrac{8x - 4}{x^2 - 1}$

In Problems 10 through 13, find the solution set. See Example 3.

10. $\dfrac{1}{3 - s} + \dfrac{s}{s - 1} + \dfrac{2}{s^2 - 4s + 3} = 0$

11. $\dfrac{2x^2}{x^2 - 4x + 3} + \dfrac{1}{x - 1} = \dfrac{9}{x - 3}$

12. $\dfrac{x + 2}{x - 1} = \dfrac{3}{x + 1} - \dfrac{6}{1 - x^2}$

13. $\dfrac{2}{x - 3} + \dfrac{x^2 - 1}{x^2 - 10x + 21} = \dfrac{12}{x - 7}$

In Problems 14 and 15, solve each inequality. Write the solution set in interval notation. See Example 4.

14. $\dfrac{x}{2} - \dfrac{1}{x} \le \dfrac{1}{2}$

15. $\dfrac{x}{12} - \dfrac{1}{x} + \dfrac{1}{12} \ge 0$

In Problems 16 through 20, find the solution set. See Example 5.

16. $x^{-2} - x^{-1} - 2 = 0$

17. $x^{-2} + 2x^{-1} - 24 = 0$

18. $y^{-2} - 3y^{-1} - 18 = 0$

19. $t^{-4} - 2t^{-2} + 1 = 0$

20. $x^{-4} - 3x^{-2} - 4 = 0$

In Problems 21 through 24, solve each problem. See Example 6.

21. A ship sails at constant speed for 24 nautical miles. If the ship's rate was 2 knots faster, it would have saved 1 hour on the trip. What was its speed?

22. A bus travels 200 miles from Atlanta at a constant speed. It returns immediately at a constant speed, 10 miles per hour slower than it went. If the total trip took 9 hours, what speed did it travel from Atlanta?

23. Alan set out on a 23-mile bike trip to Tacoma. After 22 miles his bike broke down and he walked the rest of the trip. His riding speed was 4 mph faster than his walking speed. If the total trip took 3 hours, find his walking speed.

24. Jamel sailed for 2 miles and then rowed for 5 miles. His sailing speed was 3 mph faster than his rowing speed. If the total trip took 2 hours, what was his rowing speed?

In Problems 25 and 26, solve each problem. See Example 7.

25. Carolyn's hot tub has two fill pipes. If both pipes are open, they can fill the hot tub in 4 hours. If only the smaller pipe is open, it takes 6 hours longer to fill the tub than if only the larger pipe were open. Find the time it would take each pipe to fill the tub if working alone.

26. Odette and Alex can clean their house in 2 hours working together. Working alone, it takes Odette 3 hours longer than it takes Alex working alone. How long does it take each to clean the house working alone?

Practice Exercises

In Problems 27 through 80, find the solution set.

27. $\dfrac{2x}{5} - \dfrac{2}{x} = \dfrac{1}{5}$

28. $\dfrac{y^2 + 2y - 5}{y + 1} = 1$

29. $\dfrac{-4}{3x - 1} = 2x + 3$

30. $\dfrac{x + 2}{x - 1} = x + 2$

31. $\dfrac{1}{x + 1} - \dfrac{10}{x - 2} = \dfrac{3}{2}$

32. $\dfrac{25}{y + 2} - \dfrac{9}{y - 2} = 16$

33. $\dfrac{2}{x^2 - 1} + \dfrac{3x + 6}{x^2 - x - 2} = \dfrac{8}{x^2 - 3x + 2}$

34. $\dfrac{2x}{(x - 3)(x + 1)} + \dfrac{3}{x + 1} = 1$

35. $\dfrac{5}{x + 2} - \dfrac{3}{x - 5} = \dfrac{x^2 - 8x - 6}{x^2 - 3x - 10}$

36. $\dfrac{3}{x} + \dfrac{x}{x + 4} = \dfrac{10}{x^2 + 4x}$

37. $\dfrac{1}{x} + \dfrac{5}{3x} = \dfrac{x + 2}{3}$

38. $\dfrac{1}{x - 5} + \dfrac{x}{x - 2} = \dfrac{3}{x^2 - 7x + 10}$

39. $\dfrac{2}{x + 7} + \dfrac{16}{x^2 + 6x - 7} + \dfrac{x}{x - 1} = 0$

40. $\dfrac{5}{x^2 + 3x - 4} + \dfrac{1}{x + 4} = \dfrac{x}{x - 1}$

41. $\dfrac{1}{5} - \dfrac{7}{15 - 5x} = \dfrac{x^2}{x - 3}$

42. $\dfrac{4}{y - 8} + \dfrac{2}{y^2 - 11y + 24} + \dfrac{y + 1}{y - 3} = 0$

43. $\dfrac{2}{t - 3} - \dfrac{t}{2 - t} = \dfrac{2}{t^2 - 5t + 6}$

44. $\dfrac{-3}{x + 4} + \dfrac{x}{x - 3} = \dfrac{21}{x^2 + x - 12}$

45. $\dfrac{1}{x} + \dfrac{x - 3}{x + 1} = \dfrac{16}{x^2 + x}$

46. $\dfrac{5}{w - 5} + \dfrac{w}{w + 5} = \dfrac{50}{w^2 - 25}$

47. $\dfrac{1}{x^2} + \dfrac{1}{3} = \dfrac{4}{x^2}$

48. $\dfrac{3}{t - 2} + \dfrac{t + 1}{t + 3} = \dfrac{15}{t^2 + t - 6}$

49. $\dfrac{Y}{Y + 5} = \dfrac{1}{Y - 3} - \dfrac{8}{Y^2 + 2Y - 15}$

50. $\dfrac{7}{3X + 6} + 3 = \dfrac{-8}{3X - 3}$

51. $\dfrac{-2}{x + 1} - \dfrac{5}{3 - x} = 1$

52. $\dfrac{15}{4(x - 2)} - \dfrac{7}{4(x + 2)} = 1$

53. $\dfrac{6}{x} + \dfrac{x}{x + 5} = \dfrac{30}{x^2 + 5x}$

54. $\dfrac{y}{y + 3} - \dfrac{2}{y + 1} = \dfrac{y - 3}{y^2 + 4y + 3}$

55. $\dfrac{t+3}{t-1} + \dfrac{1}{t+4} = \dfrac{2t^2+6t+12}{t^2+3t-4}$

56. $\dfrac{x}{x^2+3x+2} + \dfrac{1}{x^2+4x+3} = \dfrac{2}{x^2+5x+6}$

57. $\dfrac{1}{t^2-1} + \dfrac{2t}{2t^2-3t+1} = \dfrac{1}{2t^2+t-1}$

58. $\dfrac{3x}{x^2-x-2} + \dfrac{2}{x^2+4x+3} = \dfrac{14}{x^2+x-6}$

59. $\dfrac{18}{s^2+s-6} + \dfrac{s-1}{s^2+5s+6} = \dfrac{12}{s^2-4}$

60. $\dfrac{y}{y^2-y-2} + \dfrac{2}{y^2-2y-3} + \dfrac{2}{y^2-5y+6} = 0$

61. $\dfrac{1}{x^2-1} + \dfrac{x}{2x^2-3x+1} = \dfrac{1}{2x^2+x-1}$

62. $\dfrac{x}{x^2-x-2} - \dfrac{1}{2x^2+x-1} = \dfrac{1}{2x^2-5x+2}$

63. $\dfrac{2}{5y^2-15y-50} + \dfrac{y+1}{y^2-y-6} = \dfrac{-8}{5y^2-40y+75}$

64. $x^{-2} + 4x^{-1} + 4 = 0$

65. $x^{-2} - 3x^{-1} - 10 = 0$

66. $x^{-2} + 5x^{-1} - 6 = 0$

67. $2x^{-2} - 17x^{-1} + 21 = 0$

68. $3x^{-2} + x^{-1} = 4$

69. $4x^{-2} + 27 = 21x^{-1}$

70. $3 - 2x^{-1} = 5x^{-2}$

71. $x^{-2} - 10x^{-1} + 25 = 0$

72. $3x^{-4} - x^{-2} - 4 = 0$

73. $2x^{-4} - 11x^{-2} + 9 = 0$

74. $6x^{-4} - x^{-2} - 1 = 0$

In Problems 75 through 80, write the solution set in interval notation.

75. $\dfrac{-3}{x} - 1 < \dfrac{4}{1-x}$

76. $\dfrac{3}{x-1} < \dfrac{x}{1-x}$

77. $\dfrac{14}{x-2} - \dfrac{35}{x-5} \le \dfrac{3}{2}$

78. $\dfrac{x+1}{x-2} - \dfrac{8}{x+1} \le \dfrac{9}{x^2-x-2}$

79. $\dfrac{1}{x} + \dfrac{x+3}{7} \ge \dfrac{5}{7x}$

80. $\dfrac{20}{7} + \dfrac{x^2}{x+1} \ge \dfrac{23}{7x+7}$

In problems 81 through 86, solve each problem.

81. Igor rode his bike 3 miles and then walked for 6 miles. His walking speed was 2 mph slower than his speed on his bike. If he walked 1 hour longer than he rode his bike, find his speed on his bike.

82. Janetta ran 3 miles and then decreased her speed by 1 mph and ran 4 more miles. If it took 1 hour longer on the last 4 miles, find her original speed.

83. Joan rides her bicycle to Tacoma 74 miles away in 5 hours. Rain forces her to slow down by 6 mph for the last 10 miles. What was her original speed?

84. Marvin drove at a constant speed from Atlanta to Savannah, a distance of 260 miles, and returned immediately. He drove 13 mph faster on the return trip and shortened his time by 1 hour. What was his speed each way?

85. Working together, Cathy and Jim can prepare a gourmet Hunan feast in 6 hours. If Jim were to do it

Problem 85

alone, it would take him 9 hours longer than it would take Cathy working alone. How long would it take each to prepare the feast working alone?

86. Aslam can pour a concrete driveway in 16 hours less

than Haazim. If they work together, they can pour the driveway in 6 hours. How long does it take each to pour the driveway working alone?

Challenge Problems

In Problems 87 through 103, solve each equation.

87. $4x^{-4} - 68x^{-2} + 225 = 0$

88. $4x^{-4} + 4x^{-2} - 3 = 0$

89. $9x^{-4} + 7x^{-2} - 2 = 0$

90. $4p^{-4} + 5p^{-2} - 6 = 0$

91. $16x^{-4} - 8x^{-2} - 15 = 0$

92. $4x^{-4} + 11x^{-2} + 6 = 0$

93. $6x^{-2} - x^{-1} = 0$

94. $y^{-4} + y^{-2} = 0$

95. $\dfrac{3x}{x - 1} = \dfrac{x^2}{1 - x}$

96. $\dfrac{x}{x + 1} - \dfrac{10x}{x - 2} = \dfrac{3x}{2}$

97. $\dfrac{5x}{x^2 + 3x - 4} + \dfrac{x}{x + 4} = \dfrac{x^2}{x - 1}$

98. $\dfrac{2x^2}{x - 3} + \dfrac{x^3}{x - 2} = \dfrac{2x^2}{x^2 - 5x + 6}$

99. $\dfrac{a - 2}{x + 2} + \dfrac{x}{x - 1} = \dfrac{2a^2 - a + 2}{x^2 + x - 2}$

100. $a^2 x^{-4} + 3ax^{-2} - 4 = 0, \quad a \neq 0$

101. $(x + 1)^{-4} - 5(x + 1)^{-2} + 4 = 0$

102. $(z - 2)^{-4} - 6(z - 2)^{-2} + 9 = 0$

103. $\dfrac{2}{3\sqrt{x}} + \dfrac{\sqrt{x}}{3} = \dfrac{x^2}{3\sqrt{x}}$

104. An airplane flies between Denver and Loveland. With a tailwind of 20 mph, it arrives 4 minutes earlier than it would without the tailwind. If the cities are 40 miles apart, find the speed of the plane in still air.

105. Ron can row 30 miles down the Columbia River and back in 8 hours. If the rate of the current is 5 mph, find the rate that Ron would row in still water.

In Your Own Words

106. Explain how to solve a rational equation.

5.7   Radical Equations

TAPE 11

We learned in Section 4.5 that sometimes it is necessary to square both sides of an equation to clear radicals to solve using familiar techniques. Often, raising both sides to a power will lead to the types of higher-order equations that we are presently studying.

Never forget that raising both sides of an equation to a power may introduce extraneous solutions. We *always* must check the solutions of the squared equation to see if they are solutions of the original equation.

► **EXAMPLE** **1**

Be Careful!

Solve $\sqrt{4x + 1} + 5 = x$.

Solution

$$\sqrt{4x + 1} + 5 = x$$

First, isolate the radical.

$$\sqrt{4x + 1} = x - 5$$

Then square both sides.

$$(\sqrt{4x + 1})^2 = (x - 5)^2$$
$$4x + 1 = x^2 - 10x + 25$$

Simplify, and write in standard form.

$$0 = x^2 - 14x + 24$$
$$0 = (x - 2)(x - 12)$$
$$x = 2 \quad \text{or} \quad x = 12$$

Check 2 in the original equation.

- LS: $\sqrt{4(2) + 1} + 5 = \sqrt{9} + 5 = 3 + 5 = \boxed{8}$
- RS: $\boxed{2}$

Does *not* check.

Check 12.

- LS: $\sqrt{4(12) + 1} + 5 = \sqrt{49} + 5 = 7 + 5 = \boxed{12}$
- RS: $\boxed{12}$

Checks.

Thus the solution set is {12}.   ◄

Solving Radical Equations

1. Isolate one radical.
2. Raise both sides to the appropriate power.
3. Simplify, and isolate another radical if necessary. Continue steps 2 and 3 until the equation is free of radicals.
4. Solve the resulting equation.
5. Check all possible solutions in the original equation.
6. Write the solution set.

► **EXAMPLE 2**

Solve $\sqrt{2x - 1} - \sqrt{x - 4} = 2$.

Solution

*Be
Careful!*

$$\sqrt{2x - 1} - \sqrt{x - 4} = 2$$

$$\sqrt{2x - 1} = 2 + \sqrt{x - 4} \qquad \text{Isolate a radical}$$

$$(\sqrt{2x - 1})^2 = (2 + \sqrt{x - 4})^2 \qquad \text{Square both sides}$$

$$2x - 1 = 4 + 4\sqrt{x - 4} + (x - 4)$$

$$x - 1 = 4\sqrt{x - 4} \qquad \text{Isolate a radical}$$

$$(x - 1)^2 = (4\sqrt{x - 4})^2 \qquad \text{Square both sides}$$

$$x^2 - 2x + 1 = 16(x - 4)$$

$$x^2 - 2x + 1 = 16x - 64 \qquad \text{Distributive Property}$$

$$x^2 - 18x + 65 = 0 \qquad \text{Standard form}$$

$$(x - 5)(x - 13) = 0 \qquad \text{Factor}$$

$$x = 5 \quad \text{or} \quad x = 13 \qquad \text{Property of} \\ \text{Zero Products}$$

Check 5 in the original equation.

- LS: $\sqrt{2(5) - 1} - \sqrt{5 - 4} = \sqrt{9} - \sqrt{1} = 3 - 1 = \boxed{2}$
- RS: $\boxed{2}$

5 checks.

Check 13 in the original equation.

- LS: $\sqrt{2(13) - 1} - \sqrt{13 - 4} = \sqrt{25} - \sqrt{9} = 5 - 3 = \boxed{2}$
- RS: $\boxed{2}$

Also, 13 checks.

$$\{5, 13\} \qquad \blacktriangleleft$$

It may be necessary to raise both sides of an equation to a power greater than 2. In the following example it will be convenient to cube both sides.

► **EXAMPLE 3**

Solve $x^{1/3} = 3$.

Solution

$$x^{1/3} = 3$$

$$(x^{1/3})^3 = 3^3$$

$$x^1 = 27$$

$$x = 27 \qquad \text{27 checks by}$$

$$\{27\} \qquad \qquad \text{inspection.} \qquad \blacktriangleleft$$

Problem Set 5.7

Warm-Ups

In Problems 1 through 10, find the solution set. See Example 1.

1. $x = \sqrt{2x^2 - 3x + 2}$
2. $x + 1 = \sqrt{x + 3}$
3. $x - 2 = \sqrt{6 - 3x}$
4. $x = -1 + \sqrt{6x - 2}$
5. $x = \sqrt{2x + 6} - 3$
6. $x = 2 - \sqrt{3x + 4}$
7. $2x - 1 = \sqrt{2x^2 + 3x - 2}$
8. $3x - 1 = \sqrt{8x^2 - 5x + 7}$
9. $2x + 3 = \sqrt{8 + 7x - 2x^2}$
10. $x - 1 = \sqrt{2x^2 - 3x - 1}$

In Problems 11 through 18, solve each equation. See Example 2.

11. $\sqrt{x - 3} - 1 = \sqrt{2x - 4}$
12. $\sqrt{x + 5} = 3 + \sqrt{2x - 8}$
13. $\sqrt{3x + 1} = 1 + \sqrt{2x - 1}$
14. $\sqrt{4x + 13} = 2 + \sqrt{3x}$
15. $\sqrt{2y - 5} = \sqrt{3y + 4} - 2$
16. $\sqrt{5z + 1} = \sqrt{3z - 5} + 2$
17. $\sqrt{5x + 21} = 1 - \sqrt{3x + 16}$
18. $\sqrt{2x + 4} = \sqrt{9 - 2x} - 1$

In Problems 19 through 22, find the solution set. See Example 3.

19. $x^{1/3} = 2$
20. $x^{1/3} = -3$
21. $\sqrt[3]{2x - 1} - 3 = 0$
22. $\sqrt[3]{x^2 - 1} = 2$

Practice Exercises

In Problems 23 through 48, find the solution set.

23. $\sqrt{x + 2} = x$
24. $\sqrt{9x} = x + 2$
25. $\sqrt{16x} = x + 3$
26. $\sqrt{3x - 1} = \sqrt{x + 1}$
27. $\sqrt{4 - x} = x - 4$
28. $\sqrt{x + 3} = x - 3$
29. $\sqrt{2x} = x - 4$
30. $\sqrt{18x + 10} - x = 5$
31. $\sqrt{8x + 1} - 2 = x$
32. $x = \sqrt{2x^2 - 11x + 30}$
33. $\sqrt{z + 27} - 1 = \sqrt{2z + 20}$
34. $\sqrt{1 - 8r} + r = 2$
35. $\sqrt{5w} - \sqrt{w - 4} = 4$
36. $\sqrt{2 - 7t} - 3 = \sqrt{t + 3}$
37. $\sqrt{3 - 2x} - \sqrt{x + 4} = 2$
38. $\sqrt{4x + 1} - 2 = \sqrt{x - 1}$
39. $\sqrt{t + 9} = 2 + \sqrt{2t + 1}$
40. $\sqrt{2y - 1} - 1 = \sqrt{y - 1}$
41. $\sqrt{z - 2} + 2 = \sqrt{2z}$
42. $\sqrt{2x + 7} - 1 = \sqrt{x + 3}$
43. $\sqrt{2x + 1} + 1 = \sqrt{x + 4}$
44. $\sqrt{5w + 20} - 1 = 2\sqrt{w + 3}$
45. $\sqrt[3]{x} = 1$
46. $\sqrt[3]{x} = -2$
47. $(3x + 1)^{1/3} = 2$
48. $(x^2 + 2)^{1/3} = 3$

In Problems 49 and 50, equilibrium price is defined as the price when supply equals demand.

49. The supply in millions of a toy is predicted by the formula $S = \sqrt{11x}$, and the demand for the toy is predicted by $D = \sqrt{121 - 2x^2}$, where x is the price of the toy in dollars. Find the equilibrium price.

50. The supply in thousands of leather jackets is predicted by the formula $S = \sqrt{225x}$, and the demand for the jacket is predicted by $D = \sqrt{75,000 - 3x^2}$, where x is the price of the jacket in dollars. Find the equilibrium price.

Challenge Problems

In Problems 51 through 53, find the solution set.

51. $\sqrt{x} + \sqrt{x + 5} = \sqrt{5x + 5}$

52. $\sqrt{3 - x} - \sqrt{2 + x} = \sqrt{2x + 3}$

53. $\sqrt{4x + 2} = \sqrt{6x + 6} - \sqrt{2 - 2x}$

54. Solve $x + \sqrt{x} - 2 = 0$ two ways. First, isolate the radical and square both sides; then solve it by factoring.

In Problems 55 through 57, find the solution set.

55. $x^{2/3} - x^{1/3} - 2 = 0$

56. $x^{2/3} - 16 = 0$

57. $x^{2/5} - 2x^{1/5} - 3 = 0$

In Your Own Words

58. Explain the steps in solving an equation containing radicals.

► Chapter Summary

GLOSSARY

Quadratic equation in one variable   Equation that can be written in the form $ax^2 + bx + c = 0$. If $a > 0$, the equation is in **standard form.**

Discriminant   The value of $b^2 - 4ac$, using a, b, and c from the standard form of a quadratic equation.

If $b^2 - 4ac > 0$, the equation has two real solutions.
If $b^2 - 4ac < 0$, the equation has two complex, nonreal solutions.
If $b^2 - 4ac = 0$, the equation has one real repeated solution.

Square Root Property $X^2 = A$
is equivalent to the pair of equations
$X = \pm\sqrt{A}$

Solving a Quadratic by Completing the Square

1. Write the equation in standard form.
2. Divide both sides by the coefficient of x^2 if the coefficient is not 1.
3. Subtract the constant term from both sides.
4. Divide the coefficient of x by 2, square the result, and then add this number to both sides.

5. Factor the left side and simplify the right side.

6. Apply the Square Root Property and write the solution set.

Quadratic Formula

$ax^2 + bx + c = 0, \quad a \neq 0$

is equivalent to

$$x = \frac{-b \pm \sqrt{b^2 - 4ac}}{2a}$$

Solving a Quadratic Equation

1. Is the equation of the form $X^2 = A$? If so, use the Square Root Property.

2. Try factoring.

3. Use the Quadratic Formula

4. Use completing the square when directed to do so.

Solving Higher-degree Equations

1. Write the equation in standard form.

2. Factor, if possible.

3. Use the Property of Zero Products, the Square Root Property, or the Quadratic Formula to find the solution set.

Solving Rational Equations

1. Multiply both sides by the LCD.

2. Find possible solutions by solving the resulting equation.

3. Discard possible solutions that make any denominator have a value of zero.

4. Write the solution set.

Solving Radical Equations

1. Isolate one radical.

2. Raise both sides to the appropriate power.

3. Simplify and isolate another radical if necessary. Continue steps 2 and 3 until the equation is free of radicals.

4. Solve the resulting equation.

5. Check all possible solutions in the original equation.

6. Write the solution set.

Solving Nonlinear Inequalities

1. Solve the associated equation to find the boundary numbers.

2. Find any free boundary numbers.

3. Locate *all* the boundary numbers on a number line.

4. Determine which regions formed by the boundary numbers make the *original inequality* true by testing with one number inside the region.

5. Shade only the regions that test true.

6. Check the boundary numbers themselves.

7. Write the solution set.

CHECK-UPS

In Problems 1 through 6, solve each equation or inequality.

1. $x^2 + \frac{2}{3}x - \frac{8}{3} = 0$ Section 5.1; Example 5

2. $(x - 1)(x + 2) \leq 0$ Section 5.1; Example 10

3. $2x^2 + 16 = 0$ Section 5.3; Example 1b

4. $t^2 - t + 3 = 0$ Section 5.4; Example 3

5. $x^3 = -8$ Section 5.5; Example 3

6. $\dfrac{1}{x - 6} + \dfrac{6}{x^2 - 11x + 30} - \dfrac{x + 1}{5 - x} = 0$ Section 5.6; Example 2

7. $\sqrt{4x + 1} + 5 = x$ Section 5.7; Example 1

8. The lengths of the sides of a right triangle are three consecutive even integers. What are they?   Section 5.2; Example 5

9. Solve $2x^2 - x + 4 = 0$ by completing the square.   Section 5.3; Example 7

10. Tony entered a 17-mile race. He ran for 15 miles and hurt his leg and was forced to walk the rest of the race. His running rate was 3 mph faster than his walking rate. If the race took 2 hours to complete, find his walking speed.   Section 5.6; Example 7

► Review Problems

In Problems 1 through 58, find the solution set.

1. $12x^2 + 5x - 2 = 0$

2. $2x^2 - 50 = 0$

3. $x^2 - 7x > 0$

4. $x^2 \geq 8$

5. $3x^2 + 4x - 3 = 0$

6. $(x - 7)(x + 3)(x - 10) = 0$

7. $\dfrac{3}{x - 5} + \dfrac{x}{3(x + 2)} = \dfrac{5}{x^2 - 3x - 10}$

8. $x^{-2} + 8x^{-1} + 15 = 0$

9. $4x^2 + 12x + 9 \leq 0$

10. $6x^2 + 13x + 6 < 0$

11. $4x^2 - x + 1 = 0$

12. $(x + 3)(x - 7) = 8$

13. $(3x + 2)^2 = 9$

14. $x(x - 3)(x + 2) = 0$

15. $\sqrt{6x + 1} - 1 = \sqrt{3x + 4}$

16. $\dfrac{1}{x + 3} + \dfrac{x}{x + 1} = \dfrac{-2}{x^2 + 4x + 3}$

17. $18x^2 + 9x - 5 = 0$

18. $-x^2 = 36$

19. $4x^2 - 9 = 0$

20. $x^4 - 6x^2 + 9 = 0$

21. $\dfrac{1}{x} + \dfrac{x+1}{x^2} = \dfrac{6}{x^3}$

22. $3x^2 - 2x = 0$

23. $\dfrac{3}{x} + \dfrac{3}{x^2 - x} = \dfrac{x}{1-x}$

24. $\sqrt{5x-2} + 1 = \sqrt{10x-3}$

25. $x^3 = \dfrac{1}{8}$

26. $-x^2 + 4x + 21 = 0$

27. $3x^2 + 8x + 4 = 0$

28. $4x^2 - 6x + 9 = 0$

29. $\sqrt{x+2} - x = 0$

30. $6x^2 + 5x - 6 = 0$

31. $3x^2 - x - 2 \geq 0$

32. $\dfrac{3}{x+3} - \dfrac{x}{3-x} \leq \dfrac{7}{x^2-9}$

33. $x^4 - 50x^2 + 49 = 0$

34. $3x^2 = 27$

35. $\sqrt{2x+1} - \sqrt{x} = 1$

36. $\dfrac{5}{x+8} + \dfrac{x-3}{x+1} = \dfrac{11x-13}{x^2+9x+8}$

37. $\sqrt{2x+5} - \sqrt{x-1} = 2$

38. $6x^{-4} + x^{-2} = 1$

39. $\dfrac{3}{x^2+6x+8} + \dfrac{x+4}{x^2+3x+2} = \dfrac{5}{x^2+5x+4}$

40. $\sqrt{7x} - \sqrt{3x+4} = 2$

41. $5x^2 + 5x + 2 = 0$

42. $\dfrac{1}{x^2+7x+10} + \dfrac{7(x-3)}{x^2-3x-10} = \dfrac{29}{x^2-25}$

43. $\dfrac{1}{x+1} + \dfrac{1}{3} = 2x$

44. $(x-1)(2x+3) = 1$

45. $(x-2)^2 = 6$

46. $(x-1)^2 = 10$

47. $\sqrt{2x+7} - \sqrt{x} = 2$

48. $(x+3)(x+5) = 3$

49. $x^2(x^2-5) - 3(x^2-5) = 0$

50. $\dfrac{1}{x} + \dfrac{x+1}{x^2} = \dfrac{1}{x^3}$

51. $(x+3)^2 = -5$

52. $2\sqrt{x+1} - 1 = \sqrt{3x}$

53. $16x^4 + 9x^2 - 7 = 0$

54. $\dfrac{1}{x-1} + 6 = \dfrac{14}{x+3}$

55. $\dfrac{9}{x-1} + \dfrac{5}{3+x} \geq 2$

56. $\dfrac{2}{x+4} < \dfrac{x+1}{x+2}$

57. $(x-7)^2 = -8$

58. $(2x-5)^2 = -12$

In Problems 59 through 65, solve each problem.

59. Find two numbers whose sum is 17 and whose product is -390.

60. The perimeter of a rectangle is 84 m and its area is 425 m². Find its dimensions.

61. The hypotenuse of a right triangle is $3\sqrt{2}$ meters and one leg is 2 m. Find the length of the other leg.

62. The area of a triangle is 72 in.². If the length of the base is four times the length of the height, find the length of the height.

63. Roger drove 27 miles to Tampa on his new motorcycle. For the first 15 miles, he went 6 mph faster than he did the rest of the trip. If the trip to Tampa took 1 hour, what was his speed on the last part of the trip?

64. A missile is shot into the air. The height h above the ground is given by the formula $h = 128t - 16t^2$, where t is the time in seconds since launch. Find the time interval for which the missile is less than 112 feet above the ground.

65. A new snowplow costs $96,000. Suppose the yearly maintenance cost in dollars is given by $C = t^2 + 930t$, where t is the age of the snowplow in years. When will the maintenance cost be greater than 30% of the purchase price?

Let's Not Forget . . .

Identify the expressions that are in factored form. Factor those that are not factored.

66. $-8x^3 + 125$

67. $(2x + 5)^3$

68. $\sqrt{a}(\sqrt{a} + \sqrt{b})$

69. $16 - (x + 1)^2$

70. $x^2 + 2xy + y^2$

71. $y(c - d) + x(c - d)$

How many terms are in each expression? Which expressions have $x - y$ as a factor?

72. $x^3 - y^3$

73. $x^2 - xy + y^2$

74. $a(x - y) + b(y - x)$

75. $x - y(a + b)$

Simplify each expression, if possible, leaving only nonnegative exponents in your answer.

76. $\dfrac{4 \pm \sqrt{-12}}{2}$

77. $\dfrac{-3x^{-2}}{x^{-1}y^3}$

78. $-3^2(ab^2)^{-2}$

79. $a^{-1}b^2(a^{-2} + b)$

80. -2^{-3}

81. $\sqrt[5]{a^5 + b^5}$

Find each product.

82. $(x^{1/2} + y^{1/2})^2$

83. $(3y - 1)^3$

Reduce each, if possible.

84. $\dfrac{a(r + t) - b(r - t)}{(r + t)(r - t)}$

85. $\dfrac{x(b + c) - y(b + c)}{(b + c)(b - c)}$

The following problems can be worked using a least common denominator. Follow the directions in each and notice how the LCD is used.

86. Solve $\dfrac{x}{x + 3} - \dfrac{2}{x + 4} = \dfrac{2}{x^2 + 7x + 12}$.

87. Perform the operation indicated: $\dfrac{3}{x + 4} - \dfrac{x}{x + 3}$.

88. Simplify $\dfrac{1 + \dfrac{x}{x + 3}}{1 - \dfrac{x}{x + 4}}$.

Label each as an equation or as an expression. Solve the equations and perform the operations indicated on the expressions.

89. $(1 - 2i)(1 + i)$

90. $(x + 1)(x + 2) = 12$

91. $\dfrac{x}{x^2 - 5x - 14} - \dfrac{1}{x - 7}$

92. $\sqrt{x - 1} - \sqrt{2x + 5} = -2$

93. $\left| \dfrac{1}{4}(x - 3) \right| = 3$

94. $\dfrac{2}{x + 2} - \dfrac{2}{x + 3} = 1$

95. $\dfrac{3}{5}(x + 2) - 1 = \dfrac{1}{3}(1 - x)$

► You Decide

You have ordered 8 cubic yards of premixed concrete for your patio. You want a rectangular patio of usable size, but you also want the concrete to be thick enough to withstand outdoor conditions and ordinary wear–and–tear. The presence of an old oak tree dictates that its length must be 10 feet more than its width. You decide on the dimensions of your new patio.

► Chapter 5 Test

In Problems 1 through 5, choose the correct answer.

1. The solution set for $x^2 - 2x - 3 = 5$ is (?)

   A. $\{4, 8\}$ **B.** $\{-2, 4\}$

   C. $\{-4, 2\}$ **D.** $\{2, 8\}$

2. $\dfrac{-3 \pm \sqrt{12}}{12} = $ (?)

   A. $-\dfrac{1}{4} \pm \dfrac{\sqrt{3}}{6}$ **B.** $-\dfrac{1}{4} \pm \dfrac{\sqrt{3}}{2}$

   C. $\dfrac{-1 \pm \sqrt{3}}{2}$ **D.** $-\dfrac{1}{4} \pm \dfrac{\sqrt{3}}{3}$

3. The solution set for $2x^2 - 2x + 3 = 0$ is (?)

   A. $\left\{\dfrac{3}{2}, 1\right\}$ **B.** $\{1 \pm \sqrt{5}i\}$

   C. $\left\{\dfrac{1}{2} \pm \dfrac{\sqrt{7}}{2}\right\}$ **D.** $\left\{\dfrac{1}{2} \pm \dfrac{\sqrt{5}}{2}i\right\}$

4. The solution set for $x = \sqrt{7 - x} + 1$ is (?)

   A. $\{-2, 3\}$ **B.** $\{-3, 2\}$

   C. $\{3\}$ **D.** $\varnothing$

5. The solution set for $4x^4 - 17x^2 + 4 = 0$ is (?)

   A. $\left\{\pm\dfrac{1}{2}i, \pm 2i\right\}$ **B.** $\left\{\dfrac{1}{4}, 4\right\}$ **C.** $\left\{\pm\dfrac{1}{2}, \pm 2\right\}$

   D. None of the above

In Problems 6 through 12, find the solution set.

6. $3x^2 - 4x - 4 < 0$

7. $\dfrac{3}{x + 1} + \dfrac{2}{x + 3} = 2$

8. $(x - 3)^2 = 9$

9. $x^2 \geq 3x$

10. $x^3 - 8 = 0$

11. $\dfrac{1}{x^4} + \dfrac{3}{x^2} - 4 = 0$

12. $\sqrt{x + 1} + 2 = \sqrt{2x + 10}$

13. Find the solution set by completing the square.

$$3x^2 - 6x - 3 = 0$$

14. Find the discriminant for $2x^2 - x - 1 = 0$.
The discriminant indicates that the equation has:
   A. exactly one real solution
   B. two distinct real solutions
   C. two complex nonreal solutions
   D. no solutions

15. Martha, a cyclist, pedals 20 miles to Ventura at one speed and then decreases her speed by 10 mph for the next 30 miles to Santa Barbara. What was her original speed if the total trip took 4 hours?

6 Graphs and Functions

See Problem Set 6.4, Exercise 56.

Connections

Being visual creatures, we often say "I see," meaning "I understand." With this in mind, mathematicians in the seventeenth and eighteenth centuries set about making pictures to represent abstract mathematical ideas. Their chief tool in this pursuit was our old friend the number line, and, later, the plane formed from two number lines.

Just as each point on a number line has a coordinate, each point in a plane has a pair of coordinates. The study of such coordinates is based on the work of the French mathematician René Descartes (1596–1650).

In this chapter, we study equations of lines and their graphs. The slope of a line is defined, and we learn how to find the slope of lines from an equation or from a graph. Then we extend these ideas to functions, one of the fundamental concepts of mathematics.

6.1 Cartesian Coordinate System

TAPE 11

If two copies of the number line, one horizontal and one vertical, are placed so that they intersect at the zero point of each line, a pair of **axes** is formed. The horizontal number line is called the **x-axis** and the vertical number line is called the **y-axis.** The point where the lines intersect is called the **origin.** We call this a **rectangular coordinate system,** or a **Cartesian coordinate system,** named for Descartes.

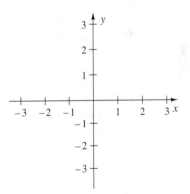

Suppose that p and q are two real numbers, written in the form

$$(p, q)$$

This is called an **ordered pair.** The order of the numbers is important.

We associate a point on a coordinate system with this ordered pair by finding the first number of the ordered pair, p, on the x-axis and the second number, q, on the y-axis. If we draw lines through these points perpendicular to the axes of the coordinate system, the lines will intersect in a single point. This unique point corresponds to the ordered pair (p, q) and is called its **graph.** To **plot** a point means to graph its ordered pair.

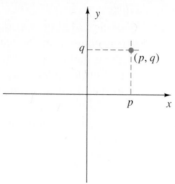

Note the graphs of the ordered pairs $(-3, 2)$, $(1, -2)$, $(-2, 1)$, and $\left(\dfrac{5}{2}, 1\right)$ on the following coordinate system.

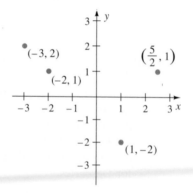

Be Careful!   Notice that the graphs of $(-2, 1)$ and $(1, -2)$ are different points. The order in which the numbers are written in an ordered pair *is important.* The coordinates of the origin are $(0, 0)$.

The first number of an ordered pair is called the **abscissa** or **first coordinate** or **x-coordinate,** and the second number, the **ordinate** or **second coordinate** or **y-coordinate.**

The four regions made by the axes are called **quadrants.** The graph of every ordered pair is in one of the four quadrants or on one of the axes.

	y-Axis	
Quadrant II		Quadrant I
x negative y positive		x positive y positive
	Origin	x-Axis
Quadrant III		Quadrant IV
x negative y negative		x positive y negative

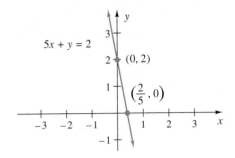

Notice that, if we let x have the value 0 and find the y value that makes the equation true, we have found where the line crosses the y-axis. This y value is called the **y-intercept.** Similarly, the **x-intercept** is the x value where the line crosses the x-axis. Find the x-intercept by giving y the value 0 and solving for x.

To Find the Intercepts of a Line

1. To find the x-intercept, replace y with 0 and solve for x.

2. To find the y-intercept, replace x with 0 and solve for y.

Do we always get two points when we find the x-intercept and the y-intercept?

► **EXAMPLE 3**

Draw the graph of $x + 2y = 0$.

Solution

To find the x-intercept, replace y with 0 and solve for x. This gives a value of 0 for x. The ordered pair (0, 0) is one solution. However, when we find the y-intercept, we get (0, 0) again!

To find another point, let y be 1 and solve for x. This gives -2 for x. Thus $(-2, 1)$ is a second ordered pair in the solution set.

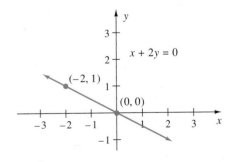

Notice that this line goes through the origin. If C is zero in the standard form $Ax + By = C$, the graph will go through the origin.

Does every line have both an x-intercept and a y-intercept?

► **EXAMPLE 4**

Draw the graph of $x = 2$.

Solution

There is no y-term in this equation. Does that mean that y is 0? No, write $x = 2$ in the standard form

$$x + 0y = 2$$

and we see that $A = 1$, $B = 0$, and $C = 2$.

Clearly, this equation is satisfied if x has the value 2 and y has any value at all! If we choose the pairs $(2, -1)$ and $(2, 1)$, we see that the equation $x = 2$ is true in both cases, giving the following graph.

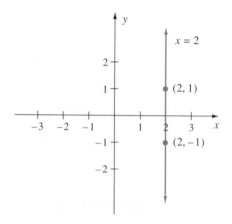

This line is called a **vertical line.**   ◄

The Graph of the Equation $x = p$

is a **vertical** line with x-intercept of p.

► **EXAMPLE 5**

Draw the graph of $2y = -3$.

Solution

In standard form, this equation is

$$0x + 2y = -3$$

Much as in the preceding example, this equation is satisfied by pairs of the form

$$\left(K, -\frac{3}{2} \right), \qquad \text{where } K \text{ is } any \text{ real number!}$$

Therefore, choosing 2 or -2 for K, we get $\left(2, -\frac{3}{2} \right)$ and $\left(-2, -\frac{3}{2} \right)$. The graph looks as follows:

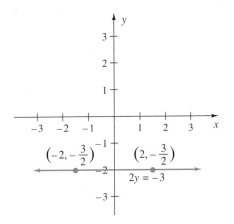

This line is called a **horizontal line.**   ◄

> ## The Graph of the Equation $y = q$
>
> is a **horizontal** line with y-intercept of q.

Linear equations in two variables are used by scientists, engineers, mathematicians, and business managers to express relationships between two variables. Usually, letters that are more appropriate to the situation are used rather than x and y.

► ## EXAMPLE 6

The distance d in miles a jogger runs at 6 mph is expressed in terms of running time t in hours by the equation $d - 6t = 0$.

(a) Graph this equation.

(b) How far has the jogger run in 2 hours?

(c) How long will it take the jogger to run 24 miles?

Solutions

(a) Let the horizontal axis be time and the vertical axis be distance. This line goes through the origin, $(0, 0)$. To find another point, let t be 1. This gives $d - 6(1) = 0$ or $d = 6$. So the point $(1, 6)$ is on the graph.

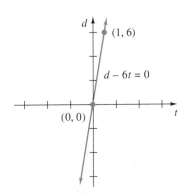

(b) If t is 2, then $d - 6(2) = 0$ or $d = 12$. The jogger has run 12 miles. On the graph this is represented by the point $(2, 12)$.

(c) If d is 24, then $24 - 6t = 0$ or $t = 4$. It will take 4 hours to run 24 miles. This corresponds to the point $(4, 24)$ on the graph.   ◄

Problem Set 6.2

Warm-Ups

In Problems 1 through 24, find the intercepts, then draw the graphs of the equations. For problems 1 through 14, see Examples 1 and 2.

1. $3x + 2y = 6$ **2.** $3x - 2y = 6$ **3.** $-3x + 2y = 6$ **4.** $-3x - 2y = 6$

5. $x + y = 1$ **6.** $x - y = 1$ **7.** $4x + y = 4$ **8.** $5x + 3y = 10$

9. $3x - 4y = 4$ **10.** $x + 3y = 1$ **11.** $2x - y - 1 = 0$ **12.** $14x - 21y = 49$

13. $10x - 15y + 25 = 0$ **14.** $x - 3y = 6$

For Problems 15 through 18, see Example 3.

15. $x + y = 0$ **16.** $x - y = 0$ **17.** $3x = -2y$ **18.** $3x = 2y$

Use the letter m for the slope. Therefore, the slope of a line containing the points (x_1, y_1) and (x_2, y_2) is given by the following formula.

The Slope Formula

The slope of the line through the points (x_1, y_1) and (x_2, y_2) is given by the formula

$$m = \frac{y_2 - y_1}{x_2 - x_1}, \qquad x_1 \neq x_2$$

► **EXAMPLE 1**

Find the slope of the line that contains the points $(1, 3)$ and $(3, 7)$.

Solution

$$m = \frac{y_2 - y_1}{x_2 - x_1}$$

Let P_1 be the point $(1, 3)$ and P_2 be the point $(3, 7)$; then x_1 is 1, y_1 is 3, x_2 is 3, and y_2 is 7, and the slope formula gives

$$m = \frac{7 - 3}{3 - 1} = \frac{4}{2} = 2$$

The slope of the line is 2. The line is graphed next.

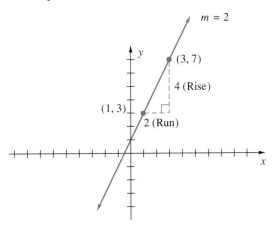

► **EXAMPLE 2**

Find the slope of the line that contains the points $(4, 3)$ and $(8, 1)$.

Solution

$$m = \frac{y_2 - y_1}{x_2 - x_1}$$

Let P_1 be (4, 3) and P_2 be (8, 1) to get

$$m = \frac{1 - 3}{8 - 4} = \frac{-2}{4}$$

$$= -\frac{1}{2}$$

The slope of the line is $-\frac{1}{2}$. The line is graphed next.

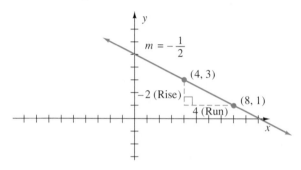

In finding the slope of the line containing the points (1, 3) and (4, 5), if P_1 is (1, 3) and P_2 is (4, 5), then using the slope formula gives

$$m = \frac{5 - 3}{4 - 1} = \frac{2}{3}$$

But if the points are turned around with P_1 as (4, 5) and (1, 3) as P_2,

$$m = \frac{3 - 5}{1 - 4} = \frac{-2}{-3} = \frac{2}{3}$$

Thus it does not matter which point is first, as long as we are consistent in the numerator and denominator.

In Examples 3 and 4, be careful with the signs.

► **EXAMPLE 3**

Find the slope of the line that contains the points $(-3, 1)$ and (3, 3).

Solution

$$m = \frac{3 - 1}{3 - (-3)} = \frac{2}{6}$$

$$= \frac{1}{3}$$

The slope is $\frac{1}{3}$. The line is graphed next.

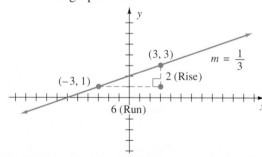

► **EXAMPLE 4**

Find the slope of the line that contains the points $(-1, 7)$ and $(2, -2)$.

Solution

$$m = \frac{-2 - 7}{2 - (-1)} = \frac{-9}{3} = -3$$

The slope is -3. The line is graphed next.

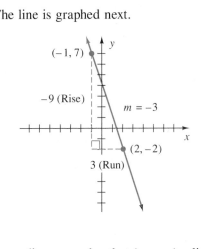

Notice in the preceding examples that *increasing* lines have *positive* slope, whereas *decreasing* lines have *negative* slope.

Let's calculate the slope of a horizontal line and a vertical line.

► **EXAMPLE 5**

Find the slope of the line containing the points $(-3, 2)$ and $(4, 2)$.

Solution

$$m = \frac{2 - 2}{4 - (-3)} = \frac{0}{7} = 0$$

The slope of the line is 0. The graph is shown next.

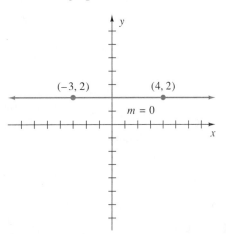

Since the y-coordinates of all the points on a horizontal line are the same, the rise is zero and so the slope is zero.

► EXAMPLE 6

Find the slope of the line containing the points $(1, 3)$ and $(1, -2)$.

Solution

$$m = \frac{-2 - 3}{1 - 1} = \frac{-5}{0}, \qquad \text{which is undefined!}$$

The slope is undefined. So the line has no slope. Its graph is shown below.

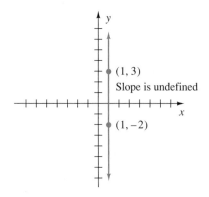

Since the x-coordinates of all the points on a vertical line are the same, the run is zero and we are unable to calculate the slope. Thus a vertical line has no slope.

Horizontal and Vertical Lines

1. Horizontal lines have equations of the form

$$y = q$$

and have a slope of zero.

2. Vertical lines have equations of the form

$$x = p$$

and have undefined slope.

Suppose that a line has slope m and y-intercept b. The equation of this line can be obtained from the slope formula. Since the y-intercept is b, the point $(0, b)$ is on the line. If (x, y) is any point on the line, the slope formula gives the following.

$$m = \frac{y - b}{x - 0} \qquad \text{Definition of slope}$$

$$mx = y - b \qquad \text{Clear fractions}$$

$$mx + b = y$$

or $\qquad\qquad\qquad y = mx + b$

We call this the **slope–intercept form.**

Slope–Intercept Form

An equation of a line with slope m and y-intercept of b is given by

$$y = mx + b$$

In Section 6.2, we graphed linear equations written in standard form, $Ax + By = C$. We can take an equation in standard form and rewrite the equation in slope–intercept form.

► EXAMPLE 7

Rewrite $5x - 6y = 18$ in slope–intercept form.

Solution

Solve $5x - 6y = 18$ for y.

$$-6y = -5x + 18$$

$$y = \frac{-5x + 18}{-6}$$

$$= \frac{5}{6}x - 3$$

This is the slope–intercept form, where m is $\dfrac{5}{6}$ and b is -3.   ◄

When a linear equation is in slope–intercept form, the slope and y-intercept can be read directly from the equation.

► EXAMPLE 8

Find the slope and y-intercept of the graph of $3x + 5y = 11$.

Solution

First, rewrite the equation in slope–intercept form by solving for y.

$$3x + 5y = 11$$

$$5y = -3x + 11$$

$$y = -\frac{3}{5}x + \frac{11}{5}$$

Now read the slope and y-intercept directly from the equation.

The slope is $-\dfrac{3}{5}$ and the y-intercept is $\dfrac{11}{5}$.   ◄

From an equation we can recognize whether the graph is a vertical or horizontal line. Then we can find the slope because we know that a vertical line has undefined slope and a horizontal line has 0 slope.

▶ **EXAMPLE 9**

Find the slope of the graph of $y = -4$.

Solution

The equation $y = -4$ is of the form $y = q$, so its graph is a horizontal line. Therefore, the slope is zero.   ◀

▶ **EXAMPLE 10**

Find the slope of the graph of $x = 16$.

Solution

This equation is of the form $x = p$, so its graph is a vertical line and its slope is undefined.   ◀

It is convenient to use the slope–intercept form when drawing the graph of a linear equation. The y-intercept gives one point immediately; then use the fact that the slope is the ratio of the rise to run to get a second point.

▶ **EXAMPLE 11**

Sketch the graph of the linear equation $y = \dfrac{2}{3}x - \dfrac{5}{2}$.

Solution

Since the equation is in slope–intercept form, read the y-intercept directly from the equation. Because b is $-\dfrac{5}{2}$, this graph crosses the y axis at $-\dfrac{5}{2}$.

The slope is $\dfrac{2}{3}$. So, working from the y-intercept of $-\dfrac{5}{2}$, we *rise* 2 while we *run* 3 (from left to right). This gives a second point on the graph, which is enough to complete the sketch.

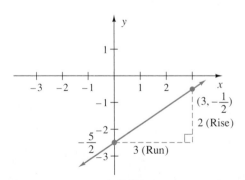

   ◀

▶ **EXAMPLE 12**

Graph the linear equation $y = -\dfrac{3}{4}x + 2$.

Solution

This equation is in slope–intercept form. So read the slope and y-intercept directly from the equation.

$$m = \frac{-3}{4} \quad \text{and} \quad b = 2$$

Notice that we always consider the *run* to be positive. So a negative slope implies a negative *rise*. Thus, working from the y-intercept of 2, rise -3 and run 4. As a negative rise is downward, go *down* 3 and *across* 4.

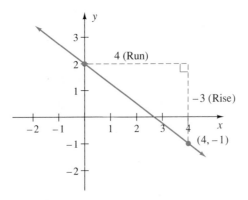

The graphs of $y = \dfrac{1}{2}x + 1$ and $y = \dfrac{1}{2}x - 1$ both have slope of $\dfrac{1}{2}$. The following graph seems to show that the lines are parallel. In fact, lines with the same slope are parallel and parallel lines have the same slope.

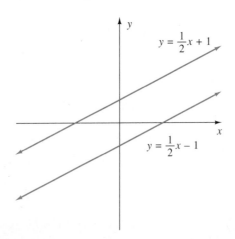

The graphs of $y = 3x + 1$ and $y = -\dfrac{1}{3}x + 2$, shown next, seem to indicate that the lines are perpendicular.

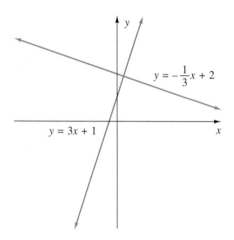

The slopes of the lines are 3 and $-\dfrac{1}{3}$. In fact, slopes of perpendicular lines are negative reciprocals of each other.

Parallel and Perpendicular Lines

Suppose that l_1 is a line with slope m_1 and l_2 is a line with slope m_2, and neither line is vertical.

1. If l_1 and l_2 are **parallel** lines, their slopes are the same; that is,

$$m_1 = m_2$$

2. If l_1 and l_2 are **perpendicular** to each other, their slopes are negative reciprocals of each other; that is,

$$m_1 m_2 = -1 \quad \text{or} \quad m_1 = -\dfrac{1}{m_2}$$

► EXAMPLE 13

Suppose that l_1 is a line with slope $\dfrac{2}{3}$. What is the slope of l_2, a line parallel to l_1, and the slope of l_3, a line perpendicular to l_1?

Solution

Since l_2 is parallel to l_1, its slope is the same as l_1. So $m_2 = \dfrac{2}{3}$. However, l_3 is perpendicular to l_1, so its slope is the negative reciprocal of the slope of l_1.

$$m_3 = -\frac{1}{m_1}$$

$$= -\frac{1}{\dfrac{2}{3}} = -1 \cdot \frac{3}{2} = -\frac{3}{2} \qquad \blacktriangleleft$$

► EXAMPLE 14

Find the slope of *any* line parallel to the graph of $2x + y = 6$.

Solution

First, find the slope of the graph of $2x + y = 6$. Write the equation in slope–intercept form.

$$y = -2x + 6$$

So the slope is -2.

Since parallel lines have the same slope, any line parallel to $2x + y = 6$ will have slope -2. $\qquad \blacktriangleleft$

► EXAMPLE 15

Suppose that L_1 is a line that is perpendicular to the graph of $y = \frac{5}{3}x - 5$. What is the slope of L_1?

Solution

Let m_1 be the slope of L_1 and m_2 be the slope of the graph of $y = \frac{5}{3}x - 5$. Then $m_2 = \frac{5}{3}$. Since L_1 is perpendicular to the line whose slope is m_2,

$$m_1 = -\frac{1}{m_2}$$

$$= -\frac{1}{\dfrac{5}{3}} = -1 \cdot \frac{3}{5} = -\frac{3}{5}$$

The slope of L_1 is $-\frac{3}{5}$. $\qquad \blacktriangleleft$

Problem Set 6.3

Warm-Ups

In Problems 1 through 20, find the slope of the line that contains both points. For Problems 1 through 6, see Examples 1 and 2.

1. $(2, 2), (3, 5)$ **2.** $(3, 2), (2, 5)$ **3.** $(2, 2), (5, 3)$

4. $(5, 2), (2, 3)$

5. $\left(\dfrac{5}{2}, \dfrac{1}{2}\right), \left(\dfrac{7}{2}, \dfrac{5}{2}\right)$

6. $\left(\dfrac{1}{3}, \dfrac{1}{2}\right), \left(\dfrac{1}{4}, \dfrac{1}{3}\right)$

For Problems 7 through 14, see Examples 3 and 4.

7. $(4, -2), (2, 2)$

8. $(-1, 4), (4, 6)$

9. $(-7, 4), (-3, 2)$

10. $(-3, 6), (-9, -3)$

11. $(-2, -6), (-4, -1)$

12. $(0, 0), (-8, -6)$

13. $(0, -9), (-4, 0)$

14. $\left(-\dfrac{1}{3}, -\dfrac{5}{4}\right), \left(\dfrac{2}{3}, \dfrac{3}{4}\right)$

For Problems 15 through 17, see Example 5.

15. $(-2, 3), (-5, 3)$

16. $\left(\dfrac{5}{7}, -\dfrac{2}{3}\right), \left(\dfrac{3}{5}, -\dfrac{2}{3}\right)$

17. $(-2, 0), (0, 0)$

For Problems 18 through 20, see Example 6.

18. $(8, -3), (8, 3)$

19. $(-3, -6), (-3, -8)$

20. $\left(\dfrac{3}{5}, \dfrac{2}{5}\right), \left(\dfrac{3}{5}, -\dfrac{2}{5}\right)$

In Problems 21 through 29, find the slope of the graph of each linear equation by rewriting the equation in slope–intercept form and reading the slope from the equation. See Examples 1 and 2.

21. $2x + 3y = 6$

22. $5x - y = 10$

23. $x + 7y = 14$

24. $2x + 3y = 5$

25. $5x - y = 7$

26. $x + 7y = 1$

27. $4x - 7y = 0$

28. $x = 2y - 3$

29. $y = x$

For Problems 30 through 32, see Examples 9 and 10.

30. $y = 4$

31. $2x = 7$

32. $4 + 3y = 0$

In Problems 33 through 38, sketch the graph of each equation by using the information from the slope–intercept form. See Examples 11 and 12.

33. $y = x - 1$

34. $y = -x + 1$

35. $y = \dfrac{1}{2}x + \dfrac{3}{2}$

36. $y = -\dfrac{1}{3}x + \dfrac{5}{3}$

37. $y = 3x - 2$

38. $y = -4x$

For Problems 39 through 44, see Examples 13 through 15.

39. Find the slope of a line parallel to the line containing the points $(2, 5)$ and $(1, -3)$.

40. Find the slope of a line perpendicular to the line containing the points $(0, -2)$ and $(-4, -5)$.

41. Find the slope of a line parallel to the line containing the points $(2, -3)$ and $(7, -3)$.

42. Find the slope of a line perpendicular to the line containing the points $(2, -3)$ and $(7, -3)$.

43. Find the slope of any line perpendicular to the graph of $4x + 3y = 24$.

44. Find the slope of a line parallel to the graph of $2x - 5y = 20$.

Practice Exercises

In Problems 45 through 66, find the slope of the line that contains both points.

45. $(4, 4), (5, 2)$ **46.** $(2, 4), (4, 5)$ **47.** $(3, 2), (6, -3)$

48. $(1, -2), (4, 4)$ **49.** $(0, 3), (6, -1)$ **50.** $(-2, 3), (1, 5)$

51. $(-2, 2), (-5, 6)$ **52.** $(-3, 5), (1, -3)$ **53.** $(-6, 5), (6, -4)$

54. $(-6, 2), (-6, 4)$ **55.** $(1, 3), (5, 3)$ **56.** $(-5, -5), (-2, -7)$

57. $(-4, -3), (-8, -1)$ **58.** $(-2, -4), (-5, -4)$ **59.** $(0, 3), (-4, 0)$

60. $(0, -3), (0, 0)$ **61.** $\left(-\frac{1}{2}, \frac{2}{3}\right), \left(\frac{1}{2}, \frac{5}{3}\right)$ **62.** $\left(\frac{1}{3}, 6\right), \left(\frac{4}{3}, 4\right)$

63. $\left(\frac{2}{5}, -\frac{7}{2}\right), \left(\frac{4}{10}, -\frac{3}{2}\right)$ **64.** $\left(\frac{5}{2}, \frac{1}{6}\right), \left(2, \frac{2}{3}\right)$ **65.** $\left(\frac{1}{5}, -\frac{2}{5}\right), \left(\frac{1}{3}, -\frac{2}{5}\right)$

66. $\left(-\frac{3}{4}, -2\right), \left(\frac{1}{2}, \frac{1}{2}\right)$

In Problems 67 through 74, find the slope of the graph of each linear equation in two ways.
(a) Find two points and use the slope formula.
(b) Rewrite in slope–intercept form and read it from the equation.

67. $3x + 4y = 12$ **68.** $4x - y = 8$ **69.** $x + 5y = 10$ **70.** $3x + 4y = 5$

71. $4x - y = 7$ **72.** $x + 5y = 1$ **73.** $3x + 7y = 0$ **74.** $x = 3y - 4$

In Problems 75 through 80, find the indicated slope.

75. Find the slope of a line parallel to the line containing the points $(1, 6)$ and $(-1, 3)$.

76. Find the slope of a line perpendicular to the line containing the points $(2, -3)$ and $(-4, 0)$.

77. Find the slope of a line parallel to the line containing the points $(5, -9)$ and $(5, -3)$.

78. Find the slope of a line perpendicular to the line containing the points $(2, -1)$ and $(2, -3)$.

79. Find the slope of any line perpendicular to the graph of $2x - 3y = 24$.

80. Find the slope of a line parallel to the graph of the equation $2x + 5y = 30$.

In Problems 81 through 86, sketch the graph of each equation by using the information from the slope–intercept form.

81. $y = x - 2$ **82.** $y = -x - 1$ **83.** $y = \frac{1}{3}x + \frac{2}{3}$

84. $y = -\frac{1}{2}x + \frac{3}{2}$ **85.** $y = -3x + 2$ **86.** $y = 4x$

Challenge Problems

In Problems 87 and 88, determine an equation for the line l_2 from the information shown on each graph.

87.

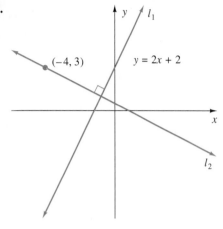

$(-4, 3)$

$y = 2x + 2$

88.

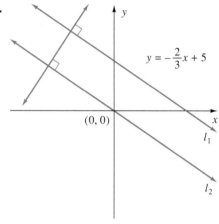

$y = -\frac{2}{3}x + 5$

$(0, 0)$

In Your Own Words

89. What is slope?

6.4 Equations of Lines

TAPE 12

In previous sections we learned that the graph of a linear equation in two variables is a line. It is reasonable to ask: Given a line, is there a linear equation with that line as its graph? The answer is yes, and we will see how to find suitable equations in this section.

Now, suppose that we are given a line and wish to find an equation having that line as its graph. Two pieces of information are needed:

1. The slope of the line
2. The coordinates of any point on the line

Substituting this information into the slope formula and simplifying will produce the desired equation.

▶ **EXAMPLE 1**

Find an equation of a line with slope 2 that contains the point $(1, -2)$, and draw its graph.

Solution

In the slope formula we let P_1 be the point $(1, -2)$ and P_2 be any point on the line, say (x, y). Since the slope is 2, the formula gives

$$2 = \frac{y - (-2)}{x - 1}$$

$$2(x - 1) = y + 2$$

$$2x - 2 = y + 2$$

$$2x - y = 4$$

The equation is $2x - y = 4$.   ◄

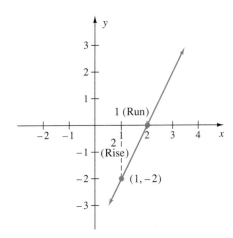

When using a given slope m, a given point (x_1, y_1) and a general point on the line (x, y), the slope formula takes the form

$$m = \frac{y - y_1}{x - x_1}$$

Multiplying both sides by $(x - x_1)$ produces the **point–slope** form,

$$m(x - x_1) = y - y_1$$

which is usually written

$$y - y_1 = m(x - x_1)$$

The Point–Slope Form

An equation of the line with slope m and containing the point (x_1, y_1) is given by

$$y - y_1 = m(x - x_1)$$

► **EXAMPLE 2**

Find an equation for the line containing the points $(-1, 3)$ and $(2, -1)$.

Solution

First, find the slope of the line containing the two points

$$m = \frac{3 - (-1)}{-1 - 2}$$

$$= -\frac{4}{3}$$

Next, pick *either* of the two given points to be (x_1, y_1) and substitute into the point–slope form of the linear equation.

$$y - y_1 = m(x - x_1)$$

$$y - (-1) = -\frac{4}{3}(x - 2) \qquad \text{Using } (2, -1) \text{ as } (x_1, y_1)$$

$$3(y + 1) = -4(x - 2)$$

$$3y + 3 = -4x + 8$$

$$4x + 3y = 5$$

This is the equation of a line in standard form. The graph of $4x + 3y = 5$ contains the points $(-1, 3)$ and $(2, -1)$.   ◄

► **EXAMPLE 3**

Find an equation for the line through the point $(4, -7)$ that is parallel to the graph of $x + 2y = 6$.

Solution

First, find the slope of the graph of the given equation by writing it in slope–intercept form.

$$x + 2y = 6$$

$$2y = -x + 6$$

$$y = -\frac{1}{2}x + 3$$

Note that the given line has slope $-\frac{1}{2}$. The line we are looking for is parallel to the line given, so its slope is the same. Because $(4, -7)$ is on our line, we have enough information to find an equation.

$$y - y_1 = m(x - x_1)$$

$$y - (-7) = -\frac{1}{2}(x - 4)$$

Multiply both sides by -2 to clear fractions.

$$-2(y + 7) = x - 4$$

$$-2y - 14 = x - 4$$

$$-10 = x + 2y \quad \text{or} \quad x + 2y = -10$$

This is the desired equation in standard form.   ◄

► ## EXAMPLE 4

Find an equation of the line through the origin perpendicular to the graph of $2x + 3y = 7$.

Solution

First, find the slope. Rewriting the equation in slope–intercept form gives us

$$2x + 3y = 7$$

$$3y = -2x + 7$$

$$y = -\frac{2}{3}x + \frac{7}{3}$$

The slope of the graph of the equation is $-\frac{2}{3}$. Therefore, the slope of a line perpendicular to this line is $\frac{3}{2}$.

Thus we are looking for a line through the origin with slope $\frac{3}{2}$. As the origin has coordinates $(0, 0)$, we have

$$y - y_1 = m(x - x_1)$$

$$y - 0 = \frac{3}{2}(x - 0)$$

$$y = \frac{3}{2}x$$

$$3x - 2y = 0$$ ◄

Forms of the Linear Equation in Two Variables

Standard Form

$$Ax + By = C, \qquad A \text{ and } B \text{ not both zero}$$

Slope–Intercept Form

$$y = mx + b$$

Point–Slope Form

$$y - y_1 = m(x - x_1)$$

Linear equations are often stated in standard form. One of the most useful forms is the slope–intercept form. The point–slope form is convenient for writing an equation of a line given a point and its slope.

> ## To Write an Equation of a Line That Has Slope
> 1. Find m, the slope of the line.
> 2. Find (x_1, y_1), the coordinates of any point on the line.
> 3. Substitute into the point–slope formula.
> $$y - y_1 = m(x - x_1)$$
> 4. Write the equation in standard or slope–intercept form, whichever is preferred.

A special case of this occurs if the slope and y-intercept are known. In this case it is easier to use the slope–intercept form.

► EXAMPLE 5

Find an equation of the line with slope -1 and y-intercept of 5.

Solution

Using the slope–intercept form, $y = mx + b$ gives
$$y = -x + 5 \qquad \blacktriangleleft$$

Since a vertical line has no slope, it must be treated in another way. It is usually best to treat horizontal and vertical lines as exceptions when writing their equations. They are easy to identify (slope of zero or undefined slope), and it is simple to write their equations.

> ## Equations of Horizontal and Vertical Lines
> 1. A line of slope 0 is *horizontal* and has the equation
> $$y = q$$
> where q is the y-coordinate of *any* point on the line.
> 2. A line with undefined slope is *vertical* and has the equation
> $$x = p$$
> where p is the x-coordinate of *any* point on the line.

► EXAMPLE 6

Find an equation of the line through $(1, 4)$ that is

(a) parallel to the x-axis;

(b) perpendicular to the x-axis.

Solutions

(a) Any line parallel to the *x*-axis is a horizontal line. All points on such a line have the same *y*-coordinate. So the equation of the line through (1, 4) parallel to the *x*-axis is *y* = 4.

(b) Any line perpendicular to the *x*-axis is a vertical line. All points on a vertical line have the same *x*-coordinate. So the equation of the line through (1, 4) and perpendicular to the *x*-axis is *x* = 1. ◄

Sometimes we use different scales on the axes when we apply these ideas about lines. Also, we use letters other than *x* and *y* as names for the axes. In applications, many times only the first quadrant has meaning for the application.

► **EXAMPLE 7**

A company knows that the price of a ticket for its scenic tour will affect the number of tickets sold. An analyst determined that the line graphed here describes the relationship between the price of a ticket for the scenic tour and the number of tickets sold.

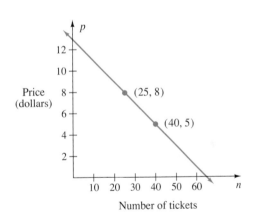

(a) Write an equation of this line.

(b) Use this equation to find the ticket price when the number of tickets sold is 10.

(c) When ticket price is $13, how many tickets are sold?

(d) Describe in words the relationship between price and number of tickets sold.

Solution

(a) We use the two given points and find the slope with the slope formula.

$$m = \frac{8 - 5}{25 - 40} = \frac{3}{-15} = -\frac{1}{5}$$

Use (40, 5) as the point and substitute into the point–slope formula. We will use n as the x-coordinate and p as the y-coordinate.

$$y - y_1 = m(x - x_1) \qquad \text{Point–Slope Formula}$$

$$p - 5 = -\frac{1}{5}(n - 40)$$

$$p = -\frac{1}{5}n + 13$$

(b) For 10 tickets, we substitute 10 for n into the equation to find p.

$$p = -\frac{1}{5}(10) + 13$$

$$= 11$$

When the price is $11, then 10 tickets will be sold.

(c) If the price is $13, we substitute 13 for p in the equation and find n.

$$13 = -\frac{1}{5}n + 13$$

$$0 = n$$

If the ticket price is $13, no tickets will be sold.

(d) As ticket sales increase, price decreases.

◄

Problem Set 6.4

Warm-Ups

In Problems 1 through 6, find an equation of the line through the point having the slope given. Sketch the graph of the line first. See Example 1.

1. $(1, 2)$; $m = 3$

2. $(-1, 1)$; $m = 2$

3. $(4, 3)$; $m = -2$

4. $(-5, -2)$; $m = -1$

5. $(4, -6)$; $m = \dfrac{1}{2}$

6. $\left(\dfrac{2}{5}, 1\right)$; $m = 1$

In Problems 7 through 12, find an equation of the line through the points. See Example 2.

7. $(2, 3), (4, 5)$

8. $(-1, 4), (2, 3)$

9. $(2, -3), (5, 0)$

10. $(0, -2), (5, 0)$

11. $(0, 0), (-1, -5)$

12. $(-1, -2), (-6, -9)$

In Problems 13 through 16, find an equation of the line satisfying the conditions stated. See Examples 3 and 4.

13. The line parallel to the graph of $4x - 2y = 13$ containing the point $(-7, 0)$.

14. The line through $(1, -5)$ parallel to the graph of $y = 6x - 5$.

15. The line through $(1, 5)$ perpendicular to the graph of $2x + 3y = 4$.

16. The line perpendicular to the graph of $y = -3x - 17$ containing the point $(4, -11)$.

In Problems 17 and 18, find an equation of the line given. See Example 5.

17. The line with y-intercept of 2 and slope $\frac{1}{2}$.

18. The line with y-intercept $\frac{3}{2}$ and slope -4.

In Problems 19 and 20, find an equation of each line. See Example 6.

19. The line through $(-4, -1)$ parallel to the x-axis.

20. The line through $(-4, -1)$ perpendicular to the x-axis.

In Problems 21 and 22, write the indicated equation. See Example 7.

21. The relationship between Fahrenheit and Celsius temperature is shown below. Write an equation of this line.

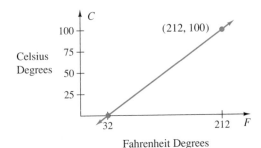

22. The marketing expenditures for Campbell Soup are graphed below. Write an equation for this line. Use the equation to find the marketing expenditures when time is 10 years.

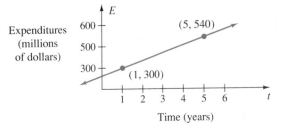

Practice Exercises

In Problems 23 through 28, find an equation of the line through the point having the slope given.

23. $(2, 1); m = 2$

24. $(1, -1); m = -2$

25. $(3, 4); m = -1$

26. $(-4, -3); m = 1$

27. $(2, -3); m = -\frac{1}{2}$

28. $\left(\frac{2}{3}, \frac{3}{5}\right); m = 1$

In Problems 29 through 34, find an equation of the line through the points.

29. $(1, 2), (3, 4)$

30. $(-2, 3), (1, 4)$

31. $(3, 0), (2, -3)$

32. $(0, 2), (-5, 0)$

33. $(-2, -4), (0, 0)$

34. $(-2, -3), (-5, -6)$

In Problems 35 through 52, find an equation of the line satisfying the conditions stated.

35. The line with y-intercept of 4 and slope $\frac{1}{3}$.

36. The line with y-intercept of 2 and slope -1.

37. The line with slope 1 and x-intercept of $-\frac{1}{2}$.

38. The line with slope -1 and x-intercept of -3.

39. The line with x-intercept of -2 and y-intercept of 3.

40. The line with x and y intercepts of 1.

41. The line through $(-1, 4)$ parallel to the graph of $y = 4x - 3$.

42. The line perpendicular to the graph of $y = -2x - 9$ containing the point $(4, -11)$.

43. The line through $(-1, 4)$ perpendicular to the graph of $x + 2y = 3$.

44. The line parallel to the graph of $5x - 3y = 13$ containing the point $(0, -7)$.

45. The line through $(2, -3)$ parallel to the y-axis.

46. The line through $(2, -3)$ perpendicular to the y-axis.

In Problems 53 through 56, answer each question.

53. Total sales for Campbell Soup are shown below. Write an equation for the line. Use the equation to find sales when time is 10 years.

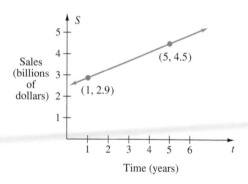

Time (years)

54. The depreciation on a boat is shown below. Write an equation of this line. Use the equation to find the depreciation when the boat is 3 years old.

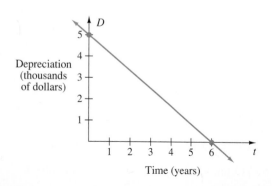

Time (years)

47. The line through the origin perpendicular to the graph of $x = -y$.

48. The line through the origin perpendicular to the line containing the points $(-5, 0)$ and $(16, 0)$.

49. The line through $(2, 0)$ parallel to the line containing the points $(-5, 3)$ and $(3, 3)$.

50. The line with x-intercept -4 and parallel to the graph of $2x + 3y = 1$.

51. The line with y-intercept 11 and perpendicular to the graph of $4x - y = 0$.

52. The line through the origin parallel to the line whose x-intercept and y-intercept are 4 and -8, respectively.

55. The Interstate Shipping Company computes the charge for shipping packages from the following graph.

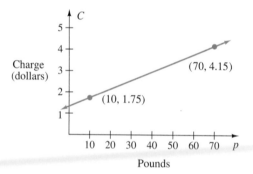

Pounds

(a) Write an equation of the line.
(b) What would the charge be for a 40-lb package?
(c) If a customer was charged $3.35, what was the weight of the package?
(d) Describe the relationship between the customer's bill and the weight of a package.

56. An increase in the number of skeletal muscle mitochondria contributes to the muscle cells' capacity to oxidize carbohydrates following training. After several studies, the Fitness Center uses the line graphed next to predict

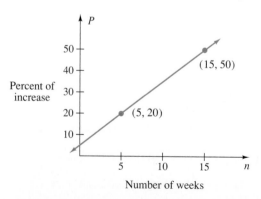

Number of weeks

the percent of increase in the number of mitochondria based on the number of weeks of training for a particular group of clients.

(a) Write an equation of this line.

(b) How many weeks of training would result in a 10% increase in mitochondria?

(c) What percent of increase in mitochondria could be expected after 4 weeks of training?

(d) Describe the relationship between the number of weeks of training and the percent of increase in mitochondria.

Challenge Problems

57. Find an equation of the line through (p, q) perpendicular to the graph of $px + qy = C$.

58. Find the x-intercept and y-intercept of the graph of $8x + 5y = 40$.

Now, divide both sides of the original equation by 40 to obtain the form

$$\frac{x}{a} + \frac{y}{b} = 1$$

This is called the *intercept form* of the linear equation. Why?

In Problems 59 through 64, find the x-intercept and y-intercept by first rewriting the given equation in intercept form.

59. $2x + 3y = 6$

60. $5x - 4y = 20$

61. $2x + 3y = 1$

62. $5x - 7y = 11$

63. $3x = 5$

64. $2y + 9 = 0$

In Your Own Words

65. Explain how to graph a line if a point on the line and the slope of the line are given.

66. Compare the advantages of the slope–intercept form and the standard form of a line.

6.5 Functions

TAPE 12

Let's consider the equation $F = \frac{9}{5}C + 32$. It is a linear equation written in slope–intercept form. Using its slope $\frac{9}{5}$ and y-intercept of 32, we can graph it.

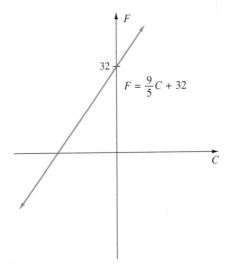

$$F = \frac{9}{5}C + 32$$

Since every point on the line represents an ordered pair of numbers, we could think of the equation as defining a set of ordered pairs. We could list some in a table that could be used to convert Fahrenheit and Celsius readings.

F	$-13°$	$-4°$	$14°$	$32°$	$59°$	$122°$
C	$-25°$	$-20°$	$-10°$	$0°$	$15°$	$50°$

Two sets of numbers result from the ordered pairs listed in the table. To help us visualize the sets as well as the pairings, we find a picture like the one below very useful.

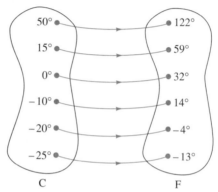

This idea of pairing numbers is very useful in mathematics. It is the central idea behind the definition of a function. The study of functions is one of the most important aspects of mathematics.

Function

A **function** is a rule that assigns to each member in one set (the **domain**) exactly one member from another set (the **range**).

Both a rule and a domain must be specified when a function is defined.

► **EXAMPLE 1**

Consider the assignments made by the following sketches. Is each a function?

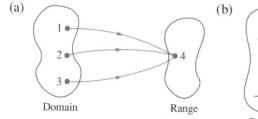

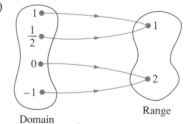

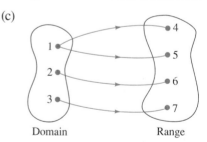

Solutions

(a) This is a function. Its rule assigns 4 to every number in the domain. To be a function, a rule must assign to each number in the domain *exactly one number* in the range. This rule does that. It does not matter that more than one number in the domain is paired with the same number in the range.

(b) This is also a function. Its domain is $\left\{-1, 0, \frac{1}{2}, 1\right\}$ and its range is $\{1, 2\}$.

It does not matter that both 1 and $\frac{1}{2}$ are paired with 1 or that both -1 and 0 are paired with 2. Each number in the domain is assigned *exactly one* number in the range.

(c) This is *not* a function. The number 1 is assigned two different numbers, 4 and 5. This violates the definition of a function, which states that *each* number in the domain must be assigned *exactly one* number in the range. ◄

There are several ways of giving the rule for a function. Using pictures as in Example 1 is one way. Sometimes the assignments are listed as ordered pairs or in tables, with the first number being from the domain.

► **EXAMPLE 2**

The domain of a function is $\{1, 2, 3, 4\}$ and the rule is: Assign to each number its square.

(a) Draw a picture to show the assignment of numbers in the domain to numbers in the range.

(b) Write the assignments as ordered pairs.

(c) Find the range.

Solutions

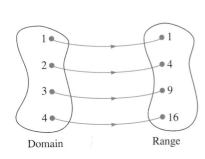

Domain   Range

(a) Draw a picture to indicate the assignment of numbers in the domain to numbers in the range.

(b) The assignments can be listed with the ordered pairs: (1, 1), (2, 4), (3, 9), (4, 16).

(c) So the range is $\{1, 4, 9, 16\}$.

Notice that *each* number in the domain is assigned *exactly one* number in the range. ◄

In mathematical applications of functions, the rule is often given using **functional notation.** We use letters such as f, g, and h to name functions.

If f is a function and x is a number in its domain, $f(x)$ is a symbol used to indicate the corresponding number in the range. We read $f(x)$ as "f of x" or "f evaluated at x." This could be represented by writing the ordered pair $(x, f(x))$ or by the following sketch.

Domain   Range

Be Careful!   NOTE: In using functional notation, remember that $f(x)$ may look like f times x, but it does *not* mean that.

The rule in Example 2 is: Assign to each number its square. Written in functional notation, this would be

$$f(x) = x^2$$

If x is a number in the domain, it is paired with $f(x)$ in the range. The illustration to the left shows the pairings.

Remember that $f(1)$ is a name for the number in the range paired with 1 from the domain. To calculate the value of $f(1)$, we use the rule $f(x) = x^2$ and replace x with 1. Thus

$$f(1) = 1^2 = 1$$

$$f(2) = 2^2 = 4$$

$$f(3) = 3^2 = 9$$

$$f(4) = 4^2 = 16$$

► **EXAMPLE 3**

If $h(x) = x^2 - 2x$, and the domain is $\mathbb{R}$, find:

(a) $h(0)$ (b) $h(-5)$

(c) $h(\sqrt{3})$ (d) $h(\pi)$

Solutions

(a) $h(0) = 0^2 - 2(0)$ Replace x with 0 in $h(x)$

$\quad\quad\;\; = 0 - 0$

$\quad\quad\;\; = 0$

(b) $h(-5) = (-5)^2 - 2(-5)$ Replace x with -5 in $h(x)$

$\quad\quad\quad\;\; = 25 + 10$

$\quad\quad\quad\;\; = 35$

(c) $h(\sqrt{3}) = (\sqrt{3})^2 - 2\sqrt{3}$ Replace x with $\sqrt{3}$ in $h(x)$

$\quad\quad\quad\;\; = 3 - 2\sqrt{3}$

(d) $h(\pi) = \pi^2 - 2\pi$ Replace x with π in $h(x)$ ◄

► **EXAMPLE 4**

If $f(x) = \sqrt{x + 1}$ with domain $x \geq -1$ and $g(x) = 3x - 4$ with domain $\mathbb{R}$, find:

(a) $f(3) + g(2)$ (b) $f(0) \cdot g(0)$

(c) $\sqrt{g(2)}$ (d) $[g(1)]^2$

Solutions

(a) $f(3) + g(2) = \sqrt{3 + 1} + 3(2) - 4$

$\quad\quad\quad\quad\;\; = 2 + 6 - 4$

$\quad\quad\quad\quad\;\; = 4$

(b) $f(0) \cdot g(0) = \sqrt{0 + 1} \cdot (3 \cdot 0 - 4)$

$\qquad = 1(-4)$

$\qquad = -4$

(c) $\sqrt{g(2)} = \sqrt{3(2) - 4}$

$\qquad = \sqrt{2}$

(d) $[g(1)]^2 = (-1)^2$

$\qquad = 1$ ◄

► EXAMPLE 5

If $g(x) = x^2 + 3$ with domain $\mathbb{R}$, find:

(a) $g(a)$ (b) $g(b)$ (c) $g(a + b)$ (d) $g(a) + g(b)$

Solutions

(a) $g(a) = a^2 + 3$ Replace x with a in $g(x)$

(b) $g(b) = b^2 + 3$ Replace x with b in $g(x)$

(c) $g(a + b) = (a + b)^2 + 3$ Replace x with $a + b$ in $g(x)$

(d) $g(a) + g(b) = (a^2 + 3) + (b^2 + 3)$

$\qquad\qquad = a^2 + b^2 + 6$ ◄

It is common practice to state the rule for a function and omit the domain. When this happens, we use the natural domain of the rule. The **natural domain** is the largest subset of real numbers for which the rule has meaning.

► EXAMPLE 6

Give the natural domain of the function $k(x) = 2x - 3$.

Solution

$k(x)$ is a real number for any value of x. So the natural domain of k is $\{x \mid x$ is a real number$\}$.   ◄

► EXAMPLE 7

Find the natural domain of the function

$$g(x) = \frac{5}{x + 3}$$

Solution

We can see that $g(x)$ is a real number for any value of x except when x is -3. $g(-3)$ is $\dfrac{5}{-3 + 3}$ or $\dfrac{5}{0}$, which is undefined. So $g(-3)$ is not a real number. Thus the natural domain of g is $\{x \mid x \neq -3\}$.   ◄

► **EXAMPLE 8**

Find the natural domain of the function $g(x) = \sqrt{x}$.

Solution

$g(x)$ is a real number for all nonnegative values of x, that is, for $x \geq 0$. [If x is -5, then $g(-5)$ is $\sqrt{-5}$, which is not a real number.] So the natural domain of g is $\{x \mid x \geq 0\}$.   ◄

► **EXAMPLE 9**

Give the natural domain of the function

$$h(x) = \sqrt{x^2 - x - 2}$$

Solution

$h(x)$ will be a real number if $x^2 - x - 2 \geq 0$. We solve the inequality by the technique of boundary numbers.

The equation $x^2 - x - 2 = 0$ becomes

$$(x - 2)(x + 1) = 0$$

and gives boundary numbers of -1 and 2.

Testing each of the regions A, B, and C, the solution set for $x^2 - x - 2 \geq 0$ is $\{x \mid x \leq -1 \text{ or } x \geq 2\}$.

So the natural domain of h is $\{x \mid x \leq -1 \text{ or } x \geq 2\}$.   ◄

A function can be evaluated only at numbers in its domain. If $f(x) = \sqrt{x}$, then $f(-2)$ is undefined since $\sqrt{-2}$ is not a real number.

Problem Set 6.5

Warm-Ups

In Problems 1 through 4, identify which rules below establish functions. Write the assignments in the pictures as ordered pairs. See Examples 1 and 2.

1.

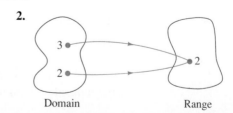

Domain   Range

2.

Domain   Range

3.

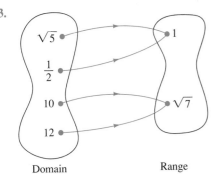

4.

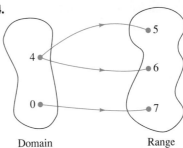

In Problems 5 through 11, let $f(x) = 2x - 5$ and $g(x) = \sqrt{x + 1}$. Find each function value. See Example 3.

5. $f(2)$ **6.** $f(-3)$ **7.** $f(0)$ **8.** $f\left(\dfrac{1}{2}\right)$

9. $g(3)$ **10.** $g(0)$ **11.** $g(-1)$

In Problems 12 through 16, let $f(x) = x^2$ and $g(x) = 2x - 3$. Find each function value. See Example 4.

12. $f(3) + g(3)$ **13.** $f(3) - g(3)$ **14.** $f(3)g(3)$

15. $\dfrac{f(3)}{g(3)}$ **16.** $(g(3))^2$

In Problems 17 through 21, let $f(x) = 2x^2 + x - 1$ and find each function value. See Example 5.

17. $f(t)$ **18.** $f(a + b)$ **19.** $f(a) + f(b)$

20. $f(2t)$ **21.** $f(\sqrt{a})$

In Problems 22 through 33, find the natural domain of each function.
For Problems 22 through 25, see Example 6.

22. $g(x) = 3x - 1$ **23.** $h(x) = x^2 - 4$ **24.** $d(x) = \sqrt[3]{x}$ **25.** $f(x) = |x|$

For Problems 26 through 29, see Example 7.

26. $r(x) = \dfrac{7}{x + 2}$ **27.** $f(x) = \dfrac{3}{x} + 2$

28. $h(x) = \dfrac{x + 7}{x - 1}$ **29.** $g(x) = \dfrac{6}{(x + 2)(x - 7)}$

For Problems 30 through 33, see Examples 8 and 9.

30. $f(x) = \sqrt{x + 2}$ **31.** $g(x) = \sqrt{x - 1}$

32. $f(x) = \sqrt{x^2 - x - 6}$ **33.** $p(x) = \sqrt{x^2 + x - 2}$

Practice Exercises

In Problems 34 through 40, find the range for each function.

34. Domain is $\{-1, 0, 1\}$; $f(x) = x^3$.

35. Domain is $\{1, 4, 7, 11, 14\}$; $f(x) = x - 7$.

36. Domain is $\{x \mid x$ is an odd natural number$\}$; $f(x) = x + 1$.

37. Domain is $\{x \mid x$ is an even natural number$\}$; $f(x) = 2x - 1$.

38. Domain is $\{x \mid x$ is an integer$\}$; $f(x) = 2x$.

39. Domain is $\{x \mid x$ is a real number$\}$; $f(x) = 11$.

40. Domain is $\{x \mid x \geq 0\}$; $f(x) = \sqrt{x}$.

In Problems 41 through 60, find the natural domain for each function.

41. $g(x) = x - 1$

42. $f(x) = \dfrac{1}{3}x + 1$

43. $h(x) = \dfrac{x}{3} - \dfrac{1}{4}$

44. $g(x) = \dfrac{1}{x + 2}$

45. $r(x) = \dfrac{3}{x - 1}$

46. $f(x) = \dfrac{1}{x} + 2$

47. $h(x) = \dfrac{x}{x - 3}$

48. $g(x) = \dfrac{3}{(x - 2)(x + 3)}$

49. $h(x) = \dfrac{x}{x + 3}$

50. $t(x) = \dfrac{x}{(x - 2)(x + 1)}$

51. $f(x) = \sqrt{x - 2}$

52. $g(x) = \sqrt{x + 1}$

53. $h(x) = x^2 - 3$

54. $f(x) = \sqrt{x^2 - 2x - 3}$

55. $d(x) = \sqrt[3]{x} + 1$

56. $p(x) = \sqrt{x^2 + 2x - 15}$

57. $g(x) = \sqrt{x^2 + 1}$

58. $f(x) = |x + 1|$

59. $f(x) = |x - 5|$

60. $w(x) = |2 - x|$

In Problems 61 through 72, let $f(x) = 3x + 1$, and find each function value.

61. $f(2)$

62. $f(-1)$

63. $f(0)$

64. $f\left(\dfrac{1}{3}\right)$

65. $f(\sqrt{3})$

66. $f(a)$

67. $f(a - b)$

68. $f(a) - f(b)$

69. $f(2t)$

70. $f(\sqrt{b})$

71. $\sqrt{f(a)}$

72. $f(-a)$

In Problems 73 through 82, let $g(x) = \sqrt{x - 1}$, and find each function value.

73. $g(6)$

74. $g(0)$

75. $g(2)$

76. $g(-2)$

77. $2g(5) - 1$

78. $g(b - 1)$

79. $g(2k + 3)$

80. $g(a^2)$

81. $[g(a)]^2$

82. $g(\pi)$

In Problems 83 through 92, let $f(x) = 1 - 2x^2$, and find each function value.

83. $f(t + h)$

84. $f(t) + f(h)$

85. $f(t + h) - f(t)$

86. $f(2t)$

87. $2f(t)$

88. $f(kt)$

89. $kf(t)$

90. $\dfrac{f(3t)}{3}$

91. $f(3t) - f(2t)$

92. $f(t) + 2f(k)$

In Problems 93 through 100, let $g(x) = \dfrac{1}{x - 1}$, and find each function value.

93. $g(0)$

94. $g(-1)$

95. $g(1)$

96. $2g(5)$

97. $\dfrac{3}{g(7)}$

98. $g\left(\dfrac{1}{2}\right)$

99. $\dfrac{g(1)}{g(2)}$

100. $g\left(\dfrac{-3}{2}\right)$

In Problems 101 through 106, let $f(x) = 2x^2$ and $g(x) = x + 5$, and find each function value.

101. $f(3) + g(3)$

102. $f(3) - g(3)$

103. $f(3)g(3)$

104. $\dfrac{f(3)}{g(3)}$

105. $f(g(3))$

106. $g(f(3))$

107. The Yellow Cab Company's fare (in dollars) for riding x miles is given by

$$F(x) = 0.25x + 1.5$$

(a) What is the fare for a 10-mile ride?

(b) What would $F(3)$ represent?

(c) How far would $3 take you?

108. The cost (in dollars) of manufacturing n radios is given by

$$C(n) = n^2 + n + 100$$

(a) What is the cost of manufacturing 1 radio?

(b) What would $C(100)$ represent?

(c) How many radios could be built for $210?

109. The number of ants on a picnic table after t minutes is given by

$$N(t) = t^3 - 2t^2 + t$$

(a) How many ants are on the picnic table after 1 minute?

(b) How many ants are on the table after 2 minutes?

(c) What does $N(10)$ represent?

Challenge Problems

110. Find the natural domain of $f(x) = \dfrac{1}{\sqrt{x^2 - x - 2}}$.

111. Find the natural domain of $f(x) = \dfrac{\sqrt{x + 1}}{\sqrt{x - 2}}$.

In Your Own Words

112. What is a function?

113. What is the domain of a function?

114. Explain the relationship between x and $f(x)$.

6.6   Graphs of Functions

TAPE 12

We have seen in previous sections that the graph of a two-variable equation yields valuable information about the equation. In this section we define the *graph of a function*. We will gain the same insights from this approach, and some new ones as well.

> **Graph of a Function**
>
> The graph of the function f is the graph of the equation
>
> $$y = f(x)$$
>
> for all x in the domain of f.

Graphs of Linear Functions

► **EXAMPLE 1**

Graph the function $g(x) = 3x - 1$.

Solution

To graph the function $g(x) = 3x - 1$, we graph the equation $y = 3x - 1$. Its graph is the line shown next.

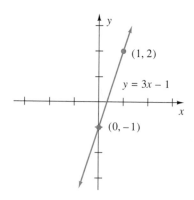

Example 1 illustrates a function whose graph is a line. We call such functions **linear functions.**

> **Linear Function**
>
> A function of the form
>
> $$f(x) = mx + b$$
>
> where m and b are real numbers, is called a **linear function.** The graph of a linear function is a line.

Mathematical models sometimes occur as linear functions.

► **EXAMPLE 2**

A manufacturer of ceramic coffee mugs has determined that his daily cost function is $C(x) = 600 + 2.5x$, where x represents the number of mugs produced. That is, if he produces 100 mugs in a day, his cost for that day is $C(100)$ dollars.

(a) Graph his daily cost function.

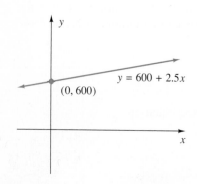

(b) What point on the graph represents the daily start-up costs (the cost if zero mugs are made)?

Solutions

(a) To graph the function, $C(x) = 600 + 2.5x$, we graph the equation $y = 600 + 2.5x$.

(b) If no mugs are made, x is 0, so the daily start-up cost is represented by the point (0, 600) on the graph. In other words, it costs him $600 a day just to open his business. ◄

The graph of a function is very helpful in determining its domain and range. The domain consists of all x-coordinates of the graph. The range consists of all possible y values. Consider the linear function f graphed next. If a value on the y-axis is chosen, is there a point on the line having it as a y-coordinate? Consider 3 on the y-axis.

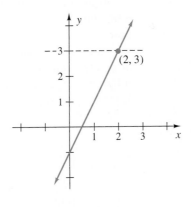

The dashed line shows that there is a point on the line with a y-coordinate of 3. Notice that (2, 3) is a point on the line. So 3 is in the range of f. Since any value on the y-axis can be chosen, the range of f is $\mathbb{R}$.

► **EXAMPLE 3**

The graph of a function g is shown next. Give its domain and range.

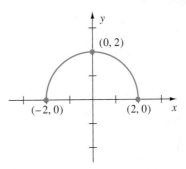

Solution

The points on the graph of g have x-coordinates that are between -2 and 2, also including -2 and 2. So the domain of f is $\{x \mid -2 \le x \le 2\}$. The y-coordinates of points on the graph of g are between 0 and 2, including 0 and 2. So the range of g is $\{y \mid 0 \le y \le 2\}$. ◄

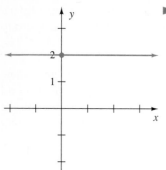

► EXAMPLE 4

The graph of *h* is at the left. Give its domain and range.

Solution

The points on the graph of *h* use every real number as an *x*-coordinate. So the domain of *h* is $\{x \mid x$ is a real number$\}$. The only *y*-coordinate for points on the graph of *h* is 2. So the range of *h* is $\{2\}$.   ◄

How can we look at a graph and determine whether or not it is the graph of a function? When will one number in the domain be paired with more than one number?

Look at the graph at the left. Draw the line $x = 1$. The line crosses the graph at points *A* and *B*. The *x*-coordinate of *A* and *B* is 1. However, *A* and *B* have different *y*-coordinates, say *a* and *b*. This means that the number 1 in the domain is paired with the two numbers, *a* and *b*. So the graph could *not* be the graph of a function.

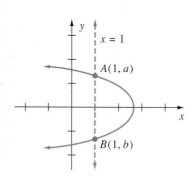

> ### Vertical Line Test
>
> If any vertical line crosses a graph in more than one point, the graph is **not** the graph of a function.

► EXAMPLE 5

Is each graph the graph of a function?

(a)

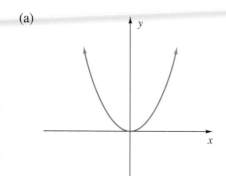

(b)

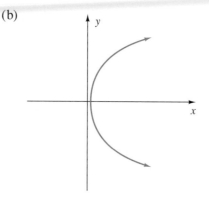

(c)

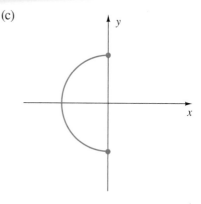

(d)

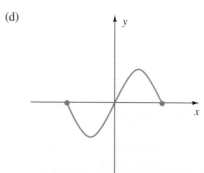

(e)

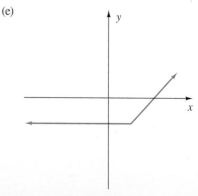

Solution

The graphs (a), (d), and (e) could be graphs of functions because they pass the vertical line test, but (b) and (c) are *not* graphs of functions because the vertical lines shown next cross each graph in more than one point.

(b)   (c)

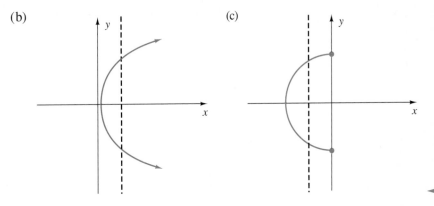

Graphs of Quadratic Functions

In Section 6.3 we learned that the graph of the first-degree equation

$$y = mx + b$$

is a line, and we have just examined functions that have linear graphs. Now, let's look at second-degree, two-variable equations of the form

$$y = ax^2 + bx + c, \qquad a \neq 0$$

where a, b, and c are constants.

First, let's graph a simple example of such an equation, $y = x^2$. We begin by making a table of x- and y-coordinates and then plotting the points.

x	y
-3	9
-2	4
-1	1
0	0
1	1
2	4
3	9

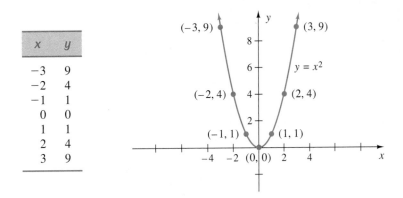

This graph is called a **parabola.** Parabolas are common in everyday life. The reflector in a flashlight is in the shape of a parabola. The cables of the George Washington Bridge hang in parabolic form, and a parabola was utilized in the design of the 200-in. telescope on Mount Palomar.

Notice that the parabola is drawn as a smooth curve. There are **no sharp places.** We make an extra effort to draw parabolas as smooth curves.

The lowest point on the parabola sketched above, $(0, 0)$, is called the **vertex.** A line through the vertex about which the parabola is symmetric is called the **axis of symmetry.** Notice that this parabola is symmetric about the y-axis.

► **EXAMPLE 6**

Graph the equation $y = x^2 + 1$ and identify the vertex and axis of symmetry.

Solution

Again we make a table of x, y values and sketch the graph.

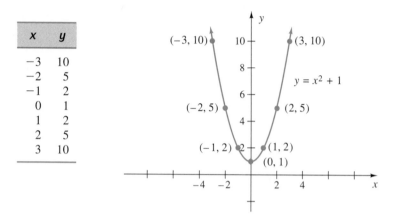

x	y
−3	10
−2	5
−1	2
0	1
1	2
2	5
3	10

By inspection, we see that the vertex is the point (0, 1), and again the axis of symmetry is the y-axis.   ◄

Compare this graph with the graph of $y = x^2$ that we drew earlier in the section. Notice that the size and shape of the two curves are just the same. The graph of $y = x^2 + 1$ is exactly the graph of $y = x^2$ shifted *up* 1 unit.

► **EXAMPLE 7**

Graph the equation, $y = x^2 - 2x - 1$.

Solution

We make a table and plot points.

x	y
−1	2
0	−1
1	−2
2	−1
3	2

Notice that all the examples of the graphs of $y = ax^2 + bx + c$ pass the vertical line test. They all qualify as possible graphs of *functions*. Since the right side of each is in the form of a quadratic polynomial, we call them **quadratic functions.**

Quadratic Functions

A function of the form

$$f(x) = ax^2 + bx + c$$

where a, b, and c are real numbers and $a \neq 0$, is called a **quadratic function.** The graph of a quadratic function is a **parabola.**

► **EXAMPLE 8**

Graph $y = 2x^2$, $y = x^2$, and $y = \dfrac{1}{4}x^2$.

Solution

As before, we graph the three curves.

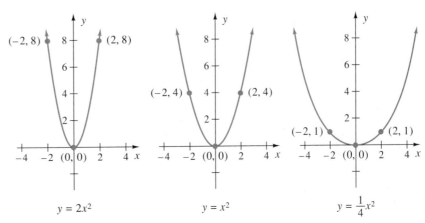

| $y = 2x^2$ | $y = x^2$ | $y = \dfrac{1}{4}x^2$ |

Compare these three parabolas. Notice the effect that the coefficient of x^2 has on the shape of the graph. The 2 makes a thinner parabola and the $\dfrac{1}{4}$ makes a wider one.   ◄

Finding the Vertex

► **EXAMPLE 9**

Graph $f(x) = -2x^2 + 1$.

Solution

Sketch the graph of $y = -2x^2 + 1$ by plotting a few points.

x	y
−1	−1
0	1
1	−1

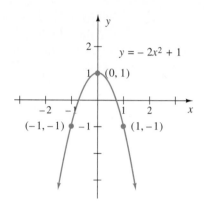

Note that this is a parabola that opens *downward* with vertex at (0, 1). It is thinner than the graph of $y = x^2$. ◄

We can now make some observations about graphing parabolas. The graph of $y = px^2 + k$ is a parabola with vertex at the point $(0, k)$. It opens downward if p is negative and it opens upward if p is positive. It is thinner than the graph of $y = x^2$ if $|p| > 1$ and wider if $|p| < 1$.

► **EXAMPLE 10**

Graph $y = (x − 1)^2$.

Solution

Plotting several points gives the following graph.

x	y
−1	4
1	0
3	4

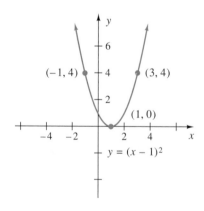

The vertex of this parabola is (1, 0). In fact, it is the graph of $y = x^2$ shifted 1 unit to the right. Notice that the axis of symmetry is the line $x = 1$. ◄

► **EXAMPLE 11**

Graph $y = −2(x − 1)^2 + 2$.

x	y
0	0
1	2
2	0

$y = -2(x-1)^2 + 2$

(1, 2)

(0, 0)   (2, 0)

Solution

Plot several points to see the parabola above. Its vertex is (1, 2). It is thinner than the graph of $y = x^2$ and it opens downward.   ◄

The graph of $y = p(x - h)^2 + k$ is a parabola with vertex at the point (h, k). Its axis of symmetry is the line $x = h$. The parabola opens upward if p is positive and downward if p is negative.

The Parabola

The graph of $y = p(x - h)^2 + k$ is a **parabola** with vertex at the point (h, k). It opens upward if p is positive and downward if p is negative. Its axis of symmetry is the line $x = h$.

► **EXAMPLE 12**

Graph $f(x) = x^2 - 2x + 3$

Solution

The graph of $y = x^2 - 2x + 3$ is a parabola. The equation must be in the form $y = p(x - h)^2 + k$ in order to find the vertex. To do this, complete the square.

$$y = x^2 - 2x + 3$$

First, group the x-terms together.

$$y = (x^2 - 2x) + 3$$

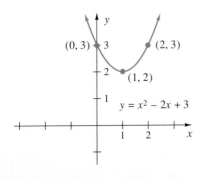

(0, 3)   3   (2, 3)

(1, 2)

$y = x^2 - 2x + 3$

Then complete the square by taking $\frac{1}{2}$ the coefficient of x and squaring it. Add and subtract this number.

$$y = (x^2 - 2x + 1) + 3 - 1$$

$$= (x - 1)^2 + 2$$

The vertex is (1, 2) and the parabola will open upward. The graph of the function is shown on the left.   ◄

► **EXAMPLE 13**

Graph $y = -2x^2 + 4x - 3$.

Solution

To graph the equation of a parabola, we write it in the form $y = p(x - h)^2 + k$. In this form we can identify the vertex and the axis of symmetry. To write the equation in the form desired, we must complete the square.

$$y = -2x^2 + 4x - 3$$

The coefficient of x^2 must be 1 for the completing the square procedure to work. We factor -2 out of the *first two terms*.

$$y = -2(x^2 - 2x) - 3$$

Now complete the square.

$$y = -2(x^2 - 2x + 1 - 1) - 3$$
$$= -2(x^2 - 2x + 1) + 2 - 3$$
$$= -2(x - 1)^2 - 1$$

The vertex is $(1, -1)$ and the parabola will open downward.

x	y
1	−1
0	−3
2	−3

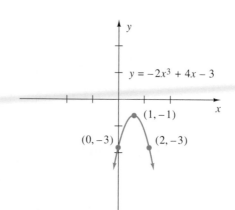

In Chapter 4 we solved quadratic equations. Now let's examine the relationship between the graph of the quadratic *function*

$$f(x) = ax^2 + bx + c$$

and the solution(s) of the quadratic *equation*

$$ax^2 + bx + c = 0$$

► **EXAMPLE 14**

Consider the quadratic function $f(x) = x^2 + 2x - 3$.

(a) Graph the function f.

(b) Solve the equation $f(x) = 0$.

Solutions

(a) We must graph the equation $y = x^2 + 2x - 3$.

$$y = (x^2 + 2x + 1 - 1) - 3 \qquad \text{Complete the square}$$
$$= (x^2 + 2x + 1) - 4$$
$$= (x + 1)^2 - 4$$

We recognize a parabola opening upward with vertex at $(-1, -4)$.

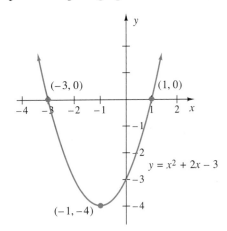

We note from the graph that the x-intercepts are -3 and 1.

(b) We solve the equation $f(x) = 0$.

$$x^2 + 2x - 3 = 0$$
$$(x - 1)(x + 3) = 0 \qquad \text{Factor}$$
$$\{-3, 1\} \qquad \text{Solution set} \qquad \blacktriangleleft$$

Notice in Example 14 that the *x-intercepts* of the graph of f are exactly the *solutions* of the equation $f(x) = 0$. That is, whenever the graph of f touches the x-axis, we have a solution to the equation $f(x) = 0$. If the solutions to the quadratic equation are complex numbers, then the parabola will not touch the x-axis.

Also note from Example 14 that the graph of f provides the solutions to various inequalities. For example, the solutions to the inequality

$$f(x) > 0$$

are those x's for which the graph of f is *above* the x-axis, or, in interval notation,

$$(-\infty, -3) \cup (1, +\infty)$$

By the same reasoning, the solutions to

$$f(x) \le 0$$

are those x's for which the graph of f is *on or below* the x-axis, or

$$[-3, 1]$$

Problem Set 6.6

Warm-Ups

In Problems 1 through 4, sketch the graph of each function. See Example 1.

1. $f(x) = 2x + 1$

2. $g(x) = 5x$

3. $h(x) = \frac{1}{2}x - 1$

4. $f(x) = -1$

For Problems 5 and 6, see Example 2.

5. A partially empty storage tank is being filled with crude oil at the rate of 20 gallons a minute. The formula for determining the number of gallons in the tank at any time during filling is the function

$$G(x) = 20x + 50$$

where $G(x)$ is the number of gallons of oil in the tank x minutes after starting.
   (a) Graph G.
   (b) What is the significance of the y-intercept?
   (c) How long will it take to fill the tank if its capacity is 210 gallons?

6. The manufacturing, marketing, and royalty costs for a certain compact disk are \$3.50 per album. The cost function for producing the album is

$$C(x) = 5000 + 3.5x$$

where x is the number of albums produced.
   (a) Graph C.
   (b) What is the significance of the y-intercept?

In Problems 7 through 12, give the domain and range for each function graphed. See Examples 3 and 4.

7.

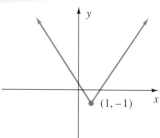

$(1, -1)$

8.

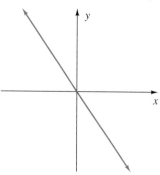

9.

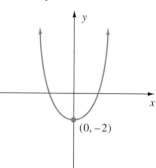

$(0, -2)$

10.

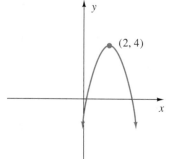

$(2, 4)$

11.

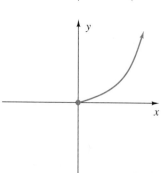

12.

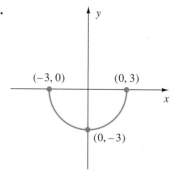

$(-3, 0)$ $(0, 3)$

$(0, -3)$

In Problems 13 through 18, determine which graphs are graphs of functions. See Example 5.

13.

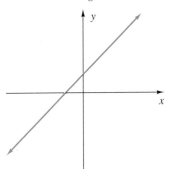

14.

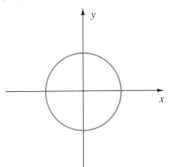

15.

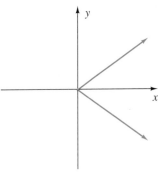

16.

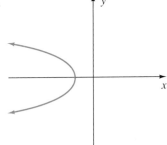

17.

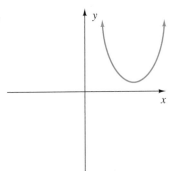

18.

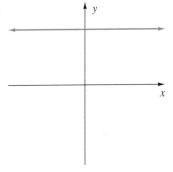

In Problems 19 through 22, identify the axis of symmetry and sketch the graph of each equation. See Example 6.

19. $y = x^2 + 3$

20. $y = x^2 - 1$

21. $y = 3x^2$

22. $y = -\frac{1}{2}x^2 + 1$

In Problems 23 through 26, sketch the graph of each equation. See Examples 7, 8 and 9.

23. $y = (x - 3)^2$

24. $y = -(x - 1)^2$

25. $y = (x + 2)^2 + 1$

26. $y = -(x + 3)^2 - 2$

In Problems 27 and 28, graph each equation. See Examples 10 and 11.

27. $y = -2x^2 + 4x$

28. $y = 3x^2 + 12x + 10$

In Problems 29 through 32, sketch the graph of each quadratic function and find the vertex. See Examples 12 and 13.

29. $g(x) = x^2 - 2x + 4$ **30.** $h(x) = x^2 + 6x + 1$ **31.** $f(x) = x^2 - 6x$ **32.** $g(x) = x^2 - 8x$

In Problems 33 and 34, use the given graphs to solve each equation. See Example 14.

33. Solve $f(x) = 0$.

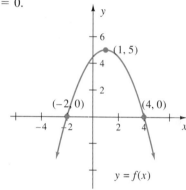

34. Solve $g(x) = 0$.

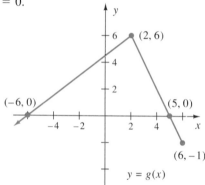

Practice Exercises

In Problems 35 through 54, sketch the graph of each function and give its domain and range. Label the vertex of parabolas.

35. $f(x) = 3x + 2$ **36.** $g(x) = 3x$ **37.** $h(x) = \dfrac{1}{4}x + 2$

38. $f(x) = 1$ **39.** $f(x) = 4x^2$ **40.** $g(x) = x^2 + 3$

41. $h(x) = \dfrac{1}{3}x^2 - 1$ **42.** $f(x) = 2 - x^2$ **43.** $f(x) = x^2 - 4$

44. $g(x) = \frac{1}{2}x^2 + 3$

45. $g(x) = (x - 3)^2$

46. $h(x) = (x + 1)^2 + 5$

47. $h(x) = \frac{1}{2}(x - 2)^2 + 1$

48. $f(x) = 3(x - 2)^2 + 4$

49. $f(x) = -(x - 1)^2 + 4$

50. $f(x) = -2(x + 3)^2 + 1$

51. $g(x) = 3x^2 + 6x + 5$

52. $h(x) = 2x^2 + 12x + 8$

53. $f(x) = -x^2 - 2x - 3$

54. $f(x) = -2x^2 + 8x - 7$

In Problems 55 through 63, answer each question.

55. Graph $f(x) = x^2 + 1$. What is the smallest y-coordinate? This number is called the *minimum* value of f.

56. Graph $f(x) = 2 - x^2$. What is the *maximum* value of f?

57. From 4 through 9 months of age a baby girl's weight is approximated by the function $f(x) = 0.16x^2 - 0.76x + 16.8$, where x represents the age of the baby in months and $f(x)$ represents the weight of the baby in pounds.

(a) Approximately how much would a 7-month-old baby weigh?

(b) Is it reasonable that this same equation would represent the weight of an adult? If this equation applied, approximately how much would a 20-year-old weigh?

58. Ms. Whitney owns and operates the Little Five Points Art Gallery. She has hired a consultant to analyze her business operation. The consultant tells her that her profit P is given by $P(x) = -x^2 + 120x$, where x is the number of paintings that she sells.

(a) How many paintings must she sell to maximize her profit?

(b) What is the maximum profit?

59. Jorge sells stereos with a price of $p = -3x + 600$. Revenue is the product of price times the number sold, x.

(a) State the revenue function.

(b) Find the maximum revenue.

60. The profit at Mary's Dress Shop is given by

$$P(x) = -(x - 250)^2 + 20,000$$

where x represents the number of dresses sold. Find how many dresses she must sell to receive a maximum profit.

61. The height (in feet) of a ball thrown upward from a building 100 ft high with an initial velocity of 96 ft/sec is given by

$$s(t) = -16t^2 + 96t + 100$$

where t is the time in seconds. Find how high up the ball will go.

62. A manufacturer of mountain bikes knows that her cost depends on the number of bicycles which are made. She determined that her cost and the number of bicycles produced are related by the function $C(x) = 0.5x^2 + $ 85, where x represents the number of bicycles her company manufactures and $C(x)$ represents the cost.

(a) What is her cost if 50 bicycles are manufactured?

(b) What is her cost if no bicycles are manufactured? Is this result reasonable?

63. Cary, the state javelin champion, can throw the javelin 200 feet. The height of the javelin is related to the distance the javelin has traveled from the foul line by the function $h(x) = -.005x^2 + x + 6$, where x represents the distance from the foul line and $h(x)$ represents the height of the javelin.

(a) Complete the table

x	$h(x)$
20	
60	
100	
140	
180	
200	

(b) How high is the javelin when it is 60 feet from the foul line?

(c) How do you interpret the meaning of the first ordered pair?

(d) How do you interpret the fact that the 2nd and 4th ordered pairs have the same y-value?

Challenge Problems

In Problems 64 through 73, graph each function. Give the domain and range for each.

64. $f(x) = \sqrt{x}$

65. $f(x) = \sqrt{x - 1}$

66. $f(x) = \sqrt{4 - x}$

67. $f(x) = -\sqrt{x}$

68. $f(x) = -\sqrt{x - 1}$

69. $f(x) = \sqrt{1 - x^2}$

70. $f(x) = x^3$ **71.** $f(x) = x^3 + 1$ **72.** $f(x) = x^3 - 1$ **73.** $f(x) = x^4$

74. Jeans Boutique discovered that the higher they marked the selling price of their designer jeans, the more pairs of jeans they sold. Sasha determined that the situation could be represented by $f(x) = .02x^2 + 1.9x - 181$ where x represents the selling price of the jeans and $f(x)$ represents the number of jeans sold.
  (a) How many pairs of jeans were sold when the price was $60 a pair?
  (b) Write an inequality which would help the boutique manager determine what the selling price should be if they want to sell at least 15 pair.

75. A farmer wishes to use at most 300 feet of fencing to enclose a rectangular plot that borders the barn using one side of the barn in lieu of additional fencing where the length is parallel to the barn.
  (a) Write an inequality for length in terms of width.
  (b) Write an inequality for the area in terms of width.
  (c) What is the largest area that can be enclosed?

In Your Own Words

76. Explain how to graph a function.

6.7 Variation

TAPE 12

The formula $C = \pi d$ relates the diameter of a circle and its circumference. The relationship between the variables C and d is such that if d gets *larger* then C also gets *larger*, and if d gets *smaller* then C also gets *smaller*. π is a constant.
  We often express such a relationship by saying that "C **varies directly** as d" or "C is **directly proportional** to d."

Direct Variation

y varies directly as x means that

$$y = kx$$

where k is a nonzero constant.

The constant k is called the **constant of proportionality.**

► **EXAMPLE 1**

y varies directly as the cube of x. Find the constant of proportionality if y is 16 when x is 2.

Solution

From the definition,

$$y = kx^3$$

Replacing y with 16 and x with 2 gives us

$$16 = k(2)^3 = 8k$$

$$2 = k$$

The constant of proportionality is 2. Thus the relationship between x and y is given by

$$y = 2x^3 \qquad \blacktriangleleft$$

Boyle's Law for the expansion of gas is $V = \dfrac{C}{P}$, where V is the volume of the gas, P is the pressure, and C is a constant. This formula relates a pressure and a volume. We can see that if P gets *larger* then V gets *smaller*, and that if P gets *smaller* then V gets *larger*.

We say that "*V* **varies inversely as** *P*" or "*V* is **inversely proportional** to *P*."

Inverse Variation

y **varies inversely as** *x* means that

$$y = \frac{k}{x}$$

where k is a nonzero constant.

► **EXAMPLE 2**

y varies inversely as the square root of x. Find the constant of proportionality if y is 15 when x is 9.

Solution

From the definition,

$$y = \frac{k}{\sqrt{x}}$$

However, y is 15 when x is 9, so

$$15 = \frac{k}{\sqrt{9}} = \frac{k}{3}$$

$$45 = k$$

Thus the constant of proportionality is 45 and the relationship between x and y is given by

$$y = \frac{45}{\sqrt{x}} \qquad \blacktriangleleft$$

> **Joint Variation**
>
> **y varies jointly as x and z** means that
> $$y = kxz$$
> where k is a nonzero constant.

We often say that **y is jointly proportional to x and z** instead of y varies jointly as x and z.

► **EXAMPLE 3**

Translate each statement into an equation.

(a) w varies jointly as y and the cube of x.

(b) x is directly proportional to y and inversely proportional to z.

Solutions

(a) Using the definition,
$$w = kx^3y$$

(b) Combining direct and inverse variation, we obtain
$$x = \frac{ky}{z} \qquad ◄$$

► **EXAMPLE 4**

Newton's law of gravitation says that the force between two particles that are a constant distance apart varies jointly as the product of the masses of the particles. If a force of 2 units exists between two particles of masses 4 and 5 units, find the force between two particles of masses 8 and 15 units.

Solution

If F is force and m_1 and m_2 are the masses,
$$F = km_1m_2$$

We find the constant of proportionality.
$$2 = k(4)(5)$$
$$\frac{1}{10} = k$$

The formula becomes $F = \frac{1}{10}m_1m_2$.

$$F = \frac{1}{10}(8)(15)$$
$$= 12$$

The force is 12 units. ◄

Problem Set 6.7

Warm-Ups

In Problems 1 through 6, translate each statement into an equation. See Examples 1, 2, and 3.

1. The area of a circle varies directly as the square of its radius.
2. The volume of a cylinder varies jointly as the square of its radius and its height.
3. The volume of a sphere is directly proportional to the cube of its radius.

4. F is inversely proportional to G.
5. u varies directly as v and inversely as w.
6. x varies jointly as the square of y and z and inversely as w.

In Problems 7 through 15, answer each question. See Example 4.

7. x varies directly as the square of y. If x is 12 when y is 2, find x when y is 5.
8. u varies inversely as the cube root of v. If u is 6 when v is 27, find u when v is 8.
9. s is jointly proportional to the square of t and g. If s is 36 when t is 2 and g is 3, find s when t is $\sqrt{2}$ and g is 32.2.
10. N is directly proportional to the square root of L and inversely proportional to M. If N is 2 when L is 16 and M is 8, find N when L is 25 and M is 2.
11. Hooke's Law says that the force required to stretch a spring is directly proportional to the distance stretched. If a force of 10 lb is required to stretch a spring 5 in., how much force will be required to stretch the spring 10 in.?
12. Ohm's Law says that the current, I, in a wire varies directly as the electromotive force, E, and inversely as the resistance, R. If I is 11 amperes when E is 110 volts and R is 10 ohms, find I if E is 220 volts and R is 11 ohms.
13. The distance a stone falls when dropped off a cliff is directly proportional to the square of time. If the stone falls 64.4 feet in 2 seconds, how far will the stone have fallen in 3 seconds?
14. The intensity of illumination, I, from a light is inversely proportional to the square of the distance, d, from the light. If the intensity is 120 candlepower at a distance of 10 ft from the light, what is the intensity 20 ft from the light?
15. The period of a pendulum varies directly as the square root of its length. If a pendulum of length 16 inches has a period of $\frac{1}{2}$ second, find the length of a pendulum with a period of $\frac{1}{4}$ second.

Practice Exercises

In Problems 16 through 21, translate each statement into an equation.

16. The distance an object falls from the Trans Am Building (ignoring air resistance) is directly proportional to the square of the time it falls.
17. The spin on a billiards ball is directly proportional to the velocity of the stroke of the cue stick.
18. The perimeter of a square varies directly as the length of a side.
19. The volume of a box varies jointly as length, width, and height.
20. t is inversely proportional to s.
21. h is directly proportional to the square root of g and inversely proportional to d.

In Problems 22 through 30, answer each question.

22. u varies directly as the cube of v. If u is 8 when v is 1, find u when v is 2.
23. x varies inversely as the square of y. If x is 2 when y is 3, find x when y is 5.
24. s is jointly proportional to t and g. If s is 10 when t is 3 and g is $\frac{5}{2}$, find s when t is 7 and g is 6.
25. N is directly proportional to the square of L and inversely proportional to the cube of M. If N is 9 when L is $\sqrt{3}$ and M is 2, find N when L is $\frac{\sqrt{2}}{2}$ and M is $\frac{3}{2}$.
26. Velocity varies directly as time. If velocity is 64 ft/sec at 2 sec, find the velocity at 4 sec.

27. If two lines are perpendicular, the slope of one is inversely proportional to the slope of the other. If two perpendicular lines have slopes of $\frac{2}{3}$ and $-\frac{3}{2}$, find the slope of a line perpendicular to a line with slope $-\frac{3}{4}$.

28. The distance traveled at a constant speed is directly proportional to time. If it takes 2 hours to go 100 miles, how many miles will be covered in 5 hours?

29. Hooke's Law (problem 11) is also true if a spring is compressed. If a force of 5 lb is required to compress a spring 2 in., how much force will be required to compress it 1 in?

30. The resistance of a wire varies directly as its length and inversely as the square of its diameter. If 50 ft of wire with a diameter of 0.01 in. has a resistance of 5 ohms, what is the resistance of 100 ft of the same kind of wire with a diameter of 0.02 in.?

In Your Own Words

31. Explain the difference in direct and inverse variation.

▶ Chapter Summary

GLOSSARY

Cartesian coordinate system   A pair of perpendicular number lines that intersect at the zero of each line. This point of intersection is called the *origin*.

Ordered pair   A pair of numbers written in the form (p, q). The first number is called the *x-coordinate* and the second number is called the *y-coordinate*.

Linear equation in two variables   An equation that can be written in the form $Ax + By = C$, where A and B are not both zero.

x-intercept of a line   The x-coordinate of the point where the line crosses the x-axis.

y-intercept of a line   The y-coordinate of the point where the line crosses the y-axis.

Graph of the function f   The graph of the equation $y = f(x)$ for x in the domain of f.

Function   A rule that assigns to each member in the **domain** exactly one member in the **range**.

Functional notation   Used in giving the rule for a function.

The **vertical line test**   Determines whether or not a graph is the graph of a function.

Linear function   Function of the form $f(x) = mx + b$, where m and b are real numbers. The graph of a linear function is a line.

Quadratic function   Function of the form $f(x) = ax^2 + bx + c$, where a, b, and c are real numbers and $a \neq 0$. The graph of a quadratic function is a **parabola.**

y **varies directly** as x means $y = kx$.

y **varies inversely** as x means $y = \dfrac{k}{x}$.

y **varies jointly** as x and z means $y = kxz$.

Distance Formula   The distance between the points (x_1, y_1) and (x_2, y_2) is given by the formula

$$d = \sqrt{(x_1 - x_2)^2 + (y_1 - y_2)^2}$$

Midpoint Formula   The midpoint between (x_1, y_1) and (x_2, y_2) is

$$\left(\frac{x_1 + x_2}{2}, \frac{y_1 + y_2}{2} \right)$$

Forms of a Linear Equation in Two Variables

1. *Standard form: $Ax + By = C$, A and B not both 0.*
2. *Slope–intercept form: $y = mx + b$.*
   m is the slope and b is the y-intercept.
3. *Point–slope form: $y - y_1 = m(x - x_1)$.*
   m is the slope and (x_1, y_1) is any point on the line.

Graphing Lines

1. Using standard form, $Ax + By = C$: Find the intercepts and plot them.
2. Using slope–intercept form, $y = mx + b$: Read the y-intercept from the equation and plot it. Read the slope and use it to draw the line.

Slope

1. The slope of the line containing the points (x_1, y_1) and (x_2, y_2) is given by the *slope formula*

$$m = \frac{\text{rise}}{\text{run}} = \frac{y_2 - y_1}{x_2 - x_1}, \qquad x_1 \neq x_2$$

2. The slope of a line can be found from the slope–intercept form of the equation of the line.

Parallel and Perpendicular Lines

1. Parallel lines have the same slope.
2. Perpendicular lines have slopes that are negative reciprocals.

Horizontal and Vertical Lines

1. $x = p$ is a line parallel to the y-axis (*vertical*). It has undefined slope.
2. $y = q$ is a line parallel to the x-axis (*horizontal*). It has zero slope.

Writing an Equation of a Line with a Slope

1. Find the slope of the line.
2. Find the coordinates of any point on the line.
3. Substitute into the point–slope form.

CHECKUPS

1. Find the distance between the points $(1, 6)$ and $(4, -2)$.

   Section 6.1; Example 2

2. Sketch the graph of each equation:
   (a) $5x + y = 2$ Section 6.2; Example 2
   (b) $x + 2y = 0$ Section 6.2; Example 3
   (c) $x = 2$ Section 6.2; Example 4
   (d) $y = \dfrac{2}{3}x - \dfrac{5}{2}$ Section 6.3; Example 11

3. Find the slope of the line that contains the points $(-1, 7)$ and $(2, -2)$

   Section 6.3; Example 4

4. Suppose that L_1 is a line perpendicular to the graph of $y = \dfrac{5}{3}x - 5$. What is the slope of L_1?

   Section 6.3; Example 15

5. Find an equation for the line containing the points $(-1, 3)$ and $(2, -1)$.

   Section 6.4; Example 2

6. If $h(x) = x^2 - 2x$, find $h(-5)$.

   Section 6.5; Example 3b

7. Find the natural domain of $g(x) = \dfrac{5}{x+3}$.   Section 6.5; Example 7

8. Graph each function.
   (a) $g(x) = 3x - 1$ Section 6.6; Example 1
   (b) $f(x) = x^2 - 2x + 3$ Section 6.6; Example 12

9. y varies inversely as the square root of x. Find the constant of proportionality if y is 15 when x is 9.   Section 6.5; Example 2

▶ Review Problems

In Problems 1 through 6, find the distance between the points and the point midway between them.

1. $(1, 1), (5, 4)$
2. $(6, 8), (2, 5)$
3. $(2, 1), (1, 3)$

4. $(1, -1), (-4, 2)$
5. $(8, -3), (9, -2)$

6. $(-6, -5), (-2, -8)$

In Problems 7 through 15, graph the equation.

7. $2x + 3y = 12$
8. $3x - 2y = 6$
9. $y = \dfrac{1}{2}x - 2$

10. $y = -x + 1$
11. $6x + y = 6$
12. $2x + 3y = 0$

13. $x = -3$
14. $5x = 9y$
15. $4y = 7$

In Problems 16 through 24, find the slope of the line that contains both points.

16. $(6, 8), (5, 3)$
17. $(5, 9), (3, 11)$
18. $(2, 6), (-2, 2)$
19. $(3, -2), (-2, 8)$
20. $(-4, -5), (-1, -7)$
21. $(-5, 0), (0, 3)$
22. $\left(\dfrac{1}{2}, \dfrac{3}{2}\right), \left(\dfrac{3}{2}, -\dfrac{1}{2}\right)$
23. $\left(-\dfrac{4}{3}, \dfrac{1}{2}\right), \left(\dfrac{1}{3}, -\dfrac{5}{2}\right)$
24. $\left(-2, -\dfrac{3}{5}\right), \left(\dfrac{1}{5}, 3\right)$

In Problems 25 through 27, find the y-intercept and the slope of the graph of the equation.

25. $2x - 3y = 6$
26. $3x = 4y - 6$
27. $2y + 3 = 0$

In Problems 28 through 36, find an equation of a line satisfying the conditions given.

28. The line through $(1, 2)$ with slope $\dfrac{3}{5}$.

29. The line through $(4, -2)$ with slope -1.

30. The line containing $(4, 1)$ and $(3, -2)$.

31. The line containing $(-9, -6)$ and $(-2, -1)$.

32. The line through $(-3, 6)$ parallel to the graph of $y = 2x - 7$.

33. The line through $(1, -7)$ parallel to the graph of $2x + 3y = 4$.

34. The line perpendicular to the line $y = -3x - 9$, containing the point $(3, -5)$.

35. The line through $(3, 1)$ perpendicular to the graph of $3x = 4y$.

36. The line through the origin parallel to the graph of the line containing the points $(7, -1)$ and $(2, 5)$.

37. Find the range of the function whose domain is $\{x \mid x$ is a positive integer$\}$; rule is $f(x) = -x$.

In Problems 38 through 51, find the natural domain of each function.

38. $f(x) = 2x - 8$

39. $f(x) = |x - 7|$

40. $f(x) = \sqrt{x + 2}$

41. $f(x) = \sqrt{x^2 - x}$

42. $f(x) = \dfrac{3}{x - 7}$

43. $f(x) = \dfrac{x + 3}{(x + 1)(x - 2)}$

If $f(x) = \dfrac{3}{x + 5}$, find:

44. $f(1)$

45. $f(-5)$

46. $f(0)$

47. $f(f(1))$

48. $f(a)$

49. $f(-3)$

50. $f(a) + f(b)$

51. $f(a^2)$

In Problems 52 through 60, sketch the graph of each function. Give the domain and range for each.

52. $f(x) = \dfrac{1}{2}x - \dfrac{5}{2}$

53. $f(x) = \dfrac{1}{4}x^2 + 3$

54. $f(x) = 5x - 2$

58. $f(x) = -\dfrac{1}{2}x^2$

59. $f(x) = x^2 - 4x + 3$

60. $f(x) = -2x^2 + 12x + 4$

61. The nautical unit of measurement *fathom* comes from a Danish word that means outstretched arms. Five fathoms is 30 feet and 10 fathoms is 60 feet. Considering these as ordered pairs, (5, 30) and (10, 60), where the first coordinate represents the number of fathoms and the second coordinate represents the associated number of feet:
   (a) Write an equation of the line in standard form that contains these two points.
   (b) According to the equation, complete the ordered pair: $(1, \square)$.
   (c) Interpret what the completed ordered pair represents.

62. Coach Moore determined that the relationship between baseball games won, or W, and baseball games lost, or L, could be expressed as $W - 3L = 4$. If the team wins 28 games, how many games would they be predicted to lose?

63. u varies directly as v. If u is 8 when v is 4, find u when v is 12.

64. x varies inversely as the fourth power of y. If x is 5 when y is $\sqrt{2}$, find x when y is $\sqrt{3}$.

65. The gravitational force between two objects varies inversely as the square of the distance between the objects. If a force of 50 force units results from two objects that are 12 units apart, how much force results from two objects that are 20 units apart?

66. The work done by a constant force is directly proportional to the distance the object is moved. If 50 foot-pounds of work is required to move an object 10 ft, how much work is done in moving the same object 35 ft?

Let's Not Forget . . .

Identify the expressions that are in factored form. Factor those that are not factored.

67. $27 + a^3$

68. $(3 - b)^3$

69. $(q + r)^2$

70. $\dfrac{a}{b} + a^2 b$

71. $(f - g)^2 - 9$

72. $a^2 bc + d^2 br$

How many terms are in each expression? Which expressions have $y - 7$ as a factor?

73. $y^3 - 7$

74. $(y - 7)^2 - x^2$

75. $2y^2 - 11y - 21$

76. $r(y - 7) - s(7 - y)$

Simplify each expression, if possible. Leave only nonnegative exponents in the answer.

77. 9^{-2}

78. -9^2

79. $(-9)^2$

80. $(-9)^{-2}$

81. $\sqrt{x^2 - 4}$

82. $\sqrt[3]{32x^3 + 64x^5}$

83. $(a^{2/3} b^{3/2})^6$

84. $\dfrac{-3a^{-4} b^2}{b^{-2}}$

Reduce, if possible.

85. $\dfrac{x^2 + y^2}{x^2 - y^2}$

86. $\dfrac{x^2 - y^2}{x^3 - y^3}$

The following problems can be worked by using a least common denominator. Follow the directions in each and notice how the LCD is used.

87. Perform the operation indicated: $\dfrac{x}{x - 1} - \dfrac{3}{1 - x}$.

88. Solve $\dfrac{3}{x - 1} + \dfrac{6}{x^2 - 1} = 1$.

89. Simplify $\dfrac{\dfrac{3}{x^2 - 1}}{\dfrac{1}{x + 1} - \dfrac{1}{x - 1}}$.

Label each problem as an expression, equation, or inequality. Solve the equations and inequalities, and perform the operations indicated on the expressions, leaving only nonnegative exponents in your answer.

90. $|5 - x| = 3$

91. $\dfrac{s - 5}{2s - 12} + \dfrac{s - 4}{12 - 2s}$

92. $(\sqrt{7} - \sqrt{5})^2$

93. $6x^2 - 5x < 6$

94. $\sqrt{x + 3} - \sqrt{x - 2} = 1$

95. $x^4 - 3x^2 - 4 = 0$

▶ You Decide

You have been negotiating with Avis, Budget, Hertz, and National rental car companies for a car to use on your upcoming trip. You have narrowed your choices to the following four cars. They are all new and fully equipped.

1. A Ford Taurus. The fee is a flat rate of $287 a week with no mileage charges.
2. An Oldsmobile 88. The fee is $18 a day plus 10¢ a mile.
3. A Chevrolet Corvette. The fee is $45 a day plus 15¢ a mile.
4. A Yugo. The fee is $10 a day plus 5¢ a mile.

Make a graph of car rental expense for one day as a function of miles traveled. Show all four cars on the same graph. What is the total fee for each of the four cars if you travel 300 miles a day, for 6 days.
Which car do you choose, and why?

▶ Chapter Test

In Problems 1 through 10, choose the correct answer.

1. The distance between the points $(2, -3)$ and $(-1, 5)$ is (?)
A. $\sqrt{65}$ B. $\sqrt{73}$
C. $\sqrt{5}$ D. $\sqrt{89}$

2. The slope of the line containing the points $(5, -1)$ and $(2, 4)$ is (?)
A. $\dfrac{5}{3}$ B. $\dfrac{3}{5}$ C. $-\dfrac{5}{3}$ D. $-\dfrac{3}{5}$

3. Which of the following is the equation of a line with slope $-\dfrac{2}{3}$ that contains the point $(3, 7)$?
A. $2x + 3y = 27$ B. $3x + 2y = 23$
C. $2x - 3y = -15$ D. $3x - 2y = -5$

4. Which of the following is an equation of the line through the origin perpendicular to the graph of $3x - 4y = -1$?
A. $4x - 3y = 0$ B. $4x + 3y = 0$
C. $3x - 4y = 0$ D. $3x + 4y = 0$

5. The slope of the line whose equation is $y = -3$ is (?)
A. $\{-3\}$ B. It has no slope.
C. 0 D. $\varnothing$

6. The statement "x varies jointly as y and z and inversely as the square of w" may be written:
A. $x = \dfrac{yz}{w^2}$ B. $x = \dfrac{yw^2}{z}$
C. $x = \dfrac{kyz}{w^2}$ D. $x = \dfrac{kyw^2}{z}$

7. If $g(x) = x^2$ and the domain of g is $\{-2, -1, 0, 1, 2\}$, the range of g is (?)
A. $\{1, 4\}$ B. $\{-4, -1, 0\}$
C. $\{0, 1, 4\}$ D. $\{-2, -1, 0, 1, 2\}$

8. If $f(x) = 1 - 2x + x^2$, then $f(-3) = $ (?)
A. -2 B. 16
C. -14 D. 13

9. The vertex of the graph of $f(x) = x^2 + 2x + 2$ is (?)
A. $(-1, 1)$ B. $(1, 1)$
C. $(1, -1)$ D. $(-1, -1)$

10. Which of the following illustrations is the graph of a function?

A. 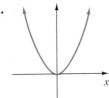   **B.** 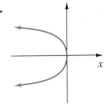   **C.** 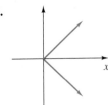   **D.**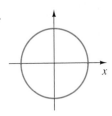

In Problems 11 through 14, sketch the graph of each equation.

11. $x + 2y = 2$

12. $2x - y = 0$

13. $x = -\dfrac{2}{3}$

14. $y = \dfrac{3}{2}x + \dfrac{1}{2}$

15. Sketch the graph of each function.
   a. $f(x) = x^2 - 1$
   b. $f(x) = -2(x - 1)^2 + 3$

16. Find the natural domain of each function.
   a. $f(x) = \sqrt{(x + 1)(x - 2)}$
   b. $g(x) = \dfrac{1}{x - 7}$

17. Write an equation of the line containing the point $(-1, 0)$ and parallel to the line through the points $(1, 2)$ and $(3, 4)$.

18. Write an equation of the line perpendicular to the x-axis containing the point $(-2, -3)$.

19. Find the distance between the points $(5, -6)$ and $(-7, -1)$.

20. Find the midpoint between the points $(5, -6)$ and $(-7, -1)$.

21. Find the slope of a line whose x-intercept is 3 and whose y-intercept is 2.

22. Find the y-intercept and the slope for the line whose equation is $5x + 3y = 7$.

23. x varies directly as the square root of y and inversely as the square of z. If x is 6 when y is 9 and z is 2, find x when y is 16 and z is 4.

7 Systems of Equations and Inequalities

See Problem Set 7.2, Exercise 27.

► Connections

Suppose we wished to cut a 13-foot length of conduit into two pieces such that one piece is 3 feet longer than the other. We are looking for two numbers whose sum is 13 and whose difference is 3. This problem could be solved with the methods developed in Chapter 2, or it could be done by trial and error. However, it seems natural to let x be the larger of the two numbers and y be the other number. Then we immediately get two equations.

$$\begin{cases} x + y = 13 \\ x - y = 3 \end{cases}$$

We call such an arrangement a system of two equations in two variables.

In this chapter we learn two analytical methods for solving systems of equations and examine some applications that can be modeled by systems of equations. We will study inequalities in two variables and learn how to graph linear systems of inequalities. Cramer's rule and matrix methods are also discussed in two optional sections.

7.1   Systems of Equations in Two Variables

TAPE 13

Let's look at the problem of finding two numbers whose sum is 13 and difference is 3. If we let x be the larger of the two numbers and y be the other number, we get two equations,

$$x + y = 13 \quad \text{and} \quad x - y = 3$$

Linked equations such as these are usually written in the form

$$\begin{cases} x + y = 13 \\ x - y = 3 \end{cases}$$

Such an arrangement is called a **system of two equations in two variables.** Notice that x stands for the same number in both equations, as does y. So a **solution** of the system is an ordered pair that satisfies both equations. Is there such an ordered pair? Is there more than one such pair? To answer these questions, let's look at the graph of these two equations on the same coordinate system.

The graphs of the two equations intersect in a single point. As the point of intersection lies on *both* lines, the corresponding ordered pair is in *both* solution sets. We conclude that the *system* of equations has exactly one solution.

We can look at the graph and guess that the lines intersect at (8, 5). This ordered pair is in the solution set of both equations, so it is the desired solution. This is the **graphical method** of solving systems of equations. As it requires judgment and estimation, it is not a very satisfactory technique. In this section we learn two analytical methods for solving such systems.

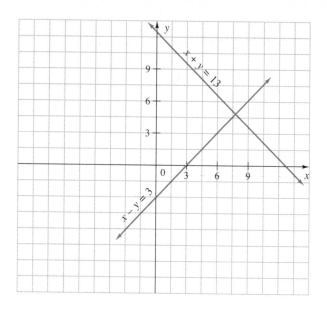

Let's look at our two equations again (numbered for reference).

$$\boxed{1} \quad \begin{cases} x + y = 13 \\ x - y = 3 \end{cases} \quad \boxed{2}$$

Suppose that we solve equation $\boxed{1}$ for x, and call it equation $\boxed{3}$

$$\boxed{3} \; x = 13 - y$$

Now, if x has the value $13 - y$ in equation $\boxed{1}$, it must have the value $13 - y$ in equation $\boxed{2}$. So we replace the x in equation $\boxed{2}$ by $(13 - y)$.

$$(13 - y) - y = 3$$

This is an equation in one variable, which can be solved by the methods learned earlier.

$$13 - 2y = 3$$

$$-2y = -10 \quad \text{or} \quad y = 5$$

This gives the y value. If we replace the y in equation $\boxed{3}$ by this value, we find the corresponding x value.

$$x = 13 - 5 = 8$$

We have found that the ordered pair $(8, 5)$ is a solution for *both* equations. Thus the solution set is

$$\{(8, 5)\}$$

This is called the **method of substitution.**

> **Method of Substitution**
>
> 1. Solve one of the equations for one of the variables.
> 2. Substitute this for the same variable in the other equation.
> 3. Solve the resulting equation.
> 4. Substitute the result back into step 1 to find the other variable.
> 5. Check solution(s), if required.
> 6. Write the solution set.

► **EXAMPLE 1**

Find the solution set for each system by substitution.

(a) $\begin{cases} 2x + y = 2 \\ 6x - 4y = -1 \end{cases}$ (b) $\begin{cases} 4x + 9y = 12 \\ 2x - 3y = 1 \end{cases}$

Solutions

(a) $\boxed{1}$ $\begin{cases} 2x + y = 2 \\ 6x - 4y = -1 \end{cases}$
      $\boxed{2}$

The most convenient variable is y in equation $\boxed{1}$. We solve that equation for y.

STEP 1   Solve one equation for one variable.

$$y = 2 - 2x$$

STEP 2   Substitute for same variable in other equation.

Replace y in equation $\boxed{2}$ by $2 - 2x$.

$$6x - 4y = -1$$
$$6x - 4(2 - 2x) = -1$$

STEP 3   Solve for the remaining variable.

$$6x - 8 + 8x = -1$$
$$14x = 7$$
$$x = \frac{7}{14} = \frac{1}{2}$$

STEP 4   Substitute back into step 1 and find the other variable.

We replace x by $\frac{1}{2}$ in the equation found in step 1 and solve.

$$y = 2 - 2x$$
$$= 2 - 2\left(\frac{1}{2}\right)$$
$$= 2 - 1 = 1$$

STEP 5 Check.

The ordered pair $\left(\frac{1}{2}, 1\right)$ checks in both equations.

STEP 6 Write the solution set.

$$\left\{\left(\frac{1}{2}, 1\right)\right\}$$

(b) $\boxed{1}$ $\begin{cases} 4x + 9y = 12 \\ 2x - 3y = 1 \end{cases}$
 $\boxed{2}$

Solve one equation for one variable.

We solve equation $\boxed{2}$ for x.

$$2x = 3y + 1$$

$$x = \frac{1}{2}(3y + 1)$$

Substitute for same variable in the other equation.

We replace x in equation $\boxed{1}$ by $\frac{1}{2}(3y + 1)$.

$$4\left[\frac{1}{2}(3y + 1)\right] + 9y = 12$$

Solve.

$$2(3y + 1) + 9y = 12$$
$$6y + 2 + 9y = 12$$
$$15y = 10$$
$$y = \frac{10}{15} = \frac{2}{3}$$

Substitute back into step 1 and solve.

We replace y by $\frac{2}{3}$ in the equation found in the first step to get

$$x = \frac{1}{2}\left(3 \cdot \frac{2}{3} + 1\right)$$

$$= \frac{1}{2}(2 + 1) = \frac{3}{2}$$

Check, if required.

We have found the ordered pair $\left(\frac{3}{2}, \frac{2}{3}\right)$, which checks in both equations.

Write the solution set.

$$\left\{\left(\frac{3}{2}, \frac{2}{3}\right)\right\}$$

◀

► **EXAMPLE 2**

Use substitution to find the solution set for

$$\begin{cases} -x + 2y = 2 \\ 3x - 6y = 1 \end{cases}$$

Solution

We solve the first equation for x, since it seems to be the most convenient.

$$-x = 2 - 2y$$

$$x = -2 + 2y$$

Putting this into the second equation gives

$$3(-2 + 2y) - 6y = 1$$

$$-6 + 6y - 6y = 1$$

$$-6 = 1$$

What has happened? The variable y is gone, and we are left with $-6 = 1$, a false statement!

Let's graph the two given equations to see what has happened.

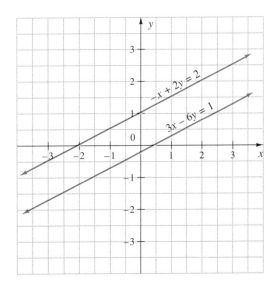

The lines look parallel. If we check the slopes, we see that the slope of each line is $\frac{1}{2}$. Because each line has the same slope, they are parallel. There is *no* ordered pair common to both lines. The solution set is the empty set.

$$\emptyset$$ ◄

We call such a system **inconsistent.** If all the variables drop out when solving a system of equations, leaving a *false* statement, the system is inconsistent and the solution set is the empty set.

What if we arrive at a statement that is always *true*?

► **EXAMPLE 3**

Solve the following system by substitution.

$$\begin{cases} x - 2y = 5 \\ 6y - 3x = -15 \end{cases}$$

Solution

We solve the first equation for x.

$$x = 2y + 5$$

Substituting that into the second gives

$$6y - 3(2y + 5) = -15$$
$$6y - 6y - 15 = -15$$
$$-15 = -15$$

This statement is always true. Let's graph these equations.

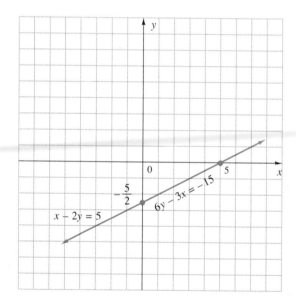

Both equations have the same line for their graph! If we look closely at the second equation, we see that it is just the first equation multiplied by -3. Thus the two equations are equivalent and have the same solution set.

Therefore, the solution set of the system is the solution set of either of the equations. In this case, the solution set is the set of ordered pairs that makes the first equation true, or

$$\{(x, y) \mid x - 2y = 5\}$$ ◄

We call such systems **dependent.**

As seen above, three situations can occur when solving systems of two linear equations in two variables.

Independent

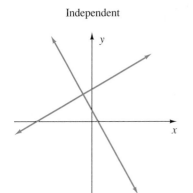

Graphs intersect
One solution

Inconsistent

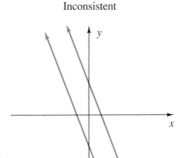

Graphs parallel
No solution

Dependent

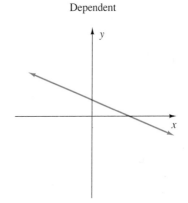

Graphs are the same
Infinite number of
solutions

A system containing one or more *nonlinear* equations is called a *nonlinear system*. The method of substitution is often the best approach to use in solving nonlinear systems.

► **EXAMPLE 4**

Solve the system by substitution.

$$\begin{cases} x^2 + y^2 = 5 \\ x + 2y = 3 \end{cases}$$

Solution

Notice that the first equation is *not* linear. However, the method of substitution will still work.

We solve the second equation for x.

$$x = 3 - 2y$$

Substitute this into the first equation, and solve for y.

$$x^2 + y^2 = 5$$
$$(3 - 2y)^2 + y^2 = 5$$
$$9 - 12y + 4y^2 + y^2 = 5$$
$$5y^2 - 12y + 4 = 0$$

This is a quadratic equation in y, which factors.

$$(5y - 2)(y - 2) = 0$$

$$y = \frac{2}{5} \quad \text{or} \quad y = 2$$

We have found two y values. We substitute each y value into the equation marked $\boxed{*}$. First, if y is 2,

$$x = 3 - 2(2)$$

$$= -1$$

We have found one ordered pair solution, $(-1, 2)$. To get another solution, replace y by $\frac{2}{5}$.

$$x = 3 - 2y$$

$$= 3 - 2\left(\frac{2}{5}\right)$$

$$= 3 - \frac{4}{5}$$

$$= \frac{11}{5}$$

Therefore, the solution set contains two ordered pairs,

$$\left\{\left(\frac{11}{5}, \frac{2}{5}\right), (-1, 2)\right\} \qquad \blacktriangleleft$$

It is instructive to look at the graphs of the two equations in the system of Example 4.

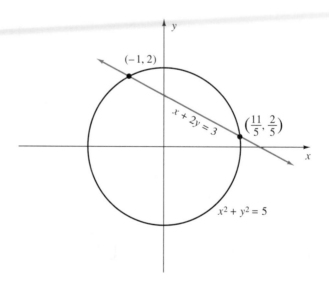

The second method for solving systems of equations works particularly well for linear systems, but may not work at all in nonlinear systems. It is called the method of **elimination.**

The addition property in Chapter 2 allows us to add two equations together. This property can be of great assistance in solving a system of equations. Some examples should clarify this.

► **EXAMPLE 5**

Solve the system by elimination.

$$\begin{cases} 3x + 2y = 2 \\ x - 2y = 14 \end{cases}$$

Solution

Notice that if we add these equations together the y terms will add to zero.

$$\begin{array}{rcl} 3x + 2y & = & 2 \\ x - 2y & = & 14 \\ \hline 4x & = & 16 \\ x & = & 4 \end{array}$$

Now we substitute this x value in either of the original equations and solve for y.

$$\begin{array}{rcl} 3(4) + 2y & = & 2 \\ 12 + 2y & = & 2 \\ 2y & = & -10 \\ y & = & -5 \end{array}$$

We have found an ordered pair, $(4, -5)$, that satisfies both equations of the system. Therefore, the solution set is

$$\{(4, -5)\} \quad \blacktriangleleft$$

► **EXAMPLE 6**

Solve the system by elimination.

$$\begin{cases} 5x - 3y = 13 \\ 2x + y = 3 \end{cases}$$

Solution

If we add these equations together, we get the equation $7x - 2y = 16$, which does not help. But note what happens if we multiply both sides of the second equation by 3 and *then* add.

$$\begin{array}{rcl} 5x - 3y & = & 13 \\ 6x + 3y & = & 9 \\ \hline 11x & = & 22 \\ x & = & 2 \end{array}$$

We now find y.

$$\begin{array}{rcl} 2(2) + y & = & 3 \\ 4 + y & = & 3 \\ y & = & -1 \end{array}$$

The solution set is

$$\{(2, -1)\} \quad \blacktriangleleft$$

So we see that, for the elimination step to work, it is sometimes necessary to multiply both sides of one of the equations by a suitable number and both sides of the other equation by another number.

Method of Elimination

1. Write both equations in standard form, $Ax + By = C$.
2. Multiply both sides of each equation by a suitable real number so that one of the variables will be eliminated by addition of the equations. (This step may not be necessary.)
3. Add the equations and solve the resulting equation.
4. Substitute the value found in step 3 into one of the original equations, and solve this equation.
5. Check the solution, if required.
6. Write the solution set.

We say that two systems are **equivalent** if they have the same solution set.

▶ **EXAMPLE 7**

Find the solution set for the following system by elimination.

$$\begin{cases} 2x + 5y = 11 \\ 3x + 7y = 15 \end{cases}$$

Solution

STEP 1   Write both equations in standard form.

Both equations are already in the form $Ax + By = C$.

STEP 2   Multiply both sides of each equation by a suitable number so that a variable will be eliminated.

One way to do this step is to multiply both sides of the first equation by 3 and both sides of the second equation by -2.

$$\begin{cases} 6x + 15y = 33 \\ -6x - 14y = -30 \end{cases}$$

This system is equivalent to the first, but notice what happens if we add the two equations together.

STEP 3   Add the equations.

$$\begin{array}{r} 6x + 15y = 33 \\ -6x - 14y = -30 \\ \hline y = 3 \end{array}$$

We have the y value.

STEP 4 Substitute the value found in step 3 into one of the original equations and solve.

We substitute 3 for y in the first of the original equations.

$$2x + 5y = 11$$
$$2x + 5(3) = 11$$
$$2x + 15 = 11$$
$$2x = -4$$
$$x = -2$$

STEP 5 Check, if required.

Since

$$2(-2) + 5(3) = -4 + 15 = 11$$

and

$$3(-2) + 7(3) = -6 + 21 = 15$$

the ordered pair $(-2, 3)$ checks in both equations.

STEP 6 Write the solution set.

$$\{(-2, 3)\}$$ ◀

► EXAMPLE 8

Solve the system by elimination.

$$\begin{cases} 9x + 12y = 18 \\ 3x + 4y = 6 \end{cases}$$

Solution

Multiply the second equation by -3.

$$\begin{cases} 9x + 12y = 18 \\ -9x - 12y = -18 \end{cases}$$

Add the two equations together.

$$9x + 12y = 18$$
$$-9x - 12y = -18$$
$$\overline{\ 0 = \ \ \ 0}$$

A true statement. As in the method of substitution, a true statement means that the system is **dependent** and the solution set is

$$\{(x, y) \mid 3x + 4y = 6\}$$ ◀

▶ **EXAMPLE 9**

Solve the following system by elimination.

$$\begin{cases} \dfrac{1}{2}x - \dfrac{1}{3}y = \dfrac{1}{6} \\ 2x - \dfrac{4}{3}y = 1 \end{cases}$$

Solution

First, we multiply both sides of the top equation by 6 and the bottom equation by 3 to clear fractions.

$$\begin{cases} 3x - 2y = 1 \\ 6x - 4y = 3 \end{cases}$$

Next we multiply the top equation by -2 to prepare for elimination.

$$\begin{cases} -6x + 4y = -2 \\ 6x - 4y = 3 \end{cases}$$

Now we add.

$$\begin{array}{r} -6x + 4y = -2 \\ 6x - 4y = 3 \\ \hline 0 = 1 \end{array}$$

which is a *false* statement, indicating an **inconsistent** system with an empty solution set,

$$\varnothing$$

◀

Many word problems are easily solved using two variables.

▶ **EXAMPLE 10**

Find two numbers whose sum is 15 if one of the numbers subtracted from twice the other number is also 15.

Solution

First, we make two assignments:

- Let x be one of the numbers.
- Let y be the other.

From the problem statement we get that the sum of x and y is 15. Also, one of the numbers, say y, subtracted from two times the other is also 15. This gives us the two equations

$$\begin{cases} x + y = 15 \\ 2x - y = 15 \end{cases}$$

We solve by elimination.

$$
\begin{aligned}
x + y &= 15 \\
2x - y &= 15 \\
\hline
3x \phantom{{}+y} &= 30 \\
x &= 10 \\
10 + y &= 15 \\
y &= 5
\end{aligned}
$$

The numbers are 5 and 10.   ◄

Problem Set 7.1

Warm-Ups

In Problems 1 through 12, use the method of substitution to find the solution set of the system. For Problems 1 through 6, see Example 1.

1. $\begin{cases} x - 2y = 0 \\ 2x + y = 5 \end{cases}$

2. $\begin{cases} x + 2y = 0 \\ 2x + y = 6 \end{cases}$

3. $\begin{cases} 2x - y = 0 \\ x + 4y = 9 \end{cases}$

4. $\begin{cases} 3x + 2y = 6 \\ 4x - y = 8 \end{cases}$

5. $\begin{cases} 2x - 5y = 4 \\ 3x - 2y = -5 \end{cases}$

6. $\begin{cases} 3x - 2y = 2 \\ 2x + 3y = -3 \end{cases}$

For Problems 7 through 9, see Examples 2 and 3.

7. $\begin{cases} x + y = 4 \\ 3x + 3y = 12 \end{cases}$

8. $\begin{cases} 3x - 2y = 6 \\ 15x - 10y = -2 \end{cases}$

9. $\begin{cases} \dfrac{1}{3}x - \dfrac{3}{4}y = 1 \\ 4x - 9y = 6 \end{cases}$

For Problems 10 through 12, see Example 4.

10. $\begin{cases} x^2 + y^2 = 25 \\ x - 3 = 0 \end{cases}$

11. $\begin{cases} x^2 - y = -1 \\ x - y = -1 \end{cases}$

12. $\begin{cases} x^2 - y^2 = 4 \\ 2x - y = -4 \end{cases}$

In Problems 13 through 27, use the method of elimination to find the solution set. For Problems 13 through 18, see Examples 5 and 6.

13. $\begin{cases} x + y = 1 \\ x - y = 3 \end{cases}$

14. $\begin{cases} 2x + 3y = 10 \\ x - 3y = -4 \end{cases}$

15. $\begin{cases} 2x - y = -5 \\ 3x + 2y = 3 \end{cases}$

16. $\begin{cases} 2x + 2y = 3 \\ -x + 6y = 2 \end{cases}$

17. $\begin{cases} 3x + 2y = 5 \\ 6x - y = 0 \end{cases}$

18. $\begin{cases} 7x - 3y = -10 \\ 5x - y = -4 \end{cases}$

For Problems 19 through 24, see Example 7.

19. $\begin{cases} 2x + 3y = 5 \\ 3x - 2y = 1 \end{cases}$

20. $\begin{cases} 4x - 5y = 2 \\ 3x + 2y = 13 \end{cases}$

21. $\begin{cases} 5x - 2y = 0 \\ 7x - 3y = -1 \end{cases}$

22. $\begin{cases} \dfrac{1}{2}x + \dfrac{1}{3}y = 11 \\ \dfrac{1}{3}x - \dfrac{1}{5}y = 1 \end{cases}$

23. $\begin{cases} \dfrac{4}{7}x - \dfrac{3}{5}y = 0 \\ -\dfrac{2}{5}x + \dfrac{1}{7}y = 0 \end{cases}$

24. $\begin{cases} -\dfrac{3}{7}x + \dfrac{2}{5}y = \dfrac{1}{35} \\ \dfrac{1}{5}x + \dfrac{7}{3}y = -\dfrac{38}{15} \end{cases}$

For Problems 25 through 27, see Examples 8 and 9.

25. $\begin{cases} 6x - 4y = 23 \\ -9x + 6y = 12 \end{cases}$

26. $\begin{cases} \dfrac{1}{3}x - \dfrac{3}{2}y = 0 \\ \dfrac{2}{5}x - \dfrac{9}{5}y = 0 \end{cases}$

27. $\begin{cases} \dfrac{2}{3}x + 8y = \dfrac{6}{7} \\ \dfrac{3}{4}x + 9y = 3 \end{cases}$

For Problems 28 through 30, see Example 10.

28. The sum of two numbers is 30. If twice one of them is three times the other, what are the numbers?

29. The difference of two numbers is 18. What are they if their sum is zero?

30. One-half the sum of two numbers equals their difference. What are the numbers if their sum is 64?

Practice Exercises

In Problems 31 through 48, use the method of substitution to find the solution set of each system.

31. $\begin{cases} x + 2y = 5 \\ 2x - y = 0 \end{cases}$

32. $\begin{cases} x - 2y = 4 \\ 2x - y = 5 \end{cases}$

33. $\begin{cases} 2x + y = 5 \\ x - 4y = -2 \end{cases}$

34. $\begin{cases} 3x - 2y = 1 \\ 4x + y = 5 \end{cases}$

35. $\begin{cases} 2x + 5y = -10 \\ 3x - 2y = 4 \end{cases}$

36. $\begin{cases} 3x - 2y = -6 \\ 2x + 3y = -4 \end{cases}$

37. $\begin{cases} 4x - 5y = -8 \\ 2x + 3y = 7 \end{cases}$

38. $\begin{cases} 2x + 6y = -2 \\ 5x + 3y = -13 \end{cases}$

39. $\begin{cases} 4x + 6y = 1 \\ 2x + 3y = 2 \end{cases}$

40. $\begin{cases} x - 6y = -1 \\ 2x - 3y = 1 \end{cases}$

41. $\begin{cases} 2x + 6y = 9 \\ 4x + y = 7 \end{cases}$

42. $\begin{cases} x + 6y = 1 \\ 3x + 18y = 3 \end{cases}$

43. $\begin{cases} 2x - y = 4 \\ 6x - 3y = 8 \end{cases}$

44. $\begin{cases} 3x + 6y = 2 \\ 6x - 9y = 2 \end{cases}$

45. $\begin{cases} x - 2y = 0 \\ 3x - 4y = 0 \end{cases}$

46. $\begin{cases} x^2 - 2y^2 = -2 \\ x - y = 1 \end{cases}$

47. $\begin{cases} x + y = 3 \\ xy = 2 \end{cases}$

48. $\begin{cases} 3x - 4y = 5 \\ \dfrac{3}{2}x - 2y = 1 \end{cases}$

In Problems 49 through 80, use the method of elimination to find the solution set of each system.

49. $\begin{cases} 2x - y = 7 \\ 3x + y = 8 \end{cases}$

50. $\begin{cases} x - 2y = 1 \\ 3x + 2y = 11 \end{cases}$

51. $\begin{cases} 3x + y = 5 \\ 4x - 2y = 10 \end{cases}$

52. $\begin{cases} 4x + 3y = 14 \\ -2x + 4y = 4 \end{cases}$

53. $\begin{cases} 2x - 3y = -5 \\ 3x + y = -2 \end{cases}$

54. $\begin{cases} 2x + 3y = 2 \\ x + 5y = -6 \end{cases}$

55. $\begin{cases} 2x - 3y = -3 \\ 3x + 2y = 15 \end{cases}$

56. $\begin{cases} 5x + 3y = 1 \\ 2x - 4y = 3 \end{cases}$

57. $\begin{cases} 7x + 4y = -1 \\ 9x - 6y = -5 \end{cases}$

58. $\begin{cases} 5x + 4y = 2 \\ 4x + 3y = 1 \end{cases}$

59. $\begin{cases} 5x - 2y = 0 \\ 3x - 7y = 0 \end{cases}$

60. $\begin{cases} 11x - 7y = 9 \\ 8x + 6y = 62 \end{cases}$

61. $\begin{cases} 3x + 4y = 0 \\ 4x + 3y = 0 \end{cases}$

62. $\begin{cases} 12x - 5y = -1 \\ 3x - 2y = -1 \end{cases}$

63. $\begin{cases} 6x - 5y = 6 \\ 9x + 7y = -20 \end{cases}$

64. $\begin{cases} 2x + 9y = -2 \\ 4x + 3y = 1 \end{cases}$

65. $\begin{cases} -4x + 15y = -2 \\ 12x + 5y = -4 \end{cases}$

66. $\begin{cases} 4x + 5y = 2 \\ 6x + 7y = 3 \end{cases}$

67. $\begin{cases} 3x - 4y = 5 \\ 4x - 7y = 9 \end{cases}$

68. $\begin{cases} 5x - 3y = 5 \\ -15x + 9y = 7 \end{cases}$

69. $\begin{cases} 17x - 11y = -1 \\ -34x + 22y = 2 \end{cases}$

70. $\begin{cases} \dfrac{1}{2}x + \dfrac{3}{4}y = \dfrac{7}{4} \\ \dfrac{1}{3}x - \dfrac{1}{6}y = \dfrac{1}{2} \end{cases}$

71. $\begin{cases} \dfrac{1}{5}x - \dfrac{2}{3}y = \dfrac{1}{15} \\ \dfrac{3}{4}x - \dfrac{5}{6}y = \dfrac{1}{3} \end{cases}$

72. $\begin{cases} \dfrac{2}{3}x - \dfrac{2}{5}y = 1 \\ \dfrac{5}{2}x - \dfrac{3}{2}y = 2 \end{cases}$

73. $\begin{cases} \dfrac{2}{7}x + \dfrac{2}{5}y = \dfrac{3}{7} \\ \dfrac{3}{5}x + \dfrac{7}{3}y = \dfrac{5}{2} \end{cases}$

74. $\begin{cases} \dfrac{6}{11}x - \dfrac{4}{7}y = 0 \\ -\dfrac{7}{4}x + \dfrac{11}{6}y = 0 \end{cases}$

75. $\begin{cases} \dfrac{3}{8}x - \dfrac{1}{9}y = 0 \\ \dfrac{9}{2}x - \dfrac{5}{3}y = 0 \end{cases}$

76. $\begin{cases} 2x + 3y^2 = 2 \\ x - y = 1 \end{cases}$

77. $\begin{cases} 3x^2 - 2y^2 = 4 \\ 2x - y^2 = 0 \end{cases}$

78. $\begin{cases} x + xy = -1 \\ xy = -6 \end{cases}$

79. $\begin{cases} x^2 + y^2 = 1 \\ x^2 - y = 1 \end{cases}$

80. $\begin{cases} x^2 + y^2 = 13 \\ x^2 - y^2 = 5 \end{cases}$

81. The difference between two numbers is 2. If three times the larger plus five times the smaller is 94, what are the numbers?

82. Three times one number is four times another. If the sum of the two numbers is 49, what are they?

83. The sum of three numbers is 68. One of the numbers is twice the average of the other two. The difference of the "other two" is 20. What are the three numbers?

Challenge Problems

In Problems 84 through 86, the systems are not linear, but the substitutions

$$u = \frac{1}{x} \quad \text{and} \quad v = \frac{1}{y}$$

will transform them into linear equations in u and v. Find the solution sets.

84. $\begin{cases} \dfrac{2}{x} + \dfrac{3}{y} = 12 \\ \dfrac{5}{x} - \dfrac{2}{y} = 11 \end{cases}$

85. $\begin{cases} \dfrac{3}{x} + \dfrac{4}{y} = 5 \\ \dfrac{6}{x} + \dfrac{5}{y} = 7 \end{cases}$

86. $\begin{cases} \dfrac{2}{x} + \dfrac{3}{y} = 0 \\ \dfrac{3}{x} + \dfrac{2}{y} = 1 \end{cases}$

In Your Own Words

87. Describe the method of *substitution* for solving two equations in two variables.

88. Describe the method of *elimination* for solving two equations in two variables.

7.2   Systems as Models for Applications

TAPE 13

Many word problems become much simpler if more than one variable is used. We follow a procedure similar to that given in Section 2.5.

> ### A Procedure to Model Word Problems
>
> **1.** Read the problem to determine what quantities are to be found.
> **2.** Assign a variable, such as x, y, or z, to represent each quantity to be found.
> **3.** Draw a figure or picture if possible. Label it.
> **4.** Reread the problem and form a model. Write a system with as many equations as there are variables.
> **5.** Solve the system of equations found in step 4.
> **6.** Chcek the solution in the original word problem. It should make sense and answer the question.
> **7.** Write an answer to the original question.

Mixture problems are particularly suited for modeling by systems of equations.

► ### EXAMPLE 1

Perry wishes to clean his boat with a 15% soap solution. How much of a 10% soap solution and a 20% soap solution should he add to a quart of pure water to make 30 quarts of a 15% solution?

Solution

STEP 1   Read the problem and determine what is to be found.

We are to find how many quarts of a 10% soap solution and a 20% soap solution are needed.

STEP 2   Assign a variable to each quantity to be found.

Let x be the number of quarts of the 10% solution.
Let y be the number of quarts of the 20% solution.

STEP 3   Draw a figure.

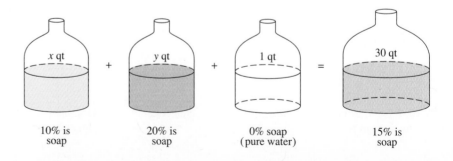

10% is   20% is   0% soap   15% is
soap   soap   (pure water)   soap

STEP 4 Form a system with as many equations as there are variables.

Because x quarts $+ y$ quarts $+ 1$ quart must be 30 quarts, our first equation is

$$x + y + 1 = 30 \qquad \text{or}$$

$\boxed{1}$
$$x + y = 29$$

Now the amount of *soap* in the 10% solution added to the amount of soap in the 20% solution must equal the amount of soap in the final solution. That is,

$$0.10x + 0.20y = 0.15(30)$$

$$10x + 20y = 15(30)$$

$$10x + 20y = 450$$

$\boxed{2}$
$$x + 2y = 45 \qquad \text{Divide by 10.}$$

This gives us the following system, which can be solved easily.

STEP 5 Solve the system.

$$\begin{cases} x + \ y = 29 \\ x + 2y = 45 \end{cases}$$

$$\begin{array}{r} -x - \ y = -29 \\ x + 2y = \ \ \ 45 \\ \hline y = \ \ \ 16 \end{array}$$

$$x + 16 = 29$$

$$x = 13$$

STEP 6 Check the solution in the original word problem.

13 quarts of the 10% solution and 16 quarts of the 20% solution make sense and answer the question.

STEP 7 Write an answer to the original question.

Perry should mix 13 quarts of the 10% solution and 16 quarts of the 20% solution with 1 quart of pure water to make his solution.   ◄

► EXAMPLE 2

Gloria bought 3 lb of raw peanuts and 2 lb of boiled peanuts from Big Bob's produce stand for $4.05. Jim bought 1 lb of raw and 4 lb of boiled for $4.65. What is the price per pound Big Bob charges for his raw and boiled peanuts?

Solution

Determine what is to be found.

We are to find the price per pound of raw peanuts and the price per pound of boiled peanuts.

Assign a variable to each quantity to be found.

Let R be the price per pound of raw peanuts and B be the price per pound of boiled peanuts.

Form as many equations as there are variables.

$$\begin{cases} 3R + 2B = 4.05 \\ R + 4B = 4.65 \end{cases}$$

Solve the system.

Using elimination, we multiply the top equation by -2 and then add.

$$\begin{array}{rcl} -6R - 4B &=& -8.10 \\ R + 4B &=& 4.65 \\ \hline -5R &=& -3.45 \\ R &=& 0.69 \\ 0.69 + 4B &=& 4.65 \\ 4B &=& 4.65 - 0.69 \\ 4B &=& 3.96 \\ B &=& 0.99 \end{array}$$

Check and then answer the question.

Big Bob charges 69 cents a pound for raw peanuts and 99 cents a pound for boiled.   ◄

Often the relationships between the lengths of sides and perimeters of geometric figures are easy to express with more than one variable.

► **EXAMPLE 3**

The perimeter of a parallelogram is 58 in. If the length of one side is 5 more than three times the length of the other side, what are the lengths of the sides?

Solution

Let x be the length of the longest side by y be the length of the shortest side.

The perimeter of a parallelogram is given by $2x + 2y$, where x and y are the lengths of the sides. Thus we have one equation: $2x + 2y = 58$.

From the relationship between the sides, we have the second equation: $x = 3y + 5$.

These two equations give us the following system:

$$\begin{cases} 2x + 2y = 58 \\ x = 3y + 5 \end{cases}$$

Since the second equation is already solved for *x*, we will use the method of substitution. We replace *x* in the first equation with its value in the second.

$$2(3y + 5) + 2y = 58$$

$$6y + 10 + 2y = 58$$

$$8y = 48$$

$$y = 6$$

Now we find *x* from the second equation.

$$x = 3(6) + 5$$

$$= 18 + 5$$

$$= 23$$

The sides of the parallelogram are 6 in. and 23 in.   ◄

Motion problems can often be simplified by using more than one variable, particularly "upstream–downstream" problems.

► EXAMPLE 4

Sarah's boat takes 1 hour to travel 5 miles upstream to the Raysville Marina, but only 20 minutes to return. What is the speed of the current in the river, and what is the average speed of Sarah's boat in still water?

Solution

We let *x* be the average speed (mph) of the boat in still water and *y* be the speed (mph) of the current. Thus Sarah's speed *upstream* is

$$x - y$$

and *downstream* is

$$x + y$$

Now we make a distance–rate–time table for each part of the trip.

	Distance	Rate	Time
Upstream	5	$x - y$	1
Downstream	5	$x + y$	$\dfrac{1}{3}$

We need time in hours, so we must change 20 min to $\dfrac{1}{3}$ hour. Since, $D = RT$, we have the two equations

$$\begin{cases} 5 = (x - y) \cdot 1 \\ 5 = (x + y) \cdot \dfrac{1}{3} \end{cases}$$

We multiply the second equation through by 3 and write the system in the usual form.

$$\begin{cases} x - y = 5 \\ x + y = 15 \end{cases}$$

We add the equations together to find the value of x.

$$2x = 20$$

$$x = 10$$

Substitute into the second equation

$$10 + y = 15$$

$$y = 5$$

Sarah's boat averages 10 mph in still water and the current in the river is 5 mph.

◀

Work problems may also be easier to solve if more than one variable is used.

► **EXAMPLE 5**

Barney and Gerald, working together, can paint a boat shed in 6 hours. But after 3 hours, Barney is called away and it takes Gerald 4 more hours to complete the job. How long would it take each to paint the shed working alone?

Solution

Let B be the number of hours it would take Barney to do the job working alone.

Let G be the number of hours it would take Gerald to do the job working alone.

If it takes Barney B hours to do the job alone, then in 1 hour Barney must do $\dfrac{1}{B}$ of the job. Since it takes Gerald G hours to do the job alone, he must do $\dfrac{1}{G}$ of the job in 1 hour. Since $\dfrac{1}{6}$ of the job is done in 1 hour when they both are working, we get the equation

$$\frac{1}{B} + \frac{1}{G} = \frac{1}{6}$$

This is not a linear equation. However, if we let $x = \dfrac{1}{B}$ and $y = \dfrac{1}{G}$, we get the linear equation

$$x + y = \frac{1}{6} \qquad \text{or}$$

①
$$6x + 6y = 1$$

When they actually do the job, Barney works 3 hours, thus doing $3\left(\dfrac{1}{B}\right)$ of the job, while Gerald works 7 hours doing $7\left(\dfrac{1}{G}\right)$ of the job. But they do one job, so we have the equation

$$\frac{3}{B} + \frac{7}{G} = 1$$

With the substitution above, we get the linear equation

②
$$3x + 7y = 1$$

and the system

$$\begin{cases} 6x + 6y = 1 \\ 3x + 7y = 1 \end{cases}$$

We multiply the second equation by -2, add, and then solve the system.

$$
\begin{array}{r}
6x + 6y = 1 \\
-6x - 14y = -2 \\
\hline
-8y = -1
\end{array}
$$

$$y = \frac{1}{8}$$

$$6x + 6y = 1$$

$$6x + 6\left(\frac{1}{8}\right) = 1$$

$$6x + \frac{3}{4} = 1$$

$$24x + 3 = 4$$

$$24x = 1$$

$$x = \frac{1}{24}$$

But $y = \dfrac{1}{8}$ means that $G = 8$, and $x = \dfrac{1}{24}$ means that $B = 24$.

It would take Gerald 8 hours to paint the boat shed alone and it would take Barney 24 hours alone. ◄

Problems that contain nonlinear statements can be very difficult to solve by any method. However, many of them are quite easy when more than one variable is used.

► **EXAMPLE 6**

The sum of two numbers is 7 and the sum of their squares is 25. What are the numbers?

Solution

Let x be one of the numbers and y the other and we have the pair of equations

$$\begin{cases} x + y = 7 \\ x^2 + y^2 = 25 \end{cases}$$

As this is a *nonlinear* system (look at the second equation), we will use the method of substitution. Solve the first equation for y.

$$y = 7 - x$$

Then substitute this into the second equation and solve.

$$x^2 + (7 - x)^2 = 25$$
$$x^2 + 49 - 14x + x^2 = 25$$
$$2x^2 - 14x + 24 = 0$$
$$x^2 - 7x + 12 = 0$$
$$(x - 3)(x - 4) = 0$$
$$x = 3, \qquad x = 4$$

If x is 3, then y is 4, and if x is 4, y is 3.
The numbers are 3 and 4.   ◄

Problem Set 7.2

Warm-Ups

In Problems 1 and 2, solve each problem. See Example 1.

1. Penny wishes to mix a tomato sauce that is 17% sugar with a sauce that is 30% sugar to obtain 26 liters of a tomato sauce that is 24% sugar. How much of each should she mix?

2. How many ounces of pure water and how many ounces of a 16% butterfat solution should Charlotte mix to obtain 32 ounces of a 10% butterfat solution?

In Problems 3 and 4, solve each problem. See Example 2.

3. Martin bought 3 heads of lettuce and 2 pounds of tomatoes at the Tasty Shopette for $4.05. Beverly bought 2 heads of lettuce and 3 pounds of tomatoes of $4.35 at the same store. What is the price of a head of lettuce and a pound of tomatoes at the Tasty Shopette?

4. Latonia paid $2.39 for three Snickers and four Mr. Goodbar candy bars, while Antonio paid $3.40 for five of each of those kinds of candy bar. What was the prices of Snickers and Mr. Goodbars?

In Problems 5 and 6, solve each problem. See Example 3.

5. The perimeter of Jon's prize-winning lawn is 54 m. It is in the shape of a rectangle with the length of one side 3 m less than twice the length of the other. What are the dimensions of the lawn?

6. When not at the lake, Colombo spends his time in a rectanglular dog pen that his master constructed with 28 yards of fencing. If the length of the pen is 2 yd longer than twice its width, what are its dimensions?

In Problems 7 and 8, solve each problem. See Example 4.

7. It takes Jim's 18-year-old outboard 3 hours to travel 24 miles downstream and 5 hours to travel 10 miles upstream from his river cottage. What is the average speed of the boat in still water, and what is the speed of the current?

8. An airplane flying with the wind travels from Sioux City to Dubuque, a distance of 500 km, in 2 hours. The return trip, against the wind, takes $2\frac{1}{2}$ hours. What is the average speed of the plane in still air, and what is the wind speed?

In Problems 9 and 10, solve each problem. See Example 5.

9. Frank and Jack working together can mow a certain field in 4 hours. One morning Frank mowed alone for 3 hours and left for the day. Jack arrived that afternoon and spent 6 hours finishing the job. How long does it take each of them, working alone, to mow the field?

10. A Texaco oil tank can be filled by two pipes in 8 hours. If the tank is filled in 10 hours when one of the pipes is shut off after 4 hours, how long will it take each of the pipes to fill the tank alone?

In Problem 11, solve each problem. See Example 6.

11. Find two numbers whose sum is 19 if the sum of their squares is 185.

Practice Exercises

In Problems 12 through 28, solve each problem.

12. The perimeter of a rectangular window is 200 cm. The length of one side is 20 cm less than 11 times the other. What are the dimensions of the window?

13. At the Farmers Market, John bought 3 pounds of grapes and 6 lemons for $3.57, and Betty bought 2 pounds of grapes and 5 lemons and paid $2.48. How much are lemones and grapes at the Farmers Market?

14. Gerry wishes to mix a 5% soap solution with an 8% soap solution to obtain 12 quarts of a 6% soap solution to wash her new Buick. How much of each should she mix?

15. Tom and Randi, working together, can mow their lawn in 1 hour and 12 minutes. Randi heads for the pool after working with Tom for 18 minutes, and it takes Tom $1\frac{1}{2}$ hours to finish the job. How long does it take each, working alone, to mow the lawn?

16. The perimeter of a triangle is 146 ft. Twice the shortest side is 9 ft more than the longest side. The longest side is 16 ft less than the sum of the other two sides. What are the lengths of the three sides of this triangle?

17. Wayne Hilton has 1200 ft of fencing and wishes to enclose a rectangular pasture that is four times as long as

it is wide. What dimensions should Wayne use for his pasture?

18. It takes a garbage scow 8 hours to travel 24 miles upstream and 8 hours to travel 88 miles downstream. What is the average speed of the scow in still water, and what is the speed of the current?

19. Carolyn walks 10 miles along the beach in the same time that Bill walks 6 miles. If Carolyn walks 1 mph less than twice Bill's rate, what is the speed at which each of them walks?

20. The sum of two positive numbers is 13, while the sum of one of them and the square of the other is 69. Find two such numbers.

21. The perimeter of an isosceles triangle is 48 inches. If the length of the shortest side is 3 inches less than either of the other two sides, what are the lengths of each of the sides of the triangle?

22. Two cars start at the same time, one from Gallop and the other from Las Cruces, 360 miles apart, and they travel toward each other. One travels 8 mph faster than the other. If they meet after 3 hours, what was the average speed of each car?

23. The perimeter of a valuable rectangular portrait is 60 in. and its area is 216 in.². What are the dimensions of a frame for this painting?

24. Tickets to the band concert were $5 for balcony and $10 for orchestra seats. If attendance at one show was 800, and if total receipts for that show were $7000, how many people bought orchestra seats?

25. Margi bought 3 leotards and 4 pairs of tights at Taffy's for $185 while Justine bought 2 leotards and 3 pairs of tights for $127.50. What is the price of a leotard and the price of a pair of tights?

26. Lafayette purchased 16 shares of Coca-Cola stock and 10 shares of Sun Trust stock for $1097.40. Norma bought 48 shares of Coca-Cola and 12 shares of Sun Trust for $2509.20. What is the price of a share of Coca-Cola stock and a share of Sun Trust stock?

27. Son Luong and Don Drew walk and run together during their lunch workout. Son burns 200 calories per hour while running and 150 calories per hour while walking. Don burns 250 calories per hour while running and 180 calories per hour walking. During one of their workouts, Son burned a total of 150 calories and Don burned a total of 185 calories. How long did they run and walk in this workout?

28. Sam's river cabin is 4 miles upstream from the landing. If the boat trip to the landing takes 30 minutes and the trip home takes 45 minutes, find the speed of the current and the speed of Sam's boat in still water.

Challenge Problems

In Problems 29 and 30, solve each problem.

29. A $\frac{1}{4}$-in. pipe and a $\frac{1}{2}$-in. pipe, working together with the drain closed, can fill a water tank in 12 minutes. With the $\frac{1}{4}$-in. pipe closed and the drain open, it takes the $\frac{1}{2}$-in. pipe 40 minutes to fill the tank. With the $\frac{1}{2}$-in. pipe closed and the drain open, it takes the $\frac{1}{4}$-in. pipe 2 hours to fill the tank. How long does it take the open drain to empty a full tank when both pipes are closed?

30. Two resistors R_1 and R_2 in a electrical circuit give a total resistance of R, where R is determined by the relationship

$$\frac{1}{R} = \frac{1}{R_1} + \frac{1}{R_2}$$

if they are hooked in parallel, and $R = R_1 + R_2$ if they are hooked in series. Suppose that two resistors give a total resitance of 12 ohms in parallel and 50 ohms in series. What is the number of ohms for each resistor?

In Your Own Words

31. Write a word problem that has the following system as its mathematical model:

$$\begin{cases} x + y = 6 \\ x - y = 3 \end{cases}$$

7.3   Linear Systems of Equations in More Than Two Variables

TAPE 13

For the same reasons that there are equations in two variables, there are equations in three, four, or even more variables. Modern high-speed computers have the ability to solve huge systems quickly.

A linear equation in three variables is an equation of the form

$$Ax + By + Cz = D, \qquad A, B, C \text{ not all zero}$$

Solution

If we add the first two equations, we get

$$2x + z = 2$$

and if we add the first and third, we get

$$4x + 2z = 4$$

We now have a system of two equations in two variables:

$$\begin{cases} 2x + z = 2 \\ 4x + 2z = 4 \end{cases}$$

If we multiply both sides of the first equation by 2, we get the system

$$\begin{cases} 4x + 2z = 4 \\ 4x + 2z = 4 \end{cases}$$

Because both of these equations are exactly the same, the system is clearly dependent, so the original system is **dependent.** (We will not write the solution set for dependent systems of three or more variables.)   ◀

Linear systems with four or more variables may be solved by an extension of the method described above. For example, a system of four linear equations in four variables may be solved by eliminating one variable from a pair of equations. Then, when the same variable is eliminated from two *other* pairs of equations, the resulting system of three equations in three variables may be solved by the three-variable method.

▶ **EXAMPLE 5**

Find the solution set.

$$\boxed{1}\quad\boxed{2}\quad\boxed{3}\quad\boxed{4}\qquad \begin{cases} x_1 + x_2 = 3 \\ 3x_2 + 3x_3 = 7 \\ x_3 + x_4 = 1 \\ 3x_1 - 3x_4 = 1 \end{cases}$$

Solution

This is an example of a system of four equations in four variables. Systems this large can be tedious to solve. However, this system has quite a few "missing terms," which will aid us in our task. For example, the variable x_4 does not occur in equation $\boxed{1}$ or $\boxed{2}$. We start by eliminating x_4 from $\boxed{3}$ and $\boxed{4}$. Multiply both sides of $\boxed{3}$ by 3; add it to $\boxed{4}$ to get $\boxed{5}$.

$$\begin{array}{l} \boxed{3} \\ \boxed{4} \\ \boxed{5} \end{array} \qquad \begin{array}{r} 3x_3 + 3x_4 = 3 \\ 3x_1 - 3x_4 = 1 \\ \hline 3x_1 + 3x_3 = 4 \end{array}$$

Now the system of equations $\boxed{1}$, $\boxed{2}$, and $\boxed{5}$ is a system of three equations in the three variables x_1, x_2, x_3.

$$\boxed{1}\quad\boxed{2}\quad\boxed{5}\qquad \begin{cases} x_1 + x_2 \phantom{{}+3x_3} = 3 \\ \phantom{x_1+{}} 3x_2 + 3x_3 = 7 \\ 3x_1 \phantom{{}+3x_2} + 3x_3 = 4 \end{cases}$$

We eliminate x_3 from this system by multiplying both sides of $\boxed{5}$ by -1 and adding it to $\boxed{2}$.

$$\begin{array}{ll}\boxed{2} & 3x_2 + 3x_3 = 7 \\ \boxed{5} & \underline{-3x_1 \phantom{{}+3x_2} - 3x_3 = -4} \\ & -3x_1 + 3x_2 \phantom{{}+3x_3} = 3 \qquad \text{or} \\ \boxed{6} & -x_1 + x_2 = 1 \end{array}$$

Finally, the system made up of equations $\boxed{1}$ and $\boxed{6}$ is a system of two equations in two variables. Adding them eliminates x_3.

$$\begin{array}{ll}\boxed{1} & x_1 + x_2 = 3 \\ \boxed{6} & \underline{-x_1 + x_2 = 1} \\ & \phantom{-x_1+{}} 2x_2 = 4 \\ & \phantom{-x_1+{}} x_2 = 2 \end{array}$$

We put this result in $\boxed{1}$ to get x_1:

$$\begin{array}{ll}\boxed{1} & x_1 + x_2 = 3 \\ & x_1 + 2 = 3 \\ & x_1 = 1 \end{array}$$

And in $\boxed{2}$ to get x_3:

$$\begin{array}{ll}\boxed{2} & 3x_2 + 3x_3 = 7 \\ & 3(2) + 3x_3 = 7 \\ & 3x_3 = 1 \\ & x_3 = \dfrac{1}{3} \end{array}$$

The value of x_3 in equation $\boxed{3}$ yields x_4.

$$\begin{array}{ll}\boxed{3} & x_3 + x_4 = 1 \\ & \dfrac{1}{3} + x_4 = 1 \\ & x_4 = \dfrac{2}{3} \end{array}$$

The solution set is

$$\left\{\left(1, 2, \frac{1}{3}, \frac{2}{3}\right)\right\}$$

Problem Set 7.3

Warm-Ups

In Problems 1 through 10, find the solution set for each system unless it is dependent.
If dependent, so indicate.
For Problems 1 through 6, see Examples 1 and 2.

1. $\begin{cases} x + y + z = 6 \\ x + 2y - z = 2 \\ 2x - 2y + z = 1 \end{cases}$
$\qquad$ **2.** $\begin{cases} 2x - y + z = 5 \\ -x - y + 2z = 4 \\ 2x + y - 3z = -5 \end{cases}$
$\qquad$ **3.** $\begin{cases} 2x - y + 2z = 2 \\ -2x + 3y + 4z = 10 \\ -2x + y - z = 0 \end{cases}$

4. $\begin{cases} x + y + z = 5 \\ 2x - 3y - 2z = 12 \\ 3x + 2y - 2z = 7 \end{cases}$
$\qquad$ **5.** $\begin{cases} x + 2y + 3z = 1 \\ 2x - 2y - 3z = -1 \\ 3x + y + 6z = 2 \end{cases}$
$\qquad$ **6.** $\begin{cases} x + y + 3z = 0 \\ 2x - 2y - z = 0 \\ 5x - 3y - z = 0 \end{cases}$

For Problems 7 through 10, see Example 3 and 4.

7. $\begin{cases} 3x + 3y - z = 5 \\ x + y + z = 5 \\ -2x - 2y + z = -3 \end{cases}$
$\qquad$ **8.** $\begin{cases} 2x + y - z = 4 \\ 4x - y + 2z = -2 \\ -2x + 2y - 3z = 6 \end{cases}$
$\qquad$ **9.** $\begin{cases} 3x - 5y - 2z = 9 \\ x - y + 2z = 1 \\ 2x - 3y = 1 \end{cases}$

10. $\begin{cases} x_1 + x_2 + x_3 = 1 \\ 3x_1 - 2x_2 + 2x_3 = 5 \\ 3x_1 - 7x_2 + x_3 = 7 \end{cases}$

Practice Exercises

In Problems 11 through 32, find the solution set for each system unless it is dependent.
If dependent, so indicate.

11. $\begin{cases} x + y + z = 6 \\ x + 2y + z = 9 \\ 2x + 2y - z = 6 \end{cases}$
$\qquad$ **12.** $\begin{cases} x + y + 3z = 0 \\ 2x - y + 2z = 1 \\ 3x + y + z = 6 \end{cases}$
$\qquad$ **13.** $\begin{cases} 3x_1 + x_2 + 2x_3 = 10 \\ -3x_1 - 2x_2 + 4x_3 = -11 \\ 2x_2 + x_3 = 2 \end{cases}$

14. $\begin{cases} 3x - 2y + 4z = 22 \\ x + y + z = 3 \\ 2x - 2y - 3z = -1 \end{cases}$
$\qquad$ **15.** $\begin{cases} 3x + 3y - z = 3 \\ 5x + y + 3z = 1 \\ 2x + 4y - 3z = 4 \end{cases}$
$\qquad$ **16.** $\begin{cases} 2x + 6y + z = 5 \\ -2x - 3y + 2z = -7 \\ x + 9y - 3z = 8 \end{cases}$

17. $\begin{cases} 2x + 2y - z = 1 \\ x + 2y - 3z = 4 \\ 5x + 6y - 5z = 3 \end{cases}$
$\qquad$ **18.** $\begin{cases} x - 2y + 3z = 9 \\ 3x - y - 2z = 6 \\ 4x + 4y + 2z = 2 \end{cases}$
$\qquad$ **19.** $\begin{cases} 3x + y - z = 8 \\ 2x - y + 2z = 3 \\ x + 2y - 3z = 5 \end{cases}$

20. $\begin{cases} x + 2y - z = -2 \\ 2x + 2y + 2z = 3 \\ 6x - 4y - 2z = 2 \end{cases}$
$\qquad$ **21.** $\begin{cases} 2x - 3y - 10z = 4 \\ 4x - 5z = 3 \\ 6y + 5z = -3 \end{cases}$
$\qquad$ **22.** $\begin{cases} 3x + y = 5 \\ 2y - 4z = 7 \\ x + y - z = 4 \end{cases}$

23. $\begin{cases} 2x + 6y = 5 \\ 3y - z = 2 \\ -3x - 3z = 1 \end{cases}$
$\qquad$ **24.** $\begin{cases} 2x + 3z = 2 \\ 5y - 9z = -3 \\ 6x + 7y = 3 \end{cases}$
$\qquad$ **25.** $\begin{cases} 2x - 3y = 6 \\ -2y - 3z = 4 \\ 3x - 2z = 0 \end{cases}$

26. $\begin{cases} 2x + 3y + 4z = 0 \\ x + y + z = 0 \\ -x + 2y + 2z = 0 \end{cases}$
$\qquad$ **27.** $\begin{cases} x + y - z = -8 \\ 4x + 5y - 6z = -2 \\ 2x + 3y - 4z = 14 \end{cases}$
$\qquad$ **28.** $\begin{cases} 2x - y + 4z = -4 \\ 4x - 5y + 7z = 1 \\ -2x + 7y - 2z = 6 \end{cases}$

29. $\begin{cases} 2x + 4y - z = 3 \\ x + y = 1 \\ 2x + 3y + z = 2 \end{cases}$

30. $\begin{cases} 2x + 3y - 3z = 0 \\ x - 3y + 2z = 5 \\ 4x - 6y - 5z = 5 \end{cases}$

31. $\begin{cases} 3x - 2y - z = 4 \\ x + 4y + 2z = -1 \\ 2x - 4y - 3z = 6 \end{cases}$

32. $\begin{cases} 3x - y = 0 \\ y + 2z = -1 \\ x - z = 0 \end{cases}$

In Problems 33 through 39, solve each problem.

33. Suppose that the sum of a certain number and a second number is 6. Further, suppose that the sum of the certain number and a third number is 7. If the sum of all three is 12, find the three numbers.

34. Find three numbers whose sum is 8 if the sum of two of them is 7 and the difference of the same two is 3.

35. Suppose that the sum of a certain number and a second number is 1. Further, suppose that the sum of the certain number and a third number is 8. If the sum of all three is 6, find the three numbers.

36. Four numbers have a sum of 9. The sum of two of them is 8 and the difference of the same two is 2. If the difference of the other two is 3, find the four numbers.

37. Terry, Saundra, and Linda made trips to the refreshment stand at the Cinema IV. Terry bought 2 small and 3 medium-sized drinks and one bag of popcorn for $5.25. Saundra bought a small drink and a bag of popcorn for $1.60, and Linda bought a medium-sized drink and 2 bags of popcorn for $2.90. What was the price of popcorn and drinks at Cinema IV?

38. The Smiths went soup shopping at the local Safeway. Kenneth bought 5 cans of tomato and 3 cans of mush-

Problem 37

room for $3.09, while Margaret bought 2 cans of chicken noodle and a can of mushroom for $1.41. If Irene bought 2 cans of tomato and 3 cans of chicken noodle for $2.19, what is the price of tomato soup, mushroom soup and chicken noodle soup at the local Safeway store?

39. The sum of four numbers is 11. Twice the first plus the third is 2, while three times the first added to twice the third is 5. If three times the second plus twice the fourth is 17, what are the four numbers?

Challenge Problems

In Problems 40 through 43, find the solution set for each system.

40. $\begin{cases} x + y + z + w = 2 \\ x - y + 2z = -1 \\ 2y + 3z - 3w = -9 \\ 2x - 3y + 2w = 6 \end{cases}$

41. $\begin{cases} x - 2y - z + 2w = -2 \\ 2x - y + z - w = 0 \\ -x + 3y - w = 4 \\ 5x - 3y = -4 \end{cases}$

42. $\begin{cases} x_1 + 2x_2 - x_3 = 2 \\ 2x_1 + x_3 + x_4 = 9 \\ x_2 - x_4 = -2 \\ 3x_1 + 4x_2 = 11 \end{cases}$

43. $\begin{cases} x_1 + x_2 = 1 \\ x_2 - x_3 = -2 \\ 2x_3 - 2x_4 = 5 \\ 2x_4 + 2x_5 = 1 \\ -2x_5 = 0 \end{cases}$

In Your Own Words

44. How do we determine if a system of equations is dependent or inconsistent?

7.4   Linear Inequalities in Two Variables

In Chapter 2 we studied inequalities in one variable. In this section we learn an extension of the method of boundary numbers that will allow us to graph linear inequalities in two variables. The graph of an inequality in two variables consists of all ordered pairs that make the statement of the inequality true. Inequalities in two variables, much like *equations* in two variables, required a coordinate system to display their graph.

Let's look at the graph of the inequality $x + y \geq 1$. The graph is the collection of all ordered pairs (x, y) that make the statement $x + y \geq 1$ true.

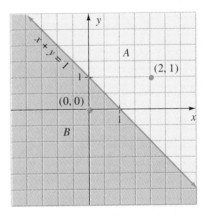

Using a technique similar to the method of boundary numbers, we graph the *equation* $x + y = 1$.

This line is called a **boundary line.** It divides the plane into two regions, marked A and B on the sketch.

Region A is tested by selecting any point in the interior of the region, say $(2, 1)$, and determining whether it makes the statement of the inequality true or false. Since $2 + 1 \geq 1$ is a true statement, the point $(2, 1)$ is in the solution set, and it follows that every point in region A is also in the solution set.

Similarly, we pick any point in region B, say $(0, 0)$, and see that the statement $0 + 0 \geq 1$ is false. Thus $(0, 0)$ is not in the solution set, which tells us that no point in region B is in the solution set.

The boundary line itself *is* in the solution set because the original statement, $x + y \geq 1$, includes equality. The solution set is graphed next.

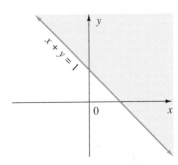

> **A Procedure for Graphing Linear Inequalities in Two Variables**
>
> **1.** Graph the boundary line:
>   (a) Draw a solid line if equality is included. ($\geq$, $\leq$)
>   (b) Draw a dashed line if equality is not included. ($>$, $<$)
> **2.** Determine which region(s) formed by the line makes the inequality true by testing with one point from inside each region.
> **3.** Shade the region(s) that makes the inequality true.

► **EXAMPLE 1**

Graph the solution set for $x - 2y > 0$.

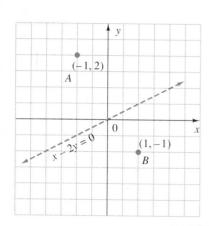

Solution

STEP 1   Graph the boundary line.

We draw the graph of the line $x - 2y = 0$.

STEP 2   Test the regions formed by the boundary line.

A table is convenient for step 2.

Region	Test Point in Region	Statement $x - 2y > 0$	Truth of Statement	Region in Solution Set?
A	$(-1, 2)$	$-1 - 4 > 0$	False	No
B	$(1, -1)$	$1 - 2(-1) > 0$	True	Yes

STEP 3   Shade the regions where the inequality is true.

Shade region *B*.   ◄

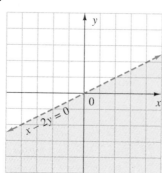

▶ **EXAMPLE 2**

Graph the solution set for $x \geq 1$ (on the x, y plane).

Solution

We graph the boundary line $x = 1$. It will be a solid line because equality is included.

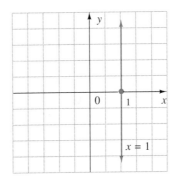

We can make a table for the next step or we can just test the regions and shade the graph.

The point $(0, 0)$ is in the region left of the boundary line, but it is not in the solution set because $0 \geq 1$ is a *false* statement. The left region *is not* shaded. The point $(2, 0)$ is in the right region and is in the solution set because $2 \geq 1$ is a *true* statement. The right region *is* shaded.   ◀

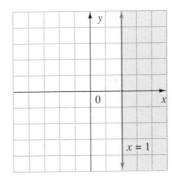

Problem Set 7.4

Warm-Ups

In Problems 1 through 10, graph the solution set on the x, y plane for each inequality.
For Problems 1 through 6, see Example 1.

1. $x - y < 2$

2. $x + 2y \leq 5$

3. $y - \dfrac{1}{2}x > 4$

4. $\frac{1}{3}y - x \geq 2$ **5.** $x - y \leq 0$ **6.** $y < 2x$

For Problems 7 through 10, see Examples 1 and 2.

7. $x \geq 3$ **8.** $y > 0$ **9.** $y \leq -1$ **10.** $x < 1$

Practice Exercises

In Problems 11 through 46, graph the solution set on the x, y plane for each inequality.

11. $x + y < 3$ **12.** $x + 5y \leq 1$ **13.** $3x + y \leq 5$

14. $x + y < 0$ **15.** $x + y \geq 0$ **16.** $y \leq 3x$

17. $y > 2x - 3$ **18.** $x < 4 - 3y$ **19.** $x \leq 1$

20. $y \geq 2$ **21.** $y < 3x + 4$ **22.** $y \geq 2x - 7$

23. $x > 0$ **24.** $y \leq 0$ **25.** $2x - 3y \leq 4$

26. $3x + 2y < 6$

27. $3x - 2y \leq 5$

28. $2x + 4y > 7$

29. $3x - y < 6$

30. $2x + y \geq 0$

31. $\frac{1}{2}x - \frac{1}{6}y \leq 1$

32. $\frac{2}{3}x + \frac{3}{4}y < \frac{1}{6}$

33. $2.4x - 3.2y \geq 0$

34. $0.3x - 0.5y \leq 0.1$

35. $3y - 6 \geq 0$

36. $2x + y < 6$

37. $y - \frac{1}{3}x < \frac{4}{3}$

38. $\frac{1}{2}y + x > 2$

39. $\frac{1}{4}x - \frac{1}{2}y < 2$

40. $\frac{1}{2}x + \frac{1}{3}y > \frac{1}{4}$

41. $2.1x - 1.2y < 1.1$

42. $0.2x - 0.3y \leq 0.2$

43. $3x \leq 6$

44. $2x - 5 > 0$

45. $6x - y > 2$

46. $2x \leq y + 3$

In Problems 47 and 48, answer each question.

47. Karen is opening a music store. She buys some guitars for $350 each and some violins for $600 each. Her budget allows her to spend no more than $20,000 on her stock of guitars and violins.

    (a) Express this information as an inequality in two variables.

    (b) Sketch a graph of the inequality.

    (c) If she buys 10 violins, how many guitars can she buy?

48. One hour of walking at 3 mph burns 210 calories, whereas 1 hour of riding a bicycle at 5.5 mph uses 380 calories. Suzanne wants to combine walking and riding into a workout routine to burn at least 800 calories. Let x represent the number of hours of walking and y represent the number of hours of bike riding. A linear inequality that represents Suzanne's combined calorie expenditure is $210x + 380y \geq 800$. If Suzanne bikes for 1 hour, how long must she walk to achieve her goal?

Problem 47

Challenge Problems

In Problems 49 through 58, graph the solution set on the x, y plane.

49. $y \leq mx; m > 0$

50. $y < mx; m < 0$

51. $y \leq mx + b; m > 0, b > 0$

52. $y \geq mx + b; m > 0, b < 0$

53. $y > mx + b; m < 0, b > 0$

54. $y < mx + b; m < 0, b < 0$

55. $y \leq a; a > 0$

56. $y \geq a; a < 0$

57. $x \geq a; a > 0$

58. $x < a; a < 0$

In Your Own Words

59. Describe the graph of the inequality $y < mx + b$.

60. Describe the graph of the inequality $x \geq k$.

7.5   Systems of Linear Inequalities in Two Variables

TAPE 14

When two or more linear inequalities are considered simultaneously, a system of linear inequalities is formed. In general, the solution of such a system is a region in the plane where *all* the inequalities are true.

► **EXAMPLE 1**

Graph the solution set of the following system of inequalities.

$$\begin{cases} x - y \le 1 \\ x + y > 0 \end{cases}$$

Solution

First we graph the inequality $x - y \le 1$, as in Section 7.4.

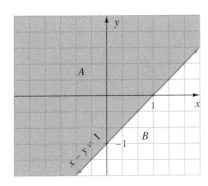

Region	Test Point in Region	Statement $x - y \le 1$	Truth of Statement	Region in Solution Set?
A	$(0, 0)$	$0 - 0 \le 1$	True	Yes
B	$(1, -1)$	$1 - (-1) \le 1$	False	No

So we see that the first inequality of the system is true for all points on, or above, the line $x - y = 1$. However, the solution set of the *system* is the set of points where *both* inequalities are true.

Next, we sketch the graph of $x + y > 0$.

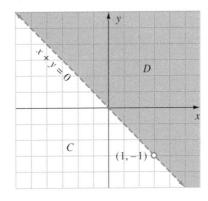

Region	Test Point in Region	Statement $x + y > 0$	Truth of Statement	Region in Solution Set?
C	$(-1, 0)$	$-1 + 0 > 0$	False	No
D	$(0, 1)$	$0 + 1 > 0$	True	Yes

The second inequality of the system is true for all points above the line $x + y = 0$. Still, we are looking for points that satisfy *both* inequalities.

Notice the picture if we graph both solution sets on the *same* coordinate system.

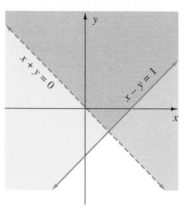

Be Careful!   The intersection of the two solution sets is the solution set for the system. Notice that the lower segment of the solid boundary line *is not* in the solution set of the inequality $x + y > 0$ and must therefore be changed from solid to dashed in the final sketch.

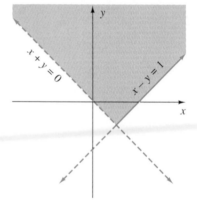

A Procedure for Graphing the Solution Set of a System of Linear Inequalities

1. Graph each inequality on the same coordinate system, lightly shading its solution set.
2. Darken the *intersection* of the lightly shaded regions.
3. Change any portion of any *solid* boundary line *not in the intersection* of the solutions from solid to dashed.

► **EXAMPLE 2**

Graph the solution set for the following system of linear inequalities.

$$\begin{cases} x - y \geq -1 \\ x + y \geq 0 \\ 3x - y < 3 \end{cases}$$

Solution

Sketching each of the inequalities as done in Section 7.4 gives the following graphs:

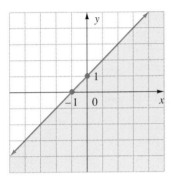

$$x - y \geq -1$$

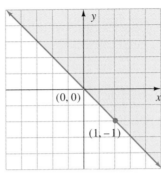

$$x + y \geq 0$$

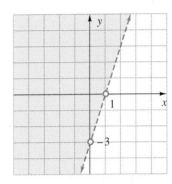

$$3x - y < 3$$

The solution set of the system is the intersection of all the shaded regions.

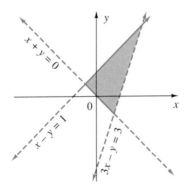

Notice the portions of the solid boundary lines that were changed from solid to dashed.   ◄

► EXAMPLE 3

The Lakeview Linen Company rents and maintains uniforms for a broad range of industrial customers. They manufacture uniforms for their service in two styles, food handling and automotive. The food-handling uniform uses 5 lb of cotton and 1 lb of polyester, while the automotive uniforms use 2 lb of cotton and 4 lb of polyester. There are 500 lb of cotton and 400 lb of polyester on hand. Write a system of inequalities that illustrate the situation and graph the system.

Solution

Let x be the number of food-handling uniforms made. Let y be the number of automotive uniforms made.

Since there are only 500 lb of cotton available and each food-handling uniform uses 5 lb and each automotive requires 2 lb, we have the inequality

$$5x + 2y \leq 500$$

Similarly, since there are 400 lb of polyester,

$$x + 4y \leq 400$$

Thus the situation is illustrated by the following system of inequalities.

$$\begin{cases} 5x + 2y \leq 500 \\ x + 4y \leq 400 \\ x \geq 0 \\ y \geq 0 \end{cases}$$

The last two inequalities are formed because x and y cannot be negative.

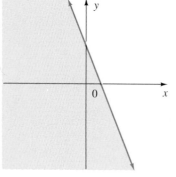

$5x + 2y \leq 500$

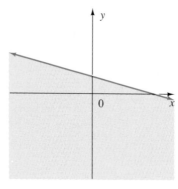

$x + 4y \leq 400$

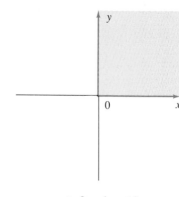

$x \geq 0$ and $y \leq 0$

The solution set of the system is the intersection of all the shaded regions.

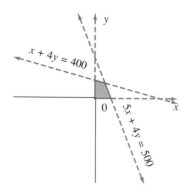

Any point (x, y) in the shaded region satisfies the system of inequalities and represents a production possibility for the Lakeside Linen Company. ◀

Problems such as this form an important branch of applied mathematics called **linear programming.**

Problem Set 7.5

Warm-Ups

In Problems 1 through 10, graph the solution set for each system on the x, y plane. For Problems 1 through 8, see Example 1.

1. $\begin{cases} x - y \le 4 \\ x + y \ge 2 \end{cases}$

2. $\begin{cases} x + 2y > 2 \\ x + y < 1 \end{cases}$

3. $\begin{cases} x + y \le 0 \\ x - y > 0 \end{cases}$

4. $\begin{cases} x - 3y > 6 \\ 2x - y \le 0 \end{cases}$

5. $\begin{cases} 3x - y < 1 \\ x + 2y > 5 \end{cases}$

6. $\begin{cases} 4x + y > 0 \\ x - 4y < 16 \end{cases}$

7. $\begin{cases} 2x + y \ge 3 \\ x - 2y > 4 \end{cases}$

8. $\begin{cases} \frac{1}{2}x + y \ge 2 \\ x - 2y < 6 \end{cases}$

For Problems 9 and 10, see Example 2.

9. $\begin{cases} x - y \ge -2 \\ x + y \le 2 \\ x - 2y \le -2 \end{cases}$

10. $\begin{cases} 2x - y > 0 \\ 2x + y < 2 \\ y > 0 \end{cases}$

Practice Exercises

In Problems 11 through 42, graph the solution set for each system on the x, y plane.

11. $\begin{cases} x - \frac{1}{3}y \le -1 \\ 3x + y \ge -9 \end{cases}$

12. $\begin{cases} 2x - 3y \le 4 \\ 2x - y > -2 \end{cases}$

13. $\begin{cases} x \ge 0 \\ y \ge 0 \end{cases}$

14. $\begin{cases} x > 0 \\ y < 0 \end{cases}$

15. $\begin{cases} x < 4 \\ y > 2 \end{cases}$

16. $\begin{cases} x > -1 \\ y > -1 \end{cases}$

17. $\begin{cases} 2x - y \le 4 \\ 2x - y \ge 5 \end{cases}$

18. $\begin{cases} x + 2y < 5 \\ x + 2y < 7 \end{cases}$

19. $\begin{cases} x + 3y \ge 6 \\ 2x + 6y \le 4 \end{cases}$

20. $\begin{cases} 4x - 3y < 15 \\ 5x + 4y > 11 \end{cases}$

21. $\begin{cases} 2x + 5y > -9 \\ 7x + 3y < 12 \end{cases}$

22. $\begin{cases} 3x + 8y \le 7 \\ 4x - 3y \le 23 \end{cases}$

23. $\begin{cases} x - 3y > 0 \\ x - 3y > 2 \end{cases}$

24. $\begin{cases} 4x + 3y > 16 \\ 4x - 5y > 0 \end{cases}$

25. $\begin{cases} x + 1 \le 0 \\ y + 3 > 0 \end{cases}$

26. $\begin{cases} x - 2 > 4 \\ 2 - y < 3 \end{cases}$

27. $\begin{cases} x - y \le 1 \\ x + 2y \le 1 \end{cases}$

28. $\begin{cases} x + y \ge 0 \\ x - y \ge 4 \end{cases}$

29. $\begin{cases} x - 2y \le 3 \\ 2x - 4y > 1 \end{cases}$

30. $\begin{cases} 2x - y > 4 \\ 6x - 3y > 0 \end{cases}$

31. $\begin{cases} 2x + y < 2 \\ x + \dfrac{1}{2}y < 0 \end{cases}$

32. $\begin{cases} 3x + y < 6 \\ x + \dfrac{1}{3}y > 4 \end{cases}$

33. $\begin{cases} x < 0 \\ y \le 0 \end{cases}$

34. $\begin{cases} x \ge 1 \\ y < 2 \end{cases}$

35. $\begin{cases} x < -3 \\ y \ge 1 \end{cases}$

36. $\begin{cases} x > -4 \\ y \ge -3 \end{cases}$

37. $\begin{cases} \dfrac{1}{2}x - y > 5 \\ x + 2y < 14 \end{cases}$

38. $\begin{cases} x + \dfrac{1}{4}y \le -2 \\ 2x - y \le -4 \end{cases}$

39. $\begin{cases} 3x + 2y < 6 \\ 3x + 2y \ge 1 \end{cases}$

40. $\begin{cases} 4x + 4y > 4 \\ x + y > 2 \end{cases}$

41. $\begin{cases} x - y > -3 \\ x - 2y \le 0 \\ 3x + y \le 3 \end{cases}$

42. $\begin{cases} x + 2y > 2 \\ x + y < 3 \\ x - y \ge -1 \end{cases}$

In Problems 43 and 44, write a system of inequalities that represents the situation and sketch the graph.

43. The Nut Boutique cans two styles of mixed nuts. A can of Deluxe Mixed Nuts contains 1 lb of cashews and 2 lb of peanuts, while a can of Premium Mixed Nuts contains 2 lb of cashews and 1 lb of peanuts. Only 30 lb of cashews and 40 lb of peanuts are available.

44. The High Point Furniture Company manufactures two types of desk–chair sets. The first type uses 10 board feet of pine and 5 board feet of oak, while the other requires 5 board feet of each. There are 800 board feet of pine and 500 board feet of oak available.

Challenge Problems

In Problems 45 through 48, graph the solution set on the x, y plane.

45. $\begin{cases} y \le mx, & m > 0 \\ y < nx, & n < 0 \end{cases}$

46. $\begin{cases} y \le mx + b, & m > 0, b > 0 \\ y \ge nx + b, & n < 0 \end{cases}$

47. $\begin{cases} y < a, & a > 0 \\ x < b, & b > 0 \end{cases}$

48. $\begin{cases} y \ge a, & a < 0 \\ x \le b, & b > 0 \end{cases}$

In Your Own Words

49. Describe the steps in graphing a system of linear inequalities.

7.6   Determinants and Cramer's Rule (Optional)

TAPE 14

A square array of numbers with a bar on each side, such as

$$\begin{vmatrix} 1 & 0 & 2 \\ -3 & 1 & 5 \\ 0 & 4 & 2 \end{vmatrix}$$

is called a **determinant.** The numbers in the array are called **elements** or **entries.** The number of rows (or columns) of a determinant is called the **order** of the determinant. The example above is a third-order determinant. The following examples are second-order and fourth-order determinants.

$$\begin{vmatrix} 2 & 0 \\ 1 & 3 \end{vmatrix}$$

Second-order
determinant

$$\begin{vmatrix} 3 & 0 & 2 & 0 \\ 0 & 9 & 1 & 4 \\ 6 & 0 & -1 & 2 \\ 5 & 1 & 0 & 1 \end{vmatrix}$$

Fourth-order
determinant

A determinant represents a number just as $\sqrt{4}$ represents a number. To evaluate any determinant, we must first learn how to evaluate a second-order determinant.

Value of a Second-order Determinant

$$\begin{vmatrix} a & b \\ c & d \end{vmatrix} = ad - bc$$

► **EXAMPLE 1**

Evaluate the following determinants.

(a) $\begin{vmatrix} 1 & 2 \\ 3 & 4 \end{vmatrix}$ (b) $\begin{vmatrix} 2 & 0 \\ 5 & 3 \end{vmatrix}$ (c) $\begin{vmatrix} 3 & -2 \\ 4 & 2 \end{vmatrix}$ (d) $\begin{vmatrix} -2 & 3 \\ -4 & 6 \end{vmatrix}$

Solutions

(a) $\begin{vmatrix} 1 & 2 \\ 3 & 4 \end{vmatrix} = 1 \cdot 4 - 2 \cdot 3 = 4 - 6 = -2$

(b) $\begin{vmatrix} 2 & 0 \\ 5 & 3 \end{vmatrix} = 2 \cdot 3 - 0 \cdot 5 = 6 - 0 = 6$

(c) $\begin{vmatrix} 3 & -2 \\ 4 & 2 \end{vmatrix} = 3 \cdot 2 - (-2) \cdot 4 = 6 + 8 = 14$

(d) $\begin{vmatrix} -2 & 3 \\ -4 & 6 \end{vmatrix} = (-2) \cdot 6 - 3 \cdot (-4) = -12 + 12 = 0$ ◄

A determinant of order higher than 2 is evaluated by reducing it in successive steps until it is a sum of second-order determinants. To do this, two more definitions are needed.

The **minor** of an element in a determinant is the determinant that remains when we delete the row and the column of the element. For example, the minor of b in the determinant

$$\begin{vmatrix} a & b & c \\ d & e & f \\ g & h & i \end{vmatrix} \quad \text{is} \quad \begin{vmatrix} d & f \\ g & i \end{vmatrix}$$

and the minor of g in the same determinant

$$\begin{vmatrix} a & b & c \\ d & e & f \\ g & h & i \end{vmatrix} \quad \text{is} \quad \begin{vmatrix} b & c \\ e & f \end{vmatrix}$$

The **cofactor** of an element is the minor of the element with the appropriate sign attached. The appropriate sign is taken from the array of signs shown next.

$$\begin{array}{ccc} + & - & + \\ - & + & - \\ + & - & + \end{array} \qquad \begin{array}{cccc} + & - & + & - \\ - & + & - & + \\ + & - & + & - \\ - & + & - & + \end{array} \qquad \begin{array}{ccccc} + & - & + & - & + \\ - & + & - & + & - \\ + & - & + & - & + \\ - & + & - & + & - \\ + & - & + & - & + \end{array}$$

Third-order   Fourth-order   Fifth-order

Notice that the signs alternate, like the squares on a checkerboard, with a $+$ in the upper-left location.

▶ **EXAMPLE 2**

Write the cofactor of b in the determinant

$$\begin{vmatrix} a & b & c \\ d & e & f \\ g & h & i \end{vmatrix}$$

Solution

The minor of b is the determinant that remains when the row and column of b are deleted (the first row and the second column).

$$\begin{vmatrix} a & b & c \\ d & e & f \\ g & h & i \end{vmatrix}$$

$$\text{Minor of } b = \begin{vmatrix} d & f \\ g & i \end{vmatrix}$$

The position of b in the array of signs is as follows:

$$\begin{array}{ccc} + & - & + \\ - & + & - \\ + & - & + \end{array}$$

The cofactor of b is the minor with that sign.

$$\text{Cofactor of } b = - \begin{vmatrix} d & f \\ g & i \end{vmatrix}$$

◄

To Evaluate a Determinant of Order Higher Than Two

1. Select a row or column to expand about.
2. For each element in the chosen row (or column), multiply the element by its cofactor.
3. The value of the determinant is the sum of the products found in step 2.

► EXAMPLE **3**

Evaluate the following determinant by expanding about the first row.

$$\begin{vmatrix} 2 & 2 & -3 \\ 2 & -1 & 1 \\ 1 & 1 & 0 \end{vmatrix}$$

Solution

The minors of the elements in the first row are

STEP 2

$$\begin{vmatrix} -1 & 1 \\ 1 & 0 \end{vmatrix} \qquad \begin{vmatrix} 2 & 1 \\ 1 & 0 \end{vmatrix} \qquad \begin{vmatrix} 2 & -1 \\ 1 & 1 \end{vmatrix}$$

First element   Second element   Third element

The cofactors of the elements in the first row are

$$+ \begin{vmatrix} -1 & 1 \\ 1 & 0 \end{vmatrix} \qquad - \begin{vmatrix} 2 & 1 \\ 1 & 0 \end{vmatrix} \qquad + \begin{vmatrix} 2 & -1 \\ 1 & 1 \end{vmatrix}$$

The products of the elements times their cofactors are

$$2 \begin{vmatrix} -1 & 1 \\ 1 & 0 \end{vmatrix} \qquad -2 \begin{vmatrix} 2 & 1 \\ 1 & 0 \end{vmatrix} \qquad -3 \begin{vmatrix} 2 & -1 \\ 1 & 1 \end{vmatrix}$$

Therefore, the value of the determinant is given by

STEP 3

$$\begin{vmatrix} 2 & 2 & -3 \\ 2 & -1 & 1 \\ 1 & 1 & 0 \end{vmatrix} = 2 \begin{vmatrix} -1 & 1 \\ 1 & 0 \end{vmatrix} - 2 \begin{vmatrix} 2 & 1 \\ 1 & 0 \end{vmatrix} - 3 \begin{vmatrix} 2 & -1 \\ 1 & 1 \end{vmatrix}$$

$$= 2(0 - 1) - 2(0 - 1) - 3(2 - (-1))$$

$$= -2 + 2 - 6 - 3$$

$$= -9$$

Notice that this determinant is easier to evaluate about the *third* row.

$$\begin{vmatrix} 2 & 2 & -3 \\ 2 & -1 & 1 \\ 1 & 1 & 0 \end{vmatrix} = 1 \begin{vmatrix} 2 & -3 \\ -1 & 1 \end{vmatrix} - 1 \begin{vmatrix} 2 & -3 \\ 2 & 1 \end{vmatrix} + 0 \begin{vmatrix} 2 & 2 \\ 2 & -1 \end{vmatrix}$$

$$= 2 - 3 - (2 + 6)$$

$$= 2 - 3 - 8 = -9 \qquad \blacktriangleleft$$

► EXAMPLE 4

Evaluate the following determinant.

$$\begin{vmatrix} 3 & 0 & 2 \\ 1 & 0 & 4 \\ 2 & 1 & 5 \end{vmatrix}$$

Solution

First, we note the two zeros in the second column and decide to expand about that column.

$$\begin{vmatrix} 3 & 0 & 2 \\ 1 & 0 & 4 \\ 2 & 1 & 5 \end{vmatrix} = -0 + 0 - 1 \begin{vmatrix} 3 & 2 \\ 1 & 4 \end{vmatrix}$$

$$= -(12 - 2) = -10 \qquad \blacktriangleleft$$

Evaluation of determinants larger than third order can be very tedious. It pays to look carefully for rows or columns containing zeros.

► EXAMPLE 5

Evaluate

$$\begin{vmatrix} 1 & 1 & 0 & 1 \\ 0 & 1 & 1 & 0 \\ 0 & 1 & 3 & 1 \\ 2 & 0 & 1 & 1 \end{vmatrix}$$

Solution

Notice that the first column has two zero entries. We decide to expand down that column.

$$\begin{vmatrix} 1 & 1 & 0 & 1 \\ 0 & 1 & 1 & 0 \\ 0 & 1 & 3 & 1 \\ 2 & 0 & 1 & 1 \end{vmatrix} = 1 \begin{vmatrix} 1 & 1 & 0 \\ 1 & 3 & 1 \\ 0 & 1 & 1 \end{vmatrix} - 0 + 0 - 2 \begin{vmatrix} 1 & 0 & 1 \\ 1 & 1 & 0 \\ 1 & 3 & 1 \end{vmatrix}$$

Now we have two third-order determinants to evaluate. We choose the first row to expand each.

$$= 1 \begin{vmatrix} 3 & 1 \\ 1 & 1 \end{vmatrix} - 1 \begin{vmatrix} 1 & 1 \\ 0 & 1 \end{vmatrix} - 2 \left(1 \begin{vmatrix} 1 & 0 \\ 3 & 1 \end{vmatrix} - 0 + 1 \begin{vmatrix} 1 & 1 \\ 1 & 3 \end{vmatrix} \right)$$

$$= 3 - 1 - (1 - 0) - 2[(1 - 0) + (3 - 1)]$$

$$= 3 - 1 - 1 - 2(1 + 2)$$

$$= 1 - 2 \cdot 3 = -5$$ ◄

Determinants are an important structure found throughout mathematics. Determinants are used in a formula for the solution of systems of linear equations called **Cramer's rule.**

Cramer's rule is a general method for solving systems of linear equations that uses determinants. Cramer's rule can be used whenever there is the same number of linear equations as there are variables.

Cramer's Rule for Two Equations in Two Variables

The system of equations

$$\begin{cases} Ax + By = P \\ Cx + Dy = Q \end{cases}$$

has the solution

$$x = \frac{\begin{vmatrix} P & B \\ Q & D \end{vmatrix}}{\begin{vmatrix} A & B \\ C & D \end{vmatrix}}, \qquad y = \frac{\begin{vmatrix} A & P \\ C & Q \end{vmatrix}}{\begin{vmatrix} A & B \\ C & D \end{vmatrix}} \qquad \text{provided that} \quad \begin{vmatrix} A & B \\ C & D \end{vmatrix} \neq 0$$

Each equation in the system must first be written in standard form. When using Cramer's rule, it is helpful to write the equations of the system carefully, lining up the same variables under one another.

Notice that the determinant in the denominator is made up of the coefficients of the variables of the original equation in their original positions. We call it the **determinant of coefficients.**

Examination of the determinant of coefficients shows that there is a *column* of numbers for each variable.

$$\downarrow \text{ column of } x\text{-coefficients}$$

$$\text{Determinant of coefficients:} \quad \begin{vmatrix} A & B \\ C & D \end{vmatrix}$$

$$\text{column of } y\text{-coefficients} \uparrow$$

The determinant in the numerator of the formula for x is the determinant of coefficients with the x column replaced by the column of constant terms. The formula for y is arranged in a similar manner.

If the determinant of coefficients is zero, the system is either inconsistent or dependent, and Cramer's rule will not work. When this occurs, solve the system by elimination or substitution.

► **EXAMPLE 6**

Use Cramer's rule to solve the system

$$\begin{cases} 3x + 2y = 4 \\ 4x - y = 2 \end{cases}$$

Solution

By Cramer's rule,

$$x = \frac{\begin{vmatrix} 4 & 2 \\ 2 & -1 \end{vmatrix}}{\begin{vmatrix} 3 & 2 \\ 4 & -1 \end{vmatrix}} = \frac{4(-1) - 2 \cdot 2}{3(-1) - 2 \cdot 4} = \frac{-4 - 4}{-3 - 8} = \frac{-8}{-11},$$

$$y = \frac{\begin{vmatrix} 3 & 4 \\ 4 & 2 \end{vmatrix}}{\begin{vmatrix} 3 & 2 \\ 4 & -1 \end{vmatrix}} = \frac{3 \cdot 2 - 4 \cdot 4}{3(-1) - 2 \cdot 4} = \frac{6 - 16}{-3 - 8} = \frac{-10}{-11}$$

So the solution set is

$$\left\{ \left(\frac{8}{11}, \frac{10}{11} \right) \right\}$$ ◄

Notice that the denominator is the same in the calculation of both variables. This observation saves a little work.

► **EXAMPLE 7**

Solve

$$\begin{cases} 2x - 3y = 4 \\ 5x - 6y = 7 \end{cases}$$

Solution

By Cramer's rule,

$$x = \frac{\begin{vmatrix} 4 & -3 \\ 7 & -6 \end{vmatrix}}{\begin{vmatrix} 2 & -3 \\ 5 & -6 \end{vmatrix}} = \frac{4(-6) - 7(-3)}{2(-6) - 5(-3)} = \frac{-24 + 21}{-12 + 15} = \frac{-3}{3} = -1$$

$$y = \frac{\begin{vmatrix} 2 & 4 \\ 5 & 7 \end{vmatrix}}{3} = \frac{2 \cdot 7 - 4 \cdot 5}{3} = \frac{14 - 20}{3} = \frac{-6}{3} = -2$$

The solution set is

$$\{(-1, -2)\}$$ ◄

Cramer's Rule for Three Equations in Three Variables

The system of equations

$$\begin{cases} Ax + By + Cz = P \\ Dx + Ey + Fz = Q \\ Gx + Hy + Iz = R \end{cases}$$

has solutions

$$x = \frac{\begin{vmatrix} P & B & C \\ Q & E & F \\ R & H & I \end{vmatrix}}{\begin{vmatrix} A & B & C \\ D & E & F \\ G & H & I \end{vmatrix}}, \quad y = \frac{\begin{vmatrix} A & P & C \\ D & Q & F \\ G & R & I \end{vmatrix}}{\begin{vmatrix} A & B & C \\ D & E & F \\ G & H & I \end{vmatrix}}, \quad z = \frac{\begin{vmatrix} A & B & P \\ D & E & Q \\ G & H & R \end{vmatrix}}{\begin{vmatrix} A & B & C \\ D & E & F \\ G & H & I \end{vmatrix}}$$

provided that $\begin{vmatrix} A & B & C \\ D & E & F \\ G & H & I \end{vmatrix} \neq 0$

► EXAMPLE 8

Find the solution set for the following system.

$$\begin{cases} 3x - y = 0 \\ y + 2z = -1 \\ x - z = 0 \end{cases}$$

Solution

First, rewrite the system with the variables "lined up" and with zero coefficients where they occur.

$$\begin{cases} 3x - y + 0z = 0 \\ 0x + y + 2z = -1 \\ x + 0y - z = 0 \end{cases}$$

Next, apply Cramer's rule.

$$x = \frac{\begin{vmatrix} 0 & -1 & 0 \\ -1 & 1 & 2 \\ 0 & 0 & -1 \end{vmatrix}}{\begin{vmatrix} 3 & -1 & 0 \\ 0 & 1 & 2 \\ 1 & 0 & -1 \end{vmatrix}} = \frac{1\begin{vmatrix} -1 & 2 \\ 0 & -1 \end{vmatrix}}{3\begin{vmatrix} 1 & 2 \\ 0 & -1 \end{vmatrix} + 1\begin{vmatrix} 0 & 2 \\ 1 & -1 \end{vmatrix}} = \frac{(-1)(-1) - 2 \cdot 0}{3(-1 - 0) + 0 - 2}$$

$$= -\frac{1}{5}$$

$$y = \frac{\begin{vmatrix} 3 & 0 & 0 \\ 0 & -1 & 2 \\ 1 & 0 & -1 \end{vmatrix}}{-5} = \frac{3\begin{vmatrix} -1 & 2 \\ 0 & -1 \end{vmatrix}}{-5} = \frac{3(1-0)}{-5} = -\frac{3}{5}$$

$$z = \frac{\begin{vmatrix} 3 & -1 & 0 \\ 0 & 1 & -1 \\ 1 & 0 & 0 \end{vmatrix}}{-5} = \frac{1\begin{vmatrix} -1 & 0 \\ 1 & -1 \end{vmatrix}}{-5} = -\frac{1}{5}$$

The solution set is

$$\left\{ \left(-\frac{1}{5}, -\frac{3}{5}, -\frac{1}{5} \right) \right\} \qquad \blacktriangleleft$$

Systems with more than three variables can be solved with Cramer's rule. There must be the same number of equations as there are variables, and all the equations must be linear. The system should be written in the standard form shown above for the three-variable case, being sure to insert zeros where necessary. The determinant of coefficients can then be formed by inspection.

Examination of the determinant of coefficients shows that there is a *column* of numbers for each variable. For example, in the three-variable case

$$\begin{array}{c} \downarrow \text{ column of } y\text{-coefficients} \\ \begin{vmatrix} A & B & C \\ D & E & F \\ G & H & I \end{vmatrix} \end{array}$$

Notice that if we were to replace the column of y-coefficients with the column of constant terms (numbers to the right of the equals sign) we would have the determinant in the numerator of the formula for y. The formulas for x and z are formed in a similar manner. By following that pattern, Cramer's rule can be applied to any linear system of n equations in n variables.

Problem Set 7.6

Warm-Ups

In Problems 1 through 12, evaluate the determinants. For Problems 1 through 6, see Example 1.

1. $\begin{vmatrix} 1 & 2 \\ 2 & 5 \end{vmatrix}$ **2.** $\begin{vmatrix} 2 & 3 \\ 4 & 5 \end{vmatrix}$ **3.** $\begin{vmatrix} 3 & 2 \\ 6 & 4 \end{vmatrix}$

4. $\begin{vmatrix} -2 & 2 \\ 3 & 5 \end{vmatrix}$ **5.** $\begin{vmatrix} 3 & 5 \\ -2 & 4 \end{vmatrix}$ **6.** $\begin{vmatrix} 7 & 5 \\ -3 & -2 \end{vmatrix}$

For Problems 7 through 12, see Examples 3 and 4.

7. $\begin{vmatrix} 1 & 1 & 2 \\ 2 & 3 & 1 \\ 1 & -1 & 1 \end{vmatrix}$ **8.** $\begin{vmatrix} 1 & -1 & 2 \\ 2 & 3 & 0 \\ 1 & 4 & 1 \end{vmatrix}$ **9.** $\begin{vmatrix} 4 & 2 & 1 \\ 0 & 0 & 3 \\ 3 & 5 & 3 \end{vmatrix}$

10. $\begin{vmatrix} 2 & 1 & 3 \\ 4 & 5 & 7 \\ 2 & 1 & 3 \end{vmatrix}$
11. $\begin{vmatrix} 0 & 1 & 0 \\ 0 & 0 & 1 \\ 2 & 0 & 0 \end{vmatrix}$
12. $\begin{vmatrix} 3 & 2 & -1 \\ 5 & 7 & 8 \\ 4 & -5 & 6 \end{vmatrix}$

In Problems 13 through 22, use Cramer's rule to solve each system. For Problems 13 through 18, work Problems 13 through 18 in Problem Set 7.1. See Examples 6 and 7. For Problems 19 through 22, work Problems 1 through 4 in Problem Set 7.3. See Example 8.

Practice Exercises

In Problems 23 through 40, evaluate the determinants.

23. $\begin{vmatrix} 2 & 1 \\ 5 & 3 \end{vmatrix}$
24. $\begin{vmatrix} 3 & 2 \\ 5 & 4 \end{vmatrix}$
25. $\begin{vmatrix} 2 & 6 \\ 3 & 9 \end{vmatrix}$
26. $\begin{vmatrix} -3 & 3 \\ 2 & 4 \end{vmatrix}$

27. $\begin{vmatrix} 4 & 5 \\ -3 & 3 \end{vmatrix}$
28. $\begin{vmatrix} -2 & -3 \\ 6 & 4 \end{vmatrix}$
29. $\begin{vmatrix} 2 & 1 & 1 \\ 1 & 3 & -1 \\ 1 & -2 & 4 \end{vmatrix}$
30. $\begin{vmatrix} 1 & 2 & 3 \\ 2 & -1 & 4 \\ 4 & 0 & 1 \end{vmatrix}$

31. $\begin{vmatrix} 1 & 2 & 0 \\ -1 & -1 & 0 \\ 3 & 4 & 2 \end{vmatrix}$
32. $\begin{vmatrix} 2 & -1 \\ -5 & -7 \end{vmatrix}$
33. $\begin{vmatrix} 2 & 0 \\ 0 & -3 \end{vmatrix}$
34. $\begin{vmatrix} 5 & 0 \\ -1 & 0 \end{vmatrix}$

35. $\begin{vmatrix} 3 & 3 & 4 \\ 2 & 2 & 1 \\ 1 & 1 & 3 \end{vmatrix}$
36. $\begin{vmatrix} 0 & 3 & 0 \\ 1 & 0 & 0 \\ 0 & 0 & 2 \end{vmatrix}$
37. $\begin{vmatrix} -2 & 3 & 5 \\ 4 & 2 & 3 \\ 7 & -3 & 2 \end{vmatrix}$
38. $\begin{vmatrix} -1 & -5 \\ 3 & -2 \end{vmatrix}$

39. $\begin{vmatrix} 4 & 0 \\ 2 & -6 \end{vmatrix}$
40. $\begin{vmatrix} 0 & 0 \\ -3 & 21 \end{vmatrix}$

For Problems 41 through 60, use Cramer's rule, if it applies, to work Problems 61 through 72 in Problem Set 7.1 and Problems 11 through 18 in Problem Set 7.3.

Challenge Problems

In Problems 61 through 63, evaluate the determinants.

61. $\begin{vmatrix} 1 & 1 & 2 & 1 \\ 2 & 1 & 0 & -3 \\ -2 & 0 & 1 & 2 \\ 4 & 0 & 5 & 1 \end{vmatrix}$
62. $\begin{vmatrix} 0 & 1 & 0 & 1 \\ 1 & 1 & 2 & 1 \\ 0 & 0 & 1 & 0 \\ 2 & 3 & 5 & 1 \end{vmatrix}$
63. $\begin{vmatrix} 1 & 1 & 0 & 0 & 0 \\ 0 & 2 & 2 & 0 & 0 \\ 0 & 0 & 3 & 3 & 0 \\ 0 & 0 & 0 & 4 & 4 \\ 5 & 0 & 0 & 0 & 5 \end{vmatrix}$

For Problems 64 through 66, use Cramer's rule to work Problems 37 through 39 in Problem Set 7.3.

In Your Own Words

67. What does a determinant represent?

68. How are third-order determinants evaluated?

7.7 Matrix Methods (Optional)

TAPE 14

A **matrix** is a rectangular array of numbers, usually written inside brackets or parentheses. For example,

$$\begin{bmatrix} 3 & -1 & 4 \\ -5 & 0 & 7 \end{bmatrix}$$

The numbers in the array are called **elements** or **entries.** The position of elements in a matrix is identified by a row number and a column number.

$$\begin{array}{c} \text{Column} \quad 1 \quad 2 \quad 3 \quad 4 \\ \qquad\qquad \downarrow \ \downarrow \quad \downarrow \quad \downarrow \\ \begin{array}{c} 1 \to \\ \text{Row} \quad 2 \to \\ 3 \to \end{array} \begin{bmatrix} -2 & 4 & -9 & 1 \\ 0 & 7 & 1 & -1 \\ 3 & 3 & 2 & -5 \end{bmatrix} \end{array}$$

The number -9 in this matrix is in row 1, column 3. The size (or dimension) of a matrix is given by the number of rows followed by the number of columns (rows *always* before columns). The matrix above has 3 rows and 4 columns, so it is a 3×4 matrix. The matrix

$$\begin{bmatrix} 3 & -1 & 4 \\ -5 & 0 & 7 \end{bmatrix}$$

is a 2×3 matrix.

There are several methods for solving systems of equations that involve matrices (the plural of "matrix"). We will examine one of them, the **method of augmented matrices.**

Suppose that we have a system of three linear equations in three variables, written in standard form.

$$Ax + By + Cz = P$$
$$Dx + Ey + Fz = Q$$
$$Gx + Hy + Iz = R$$

(Remember, some of the coefficients may be zero.) The matrix made up of the coefficients of the variables, in their proper position,

$$\begin{bmatrix} A & B & C \\ D & E & F \\ G & H & I \end{bmatrix}$$

is called the **matrix of coefficients** of the system. If we add a column on the right side of the matrix containing the constant terms of the system, we have the **augmented matrix** of the system.

$$\begin{bmatrix} A & B & C & | & P \\ D & E & F & | & Q \\ G & H & I & | & R \end{bmatrix}$$

▶ **EXAMPLE 1**

Write the augmented matrix of the system

$$\begin{cases} 2x + 3y + 4z = 5 \\ 6x + 7y + 8z = 9 \\ -x + 2y - 3z = 4 \end{cases}$$

Solution

We write the matrix by inspection.

$$\begin{bmatrix} 2 & 3 & 4 & | & 5 \\ 6 & 7 & 8 & | & 9 \\ -1 & 2 & -3 & | & 4 \end{bmatrix}$$

◀

▶ **EXAMPLE 2**

Write the augmented matrix of the system

$$\begin{cases} 2x - z = 3 \\ x + 5y = -7 \\ 6y - 5z + 4 = 0 \end{cases}$$

Solution

First, we write the system in standard form, inserting zeros where necessary.

$$\begin{cases} 2x + 0y - z = 3 \\ x + 5y + 0z = -7 \\ 0x + 6y - 5z = -4 \end{cases}$$

Now we can write the augmented matrix.

$$\begin{bmatrix} 2 & 0 & -1 & | & 3 \\ 1 & 5 & 0 & | & -7 \\ 0 & 6 & -5 & | & -4 \end{bmatrix}$$

◀

▶ **EXAMPLE 3**

Write a system of equations that corresponds to the augmented matrix

$$\begin{bmatrix} 1 & 2 & 3 & | & 4 \\ 3 & -2 & 0 & | & -1 \\ 0 & 1 & -1 & | & 0 \end{bmatrix}$$

Solution

We copy the coefficients from the augmented matrix and form the system

$$\begin{cases} 1x + 2y + 3z = 4 \\ 3x - 2y + 0z = -1 \\ 0x + 1y - 1z = 0 \end{cases}$$

which simplifies to

$$\begin{cases} x + 2y + 3z = 4 \\ 3x - 2y \quad\quad = -1 \\ \quad\quad y - z = 0 \end{cases}$$

◄

We say two augmented matrices are **equivalent** if the systems they represent have the same solution set.

Elementary Row Operations

There are three operations that can be performed on an augmented matrix that will produce an equivalent matrix.

1. Exchange any two rows.
2. Multiply the numbers of any row by the same nonzero number.
3. Multiply the numbers in any row by a real number and add the result to any *other* row.

For example, the following two augmented matrices are equivalent.

$$\begin{bmatrix} 3 & 1 & -7 & | & -2 \\ 1 & 4 & 4 & | & 3 \\ 2 & -1 & 6 & | & 0 \end{bmatrix} \qquad \begin{bmatrix} 1 & 4 & 4 & | & 3 \\ 3 & 1 & -7 & | & -2 \\ 2 & -1 & 6 & | & 0 \end{bmatrix}$$

Matrix A Matrix B

Matrix B was obtained from matrix A by exchanging rows 1 and 2. The following matrices are also equivalent.

$$\begin{bmatrix} 1 & 3 & | & -2 \\ -2 & -4 & | & 2 \end{bmatrix} \qquad \begin{bmatrix} 1 & 3 & | & -2 \\ 1 & 2 & | & -1 \end{bmatrix}$$

Matrix C Matrix D

Matrix D was obtained from matrix C by multiplying row 2 by $-\dfrac{1}{2}$.

$$\begin{bmatrix} 1 & 1 & 3 & 2 & | & 2 \\ 2 & 4 & 1 & -5 & | & 3 \\ 0 & 0 & -3 & -1 & | & -1 \\ 0 & 5 & -6 & 2 & | & 2 \end{bmatrix} \qquad \begin{bmatrix} 1 & 1 & 3 & 2 & | & 2 \\ 0 & 2 & -5 & -9 & | & -1 \\ 0 & 0 & -3 & -1 & | & -1 \\ 0 & 5 & -6 & 2 & | & 2 \end{bmatrix}$$

Matrix E Matrix F

Matrices E and F are also equivalent. Matrix F was obtained by multiplying the first row of matrix E by -2 and adding the result to the second row of matrix E.

Examine the following matrices. They are examples of matrices in **triangular form.**

$$\begin{bmatrix} 1 & 2 & -3 \\ 0 & 1 & 7 \end{bmatrix} \qquad \begin{bmatrix} 1 & -9 & 2 & 0 \\ 0 & 1 & 3 & -4 \\ 0 & 0 & 1 & 2 \end{bmatrix} \qquad \begin{bmatrix} 1 & -7 & -1 & 8 & 6 \\ 0 & 1 & 4 & -1 & 0 \\ 0 & 0 & 1 & 1 & -2 \\ 0 & 0 & 0 & 1 & 4 \end{bmatrix}$$

Notice the number 1 down the main diagonal. Notice that all elements *under* the main diagonal are zeros.

The idea of the method of augmented matrices is to form the augmented matrix of the system to be solved; then write a series of equivalent matrices, ending with a matrix in triangular form. An example will illustrate this.

► EXAMPLE 4

Solve the system

$$\begin{cases} 2x - y - z = 5 \\ x + 2y + 3z = -2 \\ 3x - 2y + z = 2 \end{cases}$$

Solution

First, we write the augmented matrix of the system.

$$\left[\begin{array}{ccc|c} 2 & -1 & -1 & 5 \\ 1 & 2 & 3 & -2 \\ 3 & -2 & 1 & 2 \end{array}\right]$$

Now, to find an equivalent matrix in the triangular form shown above, it is necessary to get a 1 in the upper-left corner. This could be done by multiplying row 1 by $\frac{1}{2}$. However, we can avoid fractions by exchanging rows 1 and 2.

$$\left[\begin{array}{ccc|c} 2 & -1 & -1 & 5 \\ 1 & 2 & 3 & -2 \\ 3 & -2 & 1 & 2 \end{array}\right] \xrightarrow[\text{rows 1 and 2}]{\text{Exchange}} \left[\begin{array}{ccc|c} 1 & 2 & 3 & -2 \\ 2 & -1 & -1 & 5 \\ 3 & -2 & 1 & 2 \end{array}\right]$$

The next step is to get the zeros in column 1. Notice that row 1 multiplied by -2 is

$$-2 \quad -4 \quad -6 \quad 4$$

If we add that to row 2, we get

$$\left[\begin{array}{ccc|c} 1 & 2 & 3 & -2 \\ 2 & -1 & -1 & 5 \\ 3 & -2 & 1 & 2 \end{array}\right] \xrightarrow[-2 \text{ and add to row 2}]{\text{Multiply row 1 by}} \left[\begin{array}{ccc|c} 1 & 2 & 3 & -2 \\ 0 & -5 & -7 & 9 \\ 3 & -2 & 1 & 2 \end{array}\right]$$

We get the zero in row 3 by a similar operation.

$$\left[\begin{array}{ccc|c} 1 & 2 & 3 & -2 \\ 0 & -5 & -7 & 9 \\ 3 & -2 & 1 & 2 \end{array}\right] \xrightarrow[-3 \text{ and add to row 3}]{\text{Multiply row 1 by}} \left[\begin{array}{ccc|c} 1 & 2 & 3 & -2 \\ 0 & -5 & -7 & 9 \\ 0 & -8 & -8 & 8 \end{array}\right]$$

Next, we need a 1 in row 2, column 2. Again we can avoid fractions by exchanging rows 2 and 3.

$$\begin{bmatrix} 1 & 2 & 3 & | & -2 \\ 0 & -5 & -7 & | & 9 \\ 0 & -8 & -8 & | & 8 \end{bmatrix} \xrightarrow[\text{rows 2 and 3}]{\text{Exchange}} \begin{bmatrix} 1 & 2 & 3 & | & -2 \\ 0 & -8 & -8 & | & 8 \\ 0 & -5 & -7 & | & 9 \end{bmatrix}$$

$$\begin{bmatrix} 1 & 2 & 3 & | & -2 \\ 0 & -8 & -8 & | & 8 \\ 0 & -5 & -7 & | & 9 \end{bmatrix} \xrightarrow[\text{2 by } \frac{-1}{8}]{\text{Multiply row}} \begin{bmatrix} 1 & 2 & 3 & | & -2 \\ 0 & 1 & 1 & | & -1 \\ 0 & -5 & -7 & | & 9 \end{bmatrix}$$

We finish column 2 by multiplying row 2 by 5 and adding it to row 3.

$$\begin{bmatrix} 1 & 2 & 3 & | & -2 \\ 0 & 1 & 1 & | & -1 \\ 0 & -5 & -7 & | & 9 \end{bmatrix} \xrightarrow[\text{5 and add to row 3}]{\text{Multiply row 2 by}} \begin{bmatrix} 1 & 2 & 3 & | & -2 \\ 0 & 1 & 1 & | & -1 \\ 0 & 0 & -2 & | & 4 \end{bmatrix}$$

Now, if we multiply row 3 by $-\dfrac{1}{2}$, we are finished.

$$\begin{bmatrix} 1 & 2 & 3 & | & -2 \\ 0 & 1 & 1 & | & -1 \\ 0 & 0 & -2 & | & 4 \end{bmatrix} \xrightarrow[\text{3 by } \frac{-1}{2}]{\text{Multiply row}} \begin{bmatrix} 1 & 2 & 3 & | & -2 \\ 0 & 1 & 1 & | & -1 \\ 0 & 0 & 1 & | & -2 \end{bmatrix}$$

The reason we wanted the triangular form can be seen when we write the system for this augmented matrix.

$$\begin{cases} x + 2y + 3z = -2 \\ y + z = -1 \\ z = -2 \end{cases}$$

We automatically have the value for z, and we can find the values for y and then x in successive steps.

$$y + z = -1$$
$$y + (-2) = -1$$
$$y = 1$$
$$x + 2y + 3z = -2$$
$$x + 2(1) + 3(-2) = -2$$
$$x + 2 - 6 = -2$$
$$x = 2$$

So we have the solution set.

$$\{(2, 1, -2)\}$$

> **The Method of Augmented Matrices**
>
> 1. Write the system in standard form, inserting zeros where necessary. Form the augmented matrix.
> 2. Using elementary row operations, change the matrix to triangular form. It is *important* to make the changes in the following order.
>   (a) Get the number 1 in row 1, column 1.
>   (b) Get zeros in column 1, under the 1 in row 1.
>   (c) Get 1 in row 2, column 2.
>   (d) Get zeros in column 2, under the 1 in row 2.
>   (e) Continue steps similar to (c) and (d) until the matrix is in triangular form.
> 3. Write the system associated with the triangular matrix.
> 4. Find the values of each of the variables in turn.
> 5. Write the solution set.

► **EXAMPLE 5**

Solve the system

$$\begin{cases} 2x + 3y = -1 \\ 3x + 2y = 1 \end{cases}$$

Solution

$$\begin{bmatrix} 2 & 3 & | & -1 \\ 3 & 2 & | & 1 \end{bmatrix} \qquad \text{the augmented matrix}$$

$$\begin{bmatrix} 2 & 3 & | & -1 \\ 3 & 2 & | & 1 \end{bmatrix} \xrightarrow[\text{1 by } \frac{1}{2}]{\text{Multiply row}} \begin{bmatrix} 1 & \frac{3}{2} & | & -\frac{1}{2} \\ 3 & 2 & | & 1 \end{bmatrix}$$

$$\begin{bmatrix} 1 & \frac{3}{2} & | & -\frac{1}{2} \\ 3 & 2 & | & 1 \end{bmatrix} \xrightarrow[\text{-3 and add to row 2}]{\text{Multiply row 1 by}} \begin{bmatrix} 1 & \frac{3}{2} & | & -\frac{1}{2} \\ 2 & -\frac{5}{2} & | & \frac{5}{2} \end{bmatrix}$$

$$\begin{bmatrix} 1 & \frac{3}{2} & | & -\frac{1}{2} \\ 0 & -\frac{5}{2} & | & \frac{5}{2} \end{bmatrix} \xrightarrow[\text{2 by } \frac{-2}{5}]{\text{Multiply row}} \begin{bmatrix} 1 & \frac{3}{2} & | & -\frac{1}{2} \\ 0 & 1 & | & -1 \end{bmatrix}$$

The matrix is in triangular form. The associated system is

$$x + \frac{3}{2}y = -\frac{1}{2}$$

$$y = -1$$

and x is obtained from

$$x + \frac{3}{2} \cdot (-1) = -\frac{1}{2}$$

$$x - \frac{3}{2} = -\frac{1}{2}$$

$$x = -\frac{1}{2} + \frac{3}{2} = \frac{2}{2}$$

$$= 1$$

The solution set is

$$\{(1, -1)\}$$

◄

► **EXAMPLE 6**

Solve

$$\begin{cases} 3x - 4y + 3z = 2 \\ x - y + z = 1 \\ x - 2y + z = 0 \end{cases}$$

Solution

Write the augmented matrix.

$$\begin{bmatrix} 3 & -4 & 3 & | & 2 \\ 1 & -1 & 1 & | & 1 \\ 1 & -2 & 1 & | & 0 \end{bmatrix} \quad \text{the augmented matrix}$$

$$\begin{bmatrix} 3 & -4 & 3 & | & 2 \\ 1 & -1 & 1 & | & 1 \\ 1 & -2 & 1 & | & 0 \end{bmatrix} \xrightarrow[\text{rows 1 and 2}]{\text{Exchange}} \begin{bmatrix} 1 & -1 & 1 & | & 1 \\ 3 & -4 & 3 & | & 2 \\ 1 & -2 & 1 & | & 0 \end{bmatrix}$$

$$\begin{bmatrix} 1 & -1 & 1 & | & 1 \\ 3 & -4 & 3 & | & 2 \\ 1 & -2 & 1 & | & 0 \end{bmatrix} \xrightarrow[-3 \text{ and add to row 2}]{\text{Multiply row 1 by}} \begin{bmatrix} 1 & -1 & 1 & | & 1 \\ 0 & -1 & 0 & | & -1 \\ 1 & -2 & 1 & | & 0 \end{bmatrix}$$

$$\begin{bmatrix} 1 & -1 & 1 & | & 1 \\ 0 & -1 & 0 & | & -1 \\ 1 & -2 & 1 & | & 0 \end{bmatrix} \xrightarrow[-1 \text{ and add to row 3}]{\text{Multiply row 1 by}} \begin{bmatrix} 1 & -1 & 1 & | & 1 \\ 0 & -1 & 0 & | & -1 \\ 0 & -1 & 0 & | & -1 \end{bmatrix}$$

$$\begin{bmatrix} 1 & -1 & 1 & | & 1 \\ 0 & -1 & 0 & | & -1 \\ 0 & -1 & 0 & | & -1 \end{bmatrix} \xrightarrow[2 \text{ by } -1]{\text{Multiply row}} \begin{bmatrix} 1 & -1 & 1 & | & 1 \\ 0 & 1 & 0 & | & 1 \\ 0 & -1 & 0 & | & -1 \end{bmatrix}$$

$$\begin{bmatrix} 1 & -1 & 1 & | & 1 \\ 0 & 1 & 0 & | & -1 \\ 0 & -1 & 0 & | & -1 \end{bmatrix} \xrightarrow[1 \text{ and add to row 3}]{\text{Multiply row 2 by}} \begin{bmatrix} 1 & -1 & 1 & | & 1 \\ 0 & 1 & 0 & | & 1 \\ 0 & 0 & 0 & | & 0 \end{bmatrix}$$

Notice the complete row of zeros, which will not allow us to finish the triangular form. This indicates that the original system was dependent.   ◄

As noted in Example 6, a complete row of zeros anywhere in the matrix signals a dependent system. A row of all zeros *except* a nonzero entry in the last column signals an inconsistent system. That is, rows anywhere in the matrix, such as

$$[0 \quad 0 \quad 0 \mid 5] \quad \text{or} \quad [0 \quad 0 \mid -2] \quad \text{or} \quad [0 \quad 0 \quad 0 \quad 0 \quad 0 \mid 19]$$

indicate inconsistent systems.

Problem Set 7.7

Warm-Ups

In Problems 1 through 6, write the augmented matrix for the system of linear equations. See Examples 1 and 2.

1. $\begin{cases} 2x + 3y = -4 \\ 5x - 6y = 1 \end{cases}$

2. $\begin{cases} 3x - y = 2 \\ x + 7y + 3 = 0 \end{cases}$

3. $\begin{cases} 2x - 3y + 4z = 1 \\ 5x - y + z = 0 \\ 7x + 6y - 2z = 5 \end{cases}$

4. $\begin{cases} x + 2y = -5 \\ 3x - 4z = 1 \\ y + 2z = 10 \end{cases}$

5. $\begin{cases} x + 2y = z \\ -x - 11 = y \\ y - 13 = 4z \end{cases}$

6. $\begin{cases} 2x + 3y - 4z + w = 17 \\ 5x + 6z - w = 8 \\ 7x - 2y - 9w = 5 \\ y + 2z + 2w = -1 \end{cases}$

In Problems 7 through 10, write the linear system associated with the matrix. See Example 3.

7. $\begin{bmatrix} 1 & 2 & | & 4 \\ 9 & -8 & | & -1 \end{bmatrix}$

8. $\begin{bmatrix} 1 & -1 & 0 & | & 7 \\ 0 & 1 & 1 & | & 1 \\ 0 & 0 & 1 & | & -4 \end{bmatrix}$

9. $\begin{bmatrix} 1 & 0 & 0 & | & 5 \\ 0 & 1 & 0 & | & -3 \\ 0 & 0 & 1 & | & -2 \end{bmatrix}$

10. $\begin{bmatrix} 1 & -2 & 3 & -4 & | & 5 \\ 4 & -1 & 1 & 0 & | & 2 \\ 2 & 1 & 0 & 0 & | & 1 \\ -1 & 1 & -1 & 1 & | & -1 \end{bmatrix}$

In Problems 11 through 16, solve the systems by the method of augmented matrices. See Examples 4, 5, and 6.

11. $\begin{cases} x + 3y = 7 \\ -2x + y = 0 \end{cases}$

12. $\begin{cases} 2x + 5y = 8 \\ x + 2y = 3 \end{cases}$

13. $\begin{cases} 2x + 4y = 2 \\ 3x + 7y = 1 \end{cases}$

14. $\begin{cases} 2x + 5y = 4 \\ 3x - 6y = 33 \end{cases}$

15. $\begin{cases} x + 2y + 2z = 3 \\ 2x + 3y + 6z = 2 \\ -x + y + z = 0 \end{cases}$

16. $\begin{cases} 2x + 5y + 2z = 9 \\ x + 3y - z = 0 \\ 2x + 3y - 3z = 1 \end{cases}$

Practice Exercises

In Problems 17 through 22, write the augmented matrix for the system of linear equations.

17. $\begin{cases} 3x + 2y = 5 \\ 6x - 5y = 3 \end{cases}$

18. $\begin{cases} 2x - 2y = 1 \\ -x + y - 2 = 0 \end{cases}$

19. $\begin{cases} 3x - 2y + 5z = 4 \\ 2x + 3y - z = 0 \\ x - y - 2z = 1 \end{cases}$

20. $\begin{cases} x - 3y = 4 \\ 2x - z = 3 \\ -y - 2z = 13 \end{cases}$

21. $\begin{cases} x + 3z = 2y \\ -x - 10 = z \\ y + 3 = 4x - z \end{cases}$

22. $\begin{cases} x - 3y + 4z + 2w = 11 \\ 4x - 5z + 4w = 7 \\ 5x - y - w = 13 \\ -x + 4y + 2z = 9 \end{cases}$

In Problems 23 through 26, write the linear system associated with the matrix.

23. $\begin{bmatrix} 2 & 1 & | & 3 \\ 6 & -5 & | & 11 \end{bmatrix}$

24. $\begin{bmatrix} 1 & 0 & -1 & | & 6 \\ 0 & 1 & -1 & | & 9 \\ 0 & 0 & 1 & | & -6 \end{bmatrix}$

25. $\begin{bmatrix} 0 & 0 & 1 & | & -3 \\ 0 & 1 & 0 & | & 4 \\ 1 & 0 & 0 & | & 2 \end{bmatrix}$

26. $\begin{bmatrix} 1 & 0 & 0 & 1 & | & -3 \\ 0 & 1 & -3 & 0 & | & 2 \\ 0 & 0 & 1 & -2 & | & -5 \\ 0 & 0 & 0 & 1 & | & 2 \end{bmatrix}$

In Problems 27 through 36, solve each system by the method of augmented matrices.

27. $\begin{cases} x - 2y = -1 \\ -3x + 8y = 5 \end{cases}$

28. $\begin{cases} -3x + y = 2 \\ x + 3y = -4 \end{cases}$

29. $\begin{cases} 3x - 9y = 3 \\ 2x - 5y = 4 \end{cases}$

30. $\begin{cases} 3x - 7y = 1 \\ -2x + 10y = 10 \end{cases}$

31. $\begin{cases} x + 3y - 2z = 1 \\ 2x + 5y - 2z = 6 \\ -2x - 4y + 3z = -1 \end{cases}$

32. $\begin{cases} 5x - 9y - 7z = 7 \\ 3x - 4y + z = 7 \\ x - 2y - z = 1 \end{cases}$

33. $\begin{cases} 2x - 6y + 4z = 0 \\ 3x + y - 4z = 10 \\ 4x + 3y - 9z = 13 \end{cases}$

34. $\begin{cases} 2x - 3y - 2z = -3 \\ 3x - 2y + 5z = 14 \\ 3x + 6y - 3z = 6 \end{cases}$

35. $\begin{cases} 3x + 9y - 6z = -15 \\ 2x + 3y + 2z = 2 \\ 5x + 8y + 14z = 13 \end{cases}$

36. $\begin{cases} -3x + 5y + 2z = 5 \\ 2x + 3y = 13 \\ 5x - 5y + 10z = -25 \end{cases}$

For Problems 37 through 56, solve the systems in Problems 61 through 72 in Problem Set 7.1 and Problems 11 through 18 in Problem Set 7.3 by the method of augmented matrices.

Challenge Problems

For Problems 57 through 60, use the method of augmented matrices to work Problems 37 through 40 in Problem Set 9.3.

In Your Own Words

61. How do we form the augmented matrix of a system of linear equations?

62. What are the elementary row operations?

▶ Chapter Summary

GLOSSARY

System of equations   Two or more equations considered simultaneously.

A **solution** of a system of equations with n variables   An ordered n-tuple of numbers that satisfies all the equations in the system.

Inconsistent system   A system of equations that has no solutions.

Dependent system   A system of equations that has an infinite number of solutions.

Boundary line   A line separating the regions of an inequality in two variables.

OPTIONAL SECTIONS

Determinant   A square array of numbers that represents a number.

Matrix   A rectangular array of numbers.

Augmented matrix   A matrix that represents a system of equations.

Systems of Equations

Graphical Method

Graph both equations and estimate the coordinates of their point(s) of intersection. This method provides an approximation to the solution(s).

Method of Substitution

1. Solve one of the equations for one of the variables.
2. Substitute this for the same variable in the other equation.
3. Solve the resulting equation.
4. Substitute the result back into step 1 to find the other variable.
5. Check solution(s), if required.
6. Write the solution set.

Method of Elimination

1. Write both equations in standard form, $Ax + By = C$.
2. Multiply both sides of each equation by a suitable real number so that one of the variables will be eliminated by addition of the equations. (This step may not be necessary.)
3. Add the equations and solve the resulting equation.
4. Substitute the value found in step 3 into one of the original equations, and solve this equation.
5. Check the solution, if required.
6. Write the solution set.

Inequalities in Two Variables

1. Graph the equation formed by replacing the inequality symbol with an equals symbol.
   (a) Draw a solid line if equality is included.   ($\geq$, $\leq$)
   (b) Draw a dashed line if equality is not included.   ($>$, $<$)

2. Determine which region(s) formed by the line makes, the inequality true by testing with one point from inside each region.

3. Shade the region(s) that make the inequality true.

Systems of Inequalities in Two Variables

1. Graph each inequality on the same coordinate system, lightly shading its solution set.

2. Darken the *intersection* of the lightly shaded regions.

3. Change any portion of any solid boundary line not in the intersection of the solutions from solid to dashed.

CHECK UPS

1. Use the method of substitution to solve

$$\begin{cases} 2x + y = 2 \\ 6x - 4y = -1 \end{cases}$$

Section 7.1; Example 1a

2. Use the method of elimination to solve

$$\begin{cases} 2x + 5y = 11 \\ 3x + 7y = 15 \end{cases}$$

Section 7.1; Example 7

3. Gloria bought 3 lb of raw peanuts and 2 lb of boiled peanuts from Big Bob's produce stand for $4.05. Jim bought 1 lb of raw and 4 lb of boiled for $4.65. What is the price per pound for raw and boiled peanuts at Big Bob's?

Section 7.2; Example 2

4. Solve the system

$$\begin{cases} x + 2y + z = 0 \\ 2x - 2y + z = -2 \\ 3x + 2y - 2z = 3 \end{cases}$$

Section 7.3; Example 2

5. Graph the solution set for $x - 2y > 0$.

Section 7.4; Example 1

6. Graph the solution set for the system of inequalities

$$\begin{cases} x - y \le 1 \\ x + y > 0 \end{cases}$$

Section 7.5; Example 1

OPTIONAL SECTIONS

7. Use Cramer's rule to solve

$$\begin{cases} 3x + 2y = 4 \\ 4x - y = 2 \end{cases}$$

Section 7.6; Example 6

8. Solve by the method of augmented matrices.

$$\begin{cases} 2x + 3y = -1 \\ 3x + 2y = 1 \end{cases}$$

Section 7.7; Example 5

► Review Problems

In Problems 1 through 6, use the method of substitution to find the solution set for each system.

1. $\begin{cases} 2x + y = 4 \\ 3x - 2y = -1 \end{cases}$

2. $\begin{cases} 2x - 3y = 7 \\ x - 2y = 4 \end{cases}$

3. $\begin{cases} 5x + 3y = 1 \\ 7x + 2y = 8 \end{cases}$

4. $\begin{cases} 6x - 3y = 2 \\ 9x + 4y = 3 \end{cases}$ **5.** $\begin{cases} x^2 - 4y^2 = 9 \\ x - y = 3 \end{cases}$ **6.** $\begin{cases} x - 6y = 5 \\ xy = 11 \end{cases}$

In Problems 7 through 25, use the method of elimination to find the solution set for each system.

7. $\begin{cases} x + 2y = 2 \\ x - 2y = 6 \end{cases}$ **8.** $\begin{cases} 3x + y = 1 \\ 4x + y = -1 \end{cases}$ **9.** $\begin{cases} 2x + 3y = 15 \\ 2x - 7y = 5 \end{cases}$

10. $\begin{cases} 2x - 3y = -5 \\ 3x + 2y = -1 \end{cases}$ **11.** $\begin{cases} 5x + 6y = -3 \\ -4x + 9y = 7 \end{cases}$ **12.** $\begin{cases} 5x + 4y = 12 \\ -6x + 8y = -8 \end{cases}$

13. $\begin{cases} 24x - 18y = -1 \\ 6x - 10y = -3 \end{cases}$ **14.** $\begin{cases} -x + 3y = 4 \\ 2x - 6y = -8 \end{cases}$

15. $\begin{cases} 5x + 4y = 2 \\ 10x + 12y = 5 \end{cases}$ **16.** $\begin{cases} x + y - z = -3 \\ x - y + z = -1 \\ x + y + z = -1 \end{cases}$ **17.** $\begin{cases} x - 2y + z = 2 \\ x + y + z = 8 \\ x - y - z = 2 \end{cases}$

18. $\begin{cases} x - y - z = 0 \\ x + 2y - z = 3 \\ x + y + 2z = 5 \end{cases}$ **19.** $\begin{cases} 2x + 3y + 5z = 0 \\ x - 2y - z = 0 \\ 3x + 2y - 2z = 0 \end{cases}$ **20.** $\begin{cases} x - 2y + z = 17 \\ 2x + 3y + 2z = 13 \\ 3x + y - z = 14 \end{cases}$

21. $\begin{cases} 2x + 5y - z = -6 \\ x - y + 3z = 18 \\ 2x + 3y - 2z = -12 \end{cases}$ **22.** $\begin{cases} 3x - y + 4z = 9 \\ x - 2y + 3z = -2 \\ 2x - y + z = 3 \end{cases}$

23. $\begin{cases} x + 3y + z = 1 \\ 2x + y + 2z = 2 \\ 3x - y + 2z = 0 \end{cases}$ **24.** $\begin{cases} 2x - 3y + 4z = 4 \\ 3x - y + z = -2 \\ x + 2y - 3z = -6 \end{cases}$ **25.** $\begin{cases} x - y + 3z = 6 \\ 2x + y - z = -3 \\ 3x - 2y + z = -4 \end{cases}$

26. The sum of two numbers is 12. Their difference is 2. Find the numbers.

27. Jane, Brenda, and Virginia bought melons at Hasty-Mart on the same day. Jane bought 3 cantaloupes, 4 honeydews, and 1 watermelon for $11.92. Brenda bought 1 cantaloupe, 1 honeydew, and 1 watermelon for $4.77. Virginia bought 1 cantaloupe and 2 watermelons for $5.17. What was the price of each type of melon?

28. Jon and Tamra know that they can paint their house in 8 hours if they work together. This year Jon painted alone for 6 hours; then Tamra finished it in 12 hours. How long would it have taken each working alone to paint the house?

29. The sum of the squares of two numbers is 58 and their difference is 10. Find one such pair of numbers.

In Problems 30 through 53, graph the solution set on the x,y plane.

30. $x - y \leq 4$ **31.** $x + 2y \geq 3$ **32.** $2x - y \geq 3$ **33.** $x - y < 0$

34. $2x + y < 1$ **35.** $y < 3x - 2$ **36.** $\begin{cases} x - y \leq 2 \\ x + y \geq 0 \end{cases}$ **37.** $\begin{cases} 2x + y < 7 \\ x - y > 5 \end{cases}$

38. $\begin{cases} 2x + y \leq 3 \\ x - 2y > 9 \end{cases}$ **39.** $\begin{cases} y \leq 2 \\ x \geq 0 \end{cases}$ **40.** $y > x$ **41.** $2x \leq y + 3$

42. $x + 2y < 3$ **43.** $y < 2x$ **44.** $\begin{cases} 2x + y < 7 \\ 4x + 2y < 14 \end{cases}$ **45.** $\begin{cases} x - y > 7 \\ x - y < 5 \end{cases}$

46. $2x + 3y \leq 6$ **47.** $2x - 3y \leq 4$ **48.** $x - 3y \leq 1$ **49.** $5x - 7y > 0$

50. $3x + 2y \leq 4$ **51.** $\frac{1}{2}x - y < 1$ **52.** $\begin{cases} 3x + 2y \leq 1 \\ x - y \leq 2 \end{cases}$ **53.** $\begin{cases} 2x - 5y \leq 5 \\ 4x - 10y \leq 0 \end{cases}$

Cramer's Rule

For Problems 54 through 72, use Cramer's rule to solve the systems in Problems 7 through 25 above.

Augmented Matrices

For Problems 73 through 91, use the method of augmented matrices to solve the systems in Problems 7 through 25 above.

Let's Not Forget . . .

Identify the expressions that are in factored form. Factor those that are not factored.

92. $125 - 45s^2$ **93.** $(4 - d)^3$

94. $(x - 2)(x^2 + 2x + 4)$ **95.** $x^3 + 8$

How many terms are in each expression? Which expressions have x + 2 as a factor?

96. $x^4 - 16\ 2$;

97. $1 + (x + 2)^3$

98. $8 + x^3\ 2$;

99. $ax - bx + 2a - 2b$

Simplify each expression, if possible. Leave only nonnegative exponents in the answer.

100. -4^{-2}

101. -2^{-4}

102. $(-3)^{-2}$

103. $(-2)^{-3}$

104. $\dfrac{-2k^{-2}}{3b^3c^{-1}}$

105. $\left(\dfrac{8x^{-6}y^9}{27y^3}\right)^{-1/3}$

Reduce, if possible.

106. $\dfrac{4 \pm \sqrt{8}}{2}$

107. $\dfrac{a - (x + b)^2}{a^2 + (x + b)^2}$

The following problems can be worked by using a least common denominator. Follow the directions in each and notice how the LCD is used.

108. Perform the operation indicated: $\dfrac{5}{x + 1} + \dfrac{4x - 1}{1 - x^2}$.

109. Solve $\dfrac{1}{1 - x} + \dfrac{2}{x^2 - 1} = 1$.

110. Simplify $\dfrac{\dfrac{1}{x + 1} - \dfrac{1}{x - 1}}{\dfrac{1}{x + 1} + \dfrac{1}{x - 1}}$.

Classify each problem as an expression, equation, or inequality. Solve the equations and inequalities, and perform the operations indicated on the expressions.

111. $|5 - x| < 3$

112. $(-x)^2 = -x^2 + 2$

113. $\sqrt[3]{8x^3 + 64x^6}$

114. $\dfrac{x^2 + 1}{x^2 - 2} + \dfrac{x^2 - 1}{2 - x^2}$

► You Decide

Your boat is docked by your fish-camp on the Arkansas River, 24 miles downstream from Morrison Bluff. The current runs a constant 4 mph on that section of the river. You have to drop some papers off at the Morrison Bluff dock tomorrow and return to the camp by noon. You decide to run your boat at a constant speed for the entire trip. What time do you leave? What speed do you run the boat?

► Chapter Text

In Problems 1 through 5, choose the correct answer.

1. The solution set for the system $\begin{cases} 3x + y = 1 \\ y = x - 1 \end{cases}$ is (?)

A. $\varnothing$

B. $\{0\}$

C. $\{(0, -1)\}$

D. $\left\{\left(\dfrac{1}{2}, -\dfrac{1}{2}\right)\right\}$

2. The solution set for the system $\begin{cases} 3x + 2y = -1 \\ 3x - 4y = 11 \end{cases}$ is (?)

A. $\{(-1, -2)\}$ **B.** $\{(1, -2)\}$

C. $\varnothing$

D. $\left\{\left(\dfrac{13}{9}, \dfrac{5}{3}\right)\right\}$

3. The solution set for the system $\begin{cases} x + y = -1 \\ y + z = 1 \\ x + z = 4 \end{cases}$ is (?)

A. Ø **B.** $\{(0, -1, 4)\}$
C. $\{2, -3, 4)\}$ **D.** $\{(1, -2, 3)\}$

4. The system $\begin{cases} x + y = -1 \\ 2x + 4y + z = 1 \\ 2y + z = 1 \end{cases}$ is best described as

(?)

A. dependent **B.** inconsistent
C. independent **D.** consistent

In Problems 6 through 9, sketch the graph.

6. $x + 2y \le 2$

7. $2x - y > 0$

5. The solution set for the system $\begin{cases} 2x + 3y = 1 \\ 4x + 6y = 0 \end{cases}$ is (?)

A. $\{(-1, 1)\}$ **B.** $\{(0, -2)\}$
C. Ø **D.** None of the above

8. $\begin{cases} x - 2y \ge -2 \\ 2x - y \ge 0 \end{cases}$

9. $\begin{cases} x + y \ge 1 \\ x < 1 \end{cases}$

10. Solve the system

$$\begin{cases} x + y = 3 \\ x^2 + y^2 = 17 \end{cases}$$

11. Use the method of *substitution* to solve the system

$$\begin{cases} 3x - 5y = 6 \\ 6x + y = 1 \end{cases}$$

12. Use the method of *elimination* to solve the system

$$\begin{cases} 7x + 2y = 3 \\ 9x + 3y = 3 \end{cases}$$

13. Find the solution set for the system

$$\begin{cases} y = 3x - 6 \\ 9x - 3y = 18 \end{cases}$$

14. Find the solution set for the system

$$\begin{cases} 3x + 2y = 1 \\ y - 3z = -10 \\ x + y + z = 3 \end{cases}$$

15. If 3 lb of bananas and 2 lb of apples cost $2.97, while one lb of apples and 5 lb of bananas cost $2.64, how much does a pound of bananas cost?

Optional Sections

16. Evaluate $\begin{vmatrix} 1 & 2 & 1 \\ 2 & 0 & -1 \\ 3 & 1 & 0 \end{vmatrix}$.

17. Use Cramer's rule to solve

$$\begin{cases} 2x + 3y = 1 \\ 3x + 2y = 2 \end{cases}$$

18. Use Cramer's rule to solve

$$\begin{cases} x + 2y + z = 0 \\ -y + z = 1 \\ 2x + 3y = 2 \end{cases}$$

19. Use the method of augmented matrices to solve

$$\begin{cases} 2x + y = 1 \\ x - 7y = -2 \end{cases}$$

20. Use the method of augmented matrices to solve

$$\begin{cases} 2x - z = 1 \\ x + y = 2 \\ x + 3y - 4z = -2 \end{cases}$$

8 Exponential and Logarithmic Functions

See Problem Set 8.3, Exercise 61.

▶ Connections

In Chapter 6 we introduced the concept of a function and learned how to use functional notation. In this chapter we expand on those ideas and investigate two of the most widely used functions, exponential and logarithmic functions. We define exponential functions and develop their connection to logarithmic functions, and then we go on to investigate the powerful properties of these functions.

Logarithms were developed by Henry Briggs and John Napier in the nineteenth century as a tool to aid in tedious computations with large numbers. Calculators and computers have made the use of logarithms for computations obsolete. However, logarithms have found many other applications and occur in such diverse roles as the Richter scale for measuring earthquake intensity and the calculation in interest rates.

8.1   Algebra of Functions

TAPE 14

In Section 6.5 we introduced the concept of a function. We continue our study of functions in this section.

If f and g are functions, we can make new functions by adding, subtracting, multiplying, and dividing the rules for f and g.

Operations with Functions

1. $(f + g)(x) = f(x) + g(x)$ Sum

2. $(f - g)(x) = f(x) - g(x)$ Difference

3. $(fg)(x) = f(x)g(x)$ Product

4. $\left(\dfrac{f}{g}\right)(x) = \dfrac{f(x)}{g(x)}, \quad g(x) \neq 0$ Quotient

---

▶ EXAMPLE 1

If $f(x) = x^2$ and $g(x) = x + 2$, find:

(a) $(f + g)(x)$ (b) $(f - g)(x)$

(c) $(fg)(x)$ (d) $\left(\dfrac{f}{g}\right)(x)$

Solutions

(a) $(f + g)(x) = f(x) + g(x)$
$$= x^2 + (x + 2)$$
$$= x^2 + x + 2$$

(b) $(f - g)(x) = f(x) - g(x)$
$$= x^2 - (x + 2)$$
$$= x^2 - x - 2$$

(c) $(fg)(x) = f(x)g(x)$
$$= x^2(x + 2)$$
$$= x^3 + 2x^2$$

(d) $\left(\dfrac{f}{g}\right)(x) = \dfrac{f(x)}{g(x)}$
$$= \dfrac{x^2}{x + 2}, \qquad x \neq -2$$

◀

► **EXAMPLE 2**

If $g(x) = \sqrt{x}$ and $h(x) = x^2$, find:

(a) $(g + h)(2)$ (b) $\left(\dfrac{h}{g}\right)(4)$

Solutions

(a) $(g + h)(2) = g(2) + h(2)$
$$= \sqrt{2} + 4$$

(b) $\left(\dfrac{h}{g}\right)(4) = \dfrac{h(4)}{g(4)}$
$$= \dfrac{4^2}{\sqrt{4}}$$
$$= \dfrac{16}{2}$$
$$= 8$$

◀

Composition of Functions

If f and g are functions, then

$$(f \circ g)(x) = f(g(x))$$

for all x in the domain of g such that $g(x)$ is in the domain of f. $f \circ g$ is called the **composition** of f with g or f **composed with** g.

► **EXAMPLE 3**

If $f(x) = 2x - 5$ and $g(x) = x + 3$, find $(f \circ g)(x)$.

Solution

$$(f \circ g)(x) = f(g(x)) \qquad \text{Definition}$$
$$= f(x + 3)$$
$$= 2(x + 3) - 5$$
$$= 2x + 6 - 5$$
$$= 2x + 1$$

Another way to evaluate $(f \circ g)(x)$ in this example is

$$(f \circ g)(x) = f(g(x))$$
$$= 2g(x) - 5$$
$$= 2(x + 3) - 5$$
$$= 2x + 6 - 5$$
$$= 2x + 1$$
◄

► **EXAMPLE 4**

If $f(x) = x^2$ and $g(x) = x - 2$, find $(f \circ g)(x)$ and $(g \circ f)(x)$.

Solution

$$(f \circ g)(x) = f(g(x))$$
$$= f(x - 2)$$
$$= (x - 2)^2$$
$$(g \circ f)(x) = g(f(x))$$
$$= g(x^2)$$
$$= x^2 - 2$$
◄

► **EXAMPLE 5**

If $f(x) = x^2$ and $g(x) = 2x - 1$, find $f(g(3))$.

Solution

$$f(g(3)) = f(2 \cdot 3 - 1)$$
$$= f(5)$$
$$= 25$$
◄

Problem Set 8.1

Warm-Ups

In Problems 1 through 4, find $(f + g)(x)$, $(f - g)(x)$, $(fg)(x)$, and $\left(\dfrac{f}{g}\right)(x)$. See Example 1.

1. $f(x) = 2x$; $g(x) = x - 5$

2. $f(x) = x - 7$; $g(x) = \dfrac{2}{3}x - 3$

3. $f(x) = 3 - x^2$; $g(x) = x + 1$

4. $f(x) = x^2 - x - 3$; $g(x) = x^2 - x$

In Problems 5 through 8, $f(x) = x - 1$ and $g(x) = 1 - x^2$. Find the value of each. See Example 2.

5. $(g - f)(0)$ 　　　　　**6.** $(fg)(-1)$ 　　　　　**7.** $\left(\dfrac{f}{g}\right)(2)$ 　　　　　**8.** $(f + g)(-2)$

In Problems 9 through 12, find $(f \circ g)(x)$ and $(g \circ f)(x)$. See Examples 3 and 4.

9. $f(x) = 5x + 8; g(x) = 7x - 4$

10. $f(x) = 6x + 9; g(x) = \dfrac{1}{4}x$

11. $f(x) = x^2 + 2; g(x) = x - 3$

12. $f(x) = x^3; g(x) = x - 1$

In Problems 13 through 16, $f(x) = x - 1$ and $g(x) = 1 - x^2$. Find the value of each. See Example 5.

13. $(f \circ g)(-1)$ 　　　　**14.** $(g \circ f)(0)$ 　　　　**15.** $(f \circ f)(1)$ 　　　　**16.** $(f \circ g)(1)$

Practice Exercises

In Problems 17 through 28, find $(f + g)(x)$, $(f - g)(x)$, $(fg)(x)$, $\left(\dfrac{f}{g}\right)(x)$, and $(f \circ g)(x)$.

17. $f(x) = 3x; g(x) = x + 8$

18. $f(x) = x + 7; g(x) = x + 4$

19. $f(x) = 4x + 2; g(x) = 3x - 9$

20. $f(x) = 6x - 1; g(x) = \dfrac{1}{2}x$

21. $f(x) = x - 3; g(x) = \dfrac{1}{3}x + 3$

22. $f(x) = x^2 + 1; g(x) = x - 5$

23. $f(x) = 2x^2 + x + 3; g(x) = 2x - 5$

24. $f(x) = 3x^2 - x - 2; g(x) = 3 - x$

25. $f(x) = 1 - x^2; g(x) = x + 3$

26. $f(x) = x^2 + 1; g(x) = 3x^2 - 1$

27. $f(x) = x^2 - x - 2; g(x) = x^2 - 2x$

28. $f(x) = x^3; g(x) = x + 1$

In Problems 29 through 36, $f(x) = 2x - 5$ and $g(x) = 4 - x^2$. Find the value of each.

29. $(g - f)(0)$ 　　　　**30.** $(fg)(-1)$ 　　　　**31.** $\left(\dfrac{f}{g}\right)(1)$ 　　　　**32.** $(f + g)(-2)$

33. $(f \circ g)(-1)$ 　　　**34.** $(g \circ f)(0)$ 　　　**35.** $(f \circ f)(1)$ 　　　**36.** $(f \circ g)(1)$

Challenge Problems

In Problems 37 and 38, find $(f \circ g)(x)$ *and* $(g \circ f)(x)$.

37. $f(x) = x + 3; g(x) = x - 3$
38. $f(x) = x^3; g(x) = \sqrt[3]{x}$
39. Is 1 in the domain of $\dfrac{f}{g}$ if $f(x) = x - 1$ and $g(x) = 1 - x^2$?

In Your Own Words

40. Explain the composition of two functions.

8.2   Inverse Functions

TAPE 15

Let's consider the following functions:

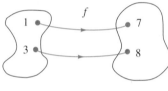

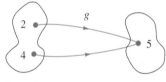

Notice that each number in the domain of f is assigned to a different number in the range of f. However, this is not true for each number in the domain of g.

> **One-to-One Function**
>
> A function is said to be **one-to-one** if every number in the *range* is paired with exactly one number in the domain.

► **EXAMPLE 1**

The following illustration describes a function. Is it one-to-one?

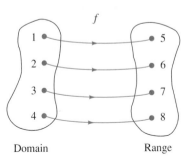

Solution

Notice that each number in the range is paired with *one* number in the domain. So f is one-to-one. ◄

► **EXAMPLE 2**

A function *f* is graphed as shown. Is *f* one-to-one?

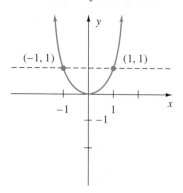

Solution

Notice that 1 is a number in the range. It is paired with −1 and 1. So *f* is *not* one-to-one.   ◄

Another way to picture Example 2 is shown below.

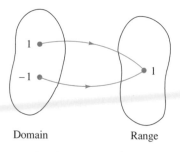

Domain   Range

Horizontal Line Test

If any horizontal line crosses the graph of a function in more than one point, the function is *not* one-to-one.

In many applications of functions, it is desirable to interchange the domain and range of a function *f*. In other words, make a function that would "reverse" what *f* does. Such a function is named f^{-1} (read "*f* inverse").

Be Careful!

NOTE: f^{-1} does *not* mean $\dfrac{1}{f}$.

Consider *f* and f^{-1} as shown next.

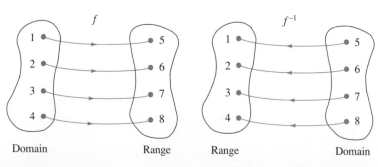

Domain   Range   Range   Domain

Notice that the domain of f^{-1} is the range of f and the range of f^{-1} is the domain of f. In other words, the domain and range are interchanged. We can see that f defines ordered pairs of (1, 5), (2, 6), (3, 7), and (4, 8), while f^{-1} defines (5, 1), (6, 2), (7, 3), and (8, 4). Notice that, in each case, the x- and y-coordinates are interchanged.

Functional notation is very useful in giving names to the numbers in the range.

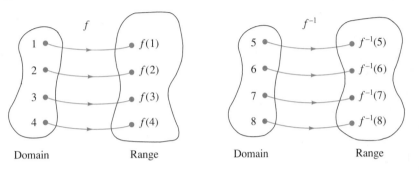

The *value* of $f(1)$ is 5 and the *value* of $f^{-1}(5)$ is 1.

To see how f^{-1} undoes what f does, consider the following. The function f assigns 1 to $f(1)$, whose value is 5.

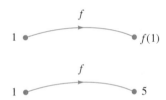

The function f^{-1} assigns 5 to $f^{-1}(5)$, whose value is 1.

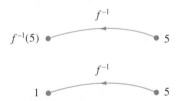

Combining these into one picture shows that f and f^{-1} undo each other.

$$f(f^{-1}(5)) = 5 \qquad f^{-1}(f(1)) = 1$$

Definition of f^{-1}

If f is a one-to-one function, f^{-1} is the function such that

$$f(f^{-1}(x)) = x \quad \text{for } x \text{ in the domain of } f^{-1}$$

and

$$f^{-1}(f(x)) = x \quad \text{for } x \text{ in the domain of } f$$

f^{-1} is called the **inverse** of f.

Notice that f must be one-to-one in order for f^{-1} to exist. There is an algorithm that finds the inverse of simple one-to-one functions.

To Find a Rule for f^{-1}

1. Replace $f(x)$ with y.
2. Swap x and y.
3. Solve this for y.
4. Replace y with $f^{-1}(x)$.

▶ **EXAMPLE 3**

If $f(x) = 2x - 3$, find $f^{-1}(x)$.

Solution

$$f(x) = 2x - 3$$
$$y = 2x - 3 \qquad \text{Replace } f(x) \text{ with } y$$
$$x = 2y - 3 \qquad \text{Swap } x \text{ and } y$$
$$x + 3 = 2y \qquad \text{Solve for } y$$
$$\frac{x + 3}{2} = y$$
$$f^{-1}(x) = \frac{x + 3}{2} \qquad \text{Replace } y \text{ with } f^{-1}(x) \qquad ◀$$

▶ **EXAMPLE 4**

If $f(x) = 5x + 4$, find a rule for f^{-1}.

Solution

$$y = 5x + 4 \qquad \text{Replace } f(x) \text{ with } y$$
$$x = 5y + 4 \qquad \text{Swap } x \text{ and } y$$
$$\frac{x - 4}{5} = y \qquad \text{Solve for } y$$
$$f^{-1}(x) = \frac{x - 4}{5} \qquad \text{Replace } y \text{ with } f^{-1}(x) \qquad ◀$$

Problem Set 8.2

Warm-Ups

In Problems 1 through 8, determine whether or not each function graphed is one-to-one.
See Examples 1 and 2.

1.

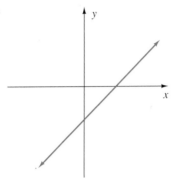

2.

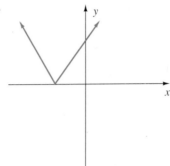

3.

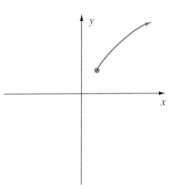

4.

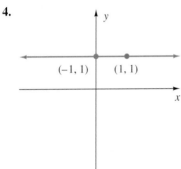

5.

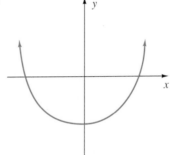

6.

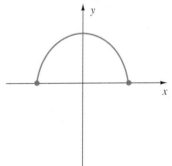

7.

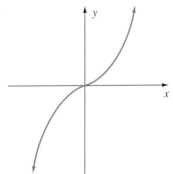

8.

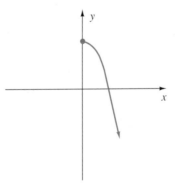

In Problems 9 through 14, find $f^{-1}(x)$. See Examples 3 and 4.

9. $f(x) = 3x - 1$ **10.** $f(x) = x - 4$ **11.** $f(x) = \dfrac{1}{2}x + 3$

12. $f(x) = \dfrac{2}{3}x + \dfrac{1}{2}$ **13.** $f(x) = x^3$ **14.** $f(x) = x^3 + 1$

Practice Exercises

In Problems 15 through 22, sketch the graph of each function and determine whether or not each function is one-to-one.

15. $f(x) = 3x + 1$ **16.** $h(x) = 5$ **17.** $f(x) = x^2 + 1$

18. $f(x) = -x^2$ **19.** $g(x) = (x + 1)^2 - 1$ **20.** $f(x) = -\dfrac{1}{2}x^2$

In Problems 21 through 26, find $f^{-1}(x)$.

21. $f(x) = 5x - 1$ **22.** $f(x) = x + 2$ **23.** $f(x) = \dfrac{1}{3}x - 1$

24. $f(x) = \dfrac{2}{5}x - \dfrac{1}{2}$ **25.** $f(x) = x^3 + 2$ **26.** $f(x) = x^3 - 1$

In Problems 27 through 30, f and f^{-1} are given. Verify that $f(f^{-1}(x)) = f^{-1}(f(x)) = x$ for each.

27. $f(x) = 2x + 1; f^{-1}(x) = \dfrac{x - 1}{2}$ **28.** $f(x) = x^3 + 1; f^{-1}(x) = \sqrt[3]{x - 1}$

29. $f(x) = x + 1; f^{-1}(x) = x - 1$ **30.** $f(x) = x^3; f^{-1}(x) = \sqrt[3]{x}$

Challenge Problems

31. Sketch the graph of $f(x) = x + 2$ and $f^{-1}(x) = x - 2$ on the same set of axes. Draw the graph of $y = x$ as a dashed line. Is there a relationship between the graphs of f, f^{-1}, and $y = x$? (HINT: Imagine folding the paper along the graph of $y = x$.)

32. The graph of g follows: Draw the graph of g^{-1}.

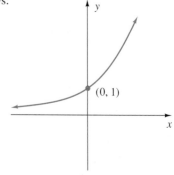

In Your Own Words

33. What is a one-to-one function?

34. If f is a one-to-one function, explain what f^{-1} is.

8.3 Exponential Functions

TAPE 15

In Section 6.6 we studied functions such as $f(x) = x^2$. In such a function, the base is a variable and the exponent is a constant. In this section we will examine another classification of important functions whose rules look much like these functions. However, the base will be a constant and the exponent a variable. For example, $f(x) = 2^x$ or $g(x) = \left(\dfrac{1}{2}\right)^x$. Such functions are called **exponential functions.**

If x is an integer, we can evaluate 2^x easily. For example, $2^4 = 16$ and $2^{-6} = \dfrac{1}{64}$. A calculator can be used to approximate other values of 2^x.

▶ **EXAMPLE 1**

Approximate each exponential to four decimal places.

(a) $2^{2/3}$ (b) $2^{1.1}$ (c) $2^{\sqrt{3}}$

Solutions

(a) $2^{2/3}$; the keystrokes $\boxed{2}\;\boxed{y^x}\;\boxed{(}\;\boxed{2}\;\boxed{\div}\;\boxed{3}\;\boxed{)}\;\boxed{=}$ should yield a display like $\boxed{1.58740105}$.

$2^{2/3} \approx 1.587$

(b) $2^{1.1} \approx 2.1436$

(c) $2^{\sqrt{3}}$; the keystrokes $\boxed{2}\;\boxed{y^x}\;\boxed{3}\;\boxed{\sqrt{}}\;\boxed{=}$ on a scientific calculator or $\boxed{2}\;\boxed{\wedge}\;\boxed{\sqrt{}}\;\boxed{3}\;\boxed{\text{ENTER}}$ on a graphing calculator should yield a display like $\boxed{3.321997085}$.

$2^{\sqrt{3}} \approx 3.3220$ ◀

▶ **EXAMPLE 2**

Graph the function $f(x) = 2^x$.

Solution

To help in graphing the function $f(x) = 2^x$, we make a table of some coordinates for $y = 2^x$, plot the points, and connect them with a smooth curve.

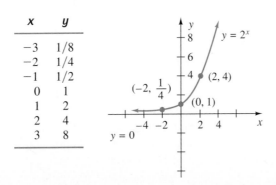

x	y
-3	$1/8$
-2	$1/4$
-1	$1/2$
0	1
1	2
2	4
3	8

Notice from the graph in Example 2 that, as x continues in the *negative* direction, the graph gets closer and closer to the x-axis, but it never touches it. The x-axis is called an **asymptote** to the graph of f.

In calculus we will see that 2^x is defined for all real numbers. Therefore, the domain of f is all real numbers. From the graph we see that the *range* of this function is all positive real numbers.

> ### Exponential Function
>
> A function of the form
>
> $$f(x) = b^x$$
>
> where $b > 0$ and $b \neq 1$, is called an **exponential function.**

We exclude $b = 1$ from the definition because the graph of $f(x) = 1$ is a horizontal line, which does not provide a useful exponential function.

► ## EXAMPLE 3

Sketch the graph of the function $g(x) = \left(\dfrac{1}{2}\right)^x$ and state its domain and range.

Solutions

Again, we make a table and sketch the graph.

x	y
-3	8
-2	4
-1	2
0	1
1	1/2
2	1/4
3	1/8

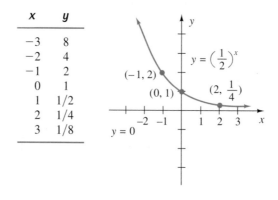

The x-axis is an asymptote. We indicate this by labeling the x-axis with $y = 0$ on the graph.   ◄

Examples 2 and 3 illustrate two types of exponential functions, **increasing** and **decreasing.** If $b > 1$, the function $f(x) = b^x$ is an *increasing function,* and if $0 < b < 1$, it is a *decreasing function.*

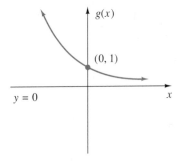

$f(x) = b^x, \; b > 1$
Increasing exponential function

$g(x) = b^x, \; 0 < b < 1$
Decreasing exponential function

Solving Some Exponential Equations

The horizontal line test indicates that an exponential function is a one-to-one function. Because of this, we have the following result.

> ## One-to-One Property of Exponential Functions
> For $b > 0$ and $b \neq 1$,
>
> if $b^x = b^y$, then $x = y$

The One-to-One Property allows us to equate exponents when the bases are the same. This property provides a tool to use when solving certain equations.

► ## EXAMPLE 4

Solve each equation.

(a) $3^x = 81$ (b) $4^x = \dfrac{1}{8}$ (c) $25^{x-7} = 625$

Solutions

(a) $3^x = 81$
$ = 3^4$
$ x = 4$ One-to-One Property of Exponential Functions
$\{4\}$

(b) $\quad 4^x = \dfrac{1}{8}$

$(2^2)^x = 2^{-3}$ $4 = 2^2$ and $2^{-3} = \dfrac{1}{8}$.

$2^{2x} = 2^{-3}$ Power of a Power
$2x = -3$ One-to-One Property

$x = -\dfrac{3}{2}$

$\left\{ -\dfrac{3}{2} \right\}$

(c) $25^{x-7} = 625$

$\quad\quad (5^2)^{x-7} = 5^4$

$\quad\quad 5^{2(x-7)} = 5^4$

$\quad\quad 2(x - 7) = 4$ One-to-One Property

$\quad\quad 2x - 14 = 4$ Distributive Property

$\quad\quad\quad 2x = 18$

$\quad\quad\quad x = 9$

$\quad\quad\quad \{9\}$ ◀

The Number e

One particular base for exponential functions occurs so often naturally in applications that it has a special name. It is an irrational number, like π. We call it e. The key $\boxed{e^x}$ can be found on any scientific calculator. It is usually a second function key. Look for it above the $\boxed{\ln}$ key. To find the value of e on a scientific calculator, press $\boxed{1}$ $\boxed{e^x}$. On a graphing calculator, press $\boxed{e^x}$ $\boxed{1}$ $\boxed{\text{ENTER}}$.

$$e \approx 2.718281828$$

▶ **EXAMPLE 5**

Approximate each exponential to four decimal places.

(a) e^4 (b) $e^{\sqrt{2}}$ (c) e^{-1}

Solutions

(a) $e^4 \approx 54.5982$ Keystrokes for scientific calculator: $\boxed{4}$ $\boxed{e^x}$

Keystrokes for graphing calculator:

$\boxed{e^x}$ $\boxed{4}$ $\boxed{\text{ENTER}}$

(b) $e^{\sqrt{2}} \approx 4.1133$ Keystrokes for scientific calculator:

$\boxed{2}$ $\boxed{\sqrt{}}$ $\boxed{e^x}$

Keystrokes for graphing calculator:

$\boxed{e^x}$ $\boxed{\sqrt{}}$ $\boxed{2}$ $\boxed{\text{ENTER}}$

(c) $e^{-1} \approx 0.3679$ Keystrokes for scientific calculator:

$\boxed{1}$ $\boxed{+/-}$ $\boxed{e^x}$

Keystrokes for graphing calculator:

$\boxed{e^x}$ $\boxed{(-)}$ $\boxed{1}$ $\boxed{\text{ENTER}}$

◀

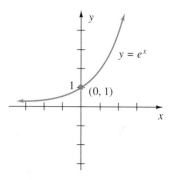

$y = e^x$

$(0, 1)$

The Natural Exponential Function

The exponential function with base e, $f(x) = e^x$, is called the **natural exponential function.** It is used in physics, engineering, business, and other fields. Because $e > 1$, the graph of the natural exponential function is increasing, like the graph of $y = 2^x$. Its domain is all real numbers and its range is $(0, +\infty)$.

Exponential functions are used as models for many applications. Some of the examples that we examined in Section 2.4 were actually exponential functions.

► **EXAMPLE 6**

The atmospheric pressure, $P(x)$ (lb/in.^{2}), at an elevation of x feet above sea level is given by the formula

$$P(x) = 14.7e^{-0.00004x}$$

Approximate (to the nearest tenth) the atmospheric pressure at an elevation of 1000 ft.

Solution

$$P(x) = 14.7e^{-0.00004x}$$

$$P(1000) = 14.7e^{-0.00004(1000)}$$

$$= 14.7e^{-0.04}$$

Approximate the value of $e^{-0.04}$ with a calculator.

$$P(1000) \approx 14.1$$

The atmospheric pressure is approximately 14.1 lb/in.^{2}.   ◄

If a quantity grows exponentially, the growth rate becomes very rapid.

► **EXAMPLE 7**

A strain of *Bacillus anthracis* bacteria initially contains 10,000 bacteria. The formula

$$A(t) = (10,000)3^{0.3t}$$

gives the number of bacteria present after t hours. Approximate to the nearest integer how many bacteria will be present after $\frac{1}{2}$ hour.

Solution

$$A(t) = (10,000)3^{0.3t}$$

$$A(0.5) = (10,000)3^{0.3(0.5)}$$

$$= (10,000)3^{0.15}$$

With a calculator, approximate the value of $3^{0.15}$.

$$A(0.5) \approx 11{,}791$$

After $\frac{1}{2}$ hour, there will be approximately 11,791 bacteria.   ◄

Problem Set 8.3

Warm-Ups

For Problems 1 through 4, answer each question. Approximate each exponential to three decimal places. See Example 1.

1. $3^{1.5}$

2. $5^{1/6}$

3. $10^{5/3}$

4. $3^{\sqrt{2}}$

In Problems 5 and 6, sketch the graph of each function. See Example 2.

5. $f(x) = 3^x$

6. $g(x) = 10^x$

In Problems 7 and 8, sketch the graph of each function. See Example 3.

7. $h(x) = \left(\dfrac{1}{3}\right)^x$

8. $F(x) = \left(\dfrac{1}{10}\right)^x$

In Problems 9 through 12, solve each equation. See Example 4.

9. $5^x = 625$

10. $7^x = \dfrac{1}{49}$

11. $6^{2x+1} = 36$

12. $9^{x-1} = 27$

In Problems 13 through 16, approximate each exponential to four decimal places. See Example 5.

13. e^3

14. $e^{\sqrt{5}}$

15. e^{-4}

16. e^{π}

For Problems 17 and 18, answer each question. See Example 6.

17. Use the formula given in Example 6 to approximate the atmospheric pressure to the nearest tenth at 1 mile (5280 ft).

18. Rabbits multiply exponentially according to the formula

$$N(t) = N_0 e^{0.3t}$$

where N_0 is the number present initially and t is the time in months. How many rabbits will there be in 3 months if 20 are present initially (to the nearest rabbit)?

In Problems 19 and 20, answer each question. See Example 7.

19. Use the formula in Example 7 to approximate the number of bacteria present after 2.5 hours.

20. The formula for calculating compound interest is

$$A = P\left(1 + \frac{r}{n}\right)^{nt}$$

where A is the amount accumulated and P is the amount invested at a rate of r compounded n times a year for t years. Approximate the amount that would accumulate if $25,000 is invested at 6.25% compounded quarterly for 10 years.

Practice Exercises

In Problems 21 through 34, approximate each exponential to four decimal places.

21. $3^{1/3}$ **22.** $5^{1/5}$ **23.** $2^{3.51}$

24. $6^{1.61}$ **25.** $5^{7/3}$ **26.** $3^{11/7}$

27. e^2 **28.** e^5 **29.** $6^{\sqrt{3}}$

30. $2^{\sqrt{19}}$ **31.** e^{-2} **32.** e^{-3}

33. $e^{\pi-2}$ **34.** $e^{2-\pi}$

In Problems 35 through 40, sketch the graph of each function.

35. $f(x) = 4^x$ **36.** $g(x) = 5^x$ **37.** $h(x) = \left(\dfrac{1}{5}\right)^x$

38. $k(x) = \left(\dfrac{1}{3}\right)^x$ **39.** $p(x) = 5^{-x}$ **40.** $q(x) = 3^{-x}$

In Problems 41 through 54, solve each equation.

41. $2^x = 64$ **42.** $6^x = 216$ **43.** $6^{x+1} = 36$

44. $2^{x-1} = 32$ **45.** $3^{2x} = 27$ **46.** $5^{3x} = 25$

47. $2^x = \dfrac{1}{8}$ **48.** $3^x = \dfrac{1}{81}$ **49.** $27^x = 9$

50. $16^x = 32$ **51.** $4^{x-1} = 8^{x+1}$ **52.** $9^{x+2} = 27^{2-x}$

53. $25^{2x-1} = 125^{3x+2}$ **54.** $216^{2-2x} = 36^{3-5x}$

In Problems 55 through 60, use a calculator to determine which expression is larger.

55. 3^5 or 5^3 **56.** 7^8 or 8^7 **57.** $3^{0.5}$ or $5^{0.3}$

58. $7^{0.8}$ or $8^{0.7}$ **59.** $3^{\sqrt{5}}$ or $5^{\sqrt{3}}$ **60.** $7^{\sqrt{8}}$ or $8^{\sqrt{7}}$

61. The population of a city is growing exponentially. The city had a population of 50,000 in 1970. If the population is given by the formula

$$P(t) = P_0 e^{0.05t}$$

where t is the number of years since 1970 and P_0 is the population in 1970, what will the population be in the year 2000 (to the nearest person)?

62. Use the formula in Problem 61 to find the population in the year 1994 (to the nearest person).

The half-life of a substance is the amount of time it takes for half of the substance to decay. The half-life of cobalt 60 is 5.2 years. The formula for the amount of cobalt 60 remaining after t years is $A(t) = A_0 \cdot 2^{-t/5.2}$, where A_0 is the amount of cobalt 60 present initially (t = 0).

63. If 100 g of cobalt 60 is present initially, approximately how many grams will remain after 10 years?

64. If 100 g of cobalt 60 is present initially, approximately how many grams will remain after 100 years?

65. The Natick Radiation Laboratory was charged with 2.5 million curies of cobalt 60 in 1962. How many curies remained in 1967?

66. How many curies of cobalt 60 will remain at the Natick Laboratory in 2002?

67. If a chain letter is sent to five people, and each person who receives a letter also sends a copy of the letter to five more people within one week, the function $f(x) = 5^x$ represents the number of letters in terms of x, the number of weeks elapsed since the first letter was sent.

(a) Complete the chart:

x	$f(x)$
1	
2	
3	
4	
5	

(b) Estimate how many weeks it will take for more than one-quarter of a million people to be involved. Check the estimate by substituting it for x in the function.

68. If a chain letter is sent to seven people, and each person who receives a letter also sends a copy of the letter to seven more people within one week, the function $f(x) = 7^x$ represents the number of letters in terms of x, the number of weeks elapsed since the first letter was sent.

(a) Complete the chart:

x	$f(x)$
1	
2	
3	
4	
5	

(b) Estimate how many weeks it will take for more than one-quarter of a million people to be involved. Check the estimate by substituting it for x in the function.

The following figure shows the amount accumulated if $30 dollars a month is deposited in a savings program that pays 7% interest compounded monthly. The graph f shows the amount accumulated if the program starts at age 20, while the graph g shows the accumulation if the program starts at age 40. Use this figure in Problems 69 through 72.

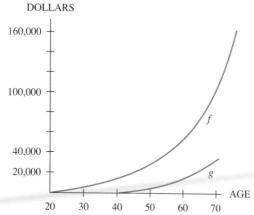

69. Estimate from the figure how much is accumulated at age 70 if the program is started at age 40.

70. Use the formula

$$A(m) = 30 \cdot \frac{(1 - I^m)}{1 - I}$$

where I is $1 + \dfrac{0.07}{12}$ and m is 360 months, to calculate the accumulation estimated in Problem 69.

71. Estimate from the figure how much is accumulated at age 70 if the program is started at age 20.

72. Use the formula in Problem 70 to calculate the accumulation estimated in Problem 71.

A well-known financial planning group advertised a 22% annual growth for a particular investment, as opposed to a 10.7% annual growth for S&P 500. The following graph appeared in the advertisement: Use this figure in Problems 73 through 80.

73. Which function shows 22%? 10.7%?

74. How much money is invested initially?

75. At 22%, in how many years will the money double? Quadruple?

76. At 10.7%, in how many years will the money double?

77. Approximate how much money has accumulated at 22% after 5 years.

78. Approximate how much money has accumulated at 10.7% after 5 years.

79. Approximate how much money has accumulated at 22% after 10 years.

80. Approximate how much money has accumulated at 10.7% after 10 years.

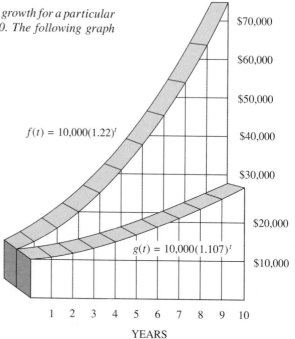

Challenge Problems

In Problems 81 through 86, sketch the graph of each function.

81. $f(x) = 2^x + 1$ **82.** $f(x) = 2^x - 1$ **83.** $f(x) = 2^{x+1}$

84. $f(x) = 2^{x-1}$ **85.** $f(x) = e^{x+1} + 2$ **86.** $f(x) = e^{x-2} - 3$

In Your Own Words

87. Discuss the graphs of $f(x) = 2^x$ and $g(x) = \left(\dfrac{1}{2}\right)^x$.

88. Compare the graphs of $g(x) = \left(\dfrac{1}{2}\right)^x$ and $h(x) = 2^{-x}$.

8.4 Logarithmic Functions

TAPE 15

Since an exponential function is a one-to-one function, it has an inverse. To find a rule for the inverse of $f(x) = b^x$, we follow the procedure developed in Section 8.2.

$$y = b^x \qquad \text{Replace } f(x) \text{ with } y.$$

$$x = b^y \qquad \text{Swap } x \text{ and } y.$$

The next step is to solve the equation $x = b^y$ for y. We do not have the tools to solve this for y, but since f has an inverse, we know such a function exists. It is called a **logarithmic function.** The name for this function is $\log_b$. ("Log" is short for "logarithm.") Continuing the procedure, we write

$$y = \log_b x \qquad \text{Solve for } y.$$

$$f^{-1}(x) = \log_b x \qquad \text{Replace } y \text{ with } f^{-1}(x).$$

We read $\log_b x$ as "logarithm of x to the base b" or, simply, "log to the base b of x." The symbol $\log_b$ is the name of a function, just as $f, g, h,$ and f^{-1} are names of functions.

> **Logarithmic Function**
>
> A function of the form
>
> $$f(x) = \log_b x, \qquad x > 0$$
>
> where $b > 0$ and $b \neq 1$, is called a **logarithmic function.**

We have seen that the statements $y = \log_b x$ and $x = b^y$ are equivalent. There are many times when the form $y = \log_b x$ (logarithmic form) must be changed to the form $x = b^y$ (exponential form) or, perhaps, the exponential form needs to be changed into the logarithmic form.

Fundamental Equivalence between Logarithms and Exponentials

If b and x are positive real numbers with $b \neq 1$, then

$$y = \log_b x \quad \text{is equivalent to} \quad b^y = x$$

$$\text{logarithmic form} \qquad\qquad \text{exponential form}$$

Notice that logarithms are really exponents. That is, when we write $y = \log_b x$ we mean that y is the exponent of b that gives x, or $b^{\log_b x} = x$.

► EXAMPLE 1

Change $2 = \log_3 9$ to exponential form.

Solution

$y = \log_b x$ is equivalent to $b^y = x$.
So $2 = \log_3 9$ is equivalent to $3^2 = 9$.   ◄

► EXAMPLE 2

Change $5^2 = 25$ to logarithmic form.

Solution

$b^y = x$ is equivalent to $y = \log_b x$.
So $5^2 = 25$ is equivalent to $2 = \log_5 25$.   ◄

► EXAMPLE 3

Solve $\log_2 x = 4$.

Solution

$\log_2 x = 4$ is an expression in logarithmic form. Writing it in exponential form gives $2^4 = x$ or $x = 16$.

$$\{16\}$$ ◄

In order to work with logarithmic functions, we must know how to find logarithms. This often involves changing forms.

► **EXAMPLE 4**

Find each logarithm, if possible.

(a) $\log_6 36$ (b) $\log_2 128$

(c) $\log_{10} 10$ (d) $\log_7 1$

(e) $\log_2 \dfrac{1}{8}$ (f) $\log_2 (-8)$

Solutions

(a) We want to find the value of $\log_6 36$. The strategy is to let $y = \log_6 36$ and then use the fundamental equivalence to find y.

$y = \log_b x$ is equivalent to $b^y = x$, so $y = \log_6 36$ is equivalent to $6^y = 36$.

$$6^y = 6^2 \qquad 36 = 6^2.$$

$$y = 2 \qquad \text{One-to-One Property of Exponentials}$$

Therefore, $\log_6 36 = 2$.

(b) $y = \log_2 128$ is equivalent to $2^y = 128$.

$$2^y = 2^7 \qquad 128 = 2^7.$$

$$y = 7 \qquad \text{One-to-One Property of Exponentials}$$

$$\log_2 128 = 7.$$

(c) $y = \log_{10} 10$ is equivalent to $10^y = 10$.

$$y = 1 \qquad \text{One-to-One Property of Exponentials}$$

$$\log_{10} 10 = 1$$

(d) $y = \log_7 1$ is equivalent to $7^y = 1$.

$$7^y = 7^0 \qquad 1 = 7^0.$$

$$y = 0 \qquad \text{One-to-One Property of Exponentials}$$

$$\log_7 1 = 0$$

(e) $y = \log_2 \dfrac{1}{8}$ is equivalent to $2^y = \dfrac{1}{8}$.

$$2^y = \frac{1}{2^3} = 2^{-3} \qquad \frac{1}{8} = 2^{-3}$$

$$y = -3 \qquad \text{One-to-One Property of Exponentials}$$

$$\log_2 \frac{1}{8} = -3$$

(f) $y = \log_2 (-8)$ is equivalent to $2^y = -8$.

But 2^y is *always* positive! There is no power of 2 that yield -8.

$\log_2 (-8)$ is undefined.   ◄

The number b in the expression $\log_b x$ is called the **base** of the logarithm. Base 10 logarithms are called **common logarithms** and are usually written omitting the base. That is, $\log x$ means $\log_{10} x$. However, as will be seen in calculus, logarithms arise *naturally* in nature and mathematical development.

These logarithms are to the base e and are called **natural logarithms.** We abbreviate natural logarithms by writing $\ln x$. In other words, $\ln x$ means $\log_e x$.

Common Logs and Natural Logs

The following two abbreviations are widely used.

$$\text{Common logarithms:} \quad \log x = \log_{10} x$$

$$\text{Natural logarithms:} \quad \ln x = \log_e x$$

► **EXAMPLE 5**

Find each common or natural logarithm.

Solutions

(a) $\log 100$ (b) $\log 0.001$ (c) $\ln e$ (d) $\ln \sqrt{e}$

(a) $\log 100 = \log_{10} 100$ Definition of common
           $y = \log_{10} 100$ logarithm
                   is equivalent to $10^y = 100$
           $10^y = 10^2$ One-to-One Property of
           $y = 2$ Exponentials
    $\log 100 = 2$

(b) $\log 0.001 = \log_{10} 0.001$ Definition of common
              $y = \log_{10} 0.001$ logarithm
                    is equivalent to $10^y = 0.001$
            $10^y = 10^{-3}$ One-to-One Property of
            $y = -3$ Exponentials
    $\log 0.001 = -3$

(c) $\ln e = \log_e e$ Definition of natural
          $y = \log_e e$ logarithm
                 is equivalent to $e^y = e$
          $y = 1$ One-to-One Property of
      $\ln e = 1$ Exponentials

(d) $\ln \sqrt{e} = \log_e \sqrt{e}$ Definition of natural
             $y = \log_e e^{1/2}$ logarithm
                    is equivalent to $e^y = e^{1/2}$

             $y = \dfrac{1}{2}$ One-to-One Property of
                                                       Exponentials
    $\ln \sqrt{e} = \dfrac{1}{2}$

◄

► **EXAMPLE 6**

Solve each equation.

(a) $\log x = 4$ (b) $\log x = -2$

Solutions

(a) $\log x = \log_{10} x$

$4 = \log_{10} x$ is equivalent to $10^4 = x$

$x = 10000$ 10000 checks in the original equation

$\{10,000\}$

(b) $-2 = \log x$ is equivalent to $10^{-2} = x$

$x = 0.01$ 0.01 checks in the original equation

$\{0.01\}$ ◀

Graphing Logarithmic Functions

Because a logarithmic function is the inverse of an exponential function, we have a general idea of its graph. However, we can use the fundamental equivalence to plot as many points as we desire.

► **EXAMPLE 7**

Graph the function $f(x) = \log_2 x$.

Solution

To graph f, we graph the equation $y = f(x)$. Since $y = \log_2 x$ is equivalent to $2^y = x$, the graph of $y = \log_2 x$ is the same as the graph of $x = 2^y$. We can make a table by choosing a value for y first and calculating the corresponding value of x.

x	y
$\frac{1}{4}$	-2
$\frac{1}{2}$	-1
1	0
2	1
4	2

Graph showing $y = \log_2 x$ with points $(2, 1)$, $(1, 0)$, $(4, 2)$, $(\frac{1}{2}, -1)$, and asymptote $y = 0$.   ◀

Logarithms have many practical uses. For example, the scale used to measure the intensity level of sound is logarithmic.

► **EXAMPLE 8**

Intensity level of sound is measured in decibels (dB) using the formula

$$\beta = 10 \log \frac{I}{I_o}$$

where $I_o = 10^{-12}$ W/m^2 and I is the intensity of the sound measured in watts per square meter. How many decibels are produced by:

(a) Busy street traffic with an intensity measured at 10^{-5} W/m^2?

(b) A lawn mower with an intensity of 3.1×10^{-3}?

Solution

(a) Substitute into the formula.

$$\beta = 10 \log \frac{I}{I_o}$$

$$= 10 \log \frac{10^{-5}}{10^{-12}}$$

$$= 10 \log 10^7$$

$$= 10 \cdot 7 = 70$$

Busy street traffic produces 70 decibels.

(b) $\beta = 10 \log \dfrac{I}{I_o}$

$$= 10 \log \frac{3.1 \times 10^{-3}}{10^{-12}}$$

$$= 10 \log (3.1 \times 10^9) \qquad \text{Approximate } \log (3.1 \times 10^9)$$
$$\text{with a calculator}$$

$$\approx 95$$

The lawn mower produces about 95 dB.

Calculator Box

Logarithms with a Calculator

Common logarithms are obtained by using the $\boxed{\log}$ key and natural logs with the $\boxed{\ln}$ key. Let's find log 3.21 and ln 345.67 to four decimal places.

Scientific Calculators:

Press $\boxed{3.21}$ $\boxed{\log}$ and read 0.506505032 on the display.

Press $\boxed{345.67}$ $\boxed{\ln}$ and read 5.845484563 on the display.

Graphing Calculators:

Press $\boxed{\log}$ $\boxed{3.21}$ $\boxed{\text{ENTER}}$ and read 0.506505032 on the display.

Press $\boxed{\ln}$ $\boxed{345.67}$ $\boxed{\text{ENTER}}$ and read 5.845484563 on the display.

$$\log 3.21 \approx 0.5065 \quad \text{and} \quad \ln 345.67 \approx 5.8455$$

Often, we wish to find a number, given its logarithm. Let's find P when $\log P = -0.09876$ and Q when $\ln Q = 24$. We will use the fundamental equivalence and the $\boxed{10^x}$ and $\boxed{e^x}$ keys.

$$\log P = -0.09876 \quad \text{is equivalent to} \quad 10^{-0.09876} = P$$

$$\ln Q = 24 \quad \text{is equivalent to} \quad e^{24} = Q$$

Scientific Calculators:

Press $\boxed{.09876}$ $\boxed{+/-}$ $\boxed{10^x}$ and read $\boxed{0.796599446}$ on the display.

Press $\boxed{24}$ $\boxed{e^x}$ and read $\boxed{2.64891221^{10}}$ on the display.

Graphing Calculators:

Press $\boxed{10^x}$ $\boxed{(-)}$ $\boxed{.09876}$ $\boxed{\text{ENTER}}$ and read $\boxed{0.796599446}$ on the display.

Press $\boxed{e^x}$ $\boxed{24}$ $\boxed{\text{ENTER}}$ and read $\boxed{2.64891221\text{E}10}$ on the display.

$$P \approx 0.7966 \quad \text{and} \quad Q \approx 2.6489 \times 10^{10}$$

Calculator Exercises

In Problems 1 through 6, approximate each logarithm to four decimal places.

1. log 98.7   **2.** ln 0.98   **3.** log 1.009

4. ln 87,953   **5.** log 0.0016   **6.** ln 234.89

In Problems 7 through 10, find x correct to three decimal places.

7. log $x = 3.2897$ **8.** log $x = -1.3648$

9. ln $x = 10.23445$ **10.** ln $x = -5.6730$

Answers

1. 1.9943   **2.** −0.0202   **3.** 0.0039   **4.** 11.3846   **5.** −2.7959

6. 5.4591   **7.** 1948.498   **8.** 0.043   **9.** 27,846.150   **10.** 0.003

Problem Set 8.4

Warm-Ups

In Problems 1 through 4, write each equation in exponential form. See Example 1.

1. $\log_2 8 = 3$ **2.** $\log_{1/2} 8 = -3$ **3.** $\log_8 64 = 2$ **4.** $\log_3 3 = 1$

In Problems 5 through 8, write each equation in logarithmic form. See Example 2.

5. $2^3 = 8$ **6.** $4^{-2} = \dfrac{1}{16}$ **7.** $\left(\dfrac{1}{3}\right)^2 = \dfrac{1}{9}$ **8.** $3^0 = 1$

In Problems 9 through 12, solve each equation. See Example 3.

9. $\log_4 x = 2$ **10.** $\log_3 81 = x$ **11.** $\log_5 x = 3$ **12.** $\log_5 x = 0$

In Problems 13 through 20, find the value of each logarithm. See Example 4.

13. $\log_5 125$ **14.** $\log_5 625$ **15.** $\log_3 (-9)$ **16.** $\log_{1/3} (-81)$

17. $\log_3 3^2$ **18.** $\log_{10} 10^3$ **19.** $\log_{100} 10,000$ **20.** $\log_2 1$

In Problems 21 through 24, find each common or natural logarithm. See Example 5.

21. $\log 0.01$　　　　　**22.** $\log 1000$　　　　　**23.** $\ln 1$　　　　　　　　**24.** $\ln e^2$

In Problems 25 and 26, solve each equation. See Example 6.

25. $\log x = 3$　　　　　**26.** $\log_3 x = -1$

In Problems 27 and 28, sketch the graph of each function. See Example 7.

27. $f(x) = \log_6 x$　　　　　　　　　　　　　**28.** $f(x) = \log_8 x$

In Problems 29 through 32, answer the question. See Example 8.

29. How many decibels are produced by a rustle of leaves with an intensity of 10^{-11} W/m²?

30. How many decibels are produced by a jet airplane 100 ft away if intensity is measured at 10^2 W/m²?

31. How many decibels are produced by conversation with an intensity of 3.1×10^{-6} W/m²?

32. How many decibels do the amplifiers at a rock concert produce if the intensity is measured at 3.2 W/m²?

Practice Exercises

In Problems 33 through 64, solve each equation.

33. $\log_5 x = 2$　　　　**34.** $\log_3 9 = x$　　　　**35.** $\log_3 27 = x$　　　　**36.** $\log_x 1 = 0$

37. $\log_2 x = 4$　　　　**38.** $\log_4 x = \dfrac{3}{2}$　　　**39.** $\log_{1/2} x = -1$　　　**40.** $\log_2 x = 5$

41. $\log_5 5 = x$　　　　**42.** $\log_{27} 3 = x$　　　**43.** $\log_3 9 = x$　　　　**44.** $\log_2 16 = 2x$

45. $\log_3 1 = x - 1$　　**46.** $\log_3 27 = x - 7$　　**47.** $\log_{10}(x + 1) = 1$　　**48.** $\log_3(2x - 1) = 2$
49. $\log_3 81 = x$　　　　**50.** $\log_5 25 = x$　　　**51.** $\log_{1/2} 4 = x$　　　**52.** $\log_{1/3} 9 = x$

53. $\log_5 5^2 = x$　　　　**54.** $\log_{10} 10 = x$　　　**55.** $\log_{100} 10 = x$　　　**56.** $\log_{25} 5 = x$

57. $\log_2 x = 1$　　　　**58.** $\log_3 3^x = 4$　　　**59.** $\log_{1/3} x = -1$　　　**60.** $\log_3 x = -1$

61. $\log_7 7 = x$　　　　**62.** $\log_{16} 4 = x$　　　**63.** $\log_{10}(x + 2) = 2$　　**64.** $\log_3(2x - 1) = 1$

In Problems 65 through 68, sketch the graph of each function, and give the domain and range for each.

65. $f(x) = \log_3 x$　　　**66.** $f(x) = \log_{10} x$　　　**67.** $f(x) = \log_5 x$　　　**68.** $f(x) = \log_4 x$

In Problems 69 through 80, approximate each expression to four decimals.

69. $\log 15$

70. $\log 3 + \log 5$

71. $\log (3 + 5)$

72. $\log \dfrac{32}{2}$

73. $\log 32 - \log 2$

74. $\log (32 - 2)$

75. $\log (0.8)^2$

76. $[\log 0.8]^2$

77. $2 \log 0.8$

78. $\log \sqrt{2}$

79. $\dfrac{1}{2} \log 2$

80. $\sqrt{\log 2}$

The power gain G in decibels between a power input to an amplifier and and the power output is given by the formula

$$G = 10 \log \frac{P_{out}}{P_{in}}$$

81. Approximate the power gain of an amplifier whose power input is 0.1 watt and whose power output is 50 watts.

82. Approximate the power gain of an amplifier whose power input is 0.01 watt and whose power output is 25 watts.

Challenge Problems

In Problems 83 through 91, sketch the graph of each function.

83. $f(x) = \log_2 x$

84. $f(x) = \log_2 (x + 1)$

85. $f(x) = \log_2 (x - 1)$

86. $f(x) = \log_2 x + 1$

87. $f(x) = \log_2 x - 1$

In Problems 88 through 91, find the domain before graphing.

88. $f(x) = \log_2 (-x)$

89. $f(x) = \log_2 |x|$

90. $f(x) = \log_2 x^2$

91. $f(x) = 2 \log_2 x$

In Your Own Words

92. Explain the relationship between a logarithmic function and an exponential function.

8.5 **Properties of Logarithms**

TAPE 15

Logarithms are very useful tools that have properties that help simplify calculations. Logarithms were even more useful before calculators were invented.

Let $\log_b M = x$ and $\log_b N = y$, where b, M, and N are positive and $b \neq 1$. Changing to exponential form gives us

$$M = b^x \quad \text{and} \quad N = b^y$$

$$MN = b^x b^y = b^{x+y} \quad \text{is equivalent to}$$

$$\log_b (MN) = x + y$$

$$\log_b (MN) = \log_b M + \log_b N$$

An important property of logarithms has been proved. This property says, *The log of a product is the sum of the logs.* It allows us to add instead of multiply. Now consider a quotient.

$$\frac{M}{N} = \frac{b^x}{b^y} = b^{x-y} \quad \text{is equivalent to}$$

$$\log_b \frac{M}{N} = x - y$$

$$\log_b \frac{M}{N} = \log_b M - \log_b N$$

This proves that the log of a quotient is the difference in the logs. This property allows us to subtract instead of divide.

Properties of Logarithms

If $M > 0$, $N > 0$, $b > 0$, $b \neq 1$, and r is real:

1. $\log_b (MN) = \log_b M + \log_b N$ Log of a product
2. $\log_b (M/N) = \log_b M - \log_b N$ Log of a quotient
3. $\log_b M^r = r \log_b M$ Log of a power
4. $\log_b b = 1$ Log of the base
5. $\log_b 1 = 0$ Log of 1

► **EXAMPLE 1**

If $\log_3 5 \approx 1.5$ and $\log_3 2 \approx 0.6$, approximate each of the following.

(a) $\log_3 10$ (b) $\log_3 5/2$ (c) $\log_3 25$
(d) $\log_3 \sqrt{2}$ (e) $\log_3 50$

Solutions

(a) $\log_3 10 = \log_3 (2)(5)$
$\qquad\qquad = \log_3 2 + \log_3 5$ Log of a product
$\qquad\qquad \approx 0.6 + 1.5 \approx 2.1$

(b) $\log_3 5/2 = \log_3 5 - \log_3 2$ Log of a quotient
$\approx 1.5 - 0.6 \approx 0.9$

(c) $\log_3 25 = \log_3 5^2$
$= 2 \log_3 5$ Log of a power
$\approx 2(1.5) \approx 3.0$

(d) $\log_3 \sqrt{2} = \log_3 2^{1/2}$

$= \frac{1}{2} \log_3 2$ Log of a power

$\approx \frac{1}{2}(0.6) \approx 0.3$

(e) $\log_3 50 = \log_3 (2)(5^2)$
$= \log_3 2 + \log_3 5^2$ Log of a product
$= \log_3 2 + 2 \log_3 5$ Log of a power
$\approx 0.6 + 2(1.5)$
$\approx 0.6 + 3.0 \approx 3.6$ ◄

To solve an equation containing logarithms, use the properties of logarithms and try to write the equation in the form

$$\log_b M = N$$

and then change to exponential form:

$$M = b^N$$

► **EXAMPLE 2**

Solve $\log_2 x - \log_2 3 = 5$.

Solution

$$\log_2 x - \log_2 3 = 5$$

$$\log_2 \frac{x}{3} = 5 \qquad \text{Log of a quotient}$$

$$\frac{x}{3} = 2^5 \qquad \text{Change to exponential form}$$

$$= 32$$

$$x = 96$$

Check 96 in the original equation.

$$\{96\} \qquad\qquad ◄$$

Since logarithmic functions are one-to-one functions, we have the following useful result.

A Property of Logarithmic Functions

If $\log_b x = \log_b y$, then $x = y$ $(x > 0, y > 0, b > 0, b \neq 1)$.

► **EXAMPLE 3**

Solve $\log_2 (3x - 1) = \log_2 8$.

Solution

$$\log_2 (3x - 1) = \log_2 8$$
$$3x - 1 = 8 \qquad \text{Property of Logarithms}$$
$$3x = 9$$
$$x = 3$$

Check 3 in the original equation.

$$\{3\}$$

◄

Logarithms are defined for positive numbers only. So we must be careful to check possible solutions in the original equation.

► **EXAMPLE 4**

Solve $\log_3 x + \log_3 (x - 2) = 1$.

Solution

$$\log_3 x + \log_3 (x - 2) = 1$$
$$\log_3 x(x - 2) = 1 \qquad \text{Log of a Product Property}$$
$$x(x - 2) = 3^1 \qquad \text{Change to exponential form}$$
$$x^2 - 2x - 3 = 0$$
$$(x - 3)(x + 1) = 0$$
$$x = 3 \quad \text{or} \quad x = -1$$

We must be careful to check before writing the solution set.
Check 3.

$$\text{LS:} \quad \log_3 3 + \log_3 (3 - 2) = \log_3 3 + \log_3 1 = \boxed{1}$$
$$\text{RS:} \quad \boxed{1}$$

Thus 3 checks and is in the solution set.
Check -1.

$\text{LS:} \quad \log_3 (-1) + \log_3 (-1 - 2)$ is undefined because we cannot find the log of a negative number. Thus -1 is not in the solution set.

$$\{3\} \qquad \text{Write the solution set} \qquad ◄$$

Logarithms enable us to solve more general exponential equations such as $3^x = 5$.

Calculator Box

Finding Logarithms to Other Bases with a Calculator

Calculators have keys for logarithms to the base e and the base 10. However, logarithms to other bases are sometimes encountered. They may be calculated using the following change-of-base formula. If b and c are permissible bases,

$$\log_c x = \frac{\log_b x}{\log_b c}$$

That is, if c is greater than zero and not equal to 1, we can find $\log_c x$ with either

$$\log_c x = \frac{\log x}{\log c} \quad \text{or} \quad \log_c x = \frac{\ln x}{\ln c}$$

Let's find $\log_2 5$ to six decimal places. We use the change of base formula $\log_2 5 = \dfrac{\ln 5}{\ln 2}$.

Scientific Calculator:

Press $\boxed{5}$ $\boxed{\ln}$ $\boxed{\div}$ $\boxed{2}$ $\boxed{\ln}$ $\boxed{=}$ and read $\boxed{2.321928095}$ on the display.

Graphing Calculator:

Press $\boxed{\ln}$ $\boxed{5}$ $\boxed{\div}$ $\boxed{\ln}$ $\boxed{2}$ $\boxed{\text{ENTER}}$ and read $\boxed{2.321928095}$ on the display.

Thus $\log_2 5 \approx 2.321928$. The $\boxed{\log}$ key instead of the $\boxed{\ln}$ key will give the same result.

Calculator Exercises

Approximate the value of each logarithm to five decimal places.

1. $\log_2 10$ **2.** $\log_2 0.876$ **3.** $\log_{16} 127.975$ **4.** $\log_2 e$

5. $\log_3 38^2$ **6.** $\log_8 \sqrt{57}$ **7.** $\log_7 68^{-2}$ **8.** $\log_{11} 12$

Answers

1. 3.32193 **2.** −0.19100 **3.** 1.74993 **4.** 1.44270

5. 6.62215 **6.** 0.97215 **7.** −4.33680 **8.** 1.03629

Because logarithms are functions, we can sometimes take the logarithm of both sides and solve such equations. We will choose 10 or e as a base because calculators use these bases. Notice how this idea works in the next example.

► **EXAMPLE 5**

Approximate the solution of each equation to the nearest thousandth.

(a) $3^x = 5$ (b) $11^t = 2^{2t+3}$

(c) $5^{2x} \cdot 6^x = 371$ (d) $e^{2k} = 244$

Solutions

(a) $3^x = 5$.

Since both sides are positive, we can take the log of both sides. We choose either log to the base 10 or base e for convenience with the calculator. In this example, we choose base 10.

$$\log 3^x = \log 5$$

$$x \log 3 = \log 5 \qquad \text{Log of a power}$$

$$x = \frac{\log 5}{\log 3}$$

$$x \approx 1.465$$

(b) In this example we will use log to the base e.

$$\ln 11^t = \ln 2^{2t+3}$$

$$t \ln 11 = (2t + 3) \ln 2 \qquad \text{Log of a power}$$

$$t \ln 11 = 2t \ln 2 + 3 \ln 2 \qquad \text{Distributive property}$$

$$t \ln 11 - 2t \ln 2 = 3 \ln 2$$

$$t(\ln 11 - 2 \ln 2) = 3 \ln 2 \qquad \text{Factor}$$

$$t = \frac{3 \ln 2}{\ln 11 - 2 \ln 2}$$

$$t \approx 2.056$$

(c) The log of a product property is useful in this example.

$$5^{2x} \cdot 6^x = 371$$

$$\log (5^{2x} \cdot 6^x) = \log 371 \qquad \text{Take the log of both sides}$$

$$\log 5^{2x} + \log 6^x = \log 371 \qquad \text{Log of a product}$$

$$2x \log 5 + x \log 6 = \log 371 \qquad \text{Log of a power}$$

$$x(2 \log 5 + \log 6) = \log 371 \qquad \text{Factor}$$

$$x = \frac{\log 371}{2 \log 5 + \log 6}$$

$$x \approx 1.181$$

(d) We choose natural logs because of the e in the problem.

$$e^{2k} = 244$$

$$\ln e^{2k} = \ln 244 \qquad \text{Log of both sides}$$

$$2k \ln e = \ln 244 \qquad \text{Log of a power}$$

$$2k = \ln 244 \qquad \text{Log of the base } (\ln e = 1)$$

$$k = \frac{1}{2} \ln 244 \approx 2.749 \qquad \blacktriangleleft$$

We solved some problems in Section 2.4 in which the variable was an exponent in the mathematical model. Then we relied on estimating answers. Now we have the tools to solve such problems without estimation.

▶ **EXAMPLE 6**

How long would it take to double an investment of $1000 at 6% interest compounded annually?

Solution

We use the compound interest formula with $A = 2000$, $P = 1000$, $r = 0.06$, and $n = 1$.

$$A = P\left(1 + \frac{r}{n}\right)^{nt}$$

$$2000 = 1000(1 + 0.06)^t$$

$$2 = 1.06^t \qquad \text{Divide both sides by 1000}$$

$$\log 2 = \log(1.06)^t \qquad \text{Take the log of both sides}$$

$$\log 2 = t \log 1.06 \qquad \text{Log of a power}$$

$$\frac{\log 2}{\log 1.06} = t$$

$$11.9 \approx t$$

The investment will double in about 12 years. $\qquad \blacktriangleleft$

Problem Set 8.5

Warm-Ups

In Problems 1 through 8, approximate each logarithm if $\log_b 2 \approx 1.32$ and $\log_b 3 \approx 1.62$. See Example 1.

1. $\log_b 6$

2. $\log_b \dfrac{3}{2}$

3. $\log_b 2^2$

4. $\log_b \sqrt{3}$

5. $\log_b \dfrac{1}{2}$

6. $\log_b 18$

7. $\log_b \dfrac{4}{3}$

8. $\log_b \sqrt{6}$

In Problems 9 through 15, solve each equation. See Example 2.

9. $\log_3 x + \log_3 2 = 2$

10. $\log_2 3x - \log_2 (x - 1) = 2$

11. $\log x + \log 2 = 0$

12. $\log_2 x - \log_2 3 = 3$

13. $\log_{16} x + \log_{16} 4 = 1$

14. $\log_3 x + \log_3 2 - \log_3 5 = 0$

15. $\log_7(x + 1) - \log_7 49 = 1$

In Problems 16 through 18, solve each equation. See Example 3.

16. $\ln(x - 1) = \ln e$

17. $\log_8 (2x + 2) = \log_8 8$

18. $\ln x - \ln e = 0$

In Problems 19 through 22, solve each equation. See Example 4.

19. $\log x + \log(x - 3) = 1$

20. $\log x(x - 3) = 1$

21. $\log_2 x^2 = 2$

22. $2 \log_2 x = 2$

In Problems 23 through 26, approximate the solution of each equation to the nearest thousandth. See Example 5.

23. $7^x = 5$

24. $8^{2v-1} = 6^v$

25. $2^x \cdot 3^{2x-5} = 588$

26. $2e^{4.11k} = 1$

In Problems 27 and 28, answer the question. See Example 6.

27. How long will it take for an investment of $10,000 at 5% interest compounded annually to double?

28. How long will it take for an investment of $20,000 at 5% interest compounded annually to double?

Practice Exercises

In Problems 29 through 36, approximate each logarithm if $\log_b 3 \approx 1.62$ and $\log_b 5 \approx 2.52$.

29. $\log_b 15$

30. $\log_b \dfrac{3}{5}$

31. $\log_b 5^2$

32. $\log_b \sqrt{5}$

33. $\log_b \dfrac{1}{3}$

34. $\log_b 45$

35. $\log_b \dfrac{9}{5}$

36. $\log_b \sqrt{15}$

In Problems 37 through 43, classify each statement as true or false. $b > 0; b \neq 1$

37. $\log_b 14 = \log_b 7 + \log_b 2$

38. $\log_b 14 = (\log_b 7)(\log_b 2)$

39. $\log_b 13 = \dfrac{\log_b 26}{\log_b 2}$

40. $\log_b 13 = \log_b 26 - \log_b 2$

41. $\log_b (17^2) = (\log_b 17)^2$

42. $\log_b (17^2) = 2 \log_b 17$

43. $\sqrt{\log_b 5} = \log_b \sqrt{5}$

In Problems 44 through 59, solve each equation.

44. $\log_5 x + \log_5 2 = 2$

45. $\log_2 5x = \log_2 (x + 1)$

46. $\log x + \log 2 = 3$

47. $\log_2 x - \log_2 3 = 4$

48. $\log_6 (2x + 2) = \log_6 6$

49. $\log_{16} 2x + \log_{16} 2 = 2$

50. $\log_5 x - \log_5 3 + \log_5 2 = 0$

51. $\log_2 (x - 1) - \log_2 16 = 1$

52. $\ln x^2 - \ln e^2 = 0$

53. $\ln(2x + 1) = \ln e$

54. $\log_2 x + \log_2 (x + 1) = 1$

55. $\log_3 |x| = 1$

56. $\log_6 x + \log_6 (x - 1) = 1$

57. $\log_2 x(x - 3) = 2$

58. $\log x + \log x = 0$

59. $\log_5 x + \log_5 (x - 4) = 1$

In Problems 60 through 67, approximate the solutions for each equation to three decimal places.

60. $2^x = 3^{x-1}$

61. $5^{x+1} = 6^x$

62. $5^{t+1} = 3^{t-1}$

63. $13^{z-2} = 17^{z+3}$

64. $3e^{5.67k} = 5$

65. $3e^{2s} = 1$

66. $23^{3x} \cdot 31^{2x} = 2$

67. $7^{12x} \cdot 2^{7x} = 6.3314 \times 10^{11}$

68. How long will it take to double an investment of $1000 at 6% interest compounded semiannually?

69. How long will it take to double an investment of $1000 at 6% interest compounded quarterly?

70. How long will it take to double an investment of $1000 at 7.5% interest compounded monthly?

71. How long will it take to triple an investment of $1000 at 7.5% interest compounded monthly?

72. Automatic Tool Products Company is considering two plans for future growth. Plan 1 costs $50,000 initially and the total cost through year t is given by the function $C_1 = 50,000\, e^{0.02t}$. Plan 2 costs only $10,000 initially, but its total cost through year t is given by $C_2 = 10,000 e^{0.1t}$. In what year will the the total cost of plan 2 catch up with the total cost of plan 1? What will the total cost of either plan be at that time?

73. As deforestation takes place on Tomandu Island, the total number of Dutch rabbits is estimated to be given by the function $R(t) = 2.81(7/8)^t$, where $R(t)$ is in millions of rabbits and t is years. The population of star-nosed moles is also decreasing as the forest shrinks, but not as rapidly as the Dutch rabbits. The estimated star-nosed mole population is given by the function

Problem 73

$M(t) = 1.52(11/12)^t$, where $M(t)$ is millions of moles and t is years. When will the two populations be equal if both estimations are correct? What will that population be? Will either population ever be zero (using the given models)?

Challenge Problems

74. Prove that $\log_b \dfrac{1}{x} = -\log_b x$.

$$\log_b \frac{1}{x} = \log_b 1 - \log_b x = -\log_b x$$

In Your Own Words

75. What properties does a logarithmic function have that other functions do not have?

► Chapter Summary

GLOSSARY

A **one-to-one** function   Each number in the range is paired with exactly one number in the domain.

The **horizontal line test**   Determines whether or not a graph is one-to-one. The linear, exponential, and logarithmic functions that we have studied are one-to-one.

The **inverse** of a function   A function that "undoes" the function. A function must be one-to-one to have an inverse. A logarithmic function is the inverse of an exponential function.

Exponential and Logarithmic Functions

Exponential $f(x) = b^x$; $b > 0$; $b \neq 1$
Logarithmic $f(x) = \log_b x$; $b > 0$; $b \neq 1$; $x > 0$

Properties of Logarithms

$$\log_b MN = \log_b M + \log_b N \qquad \text{Log of a product}$$

$$\log_b \frac{M}{N} = \log_b M - \log_b N \qquad \text{Log of a quotient}$$

$$\log_b M^r = r \log_b M \qquad \text{Log of a power}$$

$$\log_b b = 1 \qquad \text{Log of the base}$$

$$\log_b 1 = 0 \qquad \text{Log of 1}$$

Logarithmic and Exponential Forms

$x = b^y$ and is equivalent to $y = \log_b x$.

CHECKUPS

1. If $g(x) = \sqrt{x}$ and $h(x) = x^2$, find $(g + h)(2)$.   Section 8.1; Example 2a

2. If $f(x) = 2x - 5$ and $g(x) = x + 3$, find $(f \circ g)(x)$.   Section 8.1; Example 3

3. If $f(x) = 2x - 3$, find $f^{-1}(x)$.   Section 8.2; Example 3

4. Sketch the graph of each function.

   a. $g(x) = \left(\frac{1}{2}\right)^x$ Section 8.3; Example 3

   b. $f(x) = \log_2 x$ Section 8.4; Example 7

5. Solve each equation.

   a. $4^x = \frac{1}{8}$ Section 8.3; Example 4b

   b. $\log_2 x = 4$ Section 8.4; Example 3

   c. $\log_2 x - \log_2 3 = 5$ Section 8.5; Example 2

   d. $\log_3 x + \log_3 (x - 2) = 1$ Section 8.5; Example 4

► Review Problems

In Problems 1 through 16, use $f(x) = 3x - 5$ and $g(x) = x^2$ to find each.

1. $(f + g)(x)$

2. $\left(\dfrac{f}{g}\right)(x)$

3. $(f - g)(x)$

4. $(f \circ g)(x)$

5. $(fg)(x)$

6. $(g \circ f)(x)$

7. $(f + g)(3)$

8. $(f - g)(-2)$

9. $\left(\dfrac{f}{g}\right)(-1)$

10. $(fg)(0)$

11. $(f \circ g)\left(\dfrac{1}{3}\right)$

12. $(g \circ f)(1)$

13. $f(0) - g(0)$

14. $g(f(-1))$

15. $f\left(\dfrac{1}{3}\right)g\left(\dfrac{1}{3}\right)$

16. $\dfrac{g(0)}{f(0)}$

In Problems 17 through 20, find $f^{-1}(x)$.

17. $f(x) = 1 - 2x$

18. $f(x) = 5 + 3x$

19. $f(x) = x^3 - 7$

20. $f(x) = x^3 + 7$

In Problems 21 through 24, sketch the graph of each pair of equations.

21. $f(x) = 7^x$; $g(x) = \log_7 x$

22. $f(x) = 3^x$; $g(x) = \log_3 x$

23. $f(x) = 10^x$; $g(x) = \log_{10} x$

24. $f(x) = e^x$; $g(x) = \ln x$

In Problems 25 through 42, solve each equation.

25. $2^x = 128$

26. $\log_x 4 = 2$

27. $\log_{10} 10^x = 1$

28. $\log_{10} x = \log_{10} 10^3$

29. $\ln(x + e) - \ln e = 1$

30. $\ln e + \ln e^3 - \ln e^2 = x$

31. $3^{x+1} = 9$

32. $\ln x = 1$

33. $(25)^x = 625$

34. $\log_x b = 1$

35. $\log x + \log (x + 3) = 1$

36. $\log_2 x^2 = 0$

37. $\log (x + 14) = \log (2x + 7)$

38. $\log_2 (2x - 3) = \log_2 (x + 1)$

39. $\log_2 x + \log_2 (x - 1) = 1$

40. $\log_3 x + \log_3 (x + 2) = 1$

41. $\log (x + 3) - \log x = 1$

42. $\log (2x - 1) - \log (x + 1) = 0$

In Problems 43 through 50, approximate the solutions for each equation to three decimal places.

43. $11^x = 5$

44. $10^x = 4$

45. $2^{x+3} = 3^x$

46. $7^x = 4^{x+1}$

47. $e^{2x} = 5$

48. $e^x = 4$

49. $9^{2t} = 2^{t-3}$

50. $5^{2t-3} = 7^{2t-3}$

51. How long will it take to double an investment of $5000 at 6.5% interest compounded semiannually?

52. How long will it take to double an investment of $5000 at 6.5% interest compounded quarterly?

53. How long will it take to double an investment of $10,000 at 7% interest compounded monthly?

54. How long will it take to triple an investment of $10,000 at 7% interest compounded monthly?

Let's Not Forget . . .

Identify the expressions that are in factored form. Factor those that are not factored.

55. $(a + 2)^2$

56. $125 - 8a^3$

57. $4 - (a + b)^2$

58. $ab^2(a^2 + b^2)$

59. $a^3b + c^2b$

60. $(a + 2b)^3$

How many terms are in each expression? Which expressions have $a - b$ as a factor?

61. $a^3 - b^3$

62. $a - b + 7$

63. $w(a - b) + y(b - a)$

64. $(a - b)^2 - 4$

Simplify each expression, if possible, leaving only nonnegative exponents in your answer.

65. $\sqrt[5]{64x^5 - 128y^5}$

66. $(x^{1/2}y^{1/2})^4$

67. $\sqrt[3]{(x-3)^3}$

68. $(xy^{-4})^{-2}$

69. $(2^{-1} + 4^{-1})^{-1}$

70. $\dfrac{4 \pm \sqrt{-12}}{2}$

71. $\dfrac{-5x^{-3}}{y}$

72. $\left(\dfrac{16a^{-8}}{9b^4}\right)^{-1/2}$

Find each product.

73. $(3x - 1)^3$

74. $(2x - 3y)^2$

Reduce each, if possible.

75. $\dfrac{2(x-1) + a(x+1)}{4(x-1) + a(x+1)}$

76. $\dfrac{2(x-1) + a(x-1)}{4(x-1) + a(x-1)}$

The following problems can be worked by using a least common denominator. Follow the directions in each and notice how the least common denominator is used.

77. Perform the operation indicated: $\dfrac{3x}{x^2 + 6x + 9} + \dfrac{1}{x+3}$.

78. Solve $\dfrac{x}{x+3} - \dfrac{3}{x^2 + 6x + 9} = 1$.

79. Simplify $\dfrac{\dfrac{1}{x+3} - 1}{\dfrac{x+2}{x^2 + 6x + 9}}$.

Label each problem as an expression, equation, or inequality. Solve the equations and inequalities. Perform the operations indicated on the expressions, leaving only nonnegative exponents in your answer.

80. $4^x = 16$

81. $\dfrac{1}{x-2} = \dfrac{4}{2-x}$

82. $(-2)^2$

83. -2^2

84. $(\sqrt{3} + \sqrt{5})^2$

85. $2x + 3(x - 7) \le 5$

86. $\sqrt{x-5} = 4$

87. $|x + 11| = 0$

88. $(x+2)(x-7) > 0$

89. $\sqrt{48} - \sqrt{12}$

90. $\dfrac{a}{a-b} + \dfrac{b}{b-a}$

91. $a^2(a^{-3}b)^3$

92. $\dfrac{1}{x-3} < 1$

93. $ab^{-3}(ab^4 - b^5)$

94. $\sqrt{x+1} - \sqrt{x+6} = 1$

95. $|x + 7| \ge 11$

► **You Decide**

You empty the cigar box, where throughout the last few years you have been stashing odd bits of money, and find you have accumulated $7,246.31. You decide to deposit it in a local savings institution for a year to gather a bit of interest. You narrow your choice to The First National Bank and the Federal Credit Union. The First National Bank pays 5% interest on passbook savings, compounded continuously. The Credit Union also offers 5% interest and will throw in a nice set of steak knives for opening the account. However, the Credit Union only compounds the interest monthly.

Determine how much interest each account pays and decide which plan you prefer. State the reasons for your choice.

► Chapter Test

In Problems 1 through 5, circle the correct answer.

1. $\log_a P = M$ is equivalent to (?)

   A. $M^a = P$ **B.** $a^M = P$

   C. $P^a = M$ **D.** $a^P = M$

2. The best approximation for log 43.4 is (?)

   A. 3.770459441   **B.** 1.63748973

   C. 16.3748973   **D.** 37.70459441

3. If $\log x = 0.8174$, the best approximation for x is (?)

   A. -0.0875653666 **B.** -0.20162670

   C. 2.2646042   **D.** 6.5674987

4. The best approximation for $e^{\sqrt{3}}$ is (?)

   A. 5.65223367   **B.** 20.0855369

   C. 0.0549306144   **D.** 1.09861228

5. Which graph best represents the graph of $f(x) = \log_3 x$ (?)

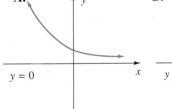
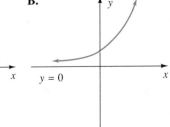
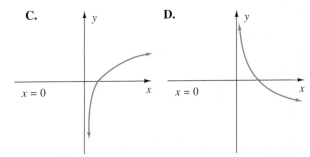

In Problems 6 and 7, evaluate each expression.

6. $\log_5 125$ **7.** $\log_2 (1/64)$

In Problems 8 through 10, suppose $f(x) = \dfrac{2}{x+3}$ and $g(x) = 2x - 1$.

8. Find $(f \circ g)(x)$.   **9.** Find $f^{-1}(x)$.   **10.** Find $(g \circ f)(-2)$.

In Problems 11 through 13, solve each equation.

11. $\log_3 27 = x$ **12.** $\log_2 x + \log_2 (x + 3) = 2$ **13.** $2^{x-6} = 64$

14. Approximate the solution to $2^{x-1} = 3^x$ to three decimal places.

15. How long will it take $35,000 invested at 6.5% compounded annually to double?

9 Conic Sections

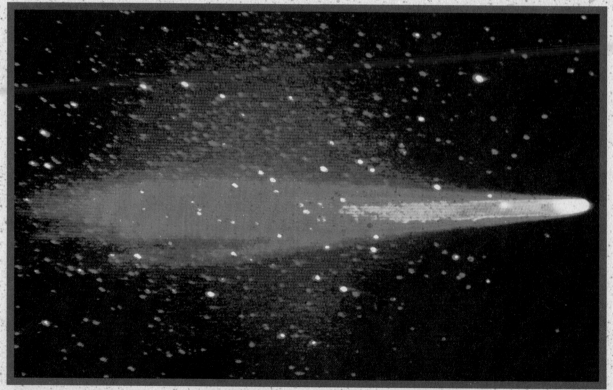

See Section 9.3, The Ellipse and the Hyperbola.

► Connections

For reasons lost in the mists of antiquity, the early Greek mathematicians spent a great deal of time studying the figures formed when a cone is sliced by a plane. In particular, Apollonius of Perga, born about 260 B.C., earned the title "The Great Geometer" from his study of these figures. The sketches below illustrate why these curves are called conic sections.

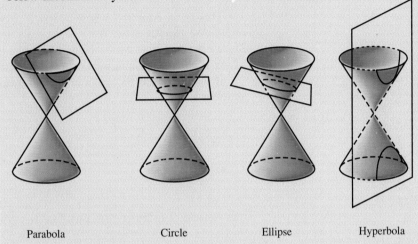

Parabola   Circle   Ellipse   Hyperbola

The conic sections have many interesting applications. For example, satellite orbits are ellipses, some comets follow hyperbolic paths as they approach the sun, and flashlight reflectors are parabolic in cross section.

This chapter begins with the study of a special figure with which we are already familiar, the circle. Next, we study the parabola and learn how to graph this conic section. In Section 9.3 we learn about the ellipse and the hyperbola and how to sketch their graphs. The last section shows the connection between the conic sections and the general second-degree equation in two variables.

9.1 The Circle

TAPE 16

In plane geometry we learned that a circle is the set of all points the same distance from a given point. The distance is called the **radius** and the given point the **center** of the circle.

Radius
Center

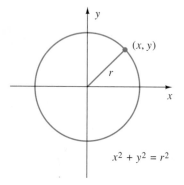

To find the basic equation for a circle with radius r, place a coordinate system so that the origin is at the center of the circle. Let (x, y) be *any* point on the circle. Then, because (x, y) must be r units from the center of the circle, the distance formula gives

$$r = \sqrt{(x - 0)^2 + (y - 0)^2}$$
$$= \sqrt{x^2 + y^2}$$
$$r^2 = x^2 + y^2$$

This is an equation whose graph is a circle of radius r and center at the origin. We usually write it $x^2 + y^2 = r^2$. It is the basic equation for a circle. Since r is a distance, it must be nonnegative. If r is 0, we have a circle of 0 radius, which is called a point circle or a degenerate circle.

If the circle has its center at the point (h, k), the basic equation is modified as follows.

The Circle

The graph of

$$(x - h)^2 + (y - k)^2 = r^2$$

is a circle of radius r with center at the point whose coordinates are (h, k).

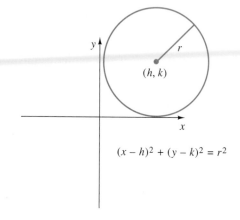

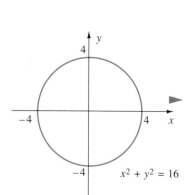

▶ **EXAMPLE 1**

Graph each equation.

(a) $x^2 + y^2 = 16$ (b) $(x - 1)^2 + (y - 2)^2 = 16$

Solutions

(a) This is of the form $x^2 + y^2 = r^2$, where r^2 is 16. Since r is a distance and must be positive, r must be 4. So this is a circle of radius 4 with center at the origin.

(b) This is of the form $(x - h)^2 + (y - k)^2 = r^2$, where h is 1, k is 2, and r is 4. Thus it is an equation of a circle of radius 4 centered at the point $(1, 2)$.

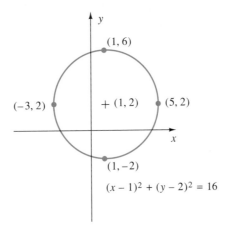

It is the graph of part (a) *shifted* so that its center is at the point (1, 2). ◄

► EXAMPLE 2

Graph the equation.

$$(x + 1)^2 + y^2 = 7$$

Solution

Notice that we can rewrite the equation as

$$[x - (-1)]^2 + (y - 0)^2 = (\sqrt{7})^2$$

which is of the form $(x - h)^2 + (y - k)^2 = r^2$, where h is -1, k is 0, and $r = \sqrt{7}$. Since $\sqrt{7} \approx 2.6$, we sketch the accompanying graph.

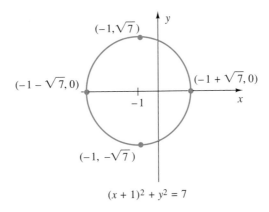

Sometimes the form of the equation is not as convenient as those in Examples 1 and 2, and we must complete the square before graphing.

► EXAMPLE 3

Sketch the graph of $x^2 + y^2 - 2x = 3$.

Solution

We note that there is *both* an x^2 term *and* an x term in the equation. We combine them to form the $(x - h)^2$ term by completing the square.

$$x^2 + y^2 - 2x = 3$$

First we group the x terms together.

$$x^2 - 2x + y^2 = 3$$

To complete the square, we divide -2 (the coefficient of x) by 2, square the result, and add this number to both sides of the equation. Since $\left(\dfrac{-2}{2}\right)^2 = 1$, we add 1 to both sides.

$$x^2 - 2x \boxed{+ 1} + y^2 = 3 \boxed{+ 1}$$
$$(x^2 - 2x + 1) + y^2 = 4$$

Now $(x^2 - 2x + 1)$ is a perfect square.

$$(x - 1)^2 + y^2 = 4$$

or, to put it in the form $(x - h)^2 + (y - k)^2 = r^2$,

$$(x - 1)^2 + (y - 0)^2 = 2^2$$

We can see that h is 1 and k is 0. So the center of the circle is $(1, 0)$ and the radius is 2.   ◄

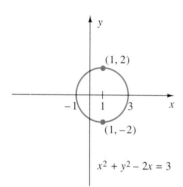

$x^2 + y^2 - 2x = 3$

Often, we must complete the square in *both* x and y.

► EXAMPLE 4

Graph each equation.

(a) $x^2 + y^2 + 4x - 6y + 12 = 0$
(b) $x^2 + y^2 - 3x + 2y = 0$

Solutions

(a) $x^2 + y^2 + 4x - 6y + 12 = 0$

We move 12 to the right side of the equation and group the x terms together and the y terms together.

$$(x^2 + 4x) + (y^2 - 6y) = -12$$

Now we complete the square twice, being careful to add the same number to both sides of the equation. We must add 4 to complete the square in x and 9 to complete the square in y.

$$(x^2 + 4x + 4) + (y^2 - 6y + 9) = -12 + 4 + 9$$
$$(x + 2)^2 + (y - 3)^2 = 1$$

To determine the values of h and k, we must be careful to write the equation in the form

$$(x - h)^2 + (y - k)^2 = r^2$$

This means that $x + 2$ must be written as $x - (-2)$.

$$[x - (-2)]^2 + (y - 3)^2 = 1^2$$

So we see that the center is $(-2, 3)$ and the radius is 1.

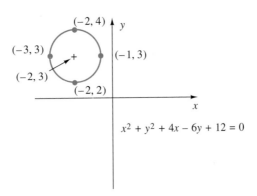

(b) $x^2 + y^2 - 3x + 2y = 0$
$(x^2 - 3x) + (y^2 + 2y) = 0$

It is necessary to add $\dfrac{9}{4}$ to complete the square in x and 1 to complete the square in y.

$$\left(x^2 - 3x + \frac{9}{4}\right) + (y^2 + 2y + 1) = 0 + \frac{9}{4} + 1$$

$$\left(x^2 - 3x + \frac{9}{4}\right) + (y^2 + 2y + 1) = \frac{13}{4}$$

$$\left(x - \frac{3}{2}\right)^2 + [y + 1]^2 = \frac{13}{4}$$

$$\left(x - \frac{3}{2}\right)^2 + [y - (-1)]^2 = \left(\frac{\sqrt{13}}{2}\right)^2$$

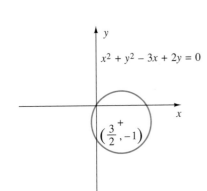

We see that the center is $\left(\dfrac{3}{2}, -1\right)$ and the radius is $\dfrac{\sqrt{13}}{2}$. To graph the circle, we must approximate $\dfrac{\sqrt{13}}{2}$.

$$\frac{\sqrt{13}}{2} \approx 1.8$$

◄

Problem Set **9.1**

Warm-Ups

In Problems 1 through 6, sketch the graph of the equation. See Examples 1 and 2.

1. $x^2 + y^2 = 4$
2. $x^2 + y^2 = 5$
3. $x^2 + (y - 1)^2 = 9$

4. $(x - 3)^2 + y^2 = 8$ **5.** $(x + 1)^2 + (y - 2)^2 = 10$ **6.** $(x - 1)^2 + (y + 1)^2 = 6$

In Problems 7 through 12, the graph of each equation is a circle. Find the center and radius; then sketch the graph. See Example 3.

7. $x^2 + y^2 - 2x = 8$ **8.** $x^2 + y^2 - 4y = 12$ **9.** $x^2 + y^2 + 8y = 4$

10. $x^2 + y^2 + 6x = -1$ **11.** $x^2 + y^2 - 10x + 16 = 0$ **12.** $x^2 + y^2 - 12y + 32 = 0$

In Problems 13 through 16, sketch the graph. See Example 4.

13. $x^2 + y^2 - 2x - 4y = 4$ **14.** $x^2 + y^2 - 6x - 8y = 0$ **15.** $x^2 + y^2 - x - 2 = 0$ **16.** $x^2 + y^2 - 3y - 4 = 0$

Practice Exercises

In Problems 17 through 58, the graph of each equation is a circle. Find the center and radius; then sketch the graph.

17. $x^2 + y^2 = 9$ **18.** $x^2 + y^2 = 6$ **19.** $x^2 + (y - 2)^2 = 16$

20. $(x - 1)^2 + y^2 = 12$

21. $(x + 3)^2 + (y + 2)^2 = 49$

22. $\left(x + \dfrac{1}{2}\right)^2 + \left(y - \dfrac{3}{2}\right)^2 = \dfrac{9}{4}$

23. $x^2 + y^2 - 4x = 5$

24. $x^2 + y^2 - 2y = 15$

25. $x^2 + y^2 - 8y = 0$

26. $x^2 + y^2 - 6x = 16$

27. $x^2 + y^2 - 6y = 0$

28. $x^2 + y^2 - 8x = 9$

29. $x^2 + y^2 + 8x = -7$

30. $x^2 + y^2 + 4y = 0$

31. $x^2 + y^2 + 2y = 7$

32. $x^2 + y^2 + 6x = 3$

33. $x^2 + y^2 - 12x - 13 = 0$

34. $x^2 + y^2 - 10y - 11 = 0$

35. $x^2 + y^2 - 4x - 2y = 11$

36. $x^2 + y^2 - 8x - 6y = 0$

37. $x^2 + y^2 - 4x + 4y - 1 = 0$

38. $x^2 + y^2 + 12x - 2y + 1 = 0$ **39.** $x^2 + y^2 + 6x - 6y + 9 = 0$ **40.** $x^2 + y^2 - 10x + 2y + 10 = 0$

41. $(x + 2)^2 + (y + 3)^2 = 25$ **42.** $\left(x + \dfrac{3}{2}\right)^2 + \left(y - \dfrac{1}{2}\right)^2 = \dfrac{9}{4}$ **43.** $x^2 + y^2 + 4x = -3$

44. $x^2 + y^2 + 2y = 0$ **45.** $x^2 + y^2 + 12x + 4y = 0$ **46.** $x^2 + y^2 + 8x + 10y - 4 = 0$

47. $x^2 + y^2 - 5x - y + 4 = 0$ **48.** $x^2 + y^2 + x - y - \dfrac{1}{2} = 0$ **49.** $x^2 + y^2 + 7x + y = 0$

50. $x^2 + y^2 + 5x + 5y + \dfrac{1}{4} = 0$ **51.** $x^2 + y^2 + 6x + 8y + 5 = 0$ **52.** $x^2 + y^2 + 2x + 6y = 0$

53. $x^2 + y^2 - y - 1 = 0$ **54.** $x^2 + y^2 - 3x - 3 = 0$ **55.** $x^2 + y^2 - x - 3y - 6 = 0$

56. $x^2 + y^2 - x + y = 0$ **57.** $x^2 + y^2 + x + 5y = 0$ **58.** $x^2 + y^2 + 7x + 7y + \dfrac{3}{4} = 0$

Challenge Problems

59. Show that the equation of a circle of radius r, centered at the point whose coordinates are (h, k), is

$$(x - h)^2 + (y - k)^2 = r^2$$

60. What is the graph of $x^2 + y^2 - 2x - 4y + 5 = 0$?

In Problems 61 through 64, the graph of each equation is a circle. Find the center and radius; then sketch the graph.

61. $2x^2 + 2y^2 - 4x - 2y + 1 = 0$

62. $2x^2 + 2y^2 + 6x - 2y - 3 = 0$

63. $4 + 2x + 4y - x^2 - y^2 = 0$

64. $5x - 3y - 2x^2 - 2y^2 + 1 = 0$

In Your Own Words

65. Describe how completing the square is used in graphing circles.

66. Would the graph of $x^2 + y^2 = -4$ be a circle? Explain.

9.2   The Parabola

TAPE 16

In Section 6.6 we learned that the graphs of second-degree equations of the form

$$y = Ax^2 + Bx + C, \qquad A \neq 0$$

are parabolas. In this section we continue our study of parabolas by graphing second-degree equations of the form

$$x = Dy^2 + Ey + F, \qquad D \neq 0$$

We begin by graphing the simplest such equation, $x = y^2$. Make a table by choosing a number for y and then calculate x.

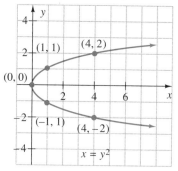

x	y
0	0
1	1
1	−1
4	2
4	−2

This is a parabola of the same general shape as the graph of $y = x^2$ except that it is lying on its side, opening to the right. In fact, if we interchange the roles of x and y (and h and k) in the equation $y = p(x - h)^2 + k$, we get similar figures, except that they open left–right instead of up–down.

The Parabolas

$$y = p(x - h)^2 + k$$

is a parabola with vertex at the point (h, k). It opens *upward* if p is positive and *downward* if p is negative. Its axis of symmetry is the line $x = h$.

$$x = p(y - k)^2 + h$$

is a parabola with vertex at the point (h, k). It opens *to the right* if p is positive and *to the left* if p is negative. Its axis of symmetry is the line $y = k$.

► EXAMPLE 1

Sketch the graph of each equation and identify the vertex of each graph.

(a) $x = y^2 + 1$ (b) $x = -2y^2$

Solutions

(a) This graph is a parabola that opens to the *right*. Since the equation is solved for x, pick numbers for y and then calculate x.

x	y
2	−1
1	0
2	1

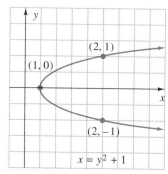

The vertex is $(1, 0)$. This graph is the same as the graph of $x = y^2$ except that it is shifted 1 unit to the right.

(b) This graph is a parabola that opens to the left. Make a table and plot the points.

x	y
−2	−1
0	0
−2	1

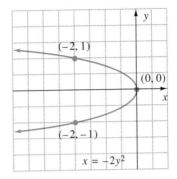

The vertex is (0, 0). This parabola is thinner than the graph of $x = y^2$.

► EXAMPLE 2

Sketch the graph of each equation. Identify the vertex.

(a) $x = \dfrac{1}{2}(y - 2)^2 + 1$ (b) $x = -(y + 2)^2$

Solutions

(a) $x = \dfrac{1}{2}(y - 2)^2 + 1$ matches the form of $x = p(y - k)^2 + h$ with p as $\dfrac{1}{2}$, k as 2, and h as 1. Thus, the parabola opens to the right and has vertex (1, 2).

x	y
3	0
3/2	1
1	2
3/2	3

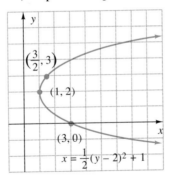

(b) $x = -(y + 2)^2$ matches the form of $x = p(y - k)^2 + h$ with p as −1, k as −2, and h as 0. So the parabola opens to the left and has vertex of (0, −2).

x	y
−1	−3
0	−2
−1	−1

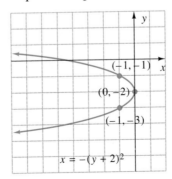

Sometimes we must complete the square to find the vertex.

► EXAMPLE 3

Graph the equation $x + 2y^2 - 8y = -5$.

Solution

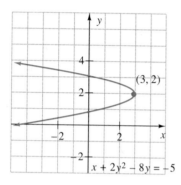

Solve the equation for x. (Note the y^2 term.)

$$x = -2y^2 + 8y - 5$$

Before completing the square, it is necessary to factor -2 from the first two terms.

$$= -2(y^2 - 4y) - 5$$

Now we can complete the square *inside the parentheses.*

$$= -2(y^2 - 4y + 4 - 4) - 5$$
$$= -2(y^2 - 4y + 4) + 8 - 5$$
$$= -2(y - 2)^2 + 3$$

This is a parabola with vertex at $(3, 2)$, opening to the left.   ◄

Problem Set 9.2

Warm-Ups

In Problems 1 through 4, sketch the graph of each equation. Label the vertex. See Example 1.

1. $x = y^2 - 1$ **2.** $x = y^2 + 2$ **3.** $x = -\dfrac{1}{2}y^2$ **4.** $x = 2y^2 + 1$

In Problems 5 through 8, sketch the graph. Label the vertex. See Example 2.

5. $x = (y + 1)^2 + 2$ **6.** $x = (y - 1)^2 - 3$ **7.** $x = -\dfrac{1}{2}(y + 1)^2$ **8.** $x = -2(y - 2)^2$

For Problems 9 through 12, sketch the graph. Label the vertex. See Example 3.

9. $x = y^2 - 3y + 2$ **10.** $y^2 + 2y = x$ **11.** $x = -2y^2 + 4y$ **12.** $x = 3y^2 + 18y + 29$

Practice Exercises

In Problems 13 through 48, the graph of each equation is a parabola. Find the vertex of each and sketch the graph.

13. $x = (y - 1)^2$ **14.** $x = -y^2 + 2$ **15.** $y = (x - 3)^2 + 2$ **16.** $y = -(x - 2)^2 - 3$

17. $x = (y + 3)^2 + 3$ **18.** $y = -2(x + 2)^2 - 1$ **19.** $y = \frac{1}{2}\left(x + \frac{2}{3}\right)^2 + \frac{1}{3}$ **20.** $x = -\frac{2}{3}\left(y + \frac{1}{2}\right)^2 - \frac{2}{3}$

21. $y = -(x + 3)^2 + 3$ **22.** $x = 2(y + 2)^2 - 1$ **23.** $x = \frac{1}{3}\left(y + \frac{3}{2}\right)^2 - \frac{1}{2}$ **24.** $y = -\left(x + \frac{1}{3}\right)^2 - \frac{4}{3}$

25. $y^2 - 2x - 4 = 0$ **26.** $x^2 + 3y = 3$ **27.** $x = -2y^2 - 8y - 8$ **28.** $y = -2x^2 - 2x - 1$

29. $3x^2 - 9x - y + 6 = 0$

30. $4y^2 - 2x - 12y = -7$

31. $y^2 - x = 0$

32. $x^2 + y = 0$

33. $x^2 - 2 = y$

34. $x = -y^2 - 1$

35. $3y^2 + 3 = x$

36. $y = -2x^2 + 2$

37. $x^2 + 2y + 4 = 0$

38. $3y^2 - x = 3$

39. $y^2 - 8y = x - 10$

40. $y = -x^2 - 2x$

41. $x^2 - x + 2 = y$

42. $y^2 + 5y = x$

43. $x = -2y^2 + 8y$

44. $y = 3x^2 + 6x + 5$

45. $y = -2x^2 + 4x - 2$

46. $x = -3y^2 - 3y - 1$

47. $3y^2 + 9y - x + 6 = 0$

48. $4x^2 - 2y + 4x = 1$

Challenge Problems

In Problems 49 through 56, the graph of each equation is a parabola. Find the vertex of each and sketch the graph.

49. $2y^2 - x - 3y + 1 = 0$ **50.** $3x^2 + 2x + y = 1$ **51.** $3x^2 - 4x - 2y = 0$ **52.** $2y^2 + 3x + 5y + 2 = 0$

53. $2x^2 + y + 3x + 2 = 0$ **54.** $2y^2 + y - 2x = 1$ **55.** $4y^2 - 5y - 2x = 0$ **56.** $2x^2 + 5y + 7x + 3 = 0$

In Your Own Words

57. Describe the procedure used to rewrite an equation of the form $y = Ax^2 + Bx + C$ into the form $y = p(x - h)^2 + k$.

9.3   The Ellipse and the Hyperbola

TAPE 16

In plane geometry, *an* **ellipse** is described as the set of all points in a plane, the sum of whose distances from two fixed points is constant. The ellipse is all points such that $d_1 + d_2 = K$, a constant.

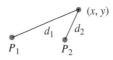

Sketching the points results in a figure like this:

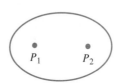

We put a coordinate system on this figure so that the points P_1 and P_2, called the *foci* (the singular is *focus*), are on the *x*-axis and the origin is halfway between the foci at the center of the ellipse.

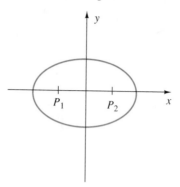

The equation whose graph is this ellipse can be written in the form

$$\frac{x^2}{a^2} + \frac{y^2}{b^2} = 1$$

where a and b are the constants shown on the following figure. The x-intercepts of the graph are at $\pm a$ and the y-intercepts are at $\pm b$.

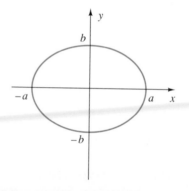

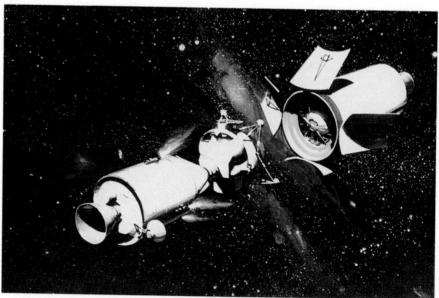

If b is larger than a, the foci lie on the y-axis and the graph looks like this:

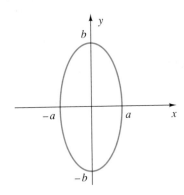

If both foci are the same point, the graph is a circle. Thus we see that a circle is a special case of an ellipse.

► **EXAMPLE 1**

Sketch the graph of the equation

$$\frac{x^2}{9} + \frac{y^2}{4} = 1$$

Solution

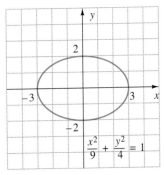

This is in the form $\dfrac{x^2}{a^2} + \dfrac{y^2}{b^2} = 1$ if a is 3 and b is 2. That is,

$$\frac{x^2}{3^2} + \frac{y^2}{2^2} = 1$$

Thus the x-intercepts are ±3 and the y-intercepts are ±2. The graph is a *smooth* oval through those points.   ◄

► **EXAMPLE 2**

Sketch the graph of the equation

$$\frac{x^2}{16} + \frac{y^2}{25} = 1$$

Solution

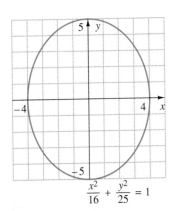

Rewriting the equation in the form $\dfrac{x^2}{4^2} + \dfrac{y^2}{5^2} = 1$ shows that a is 4 and b is 5. The x-intercepts are ±4 and the y-intercepts are ±5.   ◄

The x-intercepts, a and $-a$, and the y-intercepts, b and $-b$, are sufficient to sketch the graph of an ellipse. Find these four points; then connect them with the smoothest oval possible.

Here is the basic equation for the ellipse.

The Ellipse Centered at the Origin

The graph of the equation

$$\frac{x^2}{a^2} + \frac{y^2}{b^2} = 1$$

is an ellipse with x-intercepts a and $-a$ and y-intercepts b and $-b$, centered at the origin.

The last of the conic sections, the **hyperbola,** can be described as the set of all points the difference of whose distances from two fixed points is constant. Sketching these points results in a figure like this:

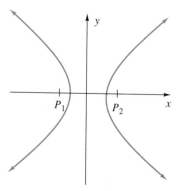

We put a coordinate system on this figure so that the foci, the points P_1 and P_2, are on the x-axis and the origin is halfway between them.

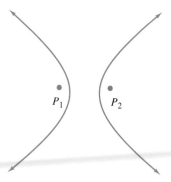

The equation whose graph is this hyperbola is

$$\frac{x^2}{a^2} - \frac{y^2}{b^2} = 1$$

where a and $-a$ are the x-intercepts as shown on the figure at left.

If we reverse the roles of x and y, we have an equation of the form

$$\frac{y^2}{b^2} - \frac{x^2}{a^2} = 1$$

where b and $-b$ are the *y-intercepts,* as shown on the following figure.

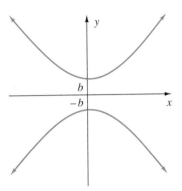

So we have hyperbolas opening left–right given by the equation

$$\frac{x^2}{a^2} - \frac{y^2}{b^2} = 1$$

and hyperbolas opening up–down given by the equation

$$\frac{y^2}{b^2} - \frac{x^2}{a^2} = 1$$

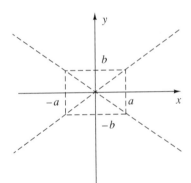

We use the values of a and b to sketch a hyperbola by drawing some construction lines. First, we draw the rectangle shown using the values of a and b. Next we draw the diagonals of the rectangle, extending them as far as necessary. The two diagonals are asymptotes of the hyperbola. That is, the farther the hyperbola is from the origin, the closer it gets to the asymptote. Using these construction lines as guides, we now sketch the hyperbola, either left–right or up–down, depending on the form of the equation.

► **EXAMPLE 3**

Sketch the graph of the equation

$$\frac{x^2}{9} - \frac{y^2}{4} = 1$$

Solution

The equation $\dfrac{x^2}{9} - \dfrac{y^2}{4} = 1$ can be written $\dfrac{x^2}{3^2} - \dfrac{y^2}{2^2} = 1$. Therefore, a is 3 and b is 2. The construction lines are drawn as follows. Noting that the equation is in the left–right form, we can then sketch the hyperbola.

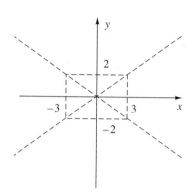

Construction Lines

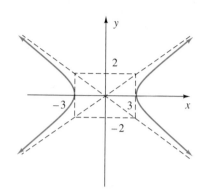

Graph of $\dfrac{x^2}{9} - \dfrac{y^2}{4} = 1$ ◄

► **EXAMPLE 4**

Sketch the graph of the equation

$$\frac{y^2}{16} - \frac{x^2}{25} = 1$$

Solution

Here a is 5 and b is 4 and we have an up–down form.

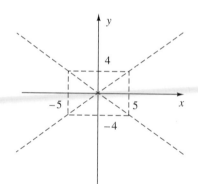

Construction Lines

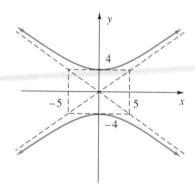

Graph of $\dfrac{y^2}{16} - \dfrac{x^2}{25} = 1$ ◄

We now have the basic equations for the hyperbola.

Hyperbolas Centered at the Origin

The graph of the equation

$$\frac{x^2}{a^2} - \frac{y^2}{b^2} = 1$$

is a hyperbola with x-intercepts a and $-a$ centered at the origin.
    The graph of the equation

$$\frac{y^2}{b^2} - \frac{x^2}{a^2} = 1$$

is a hyperbola with y-intercepts b and $-b$ centered at the origin.

Like the circle, if x is replaced with $x - h$ and y with $y - k$, the basic ellipse or hyperbola will be shifted on the coordinate system.

The Ellipse

The graph of the equation

$$\frac{(x - h)^2}{a^2} + \frac{(y - k)^2}{b^2} = 1$$

is the graph of the ellipse $\frac{x^2}{a^2} + \frac{y^2}{b^2} = 1$ shifted so that its center is at the point (h, k).

The Hyperbolas

The graph of the equation

$$\frac{(x - h)^2}{a^2} - \frac{(y - k)^2}{b^2} = 1$$

is the graph of the hyperbola $\frac{x^2}{a^2} - \frac{y^2}{b^2} = 1$ shifted so that its center is at the point (h, k).

The graph of the equation

$$\frac{(y - k)^2}{b^2} - \frac{(x - h)^2}{a^2} = 1$$

is the graph of the hyperbola $\frac{y^2}{b^2} - \frac{x^2}{a^2} = 1$ shifted so that its center is at the point (h, k).

▶ **EXAMPLE 5**

Graph the equation

$$4x^2 + y^2 - 24x + 4y + 24 = 0$$

Solution

Since this equation contains both x and x^2 terms, we complete the square to form the $(x - h)^2$ term. Because it contains y and y^2 terms, we also complete the square to obtain the $(y - k)^2$ term.

$$4x^2 + y^2 - 24x + 4y + 24 = 0$$

First, group the x and y terms and factor where necessary.

$$4(x^2 - 6x) + (y^2 + 4y) + 24 = 0$$

Now we complete the square inside each set of parentheses and simplify.

$$4(x^2 - 6x + 9 - 9) + (y^2 + 4y + 4 - 4) + 24 = 0$$

$$4(x^2 - 6x + 9) - 36 + (y^2 + 4y + 4) - 4 + 24 = 0$$

$$4(x - 3)^2 + (y + 2)^2 - 16 = 0$$

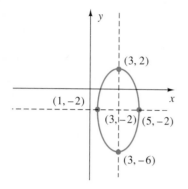

Graph of $\dfrac{(x-3)^2}{2^2} + \dfrac{(y+2)^2}{4^2} = 1$

Add 16 to both sides and then divide by 16.

$$4(x-3)^2 + (y+2)^2 = 16$$

$$\frac{4(x-3)^2}{16} + \frac{(y+2)^2}{16} = 1$$

$$\frac{(x-3)^2}{4} + \frac{(y+2)^2}{16} = 1$$

Notice that this is the ellipse $\dfrac{x^2}{2^2} + \dfrac{y^2}{4^2} = 1$ shifted so that its center is at the point $(3, -2)$. An easy way to make such a shift is to sketch a copy of the coordinate system with its origin at the point $(3, -2)$; then graph the basic ellipse on the shifted system.   ◄

► **EXAMPLE 6**

Graph the equation

$$4x^2 - 6y^2 + 32x + 12y + 22 = 0$$

Solution

Again we complete the square in both x and y.

$$4x^2 - 6y^2 + 32x + 12y + 22 = 0$$

$$4(x^2 + 8x) - 6(y^2 - 2y) + 22 = 0$$

$$4(x^2 + 8x + 16 - 16) - 6(y^2 - 2y + 1 - 1) + 22 = 0$$

$$4(x^2 + 8x + 16) - 64 - 6(y^2 - 2y + 1) + 6 + 22 = 0$$

$$4(x + 4)^2 - 6(y - 1)^2 = 36$$

$$\frac{4(x + 4)^2}{36} - \frac{6(y - 1)^2}{36} = 1$$

$$\frac{(x + 4)^2}{9} - \frac{(y - 1)^2}{6} = 1$$

We have a left–right hyperbola with $a = 3$ and $b = \sqrt{6}$, centered at the point $(-4, 1)$. We draw the basic hyperbola in its shifted position as we did the ellipse in Example 5.

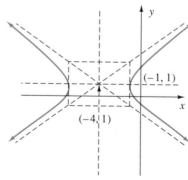

Graph of $\dfrac{(x+4)^2}{9} - \dfrac{(y-1)^2}{6} = 1$ ◄

► **EXAMPLE 7**

Graph the equation

$$81x^2 - 16y^2 - 162x + 117 = 0$$

Solution

As there is no y term in this equation, we only need to complete the square in x.

$$81x^2 - 16y^2 - 162x + 117 = 0$$
$$81(x^2 - 2x) - 16y^2 + 117 = 0$$
$$81(x^2 - 2x + 1 - 1) - 16y^2 + 117 = 0$$
$$81(x^2 - 2x + 1) - 81 - 16y^2 + 117 = 0$$
$$81(x - 1)^2 - 16y^2 = -36$$
$$\frac{81(x - 1)^2}{-36} - \frac{16y^2}{-36} = 1$$
$$-\frac{9(x - 1)^2}{4} + \frac{4y^2}{9} = 1$$

We rearrange the terms and then divide numerator and denominator of the x term by 9 and the y term by 4 to put it in the form

$$\frac{(y - k)^2}{b^2} - \frac{(x - h)^2}{a^2} = 1$$

$$\frac{y^2}{\dfrac{9}{4}} - \frac{(x - 1)^2}{\dfrac{4}{9}} = 1$$

This is an up–down hyperbola centered at $(1, 0)$, where a is $\dfrac{2}{3}$ and b is $\dfrac{3}{2}$. We are now ready to sketch the graph. Again, we move the *coordinate axes* so that they are centered at $(1, 0)$ and draw our construction lines there.   ◄

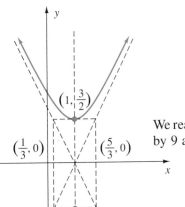

Graph of $-\dfrac{9(x - 1)^2}{4} + \dfrac{4y^2}{9} = 1$

Problem Set 9.3

Warm-Ups

In Problems 1 through 14, sketch the graph of each equation. See Examples 1 and 2.

1. $\dfrac{x^2}{36} + \dfrac{y^2}{16} = 1$

2. $\dfrac{x^2}{25} + \dfrac{y^2}{64} = 1$

3. $x^2 + \dfrac{y^2}{9} = 1$

4. $\dfrac{x^2}{49} + \dfrac{y^2}{36} = 1$

In Problems 5 through 8, sketch the graph. See Examples 3 and 4.

5. $\dfrac{x^2}{16} - \dfrac{y^2}{4} = 1$ 　　　　 **6.** $\dfrac{y^2}{9} - \dfrac{x^2}{16} = 1$ 　　　 **7.** $\dfrac{y^2}{4} - x^2 = 1$ 　　　　 **8.** $x^2 - y^2 = 1$

In Problems 9 through 12, sketch the graph. See Example 5.

9. $\dfrac{(x-1)^2}{36} + \dfrac{(y-2)^2}{49} = 1$ 　　　　　　　 **10.** $\dfrac{16(x-1)^2}{25} + \dfrac{9(y-1)^2}{4} = 1$

11. $9x^2 + 4y^2 - 18x - 8y - 23 = 0$ 　　　　 **12.** $x^2 + 9y^2 - 4x + 36y + 31 = 0$

In Problems 13 through 16, sketch the graph. See Examples 6 and 7.

13. $\dfrac{(y-1)^2}{9} - \dfrac{(x+1)^2}{4} = 1$ 　　　　　　 **14.** $\dfrac{(x+3)^2}{9} - \dfrac{(y+2)^2}{9} = 1$

15. $4x^2 - y^2 - 16x + 2y + 11 = 0$ 　　　　 **16.** $x^2 - y^2 - 2x + 2y - 4 = 0$

Practice Exercises

In Problems 17 through 48, sketch the graph of each equation.

17. $\dfrac{x^2}{36} - \dfrac{y^2}{16} = 1$ 　　　　 **18.** $\dfrac{x^2}{4} + \dfrac{y^2}{16} = 1$ 　　　　 **19.** $\dfrac{y^2}{9} + \dfrac{x^2}{16} = 1$

20. $\dfrac{y^2}{25} - \dfrac{x^2}{64} = 1$

21. $\dfrac{x^2}{4} + y^2 = 1$

22. $\dfrac{(x-1)^2}{36} - \dfrac{(y-2)^2}{49} = 1$

23. $\dfrac{(x+1)^2}{9} + \dfrac{(y-1)^2}{4} = 1$

24. $\dfrac{(y+3)^2}{9} - \dfrac{(x+2)^2}{9} = 1$

25. $\dfrac{16(x-1)^2}{25} - \dfrac{9(y-1)^2}{4} = 1$

26. $y^2 - x^2 = 1$

27. $2x^2 + y^2 = 8$

28. $x^2 - 2y^2 = 12$

29. $x^2 - 4y^2 + 16y = 0$

30. $25x^2 + 9y^2 + 50x - 200 = 0$

31. $4x^2 + 9y^2 + 32x - 36y + 64 = 0$

32. $x^2 - y^2 + 6x + 4y - 4 = 0$

33. $16x^2 + y^2 + 64x - 8y + 64 = 0$

34. $4x^2 + 5y^2 - 8x + 20y + 4 = 0$

35. $9x^2 - 36y^2 + 72y = 0$

36. $y^2 - 2x^2 = 8$

37. $x^2 - y^2 - 2x + 2y + 4 = 0$

38. $4x^2 + 9y^2 + 8x + 18y - 23 = 0$ **39.** $x^2 - 4y^2 + 16y - 32 = 0$ **40.** $9x^2 + y^2 + 36x - 4y + 31 = 0$

41. $25x^2 - 9y^2 + 50x - 200 = 0$ **42.** $4x^2 + 9y^2 + 32x + 36y + 64 = 0$

43. $x^2 - y^2 + 6x + 4y + 14 = 0$ **44.** $x^2 + 25y^2 + 10x + 50y + 25 = 0$

45. $4x^2 - 5y^2 - 8x - 20y - 36 = 0$ **46.** $5x^2 + 3y^2 - 20x + 6y + 8 = 0$

47. $9x^2 + 36y^2 + 72y + 32 = 0$ **48.** $100x^2 - 144y^2 + 600x + 1125 = 0$

Challenge Problems

In Problems 49 through 52, sketch the graph of each equation.

49. $4x^2 + 9y^2 - 4x - 6y - 34 = 0$ **50.** $64x^2 - 36y^2 + 64x + 108y - 641 = 0$

51. $9x^2 - 4y^2 - 3x - 2y + 144 = 0$ **52.** $4x^2 + 36y^2 + 12x + 36y + 9 = 0$

In Your Own Words

53. Explain how to graph $\dfrac{(x - h)^2}{a^2} + \dfrac{(y - k)^2}{b^2} = 1.$

54. Why is a circle a special case of an ellipse?

9.4 The General Second-degree Equation in Two Variables

TAPE 16

Our interest in the conic sections stems from their relationship to the general second-degree equation in two variables,

$$Ax^2 + Bxy + Cy^2 + Dx + Ey + F = 0, \qquad \text{where } A, B, \text{ and } C \text{ are not } all \ 0$$

Except for some special cases, the graphs of these equations turn out to be conic sections. The term Bxy, in the general equation, will *rotate* the conic if $B \neq 0$. We will only consider the cases where $B = 0$. Certain characteristics of the equation make some graphs parabolas, some ellipses, and some hyperbolas. It is possible to detect which conic section will be graphed by looking at the original equation.

> ### The Graph of $Ax^2 + Cy^2 + Dx + Ey + F = 0$
>
> **1.** If A or C is zero, the graph is a parabola.
>   (a) A *horizontal* axis of symmetry if $A = 0$.
>   (b) A *vertical* axis of symmetry if $C = 0$.
> **2.** If $A = C$, the graph is a circle.
> **3.** If A and C are both positive or both negative, the graph is an ellipse.
> **4.** If A and C are of opposite sign, the graph is a hyperbola.
> **5.** There are certain degenerate forms that *may* result from these equations, such as lines, points, and no graph at all.

Some examples of the degenerate forms mentioned in 5, above, are:

Equation	Graph
$x^2 + y^2 = 0$	The point $(0, 0)$
$x^2 - 1 = 0$	Two parallel lines
$x^2 - y^2 = 0$	Two intersecting lines
$x^2 + y^2 + 1 = 0$	No graph

▶ **EXAMPLE 1**

Identify each of the following as the equation of a parabola, circle, ellipse, or hyperbola. (None are degenerate.)

(a) $3x^2 - 4y^2 - 7x + 8y + 14 = 0$ (b) $5x^2 + 4y^2 + x - 10 = 0$

(c) $5x^2 + 4y + x - 10 = 0$ (d) $-3x^2 - 3y^2 + 6x + 5y - 11 = 0$

Solutions

(a) Neither the coefficient of x^2 nor the coefficient of y^2 is zero, and they are of opposite sign. Therefore, the graph of this equation is a hyperbola.

(b) The coefficient of x^2 and the coefficient of y^2 are both positive; thus the graph is an ellipse.

(c) There is no y^2 term in this equation, so its coefficient must be zero. The graph is a parabola opening left or right.

(d) The coefficients of x^2 and y^2 are equal. The graph is a circle. Of course, it is also an ellipse. ◄

Graphing Conic Sections

1. If the center (vertex, in the case of a parabola) is not at the origin, lightly sketch a coordinate system at the point where the center is shifted. Work from this "new" coordinate system.

2. **Parabola.** Plot the vertex and a point or two on either side of the vertex. Draw the smoothest "parabola shape" possible through those points.

3. **Ellipse.** Mark the x- and y-intercepts and connect them with the smoothest oval possible.

4. **Hyperbola.** Draw the construction lines lightly. Check to see if it is an up–down or left–right hyperbola, then draw both parts as smooth curves.

Note! The conic sections are all very smooth curves without sharp turns or pointy places. Draw them as evenly as possible.

Problem Set 9.4

Warm-Ups

In Problems 1 through 8, identify each equation as that of a parabola, circle, ellipse, or hyperbola. (Assume that none are degenerate.) See Example 1.

1. $x^2 - y^2 - 3x + y - 4 = 0$

2. $22x^2 + 7y^2 + 41x - 234 = 0$

3. $8x^2 + 5y^2 - 19x - 7y - 25 = 0$

4. $9y^2 + 32x - 36y + 64 = 0$

5. $3x^2 + 3y^2 + 13y = 0$

6. $2x^2 - 3y^2 + 7x + 14y - 4 = 0$

7. $x^2 - 4x + 36y + 31 = 0$

8. $64x + 8y - 16x^2 - 16y^2 + 64 = 0$

Practice Exercises

In Problems 9 through 20, identify each equation as that of a parabola, circle, ellipse, or hyperbola. (Assume that none are degenerate.)

9. $13 - 2y^2 + 3x - 7y = 0$

10. $5x^2 - 11x + 7y^2 + 9y - 44 = 0$

11. $3x^2 + 2y^2 - 2x - 2y - 14 = 0$

12. $11x^2 + 11y^2 - 13x - 5 = 0$

13. $11x^2 + 11y - 13x - 5 = 0$

14. $11x^2 - 11y^2 - 13x - 5 = 0$

15. $5x^2 + 6x - 6y^2 - 5y = 0$

16. $5x^2 + 5y - 7x + y^2 - 17 = 0$

17. $2x - 3y + 4x^2 + 5y^2 - 6 = 0$

18. $x^2 + 3x - 4y = 16 + 3y^2$

19. $7x - 8y - 4x^2 - 4y^2 = 13$

20. $17x = 21y^2$

Challenge Problems

In Problems 21 and 22, sketch the graph of each, if it exists.

21. $4x^2 + 3y^2 - 8x + 6y + 7 = 0$

22. $x^2 - 4y^2 = 0$

In Your Own Words

23. Describe the kind of equations that graph as conic sections.

▶ Chapter Summary

Second-degree Equation in Two Variables

The graph of the general second-degree equation in two variables is

$$Ax^2 + Bxy + Cy^2 + Dx + Ey + F = 0, \qquad A, B, C \text{ not all zero}$$

is a conic section or a degenerate form of a conic section.

The Conic Sections

The basic forms of the conic sections are:

1. Parabola

$y = px^2$ 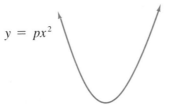 $x = py^2$

2. Ellipse

$$\frac{x^2}{a^2} + \frac{y^2}{b^2} = 1$$

3. Hyperbola

$$\frac{x^2}{a^2} - \frac{y^2}{b^2} = 1 \qquad\qquad \frac{y^2}{b^2} - \frac{x^2}{a^2} = 1$$

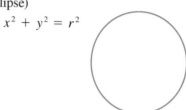

4. Circle (a special ellipse)

$$x^2 + y^2 = r^2$$

Translation of Axes   The substitution of $(x - h)$ for x and $(y - k)$ for y in one of the basic forms will shift the graph so that its center (or vertex in the case of a parabola) is at the point (h, k).

CHECKUPS

In Checkups 1 through 4, sketch the graph of each equation.

1. $(x - 1)^2 + (y - 2)^2 = 16$ Section 9.1; Example 1b

2. $x = \dfrac{1}{2}(y - 2)^2 + 1$ Section 9.2; Example 2a

3. $4x^2 + y^2 - 24x + 4y + 24 = 0$ Section 9.3; Example 5

4. $4x^2 - 6y^2 + 32x + 12y + 22 = 0$ Section 9.3; Example 6

5. Identify each of the following as the equation of
a parabola, circle, ellipse, or hyperbola.   Section 9.4; Example 1
  (a) $3x^2 - 4y^2 - 7x + 8y + 14 = 0$
  (b) $5x^2 + 4y^2 + x - 10 = 0$
  (c) $5x^2 + 4y + x - 10 = 0$
  (d) $-3x^2 - 3y^2 + 6x + 5y - 11 = 0$

► Review Problems

In Problems 1 through 10, identify the conic section.

1. $x^2 - y^2 - 2x - 6y = 0$

2. $x^2 + y^2 - 4x + 8y = 0$

3. $x^2 + 5y^2 - 10x - 13 = 0$

4. $x^2 - y - 2x = 0$

5. $2x^2 + 3y^2 - 4x + 6y = 21$

6. $x - 2y^2 - 8y - 2 = 0$

7. $3x^2 - 5y^2 - 18x - 20y - 100 = 0$

8. $5x + 10y - 5x^2 - 5y^2 + 33 = 0$

9. $2x^2 + x - y + 8 = 0$

10. $3x^2 + 4y^2 + 2x - 3y = 0$

In Problems 11 through 20, sketch the graph of each equation.

11. $\dfrac{x^2}{4} + \dfrac{y^2}{25} = 1$ **12.** $(x - 3)^2 + (y + 2)^2 = 16$ **13.** $x + 2 = 2(y - 1)^2$ **14.** $\dfrac{(y + 3)^2}{9} - \dfrac{(x + 1)^2}{36} = 1$

15. $x^2 - y^2 - 4x - 4y - 16 = 0$ **16.** $4x^2 + 5y^2 + 24x - 10y + 21 = 0$

17. $16x^2 + 9y^2 - 32x + 18y - 119 = 0$ **18.** $16x^2 - 4y^2 + 96x - 16y + 96 = 0$

19. $x^2 + y^2 - 4x + 6y + 12 = 0$ **20.** $y = 2x^2 + 8x + 7$

Let's Not Forget . . .

Identify the expressions that are in factored form. Factor those that are not factored.

21. $(2x - 1)(4x^2 + 2x + 1)$ **22.** $(3x - 2)^3$ **23.** $8 + z^3$

24. $(11x - 17)^2$ **25.** $63x^2 - 42x + 7$ **26.** $5 - 20x^2$

How many terms are in each expression? Which expressions have $x - 3$ as a factor?

27. $(x - 3)^2 - (x - 3)$ **28.** $x - 3 + y$

29. $x^2 - 6x + 9$ **30.** $x^3 - 3$

Simplify each expression, if possible, leaving only nonnegative exponents in your answer.

31. -4^{-4} **32.** $(-4)^{-1/2}$ **33.** $\dfrac{4x^2 - 9}{27 - 8x^3}$

34. $\dfrac{2^{-2} + x^{-2}}{2x^{-2}}$ **35.** $\dfrac{-3x^{-2}}{y^{-3}z^2}$ **36.** $\sqrt[5]{x^5 + y^5}$

Find each product.

37. $[\sqrt{x} + 2]^2$ **38.** $(6 - 5x)^3$ **39.** $a^2b(abc)$

Reduce, if possible.

40. $\dfrac{2(x - 7) - b(7 - x)}{(x + 7)(x - 7)}$

41. $\dfrac{x(a + b) + y(a - b)}{(a + b)(a - b)}$

The following problems can be worked by using a least common denominator. Follow the directions in each and notice how the LCD is used.

42. Perform the operation indicated: $\dfrac{x}{5 - x} + \dfrac{2}{x - 5}$.

43. Solve $\dfrac{x^2}{x^2 - 25} + \dfrac{1}{5 - x} = \dfrac{x}{5 + x}$.

44. Simplify $\dfrac{\dfrac{x + 1}{5 - x}}{\dfrac{x - 2}{x - 5}}$.

Label each as an equation or an expression. Solve the equations in one variable, graph the equations in two variables, and perform the operations indicated on the expressions.

45. $(x - 7)^2$

46. $y = (x - 7)^2$

47. $\left| \dfrac{7 - 5x}{11} \right| = 3$

48. $\dfrac{1}{2x^2 - 3x - 5} + \dfrac{1}{10 + x - 2x^2}$

49. $\dfrac{3 - 2i}{1 + i}$

50. $\sqrt{x + 1} = x - 1$

► You Decide

An approximation of population (in thousands) of Able, Pennsylvania, 61 miles east of Pittsburgh, is given by the formula

$$P = 8 + \sqrt{2t}$$

A similar approximation of the population of Baker, Pennsylvania, 11 miles west of Pittsburgh, is given by the formula,

$$P = 3 + \sqrt{10t + 1}$$

where t is the number of years since 1944 in both formulas.

What is the population of each town now? When will the populations be approximately the same? Based only on the information given, where would you prefer to live and why?

► Chapter Test

In Problems 1 through 9, circle the correct response.

1. The radius of the circle given by the equation $x^2 + y^2 - 2x + 4y = 4$ is (?)
   A. 2   **B.** 3   **C.** 9   **D.** 14

2. The vertex of the parabola given by the equation $x = y^2 - 4y + 3$ is (?)
   A. $(-1, 2)$ **B.** $(1, -2)$ **C.** $(3, -4)$ **D.** The origin

3. The center of the graph of the equation $5x^2 + 6y^2 + 50x - 24y + 119 = 0$ is (?)
   A. $(5, -2)$ **B.** $(-5, 2)$ **C.** $(10, -4)$ **D.** $(-5, 4)$

4. The graph of the equation $x^2 + y + 2x - 3 = 0$ is (?)
   A. an ellipse, but not a circle

B. a circle

C. a parabola opening upward

D. a parabola opening downward

5. The graph of the equation $2x^2 + 2y^2 + 2x - 3 = 0$ is (?)

A. an ellipse, but not a circle

B. a circle

C. a parabola opening upward

D. a parabola opening downward

6. The graph of the equation $4x^2 + 8x - 4y^2 + 8y - 4 = 0$ is (?)

A. an ellipse, but not a circle   **B.** a circle

C. a parabola   **D.** a hyperbola

7. The graph of the equation $2x^2 + 2y + y^2 = 0$ is (?)

A. an ellipse, but not a circle   **B.** a circle

C. a parabola   **D.** a hyperbola

8. The graph of the equation $2x^2 + 3y^2 - 8x + 18y + 29 = 0$ most nearly resembles (?)

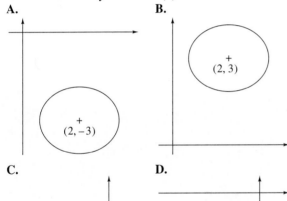

A.   **B.**

C.   **D.**

9. The graph of the equation $y^2 - x - 4y + 5 = 0$ most nearly resembles (?)

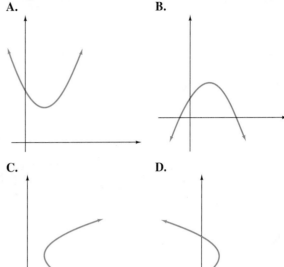

A.   **B.**

C.   **D.**

10. Find the center and radius of the circle given by the equation $x^2 + y^2 - 4y = 1$.

11. Where is the vertex of the parabola given by the equation $y = -2x^2 - 4x - 3$, and in what direction does it open?

In Problems 12 through 15, sketch the graph of the equation.

12. $x^2 - y^2 + 4x + 2y + 4 = 0$

13. $4x^2 + 9y^2 - 16x - 18y - 11 = 0$

14. $x^2 + y^2 - 6x - 6y + 9 = 0$

15. $x = y^2 - 6y + 9$

10 Other Topics

See Section 10.5, Example 2.

Connections

We complete our study of intermediate algebra with an introduction to topics that will become important in future mathematics courses.

First, we look at sequences or ordered collections of numbers such as

$$1, 2, 4, 8, 16, \ldots$$

which are used in the study of calculus.

Next, we look at sums of related numbers and introduce a very convenient notation for such sums. We use this new notation to state the Binomial Theorem.

Finally, we define permutations and combinations and learn how to compute them. These ideas form a cornerstone in the study of probability.

10.1 Sequences

TAPE 17

Ordered lists of numbers such as

$$1, 3, 5, 7, 9, \ldots$$

$$-2, 4, -8, 16, \ldots$$

$$1, \frac{1}{2}, \frac{1}{3}, \frac{1}{4}, \frac{1}{5}, \ldots$$

are called **sequences.** The three dots, called an ellipsis, mean continue on in the pattern that has been established. The expressions separated by commas are called the **terms** of the sequence. Often, we can determine the **general term** that expresses every term of the sequence. For example, the three sequences given above have general terms given by

$$2n - 1$$

$$(-2)^n$$

$$\frac{1}{n}$$

where n has the values $1, 2, 3, 4, \ldots$. This leads to the formal definition of sequence.

> ### Definition of Sequence
>
> A sequence is a function whose domain is the set of natural numbers.

Since the domain of a sequence is the set of natural numbers, we use the general term rather than the usual functional notation. Instead of writing

$$f(x) = 2x - 1, \qquad x = 1, 2, 3, \ldots$$

for the first sequence, we use n as an element of the domain and a_n instead of $f(n)$, with the understanding that n is a natural number. So we write the sequence as

$$a_n = 2n - 1$$

► **EXAMPLE 1**

Write the first five terms of the sequence $a_n = 4n + 5$.

Solution

Evaluate the terms of a sequence just like we evaluate a function. We evaluate the first five terms:

$$a_1 = 4 \cdot 1 + 5 = 9$$
$$a_2 = 4 \cdot 2 + 5 = 13$$
$$a_3 = 4 \cdot 3 + 5 = 17$$
$$a_4 = 4 \cdot 4 + 5 = 21$$
$$a_5 = 4 \cdot 5 + 5 = 25$$

And we see that the sequence is

$$9, 13, 17, 21, 25, \ldots$$ ◄

► **EXAMPLE 2**

Write the first five terms of the sequence $b_n = n(n + 1)$.

Solution

Evaluate the first five terms.

$$b_1 = 1(1 + 1) = 1 \cdot 2 = 2$$
$$b_2 = 2(2 + 1) = 2 \cdot 3 = 6$$
$$b_3 = 3(3 + 1) = 3 \cdot 4 = 12$$
$$b_4 = 4(4 + 1) = 4 \cdot 5 = 20$$
$$b_5 = 5(5 + 1) = 5 \cdot 6 = 30$$

So this sequence is

$$2, 6, 12, 20, 30, \ldots$$ ◄

Often we are given some of the terms of a sequence and are asked to provide a suitable general term.

► **EXAMPLE 3**

Find a suitable general term for each of the following sequences.

(a) $2, 6, 18, 54, 162, \ldots$ (b) $2, 5, 8, 11, 14, \ldots$

(c) $\sqrt{2}, 2, \sqrt{6}, 2\sqrt{2}, \sqrt{10}, 2\sqrt{3}, \ldots$ (d) $1, -\dfrac{1}{2}, \dfrac{1}{4}, -\dfrac{1}{8}, \dfrac{1}{16}, \ldots$

Solutions

(a) Notice that each term, after the first, is three *times* the previous term. Thus

$$a_2 = a_1 \cdot 3 = 2 \cdot 3$$

By the same reasoning

$$a_3 = a_2 \cdot 3 = 2 \cdot 3 \cdot 3 = 2 \cdot 3^2$$

$$a_4 = 2 \cdot 3^2 \cdot 3 = 2 \cdot 3^3$$

$$a_5 = 2 \cdot 3^3 \cdot 3 = 2 \cdot 3^4$$

Now we see the pattern: $a_n = 2 \cdot 3^{n-1}$.

(b) In this sequence, each term, after the first, is three *added to* the previous term. So

$$b_2 = b_1 + 3 = 2 + 3$$

$$b_3 = b_2 + 3 = 2 + 3 + 3 \quad = 2 + 2 \cdot 3$$

$$b_4 = b_3 + 3 = 2 + 2 \cdot 3 + 3 = 2 + 3 \cdot 3$$

$$b_5 = b_4 + 3 = 2 + 3 \cdot 3 + 3 = 2 + 4 \cdot 3$$

Look closely to see the pattern:

$$b_n = 2 + (n - 1) \cdot 3$$

(c) Rewrite this sequence as unsimplified square roots:

$$\sqrt{2}, \sqrt{4}, \sqrt{6}, \sqrt{8}, \sqrt{10}, \sqrt{12}, \ldots$$

We see that each term is the square root of $2n$:

$$c_n = \sqrt{2n}$$

(d) Since the terms alternate in sign, think of a negative number raised to the nth power. Since powers of 2 are in the denominators, notice that

$$d_1 = 1 = (\text{any nonzero number})^0$$

$$d_2 = -\frac{1}{2} = \left(-\frac{1}{2}\right)^1$$

$$d_3 = \frac{1}{4} = \left(-\frac{1}{2}\right)^2$$

$$d_4 = -\frac{1}{8} = \left(-\frac{1}{2}\right)^3$$

and the pattern is

$$d_n = \left(-\frac{1}{2}\right)^{n-1} \qquad \blacktriangleleft$$

A sequence such as

$$1, 4, 7, 10, 13, \ldots$$

where every term is greater than the preceding term, is called an **increasing** sequence. A sequence such as

$$1, \frac{2}{3}, \frac{4}{9}, \frac{8}{27}, \frac{16}{81}, \ldots$$

where every term is less than the preceding term, is called a **decreasing** sequence.

► **EXAMPLE 4**

Label each of the following sequences as *increasing, decreasing,* or *neither increasing nor decreasing.*

(a) $a_n = 3n - 2$ (b) $b_n = 4 - n$ (c) $c_n = 5 + (-2)^n$

Solutions

(a) A few terms of this sequence,

$$1, 4, 7, 10, \ldots$$

show that each term is 3 more than the preceding term. The sequence is *increasing.*

(b) As the value n increases by 1 in each term, the value of the term decreases by 1.

$$4, 3, 2, 1, 0, -1, \ldots$$

The sequence is *decreasing.*

(c) Notice that

$$c_1 = 5 + (-2)^1 = 5 + (-2) = 3$$
$$c_2 = 5 + (-2)^2 = 5 + 4 = 9$$
$$c_3 = 5 + (-2)^3 = 5 - 8 = -3$$
$$c_4 = 5 + (-2)^4 = 5 + 16 = 21$$
$$c_5 = 5 + (-2)^5 = 5 - 32 = -27$$

and the sequence is *neither increasing nor decreasing.*

$$3, 9, -3, 21, -27, \ldots$$ ◄

Any sequence of the form

$$a_n = a_1 + (n - 1)d$$

is called an **arithmetic sequence** or **arithmetic progression.** The number d is called its **common difference.** Some examples of arithmetic sequences are

$1, 5, 9, 13, 17, \ldots$	Common difference is 4.
$-3, 6, 15, 24, \ldots$	Common difference is 9.
$2, -1, -4, -7, \ldots$	Common difference is -3.

We could write these sequences as

$$a_n = 1 + (n - 1) \cdot 4$$
$$b_n = -3 + (n - 1) \cdot 9$$
$$c_n = 2 + (n - 1)(-3)$$

Notice that d is constant for each arithmetic sequence.

► **EXAMPLE 5**

For each of the following *arithmetic sequences,* find the common difference and write each in the form $a_n = a_1 + (n - 1)d$.

(a) 1, 3, 5, 7, 9, . . .   (b) -2, 6, 14, 22, . . .

(c) $3, \dfrac{7}{2}, 4, \dfrac{9}{2}, 5, \dfrac{11}{2}, \ldots$ (d) $-4, -7, -10, -13, \ldots$

Solutions

If the sequence is an arithmetic sequence, find d by subtracting any term from the following term.

(a) $d = a_2 - a_1 = 3 - 1 = 2$. So, because a_1 is 1,

$$a_n = 1 + (n - 1) \cdot 2$$

(b) $d = 6 - (-2) = 8$.

$$a_n = -2 + (n - 1) \cdot 8$$

(c) $d = \dfrac{7}{2} - 3 = \dfrac{7 - 6}{2} = \dfrac{1}{2}$.

$$a_n = 3 + (n - 1) \cdot \dfrac{1}{2}$$

(d) $d = -7 - (-4) = -7 + 4 = -3$.

$$a_n = -4 + (n - 1)(-3)$$ ◄

Any sequence of the form

$$a_n = a_1 r^{n-1}$$

is called a **geometric sequence** or **geometric progression**. The number r is called its **common ratio**. Some examples of geometric sequences are

1, 2, 4, 8, 16, . . .   Common ratio is 2.

2, 10, 50, 250, . . .   Common ratio is 5.

$4, -2, 1, -\dfrac{1}{2}, \dfrac{1}{4}, \ldots$ Common ratio is $-\dfrac{1}{2}$.

We could write these sequences as

$$a_n = 1 \cdot 2^{n-1} \qquad \text{(or simply } a_n = 2^{n-1})$$

$$b_n = 2 \cdot 5^{n-1}$$

$$c_n = 4\left(-\frac{1}{2}\right)^{n-1}$$

The common ratio r is a constant for each geometric sequence.

► **EXAMPLE 6**

For each of the following *geometric sequences,* find the common ratio and write each in the form $a_n = a_1 r^{n-1}$.

(a) 2, 4, 8, 16, . . .   (b) 4, 12, 36, 108, . . .   (c) 3, -2, $\dfrac{4}{3}$, $-\dfrac{8}{9}$, $\dfrac{16}{27}$, . . .

Solutions

If a sequence is a geometric sequence, find the common ratio by dividing any term into the following term.

(a) $r = \dfrac{a_2}{a_1} = \dfrac{4}{2} = 2$. Therefore, the common ratio is 2.

$$a_n = 2 \cdot 2^{n-1} \qquad \text{(or simply } a_n = 2^n)$$

(b) $r = \dfrac{12}{4} = 3$.

$$a_n = 4 \cdot 3^{n-1}$$

(c) $r = \dfrac{-2}{3} = -\dfrac{2}{3}$.

$$a_n = 3\left(-\frac{2}{3}\right)^{n-1} \qquad \blacktriangleleft$$

Problem Set 10.1

Warm-Ups

In Problems 1 through 9, write the first five terms of each sequence. See Examples 1 and 2.

1. $a_n = n$

2. $a_n = 2n + 1$

3. $a_n = 2(n + 1)$

4. $b_n = 14 - 3n$

5. $c_n = n^2$

6. $h_n = 2(n - 1)^2$

7. $a_n = 3$

8. $z_n = (-1)^{n-1} n^3$

9. $k_n = (-1)^n (1 - n)$

In Problems 10 through 15, find a suitable general term for each sequence. See Example 3.

10. 2, 4, 6, 8, . . .

11. 1, 4, 9, 16, . . .

12. 2, 5, 8, 11, . . .

13. -1, 2, -3, 4, . . .

14. 1, 9, 25, 49, . . .

15. 2, -4, 8, -16, . . .

In Problems 16 through 21, label each sequence as increasing, decreasing, or neither increasing nor decreasing. See Example 4.

16. $a_n = 3 + 2n$

17. $a_n = 3 - 2n$

18. $b_n = n^{-2}$

19. $a_n = (-n)^{-1}$

20. $v_n = (-n)^{-n}$

21. $k_n = 1 - n^{-1}$

Problems 22 through 27 are arithmetic sequences. In each, find the common difference and write in the form $a_n = a_1 + (n - 1)d$. See Example 5.

22. $3, 5, 7, \ldots$

23. $2, 3, 4, \ldots$

24. $1, 5, 9, \ldots$

25. $5, 15, 25, \ldots$

26. $0, -2, -4, \ldots$

27. $\dfrac{1}{2}, 0, -\dfrac{1}{2}, \ldots$

Problems 28 through 33 are geometric sequences. In each, find the common ratio and write in the form $a_n = a_1 r^{n-1}$. See Example 6.

28. $3, 6, 12, \ldots$

29. $1, 5, 25, \ldots$

30. $3, 9, 27, \ldots$

31. $2, -4, 8, \ldots$

32. $4, 2, 1, \ldots$

33. $1, -\dfrac{1}{4}, \dfrac{1}{16}, \ldots$

Practice Exercises

In Problems 34 through 42, write the first five terms of each sequence.

34. $a_n = n + 1$

35. $a_n = 3n - 1$

36. $a_n = 2(n - 1)$

37. $A_n = 8 - 2n$

38. $b_n = n^2 - 1$

39. $h_n = n^2 - n + 1$

40. $c_n = -4$

41. $j_n = (-1)^{n-1}n^2$

42. $a_n = (-1)^n(n - n^2)$

In Problems 43 through 48, find a suitable general term for the sequence.

43. $4, 6, 8, 10, \ldots$

44. $0, 3, 8, 15, \ldots$

45. $-1, 3, 7, 11, \ldots$

46. $1, -2, 3, -4, \ldots$

47. $0, 1, 8, 27, \ldots$

48. $-1, 2, -4, 8, \ldots$

In Problems 49 through 54, label each sequence as increasing, decreasing, or neither increasing nor decreasing.

49. $a_n = -4 + n$

50. $a_n = -4 - n$

51. $b_n = n^{-2} - 2$

52. $c_n = 2 - n^{-2}$

53. $d_n = (-2)^{-2n}$

54. $k_n = (-1 - n)^{-1}$

Problems 55 through 60 are arithmetic sequences. In each, find the common difference and write in the form $a_n = a_1 + (n - 1)d$.

55. $1, 4, 7, \ldots$

56. $-1, 4, 9, \ldots$

57. $10, 8, 6, \ldots$

58. $0, 11, 22, \ldots$

59. $-\dfrac{1}{2}, 0, \dfrac{1}{2}, \ldots$

60. $\dfrac{2}{3}, \dfrac{1}{3}, 0, \ldots$

Problems 61 through 66 are geometric sequences. In each, find the common ratio and write in the form $a_n = a_1 r^{n-1}$.

61. $1, 3, 9, \ldots$

62. $2, 4, 8, \ldots$

63. $2, -6, 18, \ldots$

64. $5, 25, 125, \ldots$

65. $27, 9, 3, \ldots$

66. $128, -64, 32, \ldots$

In Your Own Words

67. What is a sequence?

10.2 Summation Notation

TAPE 17

Ellen Adams's new job pays $1000 for the first month and includes a $10 raise each month. How much will Ellen earn the first year?

Ellen will earn the sum of 12 monthly payments as follows:

$$S = (1000) + (1000 + 10) + (1000 + 2 \cdot 10) + \cdots + (1000 + 11 \cdot 10)$$

Notice that the first two payments can be written as $(1000 + 0 \cdot 10)$ and $(1000 + 1 \cdot 10)$. Therefore, Ellen will get 12 monthly payments of the form $(1000 + j \cdot 10)$, where j has values $0, 1, 2, \ldots, 11$. Such sums can be written with a convenient shorthand called *summation notation*.

$$S = \sum_{j=0}^{11} (1000 + j \cdot 10)$$

(Ellen will earn $12,660.)

The symbol Σ is the Greek capital letter sigma, which is used to mean sum. The assignment, $j = 0$, under the sigma, is the first value of the variable j to be used in the sum. The integer 11, at the top of the sigma, is the final value of the variable j to be used in the sum. There is a term in the sum for every integer value of j from 0 to 11. The variable j is called the **index.**

► EXAMPLE 1

Write out the terms of the following sum.

$$\sum_{j=1}^{6} (2j - 1)$$

Solution

$$\sum_{j=1}^{6} (2j - 1) = (2 \cdot 1 - 1) + (2 \cdot 2 - 1) + (2 \cdot 3 - 1)$$
$$+ (2 \cdot 4 - 1) + (2 \cdot 5 - 1) + (2 \cdot 6 - 1)$$
$$= 1 + 3 + 5 + 7 + 9 + 11 \qquad \blacktriangleleft$$

► EXAMPLE 2

Write out the terms of the following sum.

$$\sum_{j=2}^{5} \frac{j - 1}{j + 1}$$

Solution

$$\sum_{j=2}^{5} \frac{j - 1}{j + 1} = \frac{2 - 1}{2 + 1} + \frac{3 - 1}{3 + 1} + \frac{4 - 1}{4 + 1} + \frac{5 - 1}{5 + 1}$$
$$= \frac{1}{3} + \frac{2}{4} + \frac{3}{5} + \frac{4}{6} \qquad \blacktriangleleft$$

► **EXAMPLE 3**

Write out the terms of the following sum.

$$\sum_{j=1}^{5} 2^{j-1} \cdot j$$

Solution

$$\sum_{j=1}^{5} 2^{j-1} \cdot j = 2^0 \cdot 1 + 2^1 \cdot 2 + 2^2 \cdot 3 + 2^3 \cdot 4 + 2^4 \cdot 5$$

$$= 1 + 2 \cdot 2 + 4 \cdot 3 + 8 \cdot 4 + 16 \cdot 5$$

$$= 1 + 4 + 12 + 32 + 80 \qquad ◄$$

► **EXAMPLE 4**

Write out the terms of the following sum.

$$\sum_{j=0}^{3} (-1)^j X^j$$

Solution

$$\sum_{j=0}^{3} (-1)^j X^j = (-1)^0 X^0 + (-1)^1 X^1 + (-1)^2 X^2 + (-1)^3 X^3$$

$$= 1 \cdot 1 + (-1)X + 1 \cdot X^2 + (-1)X^3$$

$$= 1 - X + X^2 - X^3$$

In this example, notice how the factor $(-1)^j$ caused the signs to alternate. $(-1)^j$ and $(-1)^{j-1}$ are often used for this purpose. ◄

► **EXAMPLE 5**

Write the following sum in summation notation.

$$2 + 4 + 6 + 8 + 10 + 12 + 14 + 16$$

Solution

Notice that the terms can be written

$$2 \cdot 1 + 2 \cdot 2 + 2 \cdot 3 + 2 \cdot 4 + \cdots + 2 \cdot 8$$

So we can write

$$\sum_{j=1}^{8} 2j \qquad ◄$$

► **EXAMPLE 6**

Write the following sum in summation notation.

$$x + 2x^2 + 3x^3 + 4x^4 + 5x^5$$

Solution

This sum is of the form jx^j as j goes from 1 to 5.

$$\sum_{j=1}^{5} jx^j \qquad ◄$$

Problem Set 10.2

Warm-Ups

In Problems 1 through 9, write out the terms of each sum. For Problems 1 through 6, see Examples 1 through 3.

1. $\displaystyle\sum_{j=1}^{5} 2j$

2. $\displaystyle\sum_{j=1}^{6} 3(j-1)$

3. $\displaystyle\sum_{j=0}^{7} j^2$

4. $\displaystyle\sum_{j=1}^{7} 2^j$

5. $\displaystyle\sum_{j=0}^{4} (3-j)^2$

6. $\displaystyle\sum_{j=0}^{5} (j-1)^2$

For Problems 7 through 9, see Example 4.

7. $\displaystyle\sum_{j=1}^{6} (-1)^{j-1}j$

8. $\displaystyle\sum_{j=1}^{5} (-1)^j 2^j$

9. $\displaystyle\sum_{j=1}^{5} \frac{j-1}{j+1}(-1)^{j+1}$

In Problems 10 through 15, write each sum in summation notation. For problems 10 through 12, see Example 5.

10. $1 + 2 + 3 + 4 + 5 + 6 + 7 + 8$

11. $1 - 3 + 5 - 7 + 9 - 11$

12. $1 - \dfrac{2}{3} + \dfrac{3}{5} - \dfrac{4}{7} + \dfrac{5}{9} - \dfrac{6}{11} + \dfrac{7}{13} - \dfrac{8}{15}$

For Problems 13 through 15, see Example 6.

13. $1 + \dfrac{x}{2} + \dfrac{x^2}{3} + \dfrac{x^3}{4} + \dfrac{x^4}{5} + \dfrac{x^5}{6} + \dfrac{x^6}{7}$

14. $A_1 + A_2 + A_3 + \cdots + A_{100}$

15. $1 - \sqrt{2}x^2 + \sqrt{3}x^4 - 2x^6 + \sqrt{5}x^8 - \sqrt{6}x^{10} + \sqrt{7}x^{12}$

Practice Exercises

In Problems 16 through 24, write out the terms of each sum.

16. $\displaystyle\sum_{j=1}^{6} 3j$

17. $\displaystyle\sum_{j=0}^{4} 2(j-5)$

18. $\displaystyle\sum_{j=0}^{7} (j^2 - 1)$

19. $\displaystyle\sum_{j=1}^{7} 2^{j-1}$

20. $\displaystyle\sum_{j=0}^{4} (3-2j)^2$

21. $\displaystyle\sum_{j=0}^{5} (j-1)^3$

22. $\displaystyle\sum_{j=1}^{6} (-1)^{j-1}Q_j$

23. $\displaystyle\sum_{j=1}^{5} (-1)^j 2^{j-1}$

24. $\displaystyle\sum_{j=1}^{5} \frac{2j-1}{2j+1}$

In Problems 25 through 30, write each sum in summation notation.

25. $3 + 4 + 5 + 6 + 7 + 8 + 9 + 10 + 11 + 12 + 13 + 14$

26. $2 - 4 + 6 - 8 + 10 - 12 + 14 - 16 + 18$

27. $1 + \dfrac{x^2}{3} + \dfrac{x^4}{5} + \dfrac{x^6}{7} + \dfrac{x^8}{9}$

28. $A_0 - A_1 + A_2 - A_3 + \cdots + A_{52}$

29. $-\pi + \dfrac{1 - \pi}{10} + \dfrac{2 - \pi}{100} + \dfrac{3 - \pi}{1000} + \dfrac{4 - \pi}{10,000} + \dfrac{5 - \pi}{100,000} + \dfrac{6 - \pi}{1,000,000}$

30. $1 + 2 + 4 + 8 + \cdots + 2048$

Challenge Problems

31. Write out the terms of $\displaystyle\sum_{i=1}^{5} i$, $\displaystyle\sum_{j=1}^{5} j$, and $\displaystyle\sum_{k=1}^{5} k$.

32. Why is the index in summation notation sometimes called a "dummy"?

33. Compare $\displaystyle\sum_{j=1}^{4} 5A_j$ with $5 \displaystyle\sum_{j=1}^{4} A_j$. Write a property of summation notation.

34. Compare $\displaystyle\sum_{j=1}^{4} (A_j + B_j)$ with $\displaystyle\sum_{j=1}^{4} A_j + \displaystyle\sum_{j=1}^{4} B_j$. Write another property of summation notation.

In Your Own Words

35. Why is summation notation useful?

36. Explain how the index works in summation notation.

10.3 Factorials and Binomial Coefficients

TAPE 17

Just as x^4 is shorthand notation for the product $x \cdot x \cdot x \cdot x$, shorthand notation exists for products such as $1 \cdot 2 \cdot 3 \cdot 4$ and $1 \cdot 2 \cdot 3 \cdot 4 \cdot 5 \cdot 6 \cdot 7$. Such products are called **factorials.** Write them as follows:

$$4! = 1 \cdot 2 \cdot 3 \cdot 4$$

$$7! = 1 \cdot 2 \cdot 3 \cdot 4 \cdot 5 \cdot 6 \cdot 7$$

Read 4! as "four factorial," $K!$ as "K factorial," and $(N - 1)!$ as "N minus one factorial."

Factorial Definitions

For N a natural number,

$$N! = 1 \cdot 2 \cdot 3 \cdots (N - 1) \cdot N$$

$$0! = 1$$

Factorials grow quickly, as illustrated below.

$$0! = 1 \qquad\qquad\qquad = 1$$
$$1! = 1 \qquad\qquad\qquad = 1$$
$$2! = 1 \cdot 2 \qquad\qquad\qquad = 2$$
$$3! = 1 \cdot 2 \cdot 3 \qquad\qquad\qquad = 6$$
$$4! = 1 \cdot 2 \cdot 3 \cdot 4 \qquad\qquad\qquad = 24$$
$$5! = 1 \cdot 2 \cdot 3 \cdot 4 \cdot 5 \qquad\qquad\qquad = 120$$
$$\vdots$$
$$9! = 1 \cdot 2 \cdot 3 \cdot 4 \cdot 5 \cdot 6 \cdot 7 \cdot 8 \cdot 9 \qquad = 362,880$$
$$10! = 1 \cdot 2 \cdot 3 \cdot 4 \cdot 5 \cdot 6 \cdot 7 \cdot 8 \cdot 9 \cdot 10 = 3,628,800$$

Notice from the table above that 5! is $4! \cdot 5$ and 10! is $9! \cdot 10$.

Property of Factorials

If N is a natural number,

$$N! = (N - 1)! \cdot N$$

► **EXAMPLE 1**

Calculate 11!.

Solution

By the Property of Factorials,

$$11! = 10! \cdot 11$$

From the table, 10! is 3,628,800. So

$$11! = (3,628,800) \cdot 11$$
$$= 39,916,800$$

◄

Quotients of factorials are interesting and quite common. They often simplify with vast cancellations.

► **EXAMPLE 2**

Evaluate each expression.

(a) $\dfrac{16!}{17!}$ (b) $\dfrac{20!}{18!}$ (c) $\dfrac{10!}{5! \cdot 5!}$ (d) $\dfrac{8!}{8! \cdot 0!}$

Solutions

(a) By the Property of Factorials, 17! is $16! \cdot 17$, so

$$\frac{16!}{17!} = \frac{16!}{16! \cdot 17} = \frac{1}{17}$$

Note that 16 factors were divided out from the numerator and denominator!

(b) Write 20! as $18! \cdot 19 \cdot 20$.

$$\frac{20!}{18!} = \frac{18! \cdot 19 \cdot 20}{18!} = 19 \cdot 20 = 380$$

(c) By definition,

$$\frac{10!}{5! \cdot 5!} = \frac{1 \cdot 2 \cdot 3 \cdot 4 \cdot 5 \cdot 6 \cdot 7 \cdot 8 \cdot 9 \cdot 10}{(1 \cdot 2 \cdot 3 \cdot 4 \cdot 5) \cdot (1 \cdot 2 \cdot 3 \cdot 4 \cdot 5)}$$

Divide out the first five factors, leaving

$$= \frac{6 \cdot 7 \cdot 8 \cdot 9 \cdot 10}{1 \cdot 2 \cdot 3 \cdot 4 \cdot 5}$$

$$= \frac{7 \cdot 2 \cdot 9 \cdot 2}{1} \qquad \text{Reduce}$$

$$= 252$$

(d) The 8!s divide out.

$$\frac{8!}{8! \cdot 0!} = \frac{1}{0!}$$

Remember, 0! is 1, not 0.

$$= \frac{1}{1} = 1 \qquad \blacktriangleleft$$

Factorial expressions of the form $\dfrac{7!}{3! \cdot 4!}$ or $\dfrac{10!}{8! \cdot 2!}$ are quite common in mathematics. In the next section such expressions are called **binomial coefficients** and are written

$$\binom{7}{3} = \frac{7!}{3! \cdot 4!}$$

$$\binom{10}{8} = \frac{10!}{8! \cdot 2!}$$

The Binomial Coefficients

For n and r nonnegative integers, $n \geq r$,

$$\binom{n}{r} = \frac{n!}{r! \cdot (n - r)!}$$

► EXAMPLE 3

Calculate each binomial coefficient.

(a) $\binom{7}{5}$ (b) $\binom{9}{3}$ (c) $\binom{8}{8}$ (d) $\binom{11}{0}$

Solutions

(a) $\binom{7}{5} = \dfrac{7!}{5! \cdot (7-5)!} = \dfrac{7!}{5! \cdot 2!}$

$= \dfrac{1 \cdot 2 \cdot 3 \cdot 4 \cdot 5 \cdot 6 \cdot 7}{(1 \cdot 2 \cdot 3 \cdot 4 \cdot 5)(1 \cdot 2)} = \dfrac{6 \cdot 7}{1 \cdot 2} = 21$

(b) $\binom{9}{3} = \dfrac{9!}{3! \cdot (9-3)!} = \dfrac{9!}{3! \cdot 6!}$

$= \dfrac{1 \cdot 2 \cdot 3 \cdot 4 \cdot 5 \cdot 6 \cdot 7 \cdot 8 \cdot 9}{(1 \cdot 2 \cdot 3)(1 \cdot 2 \cdot 3 \cdot 4 \cdot 5 \cdot 6)} = \dfrac{7 \cdot 8 \cdot 9}{1 \cdot 2 \cdot 3} = 7 \cdot 4 \cdot 3 = 84$

(c) $\binom{8}{8} = \dfrac{8!}{8! \cdot (8-8)!} = \dfrac{8!}{8! \cdot 0!} = \dfrac{1}{0!} = \dfrac{1}{1} = 1$

(d) $\binom{11}{0} = \dfrac{11!}{0! \cdot (11-0)!} = \dfrac{11!}{0! \cdot 11!} = 1$

Calculator Box

Factorials with a Calculator

Most scientific calculators have a factorial capability. Look for a | n! | or | x! | key. With such a key, we can find 8! easily.

Scientific Calculator:

Press | 8 | | n! | and read `40320` on the display.

Graphing Calculator:

Look for a math menu that has an n! or x! entry. For example, with the CASIO Fx7700, press | 8 | | MATH | | F2 | | F1 | | EXE | or with the TI-81 press | 8 | | MATH | | 5 | | ENTER | and read `40320` on the display.
8! = 40,320

Calculator Exercises

1. 12!   **2.** 0!   **3.** 9!   **4.** $\dfrac{20!}{18!}$ **5.** $\dfrac{6!}{2!4!}$

Answers

1. 479,001,600   **2.** 1   **3.** 362,880   **4.** 380   **5.** 15

Problem Set 10.3

Warm-Ups

In Problems 1 through 24, evaluate each expression.
For Problems 1 through 4, see Example 1.

1. $6!$
2. $8!$
3. $12!$
4. $10!$

For problems 5 through 12, see Example 2.

5. $\dfrac{14!}{13!}$
6. $\dfrac{29!}{31!}$
7. $\dfrac{8!}{4! \cdot 4!}$
8. $\dfrac{8!}{3! \cdot 5!}$

9. $\dfrac{8!}{2! \cdot 6!}$
10. $\dfrac{8!}{1! \cdot 7!}$
11. $\dfrac{20!}{1! \cdot 19!}$
12. $\dfrac{20!}{2! \cdot 18!}$

For Problems 13 through 24, see Example 3.

13. $\dbinom{8}{5}$
14. $\dbinom{5}{2}$
15. $\dbinom{5}{3}$
16. $\dbinom{6}{0}$

17. $\dbinom{6}{1}$
18. $\dbinom{6}{2}$
19. $\dbinom{6}{3}$
20. $\dbinom{6}{4}$

21. $\dbinom{6}{5}$
22. $\dbinom{6}{6}$
23. $\dbinom{500}{0}$
24. $\dbinom{69}{69}$

Practice Exercises

In Problems 25 through 48, evaluate each expression.

25. $7!$
26. $13!$
27. $11!$
28. $5!$

29. $\dfrac{17!}{18!}$
30. $\dfrac{29!}{27!}$
31. $\dfrac{9!}{4! \cdot 5!}$
32. $\dfrac{9!}{3! \cdot 6!}$

33. $\dfrac{9!}{2! \cdot 7!}$
34. $\dfrac{9!}{1! \cdot 8!}$
35. $\dfrac{30!}{1! \cdot 29!}$
36. $\dfrac{30!}{2! \cdot 28!}$

37. $\dbinom{9}{6}$
38. $\dbinom{4}{3}$
39. $\dbinom{4}{1}$
40. $\dbinom{7}{0}$

41. $\dbinom{7}{1}$
42. $\dbinom{7}{2}$
43. $\dbinom{7}{3}$
44. $\dbinom{7}{4}$

45. $\dbinom{7}{5}$
46. $\dbinom{7}{6}$
47. $\dbinom{7}{7}$
48. $\dbinom{39}{38}$

Challenge Problems

49. $\dbinom{7}{2}$ has the same value as $\dbinom{7}{5}$, and $\dbinom{6}{1}$ has the same value as $\dbinom{6}{5}$. Use the definition of binomial coefficient to show that $\dbinom{n}{r} = \dbinom{n}{n-r}$.

In Your Own Words

50. Explain $n!$ without using a formula.

10.4 **The Binomial Theorem**

TAPE 17

In Chapter 1 we learned that $(a + b)^2$ is $a^2 + 2ab + b^2$. Let's look at a few powers of the binomial $(a + b)$.

$$(a + b)^0 = \qquad\qquad 1$$
$$(a + b)^1 = \qquad\qquad a + b$$
$$(a + b)^2 = \qquad\qquad a^2 + 2ab + b^2$$
$$(a + b)^3 = \qquad\qquad a^3 + 3a^2b + 3ab^2 + b^3$$
$$(a + b)^4 = \qquad\qquad a^4 + 4a^3b + 6a^2b^2 + 4ab^3 + b^4$$
$$(a + b)^5 = \qquad a^5 + 5a^4b + 10a^3b^2 + 10a^2b^3 + 5ab^4 + b^5$$

The Binomial Theorem provides a general formula for the expansion of $(a + b)^n$, for n a natural number. Some of the formula is quite easy to see. The powers of a, for example, start at n and decrease by 1 for each term until the power becomes zero (remember, $a^0 = 1$). Similarly, b starts with an exponent of zero and increases by 1 until b^n is reached. Except for coefficients, $(a + b)^7$ can be written

$$\underline{\quad}a^7 + \underline{\quad}a^6b + \underline{\quad}a^5b^2 + \underline{\quad}a^4b^3 + \underline{\quad}a^3b^4 + \underline{\quad}a^2b^5$$
$$+ \underline{\quad}ab^6 + \underline{\quad}b^7$$

In Section 1.4 we found that the coefficients are the seventh row of Pascal's triangle.

```
0th row →                          1
1st row →                      1       1
                            1      2      1
                         1     3      3     1
                      1     4     6      4     1
                   1     5    10     10     5     1
                1     6    15    20    15     6     1
7th row →    1     7    21    35    35    21     7    1
```

However, for large values of n, this is an awkward way to find these coefficients. Let's look at some binomial coefficients from Section 10.3.

$$\binom{7}{0} = \frac{7!}{0! \cdot 7!} = \frac{1}{0!} = 1$$

$$\binom{7}{1} = \frac{7!}{1! \cdot 6!} = \frac{6! \cdot 7}{6!} = 7$$

$$\binom{7}{2} = \frac{7!}{2! \cdot 5!} = \frac{1 \cdot 2 \cdot 3 \cdot 4 \cdot 5 \cdot 6 \cdot 7}{(1 \cdot 2)(1 \cdot 2 \cdot 3 \cdot 4 \cdot 5)} = \frac{6 \cdot 7}{1 \cdot 2} = 21$$

Continuing, we have

$$\binom{7}{3} = 35, \qquad \binom{7}{4} = 35, \qquad \binom{7}{5} = 21, \qquad \binom{7}{6} = 7, \qquad \binom{7}{7} = 1$$

Notice these are exactly the elements on the seventh row of Pascal's triangle. That is, the expansion of $(a + b)^7$ could be written as

$$\binom{7}{0}a^7 + \binom{7}{1}a^6b + \binom{7}{2}a^5b^2 + \binom{7}{3}a^4b^3 + \binom{7}{4}a^3b^4$$
$$+ \binom{7}{5}a^2b^5 + \binom{7}{6}ab^6 + \binom{7}{7}b^7$$

Remembering that $x^0 = 1$ and $x^1 = x$ and using summation notation, we have

$$(a + b)^7 = \sum_{j=0}^{7} \binom{7}{j}a^{7-j}b^j$$

The Binomial Theorem

For n a natural number,

$$(a + b)^n = \sum_{j=0}^{n} \binom{n}{j}a^{n-j}b^j$$

▶ **EXAMPLE 1**

Write the first four terms in the expansion of $(x + y)^{11}$.

Solution

By the Binomial Theorem,

$$(x + y)^{11} = \sum_{j=0}^{11} \binom{11}{j}x^{11-j}y^j$$

So the first four terms are

$$\binom{11}{0}x^{11}y^0 + \binom{11}{1}x^{10}y^1 + \binom{11}{2}x^9y^2 + \binom{11}{3}x^8y^3$$

Now

$$\binom{11}{0} = \frac{11!}{0! \cdot 11!} = 1$$

$$\binom{11}{1} = \frac{11!}{1! \cdot 10!} = \frac{10! \cdot 11}{1! \cdot 10!} = \frac{11}{1!} = 11$$

$$\binom{11}{2} = \frac{11!}{2! \cdot 9!} = \frac{1 \cdot 2 \cdot 3 \cdot 4 \cdot 5 \cdot 6 \cdot 7 \cdot 8 \cdot 9 \cdot 10 \cdot 11}{(1 \cdot 2)(1 \cdot 2 \cdot 3 \cdot 4 \cdot 5 \cdot 6 \cdot 7 \cdot 8 \cdot 9)}$$

$$= \frac{10 \cdot 11}{1 \cdot 2} = 55$$

$$\binom{11}{3} = \frac{11!}{3! \cdot 8!} = \frac{1 \cdot 2 \cdot 3 \cdot 4 \cdot 5 \cdot 6 \cdot 7 \cdot 8 \cdot 9 \cdot 10 \cdot 11}{(1 \cdot 2 \cdot 3)(1 \cdot 2 \cdot 3 \cdot 4 \cdot 5 \cdot 6 \cdot 7 \cdot 8)}$$

$$= \frac{9 \cdot 10 \cdot 11}{1 \cdot 2 \cdot 3} = 165$$

So the first four terms of $(x + y)^{11}$ are

$$1 \cdot x^{11}y^0 + 11x^{10}y^1 + 55x^9y^2 + 165x^8y^3 \qquad \text{or}$$

$$x^{11} + 11x^{10}y + 55x^9y^2 + 165x^8y^3$$

◀

▶ **EXAMPLE 2**

Find the fourteenth term of $(2 + z)^{16}$.

Solution

From the Binomial Theorem,

$$(2 + z)^{16} = \sum_{j=0}^{16} \binom{16}{j} 2^{16-j} z^j$$

The fourteenth term is when j has the value 13.

$$\binom{16}{13} 2^{16-13} z^{13}$$

But

$$\binom{16}{13} = \frac{16!}{13! \cdot 3!} = \frac{14 \cdot 15 \cdot 16}{1 \cdot 2 \cdot 3} = 7 \cdot 5 \cdot 16 = 560$$

and

$$2^{16-13} = 2^3 = 8$$

so the fourteenth term is

$$560 \cdot 8z^{13} \quad \text{or} \quad 4480z^{13}$$

◀

Problem Set 10.4

Warm-Ups

In Problems 1 through 6, find the first three terms of the binomial expansion expansion. See Example 1.

1. $(a + b)^{12}$

2. $(x + y)^{20}$

3. $(s + 2)^{15}$

4. $(2 + t)^9$

5. $(2a + 3b)^6$

6. $(x^2 + 1)^{25}$

In Problems 7 through 10, see Example 2.

7. Find the fifth term of $(a + b)^{13}$.

8. Find the ninth term of $(x + 1)^{12}$.

9. Find the fourth term of $\left(3x + \dfrac{1}{2}\right)^{11}$.

10. Find the middle term of $\left(\dfrac{x}{2} + \dfrac{1}{3}\right)^{10}$.

Practice Exercises

In Problems 11 through 22, use the Binomial Theorem to expand each binomial.

11. $(a + b)^6$

12. $(x + y)^8$

13. $(x + 1)^9$
14. $(A + 2)^7$
15. $(2x + y)^6$
16. $(3x + 2y)^5$
17. $(a + b)^9$
18. $(x + 1)^8$
19. $(x + 1)^9$
20. $(A + 2)^6$
21. $(2x + y)^7$
22. $(2x + 3y)^6$

In Problems 23 through 34, find the last three terms of the binomial expansion.

23. $(a + b)^{13}$ **24.** $(x + y)^{21}$
25. $(s + 3)^5$ **26.** $(3 + t)^{17}$
27. $(3x + 2y)^7$ **28.** $(x^2 + y^2)^{30}$
29. $(a + b)^{12}$ **30.** $(x + y)^{24}$
31. $(s + 3)^6$ **32.** $(2 + t)^{19}$
33. $(2x + 3y)^7$ **34.** $(2x^2 + y^2)^{23}$

In Problems 35 through 40, find the first two terms of the binomial expansion.

35. $(a + b)^{50}$ **36.** $(x + 1)^{100}$ **37.** $(s + 3)^{500}$
38. $(a + b)^{150}$ **39.** $(1 + x)^{100}$ **40.** $(s + 5)^{400}$

41. Find the fourth term of $\left(\dfrac{x}{2} + \dfrac{1}{3}\right)^{12}$

42. Find the middle term of $\left(\dfrac{x}{3} + 1\right)^{12}$

Challenge Problems

43. Notice that $1 = 1^n = \left(\dfrac{1}{2} + \dfrac{1}{2}\right)^n$. Use the Binomial Theorem to expand $\left(\dfrac{1}{2} + \dfrac{1}{2}\right)^6$; then show that it equals 1.

44. As $1.02 = 1 + 0.02$, powers of 1.02 can be found with the Binomial Theorem. Use this technique to find the exact value of $(1.02)^5$.

45. The expansion of $(a - b)^n$ is the same as the expansion of $(a + b)^n$, except that the signs alternate, starting with positive. Write a form of the binomial theorem for $(a - b)^n$.

In Problems 46 through 51, use the form of the Binomial Theorem developed in Problem 45 to expand each binomial.

46. $(a - b)^7$
47. $(x - y)^6$
48. $(x - 1)^9$
49. $(q - 2)^8$
50. $(2x - 3y)^5$
51. $(3x^2 - 2y^2)^4$

52. Use the results of Problem 45 with the technique of Problem 44 to find the exact value of $(0.98)^5$.

In Your Own Words

53. State the Binomial Theorem without using a formula.

10.5 Permutations and Combinations

TAPE 18

In this section we examine two methods of counting the number of ways that various events can occur.

An arrangement of a collection of objects is called a **permutation.** Suppose that three books were to be arranged on a bookshelf. How many different ways could this be done? If the books are labeled *A*, *B*, and *C*, the possibilities are:

Since there are no other possibilities, we conclude that there are six different ways to arrange the three books. Now, suppose that there is room for only two books on our bookshelf. Again, we list the various ways to place the books.

There are six ways to do this task. We say that there are six permutations of *three things taken two at a time.* A common notation used for this is $P(3, 2)$. Because there are also six permutations of three things taken three at a time (the first arrangement), we see that

$$P(3, 3) = 6 \quad \text{and} \quad P(3, 2) = 6$$

If there were only one space on our shelf, we could put either book *A*, book *B*, or book *C* in that space. Thus

$$P(3, 1) = 3$$

Now suppose that there are 12 books and only room for 8.

If we start listing the possibilities we will soon tire of the tasks (or run out of paper). Let's consider the 8 spaces. Any one of the 12 books can be placed in the first space, so there are 12 ways to fill it. Suppose that we put book *G* there.

Now there are only 11 books left to fill the next space. Thus there are $12 \cdot 11$ ways to fill the first two spaces. Suppose that book *D* is chosen for the second space.

Now there are only 10 books left to fill the third space. So there are $12 \cdot 11 \cdot 10$ ways to fill the first three spaces. If we continue placing books until we have one space left,

$$\boxed{G}\boxed{D}\boxed{H}\boxed{F}\boxed{A}\boxed{J}\boxed{B}\boxed{}$$

we find that we have 5 books left to fill the last space, so we can fill it 5 different ways. Therefore, there are

$$12 \cdot 11 \cdot 10 \cdot 9 \cdot 8 \cdot 7 \cdot 6 \cdot 5 = 19{,}958{,}400$$

different ways to place 12 books in 8 spaces. That is,

$$P(12, 8) = 19{,}958{,}400$$

We extend these ideas to a formula for the number of permutations of n things taken r at a time.

$$P(n, r) = n \cdot (n - 1) \cdot (n - 2) \cdots (n - (r - 1)), \qquad 0 \le r \le n$$

Two other commonly used notations for the number of permutations of n things taken r at a time are $_nP_r$ and P_r^n.

The factorial notation, introduced in Section 10.3, is very useful for expressing this formula. However, it is more convenient to rearrange the factors using the Commutative Property for Multiplication.

Factorial Definitions

For n a natural number:

1. $n! = n \cdot (n - 1) \cdots 4 \cdot 3 \cdot 2 \cdot 1$

2. $0! = 1$

The formula for $P(n, r)$ can now be written using factorials. To see how, look at $P(12, 8)$ *again*.

$$P(12, 8) = 12 \cdot 11 \cdot 10 \cdot 9 \cdot 8 \cdot 7 \cdot 6 \cdot 5$$

If we multiply and divide by $4 \cdot 3 \cdot 2 \cdot 1$, we get

$$P(12, 8) = \frac{12 \cdot 11 \cdot 10 \cdot 9 \cdot 8 \cdot 7 \cdot 6 \cdot 5 \cdot 4 \cdot 3 \cdot 2 \cdot 1}{4 \cdot 3 \cdot 2 \cdot 1}$$

or

$$P(12, 8) = \frac{12!}{4!}$$

$$P(12, 8) = \frac{12!}{(12 - 8)!}$$

We now have a general formula for permutations:

> **Number of Permutations of *n* Things Taken *r* at a Time**
>
> For integers *n* and *r*, where $0 \leq r \leq n$,
>
> $$P(n, r) = \frac{n!}{(n - r)!}$$

► **EXAMPLE 1**

Find

(a) $P(6, 2)$ (b) $P(5, 3)$

(c) $P(8, 0)$ (d) $P(7, 7)$

Solutions

(a) $P(6, 2) = \dfrac{6!}{(6 - 2)!} = \dfrac{6!}{4!}$

$= \dfrac{6 \cdot 5 \cdot 4 \cdot 3 \cdot 2 \cdot 1}{4 \cdot 3 \cdot 2 \cdot 1} = 6 \cdot 5 = 30$

(b) $P(5, 3) = \dfrac{5!}{(5 - 3)!} = \dfrac{5!}{2!}$

$= \dfrac{5 \cdot 4 \cdot 3 \cdot 2 \cdot 1}{2 \cdot 1} = 5 \cdot 4 \cdot 3 = 60$

(c) $P(8, 0) = \dfrac{8!}{(8 - 0)!} = \dfrac{8!}{8!} = 1$

(d) $P(7, 7) = \dfrac{7!}{(7 - 7)!} = \dfrac{7!}{0!}$ Remember, $0! = 1$, not 0

$= \dfrac{7 \cdot 6 \cdot 5 \cdot 4 \cdot 3 \cdot 2 \cdot 1}{1} = 5040$ ◄

► **EXAMPLE 2**

If 9 horses are entered in The Motor City Handicap, how many different win, place, and show results are possible? (Win is first place, place is second place, and show is third place.)

Solution

As each possible selection of first-, second-, and third-place horses is a permutation of 9 horses taken 3 at a time, we must compute $P(9, 3)$.

$$P(9, 3) = \frac{9!}{(9 - 3)!} = \frac{9!}{6!}$$

$$= 9 \cdot 8 \cdot 7 = 504$$

There are 504 different win, place, and show possibilities.   ◄

Suppose that we have 5 books and only have room for 3 of them on the bookshelf, and it does not matter in what order they are placed. This is not a permutation problem, as the *order* of an arrangement is part of a permutation. In this problem the collection of books *A C D* is the same as the collection *D A C*. We call such collections **combinations** and use the notation $C(5, 3)$ for the number of combinations of 5 things taken 3 at a time. We list the different combinations of the 5 books. *A, B, C, D,* and *E,* taken 3 at a time.

$$A\,B\,C \qquad A\,B\,D \qquad A\,B\,E \qquad A\,C\,D \qquad A\,C\,E$$
$$A\,D\,E \qquad B\,C\,D \qquad B\,C\,E \qquad B\,D\,E \qquad C\,D\,E$$

Any other combination is an arrangement of 1 of the 10 combinations listed. Therefore, $C(5, 3) = 10$.

The formula for combinations also involves factorials.

Number of Combinations of *n* Things Taken *r* at a Time

For integers *n* and *r*, where $0 \le r \le n$,

$$C(n, r) = \frac{n!}{(n - r)! \cdot r!}$$

Notice that this is *exactly* the formula for the binomial coefficients that was developed in Section 10.3.

$$C(n, r) = \binom{n}{r}$$

Other commonly used notations for the number of combinations of *n* things taken *r* at a time are $_nC_r$ and C_r^n.

▶ **EXAMPLE 3**

Find

(a) $C(6, 2)$ (b) $C(5, 3)$ (c) $C(8, 0)$ (d) $C(7, 7)$

Solutions

(a) $C(6, 2) = \dfrac{6!}{(6 - 2)! \cdot 2!} = \dfrac{6!}{4! \cdot 2!}$

$= \dfrac{1 \cdot 2 \cdot 3 \cdot 4 \cdot 5 \cdot 6}{1 \cdot 2 \cdot 3 \cdot 4 \cdot 1 \cdot 2} = 5 \cdot 3 = 15$

(b) $C(5, 3) = \dfrac{5!}{(5 - 3)! \cdot 3!} = \dfrac{5!}{2! \cdot 3!}$

$= \dfrac{1 \cdot 2 \cdot 3 \cdot 4 \cdot 5}{1 \cdot 2 \cdot 1 \cdot 2 \cdot 3} = 2 \cdot 5 = 10$

(c) $C(8, 0) = \dfrac{8!}{(8 - 0)! \cdot 0!} = \dfrac{8!}{8! \cdot 0!}$

$= \dfrac{1}{0!} = \dfrac{1}{1} = 1$

(d) $C(7, 7) = \dfrac{7!}{(7 - 7)! \cdot 7!} = \dfrac{7!}{0! \cdot 7!}$

$= 1$

◀

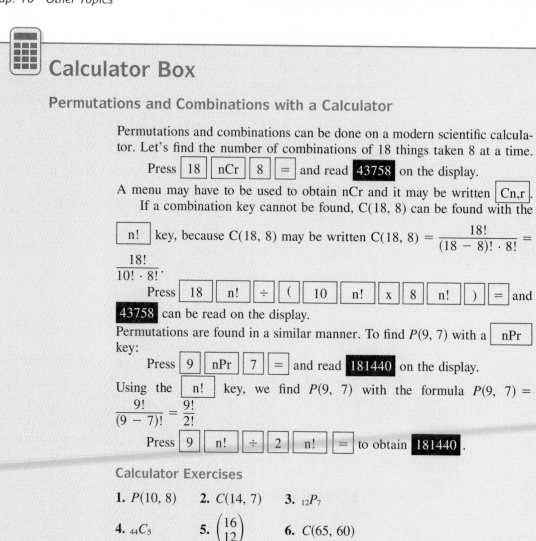

Calculator Box

Permutations and Combinations with a Calculator

Permutations and combinations can be done on a modern scientific calculator. Let's find the number of combinations of 18 things taken 8 at a time.

Press 18 nCr 8 = and read 43758 on the display.

A menu may have to be used to obtain nCr and it may be written Cn,r.

If a combination key cannot be found, C(18, 8) can be found with the

n! key, because C(18, 8) may be written $C(18, 8) = \dfrac{18!}{(18-8)! \cdot 8!} = \dfrac{18!}{10! \cdot 8!}$.

Press 18 n! ÷ (10 n! x 8 n!) = and 43758 can be read on the display.

Permutations are found in a similar manner. To find $P(9, 7)$ with a nPr key:

Press 9 nPr 7 = and read 181440 on the display.

Using the n! key, we find $P(9, 7)$ with the formula $P(9, 7) = \dfrac{9!}{(9-7)!} = \dfrac{9!}{2!}$.

Press 9 n! ÷ 2 n! = to obtain 181440 .

Calculator Exercises

1. $P(10, 8)$ **2.** $C(14, 7)$ **3.** $_{12}P_7$

4. $_{44}C_5$ **5.** $\binom{16}{12}$ **6.** $C(65, 60)$

Answers

1. 1,814,400   **2.** 3432   **3.** 3,991,680

4. 1,086,008   **5.** 1820   **6.** 8,259,888

▶ EXAMPLE 4

The Clarkston Womens club has 16 members. A steering committee made up of 4 members is to be formed. How many different ways can this be done?

Solution

Because there is no particular order to this committee, the number of different committees is the number of combinations of 16 things taken 4 at a time.

$$C(16, 4) = \frac{16!}{(16 - 4)! \cdot 4!} = \frac{16!}{12! \cdot 4!}$$

$$= \frac{1 \cdot 2 \cdot 3 \cdots 12 \cdot 13 \cdot 14 \cdot 15 \cdot 16}{1 \cdot 2 \cdot 3 \cdots 12 \cdot 1 \cdot 2 \cdot 3 \cdot 4}$$

$$= 13 \cdot 7 \cdot 5 \cdot 4 = 1820$$

There are 1820 different ways to form the steering committee. ◄

► **EXAMPLE 5**

How many different 5-card poker hands can be dealt from an ordinary deck of 52 cards?

Solution

Since the order in which the cards are dealt does not matter, we need to find the number of combinations of 52 things taken 5 at a time.

$$C(52, 5) = \frac{52!}{(52 - 5)! \cdot 5!} = \frac{52!}{47! \cdot 5!}$$

$$= \frac{1 \cdot 2 \cdot 3 \cdots 47 \cdot 48 \cdot 49 \cdot 50 \cdot 51 \cdot 52}{1 \cdot 2 \cdot 3 \cdots 47 \cdot 1 \cdot 2 \cdot 3 \cdot 4 \cdot 5}$$

$$= 2 \cdot 49 \cdot 10 \cdot 51 \cdot 52 = 2{,}598{,}960$$

There are 2,598,960 different 5-card poker hands. ◄

Problem Set 10.5

Warm-Ups

In Problems 1 through 14, evaluate each expression. For Problems 1 through 7, see Example 1.

1. $P(4, 2)$ **2.** $P(5, 0)$ **3.** $P(5, 1)$ **4.** $P(5, 2)$

5. $P(5, 3)$ **6.** $P(5, 4)$ **7.** $P(5, 5)$

For Problems 8 through 14, see Example 3.

8. $C(4, 2)$ **9.** $C(9, 0)$ **10.** $C(9, 1)$ **11.** $C(9, 2)$

12. $C(9, 3)$ **13.** $C(9, 9)$ **14.** $C(9, 8)$

For Problems 15 through 18, see Examples 2 and 4.

15. How many different three-letter "words" can be made with the letters of the word TIMERS?

16. How many different ways can first through fourth place be determined in the Kentucky Derby if 11 horses run?

17. Seven players are to be chosen from a roster of 13. In how many ways can this be done?

18. The local Ace Hardware carries 14 different shades of interior latex point. How many different combinations of four different shades can be selected?

Practice Exercises

In Problems 19 through 42, evaluate each expression.

19. $P(8, 4)$ **20.** $P(7, 5)$ **21.** $P(6, 0)$

22. $P(6, 1)$ **23.** $P(6, 2)$ **24.** $P(6, 3)$

25. $P(6, 4)$ **26.** $P(6, 5)$ **27.** $P(6, 6)$

28. $C(8, 4)$ **29.** $C(7, 5)$ **30.** $C(9, 5)$

31. $C(8, 0)$ **32.** $C(8, 1)$ **33.** $C(8, 2)$

34. $C(8, 3)$ **35.** $C(8, 8)$ **36.** $C(8, 7)$

37. $C(8, 6)$ **38.** $C(8, 5)$ **39.** $P(14, 4)$

40. $P(14, 10)$ **41.** $C(14, 4)$ **42.** $C(14, 10)$

43. How many different four-letter "words" can be made with the letters of the word OUTFIELD?

44. The officers of the Clean Air Action Coalition are president, vice-president, and secretary–treasurer. No person is allowed to hold more than one office. If there are 28 members, how many ways can a slate of officers be selected?

45. Nine players are to be chosen from a roster of 16. In how many ways can this be done?

46. In how many ways can 6 algebra books be taken from a pile of 13 algebra books?

Challenge Problems

In Problems 47 through 55, suppose that N is an integer greater than 1. Simplify each expression.

47. $P(N, N)$ **48.** $P(N, 0)$ **49.** $P(N, N - 2)$ **50.** $C(N, N)$

51. $C(N, 0)$ **52.** $C(N, N - 2)$ **53.** $C(N + 2, N)$ **54.** $P(N, 1)$

55. $C(N + 1, N - 1)$

In Your Own Words

56. Describe permutations without using a formula.

57. Describe combinations without using a formula.

► Chapter Summary

GLOSSARY

Sequence   A function whose domain is the set of natural numbers.

Terms of a sequence   The numbers in the range of the function that defines the sequence.

Arithmetic sequence   A sequence with a general term in the form $a_1 + (n - 1)d$.

Geometric sequence   A sequence with a general term in the form $a_1 r^{n-1}$.

Increasing sequence   A sequence in which every term of the sequence is larger than the preceding term.

Decreasing sequence   A sequence in which every term of the sequence is smaller than the preceding term.

Permutation   An arrangement of objects in a particular order.

Combination   An arrangement of objects where order is not important.

Summation Notation $A_1 + A_2 + A_3 + \cdots + A_n = \displaystyle\sum_{j=1}^{n} A_j$

Factorial Notation $1 \cdot 2 \cdot 3 \cdots (n - 1) \cdot n = n!$

Binomial Coefficients $\dbinom{n}{r} = \dfrac{n!}{r! \cdot (n - r)!}$, n, r nonnegative integers, with $n \geq r$

Binomial Theorem $(a + b)^n = \displaystyle\sum_{j=0}^{n} \binom{n}{j} a^{n-j} b^{j}$

CHECKUPS

1. Write the first five terms of the sequence
$$a_n = 4n + 5$$
Section 10.1; Example 1

2. Find a suitable general term for the sequence
$$1, -\frac{1}{2}, \frac{1}{4}, -\frac{1}{8}, \frac{1}{16}, \ldots$$
Section 10.1; Example 3d

3. Label the sequence as increasing, decreasing, or neither.
$$a_n = 4 - n$$
Section 10.1; Example 4b

4. Write out the terms of each sum.

   a. $\displaystyle\sum_{j=1}^{6} (2j - 1)$ Section 10.2; Example 1

   b. $\displaystyle\sum_{j=0}^{3} (-1)^j X^j$ Section 10.2; Example 4

5. Write the following sum in summation notation:
$$2 + 4 + 6 + 8 + 10 + 12 + 14 + 16$$
Section 10.2; Example 5

6. Evaluate $\dfrac{10!}{5! \cdot 5!}$.   Section 10.3; Example 2c

7. Calculate $\dbinom{7}{5}$.   Section 10.3; Example 3a

8. Write the first four terms of $(x + y)^{11}$.   Section 10.4; Example 1

9. Find $P(6, 2)$.   Section 10.5; Example 1a

10. Find $C(6, 2)$.   Section 10.5; Example 3a

► Review Problems

In Problems 1 through 6, write the first five terms of each sequence.

1. $a_n = 3n + 1$

2. $a_n = 2^n$

3. $a_n = 3 \cdot 5^{n-1}$

4. $b_n = 11 + 5n$

5. $c_n = (-1)^n \cdot 3n$

6. $d_n = (-1)^{n-1} A_n x^n$

In Problems 7 through 9, label each sequence as increasing, decreasing, or neither increasing nor decreasing.

7. $a_n = 10 + 3n$

8. $b_n = 10 - 3n$

9. $c_n = 3 - n^{-1}$

Problems 10 through 12 are arithmetic progressions. In each, find the common difference and write the next term.

10. $7, 9, 11, \ldots$

11. $-2, 0, 2, \ldots$

12. $5, \dfrac{13}{2}, 8, \ldots$

Problems 13 through 15 are geometric progressions. In each, find the common ratio and write the next term.

13. $1, 4, 16, \ldots$

14. $16, -8, 4, \ldots$

15. $40, 60, 90, \ldots$

In Problems 16 through 21, write out the terms of each sum.

16. $\displaystyle\sum_{j=0}^{5} (2j + 1)$

17. $\displaystyle\sum_{j=1}^{6} (1 - j)^2$

18. $\displaystyle\sum_{j=1}^{7} (2^j + j^2)$

19. $\displaystyle\sum_{j=0}^{5} (2j)^2$

20. $\displaystyle\sum_{j=1}^{6} (j - 6)^2$

21. $\displaystyle\sum_{j=1}^{4} (-1)^{j+1} j^j$

In Problems 22 through 24, write each sum in summation notation.

22. $1 + 3 + 5 + 7 + 9 + 11 + 13 + 15$

23. $1 + 2k + 3k^2 + 4k^3 + 5k^4 + 6k^5 + 7k^6$

24. $\dfrac{1}{2} - \dfrac{2}{3} + \dfrac{3}{4} - \dfrac{4}{5} + \dfrac{5}{6} - \dfrac{6}{7}$

In Problems 25 through 33, evaluate each expression.

25. $9!$

26. $3! \cdot 5!$

27. $0! \cdot 1! \cdot 2! \cdot 3!$

28. $\dfrac{14!}{13!}$

29. $\dfrac{41!}{39!}$

30. $\dfrac{10!}{5!}$

31. $\dbinom{9}{3}$

32. $\dbinom{17}{15}$

33. $\dbinom{10}{5}$

In Problems 34 through 39, use the Binomial Theorem to expand each binomial.

34. $(a + b)^5$

35. $(x + 1)^7$

36. $(y + 2)^8$

37. $(2 + 3y)^6$

38. $(4x + 3y)^4$

39. $(1 + s)^{10}$

40. Find the third term of $(a + b)^{11}$.

41. Find the first three terms of $(2 + 5t)^8$.

42. Find the fifteenth term of $(2 + k)^{17}$.

43. Find the middle term of $(a + b)^{14}$.

44. Simplify $P(6, 5)$.

45. Simplify $C(10, 4)$.

46. How many different ways can six balls be drawn from an urn containing 11 numbered balls if their order *is not* important?

47. How many different ways can six balls be drawn from an urn containing 11 numbered balls if their order *is* important?

Let's Not Forget . . .

Identify the expressions that are in factored form. Factor those that are not factored.

48. $(2x - 1)^5$

49. $8x^3 - 27$

50. $96x^2 - 4x - 15$

51. $(2x - 3)^3$

52. $63 + 7x^3$

53. $x^3 - 3x^2 + 3x - 1$

How many terms are in each expression? Which expressions have $7 - y$ as a factor?

54. $49 - y^2$

55. $t(7 - y) + y - 7$

56. $7 - 14y + y^2$

Simplify each expression, if possible, leaving only nonnegative exponents in your answer.

57. -2^{-4}

58. $\dfrac{-2x^{-3}}{y^2 z^{-2}}$

59. $\dfrac{16x^2 - 25}{125 - 64x^3}$

60. $|3x - 2|$

61. $\dfrac{2^{-3} + 8}{x^{-3} + 2^{-2}}$

62. $\sqrt[3]{(a + b)^6}$

Find each product.

63. $(3x + 2y)^3$

64. $(2\sqrt{2} + \sqrt{3})^2$

Reduce, if possible.

65. $\dfrac{-14 + \sqrt{-28}}{14}$

66. $\dfrac{z - w - r(w - z)}{w - z}$

The following problems can be worked by using a least common denominator. Follow the directions in each problem and notice how the LCD is used.

67. Perform the operation indicated: $\dfrac{1}{x} + \dfrac{1}{x^2}$.

68. Solve $\dfrac{4}{x} - \dfrac{1}{x^2} = 0$.

69. Simplify $\dfrac{\dfrac{1}{x} - 1}{\dfrac{1}{x^2} - 1}$.

Label each as an equation or an expression. Solve the equations in one variable, graph the equations in two variables, and perform the operations indicated on the expressions.

70. $\sqrt{4x - 3} = 2x - 3$

71. $\dfrac{1 - \dfrac{x}{x + 1}}{1 + \dfrac{1}{x + 1}}$

72. $(2x + y)^2$

73. $\left|\dfrac{2 - 3x}{4}\right| = -3$

74. $\dfrac{1}{2x^2 + 3x - 2} - \dfrac{1}{4x^2 - 4x + 1}$

75. $\dfrac{i}{2 - 3i}$

► You Decide

You are having a few friends over for an informal get-together. You want exactly 3 lbs of a party mix of nuts and other crunchies. Suppose your choices are:

Peanuts	@ $2.00 lb
Pecans	3.00 lb
Almonds	4.00 lb
Cashews	5.00 lb
Pretzel bits	0.80 lb
Sesame sticks	1.00 lb

and you want at least three different ingredients in the mix. Your budget allows only $8.00 for party mix. What combination do you choose and why?

► Chapter Test

In Problems 1 through 9, choose the correct response.

1. The sequence $a_n = (-1)^n n^2$ can be written(?)

   A. $1, -4, 9, \ldots$
   B. $-1, -4, -9, \ldots$
   C. $1, 4, 9, \ldots$
   D. $-1, 4, -9, \ldots$

2. The sequence $a_n = 3 \cdot 2^{n-1}$ is(?)

   A. an increasing, arithmetic progression
   B. an increasing, geometric progression
   C. a decreasing, arithmetic progression
   D. a decreasing, geometric progression

3. $\displaystyle\sum_{j=0}^{4} j \cdot (j + 1)$ is another way of writing the sum(?)

   A. $0 \cdot 1 + 1 \cdot 2 + 2 \cdot 3 + 3 \cdot 4$
   B. $0 \cdot 1 + 2 \cdot 3 + 4 \cdot 5$
   C. $0 + 1 + 2 + 3 + 4 + 5$
   D. $0 \cdot 1 + 1 \cdot 2 + 2 \cdot 3 + 3 \cdot 4 + 4 \cdot 5$

4. $\displaystyle\sum_{j=1}^{6} (-1)^{j+1} j = $ (?)

   A. 21   **B.** -3
   C. 3   **D.** -21

5. $6! = $ (?)

   A. 21   **B.** 30
   C. 36   **D.** 720

6. $\dfrac{12!}{10!} = $ (?)

   A. $\dfrac{6!}{5!}$ **B.** 132

   C. $2!$ **D.** $\dfrac{6}{5}$

7. $\dbinom{9}{3} = $ (?)

   A. 84   **B.** $6!$

   C. $7 \cdot 8 \cdot 9$ **D.** $\dfrac{1}{2}$

8. $(x + 2)^5 = $ (?)

   A. $x^5 + 32$
   B. $x^5 + 5x^4 + 10x^3 + 10x^2 + 5x + 2$
   C. $x^5 + 10x^4 + 40x^3 + 80x^2 + 80x + 32$
   D. $5x + 10$

9. The middle term of the expansion of $(a + b)^{12}$ is(?)

   A. $a^6 b^6$ **B.** $7 \cdot 8 \cdot 9 \cdot 10 \cdot 11 \cdot 12 a^6 b^6$
   C. $6! a^6 b^6$ **D.** $924 a^6 b^6$

10. Find a suitable general term for the sequence 4, 7, 10, 13, ...

11. Write the first five terms of the sequence $a_n = (-1)^{n+1} 2^n$.

12. Evaluate $\displaystyle\sum_{j=1}^{5} (2j - 1)$.

13. How many different five-letter "words" can be made from the letters of the word COMPUTER?

14. Use the Binomial Theorem to expand $(3 + t)^4$.

15. Find the last three terms of the expansion of $(a + b)^{18}$.

Appendix

An Introduction to the Graphing Calculator

The purpose of this appendix is to help the algebra student that has a graphing calculator become acquainted with this new tool. Although these new machines have great capabilities, it is the *student–calculator team* that must provide the results. This material is provided to get that team started on the road to a successful partnership. Although this appendix is fairly complete, it does not replace the instruction manual provided with the machine.

Several excellent graphing calculators are on the market. These instructions are designed to be useful no matter which graphing calculator is used. However, keystroke examples are given and keystrokes differ from one calculator to another. We have selected the most commonly seen machines for detailed instructions.

Texas Instruments. The TI–81 or TI–82 models are the most commonly seen graphing calculators in the algebra classroom as of this writing. Keystroke examples are provided for the TI–81. Where differences exist, they are noted, and keystrokes for the TI–82 are also supplied.

Casio. The Casio *fx*-7000 series graphing calculators are also in common usage. Keystroke examples are provided for the various *fx*-7700 models. These keystrokes are identical to those used on the *fx*-8700 and very similar to those used on other *fx*-7000 models.

Sharp. The Sharp EL-9200 and EL-9300 are recently released graphing calculator models. Keystroke examples are provided for the EL-9300. These keystrokes are identical to those used on the EL-9200.

Hewlett–Packard. The Hewlett–Packard graphing calculators are very powerful machines designed for use by working engineers and scientists. Since the HP series calculators are seldom seen in the algebra classroom and their use requires learning reverse Polish notation, they are not discussed in this appendix. However, students who have a Hewlett–Packard graphing calculator should not be disappointed. They are excellent machines and come with a complete instruction manual.

The graphing calculator is so named because of the powerful graphing capabilities of the machine. However, the ability to graph functions *is only one* of its useful features. Before we do any graphing, let's examine several other features common to all graphing calculators.

Modes

Graphing calculators have several modes. Although we may not use some of the modes, it is important to know how to get out of a mode in case we accidentally get into one. We will use the *computational* and *graphing* modes.

TI–81

Press $\boxed{\text{MODE}}$ and a list of modes will appear. The ones that are darkened are in effect. We can change these by using the arrows to move the cursor and then press $\boxed{\text{ENTER}}$.

CASIO fx–7700

Press $\boxed{\text{MODE}}$ $\boxed{+}$ to get into the computational mode. Press $\boxed{\text{M Disp}}$ and hold it down to see what modes are in effect.

Sharp El–9300

The four keys to the left of the $\boxed{\text{ON}}$ key are the mode keys. The first key is for the computational mode, and the second is for the graphing mode.

Expressions Displayed before Evaluation

Let's evaluate the expression

$$1.75^2 - \sqrt{\pi + 15}$$

We turn our calculators on and enter the expression just as we see it. We need to find keys for squaring, square root, and π. On each calculator, at least one of these is a second function. That is, the icon for the operation is written *above* a key, not *on* it. To get second function operations, we must first press a $\boxed{\text{SHIFT}}$ or $\boxed{\text{2nd}}$ key. In this appendix, we indicate keystrokes with a box like $\boxed{x^2}$ and second-function keystrokes with a double box like $\boxed{\boxed{\sqrt{}}}$. Although entering numbers may require many separate keystrokes, we will show them in a single box, like $\boxed{1.75}$. Let's enter the preceding expression.

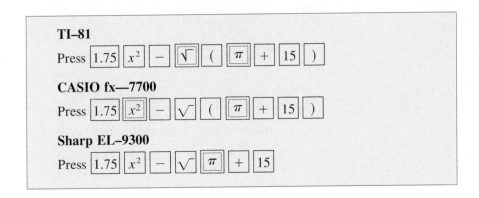

Notice that the expression to be evaluated is now displayed on the screen of the calculator. This is an important difference between graphing calculators and most nongraphing scientific calculators, which perform each operation when the key is pressed. To evaluate the expression, press $\boxed{\text{ENTER}}$ with the TI or Sharp, or $\boxed{\text{EXE}}$ with the Casio. The answer, approximately -1.196794854, should be displayed.

Editing

Suppose we wanted 14 under the square root instead of 15. That is, we wish to calculate

$$1.75^2 - \sqrt{\pi + 14}$$

We use the editing keys found at the right of the keypad, just under the top row of keys.

We refer to these keys as right arrow, left arrow, up arrow, and down arrow.

Press the up arrow on the TI–81, the left or right arrow on the Casio, and any arrow on the Sharp. Notice that the expression $1.75^2 - \sqrt{\pi + 15}$ appears on the display again. Press the left arrow until the cursor is blinking on top of the 5 in the number 15. Press $\boxed{4}$ and the editing is complete. (The Sharp will *insert* the 4, instead of writing over the 5. Press $\boxed{\text{DEL}}$ to delete the 5.) To evaluate the expression, press $\boxed{\text{ENTER}}$ with the TI or Sharp, or $\boxed{\text{EXE}}$ with the Casio. The answer, approximately -1.077740652, should be displayed.

Experiment with the editing arrows and also look for $\boxed{\text{DEL}}$ and $\boxed{\text{INS}}$ keys. On the Casio and the Sharp, the $\boxed{\text{INS}}$ is a second-function key above the $\boxed{\text{DEL}}$ key. These keys are used to delete or insert one keystroke at a time.

Negation, Subtraction, and Exponentiation

Let's enter the expression

$$-6 - 3.1^4$$

We have two keys with minus signs. The *negation* key, $\boxed{(-)}$, is used to indicate *negative* 6, while the *subtraction* key, $\boxed{-}$, is used for subtraction. To enter the exponent, we use the $\boxed{\wedge}$ key on the TI, the $\boxed{x^y}$ key on the Casio, and the $\boxed{a^b}$ key on the Sharp.

TI–81

Press $\boxed{(-)}$ $\boxed{6}$ $\boxed{-}$ $\boxed{3.1}$ $\boxed{\wedge}$ $\boxed{4}$ $\boxed{\text{ENTER}}$

CASIO fx–7700

Press $\boxed{(-)}$ $\boxed{6}$ $\boxed{-}$ $\boxed{3.1}$ $\boxed{x^y}$ $\boxed{4}$ $\boxed{\text{EXE}}$

Sharp EL–9300

Press $\boxed{(-)}$ $\boxed{6}$ $\boxed{-}$ $\boxed{3.1}$ $\boxed{a^b}$ $\boxed{4}$ $\boxed{\text{ENTER}}$

We read $\boxed{-98.3521}$ on the display.

Saving Numbers in Memory and Recalling from Memory

Each of these calculators has a memory for every letter in the alphabet. Look for a $\boxed{\text{STO}}$ key on the TI and the Sharp and a $\boxed{\rightarrow}$ key on the Casio. Notice that many keys on the keypad have letters of the alphabet written above them. These are memory locations. Let's store 41 in memory B.

TI–81

Press $\boxed{41}$ $\boxed{\text{STO} \triangleright}$ $\boxed{\text{B}}$ $\boxed{\text{ENTER}}$

TI–82

Press $\boxed{41}$ $\boxed{\text{STO} \triangleright}$ $\boxed{\text{ALPHA}}$ $\boxed{\text{B}}$ $\boxed{\text{ENTER}}$

CASIO fx–7700

Press $\boxed{41}$ $\boxed{\rightarrow}$ $\boxed{\text{ALPHA}}$ $\boxed{\text{B}}$ $\boxed{\text{EXE}}$

Sharp EL–9300

Press $\boxed{41}$ $\boxed{\text{STO}}$ $\boxed{\text{B}}$

Notice that both the TI–81 and the Sharp *automatically* go into alphabetic mode following ⎢STO⎥. Sharp users should also note that it is unnecessary to press ⎢ENTER⎥ when storing a number. To recall a number, we simply press ⎢ALPHA⎥ and the letter. Let's evaluate 37 + B. (41 is stored in B currently.) The keystrokes

⎢37⎥ ⎢+⎥ ⎢ALPHA⎥ ⎢B⎥ ⎢ENTER⎥ (or ⎢EXE⎥ on the Casio)

yield 37 + 41 or 78.

Each calculator stores the last answer in ⎢ANS⎥. It is a second-function key on both the TI–81 and the Sharp. Press ⎢ANS⎥ and Ans will appear on the display.

⎢10⎥ ⎢+⎥ ⎢ANS⎥ ⎢ENTER⎥ (⎢EXE⎥ on the Casio)

will add 10 to 78.

Variables and Implied Multiplication

The key ⎢X | T⎥ on the TI and the key ⎢X,θ,T⎥ on the Casio and Sharp supply the variables x, θ, and T used in expressions and for graphing functions. The variable given depends on the current mode of the calculator. All the calculators we are discussing use implied multiplication in a natural manner. That is, to find 37 *times* B, we simply enter 37B by pressing ⎢37⎥ ⎢ALPHA⎥ ⎢B⎥ and ⎢ENTER⎥ or ⎢EXE⎥ and read 1517 on the display (assuming that we still have 41 in B).

To evaluate the polynomial $2x^2 - 5x + 17$, when x is 11, we first store 11 in x. (When storing in x using the ⎢X | T⎥ or ⎢X,θ,T⎥ key, *do not* precede the key with ⎢ALPHA⎥.) We then enter the polynomial just as it is written.

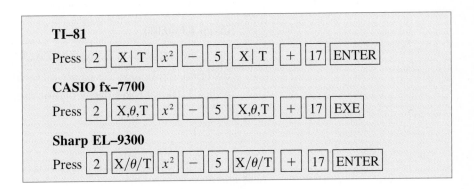

TI–81

Press ⎢2⎥ ⎢X | T⎥ ⎢x^2⎥ ⎢−⎥ ⎢5⎥ ⎢X | T⎥ ⎢+⎥ ⎢17⎥ ⎢ENTER⎥

CASIO fx–7700

Press ⎢2⎥ ⎢X,θ,T⎥ ⎢x^2⎥ ⎢−⎥ ⎢5⎥ ⎢X,θ,T⎥ ⎢+⎥ ⎢17⎥ ⎢EXE⎥

Sharp EL–9300

Press ⎢2⎥ ⎢X/θ/T⎥ ⎢x^2⎥ ⎢−⎥ ⎢5⎥ ⎢X/θ/T⎥ ⎢+⎥ ⎢17⎥ ⎢ENTER⎥

We read the value of the polynomial, 204, on the display.

Using Function Memory

Graphing calculators have memory capability for entire expressions, usually called function memory. Let's calculate the volume of several cones with the radius and height as follows:

r	h
24.5	42.1
4.3	6.4

The formula for the volume of a cone is $V = \dfrac{1}{3}\pi r^2 h$. First, we store $\pi R^2 H \div 3$ in the *function memory*.

TI–81

Press $\boxed{Y =}$. A menu appears with the cursor blinking after $:Y_1=$. Enter the expression using the $\boxed{\text{ALPHA}}$ key for R and H. Press $\boxed{\text{QUIT}}$ to to exit the Y = screen.

CASIO fx–7700

Enter the expression using the $\boxed{\text{ALPHA}}$ key for R and H. Press $\boxed{\boxed{F}\ \text{MEM}}$ at the lower left of the keypad. Notice the function memory *status line* that appears at the bottom of the display.

$$\boxed{\text{STO} \mid \text{RCL} \mid \text{fn} \mid \text{LIST}}$$

Just under the display are six keys marked F1 through F6. They apply to the items in the status line. We wish to *store* in f1, so we press $\boxed{\text{F1}}$ $\boxed{1}$. (F1 is to *store* and the second 1 indicates the storage *location*.) The function memory is displayed with the expression in f1. Press $\boxed{\text{AC}}$ to exit the function memory.

Sharp EL–9300

Enter the expression using the $\boxed{\text{ALPHA}}$ key for R and H. Press $\boxed{\text{ENTER}}$ and ignore the result. (The Sharp automatically remembers the last 110 keystrokes in calculation mode, even if the machine is turned off.)

We wish to evaluate the expression several times. We store the numbers 24.5 in R and 42.1 in H.

> **TI-81**
>
> Press ⌈Y-VARS⌉ for a listing of function names. Our function is Y_1, so we press ⌈1⌉. Now Y_1 is displayed. Press ⌈ENTER⌉ and the volume is calculated.
>
> **CASIO fx-7700**
>
> The status line should still be showing. If not, press ⌈F⌉ MEM .
>
> Press ⌈F2⌉ ⌈1⌉. (F2 is to *recall* the expression, and the 1 indicates the storage location.) Press ⌈EXE⌉ to calculate the volume.
>
> **Sharp EL-9300**
>
> Press ⌈△⌉ (⌈2ndF⌉ followed by the up arrow) until $\pi R^2 H/3$ is visible. Then press ⌈ENTER⌉ to calculate the volume.

The result, 26,463.2319, should be displayed. Enter the next numbers for R and for H and repeat the procedure. (The second answer is about 123.9211694.)

Evaluating the Quadratic Formula

Approximating the solutions of quadratic equations with the quadratic formula is a good example of formula evaluation. In all cases, the solutions to the equation $ax^2 + bx + c = 0$ are given by the formula $x = \dfrac{-b \pm \sqrt{b^2 - 4ac}}{2a}$. Some solutions may be complex numbers.

Because the TI and the Casio will not evaluate the square root of a negative number, we will calculate the discriminant first with those calculators. We put the discriminant, $b^2 - 4ac$, in function memory.

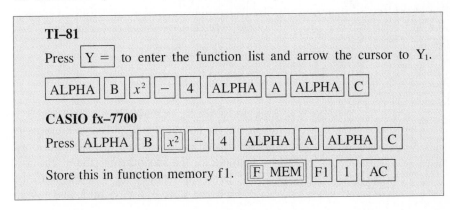

> **TI–81**
>
> Press ⌈Y =⌉ to enter the function list and arrow the cursor to Y_1.
>
> ⌈ALPHA⌉ ⌈B⌉ ⌈x^2⌉ ⌈−⌉ ⌈4⌉ ⌈ALPHA⌉ ⌈A⌉ ⌈ALPHA⌉ ⌈C⌉
>
> **CASIO fx–7700**
>
> Press ⌈ALPHA⌉ ⌈B⌉ ⌈x^2⌉ ⌈−⌉ ⌈4⌉ ⌈ALPHA⌉ ⌈A⌉ ⌈ALPHA⌉ ⌈C⌉
>
> Store this in function memory f1. ⌈F⌉ MEM ⌈F1⌉ ⌈1⌉ ⌈AC⌉

Next, we enter the quadratic formula using the stored discriminant.

TI–81

We enter the quadratic formula as Y_2. Notice how we use the discriminant stored in Y_1. We press $\boxed{Y =}$ to enter the function list and arrow the cursor to Y_2. Enter

$\boxed{(}$ $\boxed{(-)}$ $\boxed{ALPHA}$ $\boxed{B}$ $\boxed{+}$ $\boxed{\sqrt{}}$ $\boxed{\text{Y-VARS}}$ $\boxed{1}$ $\boxed{)}$ $\boxed{\div}$ $\boxed{(}$ $\boxed{2}$

$\boxed{ALPHA}$ $\boxed{A}$ $\boxed{)}$ $\boxed{QUIT}$

CASIO fx–7700

We enter the quadratic formula as f2. Notice how we use the discriminant stored in f1.

$\boxed{(}$ $\boxed{(-)}$ $\boxed{ALPHA}$ $\boxed{B}$ $\boxed{+}$ $\boxed{\sqrt{}}$ $\boxed{(}$ $\boxed{\text{F MEM}}$ $\boxed{F2}$ $\boxed{1}$ $\boxed{)}$ $\boxed{)}$

$\boxed{\div}$ $\boxed{(}$ $\boxed{2}$ $\boxed{ALPHA}$ $\boxed{A}$ $\boxed{)}$

We store this in function memory, f2. $\boxed{\text{F MEM}}$ $\boxed{F1}$ $\boxed{2}$ $\boxed{AC}$

Sharp EL–9300

We put the calculator into *complex mode* by pressing

$\boxed{\genfrac{}{}{0pt}{}{+\ -}{\times\ \div}}$ $\boxed{MENU}$ $\boxed{4}$.

Then we enter the quadratic formula.

$\boxed{(}$ $\boxed{(-)}$ $\boxed{ALPHA}$ $\boxed{B}$ $\boxed{+}$ $\boxed{\sqrt{}}$ $\boxed{ALPHA}$ $\boxed{B}$ $\boxed{x^2}$ $\boxed{-}$ $\boxed{4}$

$\boxed{ALPHA}$ $\boxed{A}$ $\boxed{ALPHA}$ $\boxed{\blacktriangleright}$ $\boxed{C}$ $\boxed{)}$ $\boxed{\div}$ $\boxed{2}$ $\boxed{ALPHA}$ $\boxed{A}$

$\boxed{ENTER}$

(If an error message occurs here, it's because $A = 0$. We ignore it by pressing $\boxed{CL}$.)

Notice that we used the $+$ case of the $\pm$ in the formula. Now we need only to put the numbers *a, b,* and *c* into memory locations A, B, and C and then recall the formula and evaluate it if the solutions are real numbers. We then edit the formula for the second root (the $-$ case of the $\pm$).

Let's approximate the solutions of the quadratic equation $2x^2 + 3x - 4 = 0$. First, we store 2 in A, 3 in B, and -4 in C.

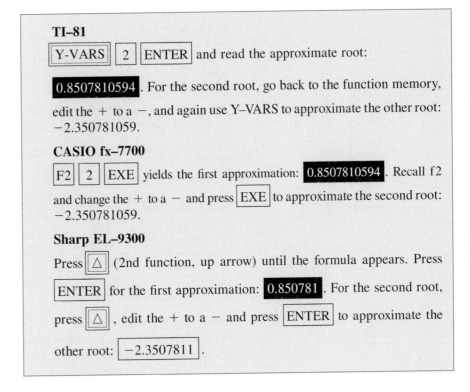

TI–81

Y-VARS 2 ENTER and read the approximate root:

0.8507810594 . For the second root, go back to the function memory, edit the + to a −, and again use Y–VARS to approximate the other root: −2.350781059.

CASIO fx–7700

F2 2 EXE yields the first approximation: 0.8507810594 . Recall f2 and change the + to a − and press EXE to approximate the second root: −2.350781059.

Sharp EL–9300

Press △ (2nd function, up arrow) until the formula appears. Press ENTER for the first approximation: 0.850781 . For the second root, press △ , edit the + to a − and press ENTER to approximate the other root: −2.3507811 .

Now let's repeat the procedure for the equation $x^2 - 2x + 2 = 0$. Store 1, −2, and 2 in A, B, C, respectively.

TI–81 and CASIO fx–7700

First we evaluate the discriminant. We get −4. The *negative* discriminant signals complex roots. The quadratic formula can be written as $x = \dfrac{-b \pm \sqrt{D}}{2a}$, where D is the discriminant. The solutions are $1 \pm i$. (Paper and pencil may be necessary for calculations.)

Sharp EL–9300

Press △ (2nd function, up arrow) until the formula appears. Press ENTER for the first solution, $1 - i$. For the second root, press △ , edit the − to a +, and press ENTER for the other root, $1 + i$.

Graphing

Now let's explore using the graphing calculator to draw graphs. We must be in graphing mode.

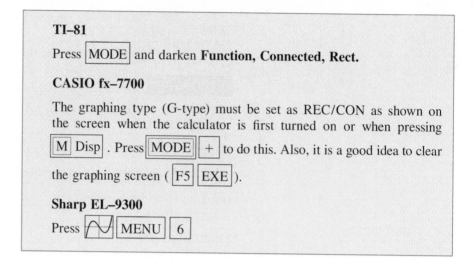

TI–81

Press MODE and darken **Function, Connected, Rect.**

CASIO fx–7700

The graphing type (G-type) must be set as REC/CON as shown on the screen when the calculator is first turned on or when pressing M Disp . Press MODE + to do this. Also, it is a good idea to clear the graphing screen (F5 EXE).

Sharp EL–9300

Press ⌐⌐⌐ MENU 6

Setting the Window

The display on a graphing calculator is called a window or viewing rectangle. Since the display shows only a portion of a coordinate system, our first consideration is to set an appropriate window. Imagine the display as a rectangle that has a set of axes at its center. The size of the rectangle is fixed by setting minimum and maximum *x* and *y* coordinates.

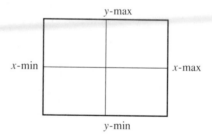

Graphing calculators have a standard window that is often good for starting a graph. It is called the *default setting*. Let's set the range at the default setting.

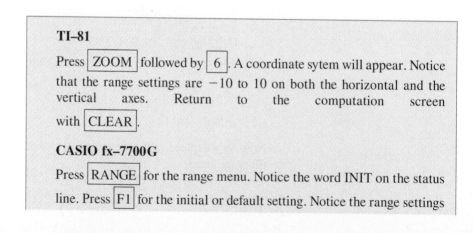

TI–81

Press ZOOM followed by 6 . A coordinate sytem will appear. Notice that the range settings are −10 to 10 on both the horizontal and the vertical axes. Return to the computation screen with CLEAR .

CASIO fx–7700G

Press RANGE for the range menu. Notice the word INIT on the status line. Press F1 for the initial or default setting. Notice the range settings on the menu. They are -4.7 to 4.7 on the x-axis and -3.1 to 3.1 on the y-axis. Press $\boxed{\text{RANGE}}$ $\boxed{\text{RANGE}}$ to return to the computation screen.

Sharp EL–9300

Press $\boxed{\text{RANGE}}$ $\boxed{\text{MENU}}$ for the range menu. (If you have left the graphing mode, first press $\boxed{\sim}$.) Notice that DEFLT is darkened.

Press $\boxed{\text{ENTER}}$ for the default range setting. Arrow down to see the default range settings. They are -4.7 to 4.7 on the x-axis and -3.1 to 3.1 on the y axis. Press $\boxed{\text{QUIT}}$ until Y1 appears.

Now let's graph $y = -2x + 1$ on the default window.

TI–81

Press $\boxed{\text{Y} =}$ and enter $\boxed{(-)}$ $\boxed{2}$ $\boxed{\text{X} \mid \text{T}}$ $\boxed{+}$ $\boxed{1}$ in Y_1. Delete any left-over symbols. Then press $\boxed{\text{GRAPH}}$.

CASIO fx–7700

To enter and graph the equation, press $\boxed{\text{GRAPH}}$ $\boxed{(-)}$ $\boxed{2}$ $\boxed{\text{X},\theta,\text{T}}$ $\boxed{+}$ $\boxed{1}$ $\boxed{\text{EXE}}$. The function memory can also be used. Store $-2x + 1$ in function memory f1. Then press $\boxed{\text{GRAPH}}$ $\boxed{\text{F} \text{ MEM}}$ $\boxed{\text{F2}}$ $\boxed{1}$ $\boxed{\text{EXE}}$. (F2 will recall memory location 1.)

Sharp EL–9300

Press $\boxed{(-)}$ $\boxed{2}$ $\boxed{\text{X}/\theta/\text{T}}$ $\boxed{+}$ $\boxed{1}$ $\boxed{\sim}$

We see the graph shown below.

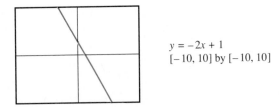

$y = -2x + 1$
$[-10, 10]$ by $[-10, 10]$

The $\boxed{\text{RANGE}}$ key allows us to change the size of the viewing rectangle. Let's set the viewing rectangle at $[-2, 2]$ for both x and y.

TI–81

Press the ⌈RANGE⌉ key. Use the editing arrows and number keys to set X_{min} to -2 and X_{max} to 2. Be careful to use the ⌈(−)⌉ key to enter a negative number. Set Y_{min} and Y_{max} in a similar manner. To quit the range screen, press ⌈QUIT⌉. To graph the line on the new coordinate system, press ⌈GRAPH⌉.

CASIO fx–7700

Press the ⌈RANGE⌉ key. Set X_{min} and press ⌈EXE⌉. Set X_{max}, Y_{min}, and Y_{max} the same way. To exit the range screen, press ⌈RANGE⌉ as many times as necessary. Notice that there are two range screens, one for x and y coordinates and one for polar coordinates. To graph the line on the new coordinate system, press ⌈GRAPH⌉ and reenter the equation.

Sharp EL–9300

In the graphing mode, press ⌈RANGE⌉. With the cursor on X_{min}, press ⌈2⌉ ⌈ENTER⌉. This sets the minimum x coordinate. Set X_{max}, Y_{min}, and Y_{max} in the same manner. To graph the line on the new coordinate system, press ⌈∿⌉.

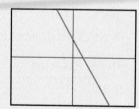

$y = -2x + 1$
$[-2, 2]$ by $[-2, 2]$

Two Graphs on the Same Screen

Let's graph $y = -2x + 1$ and $y = x$ on the same screen.

TI–81

We have already entered $y = -2x + 1$ as Y_1. Enter $y = x$ as Y_2. Notice that the equals sign is darkened. The TI will graph each function that is darkened. To turn off the graph, place the cursor on the equal sign and press ⌈ENTER⌉. Notice that the equal sign is no longer darkened. To turn on a graph, repeat the procedure. With both equal signs darkened, press ⌈GRAPH⌉.

CASIO fx–7700

Press $\boxed{\text{GRAPH}}$ $\boxed{(-)}$ $\boxed{2}$ $\boxed{\text{X,}\theta\text{,T}}$ $\boxed{+}$ $\boxed{1}$ $\boxed{\leftarrow}$ $\boxed{\text{GRAPH}}$ $\boxed{\text{X,}\theta\text{,T}}$ $\boxed{\text{EXE}}$. Notice that the $\boxed{\leftarrow}$ (second function key above EXE) moves the cursor to the next line on the display. This connects the two graph commands and is called a *multistatement*. It is a good idea to always clear the graph screen after graphing. Press $\boxed{\text{F5}}$ $\boxed{\text{EXE}}$.

Sharp EL–9300

Press $\boxed{\text{QUIT}}$. Enter $y = x$ as Y_2. The Sharp works like the TI. Press $\boxed{\text{2ndF}}$ and the up and down arrows to move from Y_1 to Y_2. Press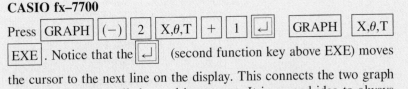

Finding the Intercepts and Vertex of a Parabola

Not only can a graphing calculator draw graphs, but it can also approximate coordinates of points on a graph. The $\boxed{\text{TRACE}}$ key on the TI and Casio allows us to move a point along a graph and shows the current coordinates of the point as it moves, and the Sharp will automatically approximate certain key points on a graph.

Let's approximate the x intercepts and vertex of the graph of the equation $y = x^2 - x - 1$. The graph of this equation is a parabola that opens upward. The intercepts and vertex can be found algebraically, but let's see what the calculator can find. We must choose a window that will contain the vertex and intercepts. Let's start with the window at $[-5, 5]$ for both x and y. We can change it later if necessary. Now, we enter the function and $\boxed{\text{GRAPH}}$ it. Let's call the negative x intercept x_1 and the positive one x_2.

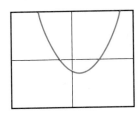

$y = x^2 - x - 1$
$[-5, 5]$ by $[-5, 5]$

Both the TI and the Casio have trace keys. With either of those, press $\boxed{\text{TRACE}}$ and notice that a blinking cursor appears. Also, the coordinates of the cursor point are written at the bottom of the screen. Use the left and right arrow keys to move the cursor along the parabola. Since the y coordinate of the x intercepts is 0, we find the y coordinates closest to 0.

TI–81

$x_1 \approx -.5789474$ $y_1 \approx -0.0858726$
$x_2 \approx 1.6315789$ $y_1 \approx 0.03047091$

Move the cursor to the vertex. To find the lowest point, compare the y coordinates as the cursor moves along the graph. Find the smallest y coordinate. The smallest y coordinate is approximately -1.249307. So the approximate coordinates of the vertex are $(0.47, -1.25)$.

CASIO fx–7700

$x_1 \approx -0.638297$ $y_1 \approx 0.045722$
$x_2 \approx 1.5957446$ $y_1 \approx -0.049343$

The smallest y coordinate is approximately -1.248981. So the approximate coordinates of the vertex are $(0.53, -1.25)$.

Sharp EL–9300

With the graph on the screen, we press $\boxed{\text{JUMP}}$ and a menu appears. Press $\boxed{4}$ for the x-intercept and the graph reappears with $x = -0.618033$, $y = 0$ underneath. The leftmost x-intercept is approximately -0.618. We repeat the same keystrokes for the other x-intercept. The calculator shows approximately 1.618. To find the vertex, we press $\boxed{\text{JUMP}}$ $\boxed{2}$ to find the minimum (low point) of the graph of the function. The approximation given is $x = 0.500007$, $y = -1.25$. So the approximate coordinates of the vertex are $(0.50, -1.25)$.

The x intercepts are solutions to the equation $x^2 - x - 1 = 0$. With the quadratic formula, we find them to be approximately -0.6180 and 1.6180.

The $\boxed{\text{TRACE}}$ feature depends on the dimensions of the window for its accuracy. Set the range at $[-2, 5]$ for both x and y and redraw the graph. Use $\boxed{\text{TRACE}}$ again to approximate the vertex. Now we find that the coordinates of the vertex are approximately $(0.50526316, -1.249972)$. Since by completing the square we know that the vertex is $(0.5, -1.25)$, this window gives a better approximation.

Box and Zoom

Graphing calculators have a zoom feature that allows us to "zoom in on" an interesting part of a graph. First, graph $y = x^2 - 2.75x + 2.02$ on the default window. Does the parabola touch the x-axis?

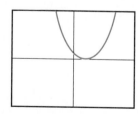

$y = x^2 - 2.75x + 2.02$
$[-10, 10]$ by $[-10, 10]$

We can zoom in on the vertex and see!

TI–81

Press $\boxed{\text{ZOOM}}$ and a menu appears. Select **1: Box** by pressing $\boxed{1}$ or $\boxed{\text{ENTER}}$. Notice that this also returns us to the graph. Next we make a box with the editing arrows. Press the right arrow several times and then the up arrow a few times and notice a + moving like a cursor.

Position the + approximately at the point $(1, 0.5)$. Press $\boxed{\text{ENTER}}$. One corner of the box is fixed. We now fix the corner diagonally across. Arrow down and across to about $(2, -0.5)$. We see a little box. Press $\boxed{\text{ENTER}}$ and the full screen becomes that box!

CASIO fx–7700G

Press $\boxed{\text{F2}}$ to bring up the zoom menu. Press $\boxed{\text{F1}}$ to select the box feature. The box feature works exactly like the one on the TI–81.

Sharp EL–9300

Press $\boxed{\text{ZOOM}}$ $\boxed{1}$ to select the box feature. It works exactly like the one on the TI–81.

The $\boxed{\text{TRACE}}$ feature can be used to estimate coordinates of points. Accuracy can be increased by boxing and zooming again and again. The vertex has coordinates about $(1.375004, 0.129375)$.

Answers to Selected Problems

PROBLEM SET 0.1

Warm-Ups

1. T   **2.** T   **3.** T   **4.** T   **5.** F   **6.** F   **7.** F   **8.** T   **9.** F
10. F   **11.** $-2, 0, 2, 4, 6, 8$ **12.** 2   **13.** $\{-2, 0, 2\}, \{-2, 0\}, \{-2, 2\}, \{0, 2\}, \{-2\}, \{0\}, \{2\}, \emptyset$
14. 1, 2, 3, 4, 5, 6, 7, 8, 9   **15.** red, white, blue   **16.** 5, 10, 15, . . .
17. a, b, c, d, e, f, g, h, i, j   **18.** $\{x \mid x$ is one of the last 4 letters in alphabet$\}$
19. $\{x \mid x$ is an even integer between 0 and 14$\}$ **20.** $\{(x \mid x$ is an odd integer$\}$
21. $\{(x \mid x$ is a day of the week$\}$ **22.** $0.\overline{3}$ **23.** 0.750 **24.** $0.1\overline{6}$ **25.** $0.625\overline{0}$ **26.** $0.\overline{285714}$
27. Rational   **28.** Rational   **29.** Irrational   **30.** Irrational   **31.** Rational   **32.** Rational

Practice Exercises

33. F   **35.** T   **37.** T   **39.** T   **41.** T   **43.** $\frac{1}{2}, 2, \frac{9}{4}, 5$ **45.** $\left\{\frac{9}{4}, 5\right\}, \left\{\frac{9}{4}\right\}, \{5\}, \emptyset$ **47.** Friday

49. q, r, s, t, u, v, w, x, y, z   **51.** $\{x \mid x$ is an integer between -3 and $3\}$
53. $\{x \mid x$ is a suit in a deck of cards$\}$ **55.** Rational   **57.** Irrational   **59.** Rational   **61.** $0.6\overline{0}$
63. $0.875\overline{0}$

Challenge Problems

65. $\frac{3}{10}$ **67.** $\frac{655}{1000}$

PROBLEM SET 0.2

Warm-Ups

1. $-\sqrt{3},$ ┼┼┼ $-\sqrt{3}\ \ 0\ \ \sqrt{3}$ **2.** 4, ┼┼┼ $-4\ \ 0\ \ 4$ **3.** $\frac{1}{3},$ ┼┼┼ $-\frac{1}{3}\ \ 0\ \ \frac{1}{3}$ **4.** $-\frac{11}{3},$ ┼┼┼ $-\frac{11}{3}\ \ 0\ \ \frac{11}{3}$

5. 0,   **6.** $-p$,   **7.** $-q$,   **8.** $\frac{1}{4}$ **9.** $\frac{1}{7}$ **10.** $\frac{3}{5}$

11. $\frac{3}{2}$ **12.** $\frac{4}{3}$ **13.** $\frac{5}{8}$ **14.** Commutative   **15.** Reflexive   **16.** Symmetric   **17.** Associative

18. Commutative   **19.** Additive Identity   **20.** Additive Inverse   **21.** Multiplicative Identity

22. Commutative   **23.** Multiplicative Inverse   **24.** Multiplication by 0   **25.** Distributive   **26.** Transitive

27. Substitution   **28.** Associative   **29.** (a) Commutative Property (b) Distributive Property (c) Addition

(d) Commutative Property   **30.** $y - x$

Practice Exercises

31. Commutative   **33.** Symmetric   **35.** Distribution   **37.** Commutative   **39.** Additive Inverse

41. Multiplication by zero   **43.** Transitive   **45.** Associative   **47.** 2,

49. $-\frac{2}{3}$,   **51.** p,   **53.** 1

55. 4   **57.** $\frac{7}{3}$ **59.** (1) Associative Property (2) Addition

Challenge Problems

61.

PROBLEM SET 0.3

Warm-Ups

1. 46   **2.** 46   **3.** 4   **4.** 38   **5.** 0   **6.** -4 **7.** 38   **8.** -13 **9.** -13 **10.** 38   **11.** -60

12. -80 **13.** a **14.** $-b$ **15.** Won't simplify   **16.** d^2 **17.** a **18.** $-b$ **19.** $|c|$ **20.** $|d|$

21. ; 8 units   **22.** ; 8 units   **23.** ; 2 units

24. ; $\pi - \sqrt{7}$ units   **25.** ; $\sqrt{5} - \sqrt{3}$ units

26. x and y can be anywhere on a number line, $|x - y|$ units.

27. $5|y|$ **28.** $14|z|$ **29.** $-53|k|$ **30.** $-2|x|$ **31.** $\frac{6}{|z|}$

Practice Exercises

33. 16   **35.** 3   **37.** 13   **39.** -47 **41.** -3 **43.** -108 **45.** $21|x|$ **47.** $-81|g|$ **49.** $\frac{|u|}{9}$

51. Won't simplify   **53.** r **55.** $-s$ **57.** r **59.** $-r$ **61.** $-r$ **63.** $|-(-3)| = -3$, False

65. $|-(-3)| = -(-3)$, True   **67.** $|7| = 7$, True   **69.** 8 units;

71. 4 units;   **73.** $\sqrt{7} - \sqrt{5}$ units;   **75.** 13   **77.** -2 **79.** $\sqrt{2} + \pi$

Challenge Problems

81. $\pi - 3$ **83.** $\pi + 3$

PROBLEM SET 0.4

Warm-Ups

1. 9   **2.** 49   **3.** -4 **4.** -18 **5.** Undefined   **6.** 0   **7.** 2   **8.** $-\dfrac{5}{3}$ **9.** $\dfrac{3}{2}$ **10.** 13

11. 5   **12.** -1 **13.** 1   **14.** -3 **15.** $-\dfrac{1}{7}$ **16.** 3   **17.** -10 **18.** 2   **19.** -17 **20.** -28

21. 81   **22.** 9   **23.** 12   **24.** $\dfrac{1}{30}$

Practice Exercises

25. -15 **27.** 42   **29.** -4 **31.** -8 **33.** -11 **35.** -7 **37.** 42   **39.** $-\dfrac{4}{27}$ **41.** -24

43. $\dfrac{25}{16}$ **45.** $-\dfrac{10}{3}$ **47.** 15   **49.** 1   **51.** -6 **53.** -1 **55.** 13

Challenge Problems

57. $\dfrac{1}{2}$

PROBLEM SET 0.5

Warm-Ups

1. $\angle a = 45°$, $\angle b = 135°$, $\angle c = 135°$, $\angle d = 45°$, $\angle e = 135°$, $\angle f = 45°$, $\angle g = 45°$, $\angle h = 135°$, $\angle i = 90°$, $\angle j = 90°$, $\angle k = 90°$, $\angle l = 90°$, $\angle m = 135°$, $\angle n = 45°$, $\angle o = 45°$, $\angle p = 135°$, $\angle q = 45°$, $\angle r = 135°$, $\angle s = 135°$, $\angle t = 45°$ **2.** 58°   **3.** 120°   **4.** 95°   **5.** 30°   **6.** $5^2 = 25$; $3^2 + 4^2 = 25$
7. (a) Two angles of one triangle are equal to two angles of the other triangle. **(b)** Two angles of one triangle are equal to two angles of the other triangle.

Practice Exercises

9. F   **11.** F   **13.** T   **15.** F   **17.** F   **19.** T   **21.** 50°   **23.** 140°   **25.** 68°
27. $10^2 = 100$; $6^2 + 8^2 = 100$ **29.** Corresponding sides are proportional.   **31.** T   **33.** T
35. F   **37.** T   **39.** T

Challenge Problems

41. $(n - 2)180°$

PROBLEM SET 0.6

Warm-Ups

1. 12 in.   **2.** 47 m   **3.** 34 cm   **4.** 28 ft   **5.** 74 yd   **6.** 28π mm   **7.** 360 ft   **8.** 700 ft
9. 28 m²   **10.** $16 - 4\pi$ in.²   **11.** 60 cm²   **12.** 2500π in.²   **13.** 125 cubic units   **14.** 696.91 in.³
15. 5250 ft³   **16.** 14,137 mm³   **17.** 4.6 in.³   **18.** 19.9 in.³
19. 201,061,930 mi²   **20.** 35,186 ft²   **21.** $325\frac{1}{2}$ in.²   **22.** 306,305 in.²

Practice Exercises

23. 7.5 in.   **25.** 62 ft   **27.** 75.06 yd   **29.** 25,133 mi   **31.** $\frac{9}{8}\pi$ in.²   **33.** $12\frac{19}{27}$ in.³   **35.** $28\frac{7}{8}$ in.³

37. 754 yd³   **39.** 268,000,000,000 mi³   **41.** $\frac{3}{2}$ ft³   **43.** 412,334 in.²   **45.** 2 ft³

Challenge Problems

47. 174 ft²

REVIEW PROBLEMS CHAPTER 0

1. 35   **3.** $-\frac{21}{13}$ **5.** 1   **7.** -41 **9.** 7   **11.** 7   **13.** 56   **15.** Commutative   **17.** Distributive
19. Multiplicative identity   **21.** Irrational   **23.** Rational   **25.** Rational   **27.** c **29.** 1, 3, 5
31. -20 **33.** -25 **35.** -25 **37.** $-b$ **39.** a **41.** The perimeter is 40 inches. The area is 104
square inches.   **43.** The circumference is 5π meters. The area is 6.25π square meters.   **45.** The volume is
12π cubic units. The surface area is 24π square units.   **47.** The volume is $\frac{500\pi}{3}$ cubic units. The surface
area is 100π square units.   **49.** Corresponding sides are proportional.

CHAPTER 0 TEST

1. B   **2.** D   **3.** B   **4.** C   **5.** C   **6.** D   **7.** C   **8.** C   **9.** B   **10.** 62   **11.** $\frac{1}{27}$ **12.** -1
13. -21 **14.** 0   **15.** 1   **16.** 18   **17.** $-7 + 6$ **18.** $x(6) + x(y)$ **19.** 0   **20.** $(x + t)2$
21. $\angle a = 130°$, $\angle b = 50°$, $\angle c = 100°$, $\angle d = 30°$, $\angle e = 150°$, $\angle f = 50°$, $\angle g = 130°$
22. The volume is 192π cubic feet.   **23.** The amount of fencing needed is 100 meters.
24. The surface area of the ball is 49π square centimeters.   **25.** The area is 308 square yards.

▶ Chapter 1

PROBLEM SET 1.1

Warm-Ups

1. Base 2; 16   **2.** Base -4; 16   **3.** Base 4; -16 **4.** Base -4; -64

5. Base 4; -64 **6.** Base $-\frac{1}{5}; \frac{3}{25}$ **7.** Base -2; -12 **8.** Base -2; 8   **9.** 1   **10.** $-\frac{1}{16}$

11. 25   **12.** 2   **13.** $\frac{1}{8}$ **14.** $\frac{1}{6}$ **15.** $\frac{1}{16}$ **16.** $-\frac{1}{16}$ **17.** $-\frac{1}{8}$ **18.** $-\frac{2}{x^3}$ **19.** 2   **20.** 1

21. $(x + 4)^6$ **22.** 64   **23.** $\frac{8x^3}{y^6}$ **24.** $\frac{1}{x^3y^2}$ **25.** $-8t^6$ **26.** $\frac{1}{xy}$ **27.** $\frac{1}{(c - a)^2}$ **28.** $\frac{1}{(2 - y)^6}$

29. $\frac{y^{15}}{x^9}$ **30.** Cannot be simplified.   **31.** 324   **32.** 49   **33.** $\frac{1}{24}$ **34.** 75   **35.** $\frac{25}{6}$

36. $\frac{-2}{x^2y}$ **37.** $\frac{9}{4}$ **38.** $-\frac{3}{4}$ **39.** $\frac{3}{2}$ **40.** $\frac{13}{5}$ **41.** 2.1367×10^{-8} **42.** -1.2345×10^{-6}

43. -3.2×10^4 **44.** 7.7722×10^{13} **45.** 160,900   **46.** 0.0000543   **47.** 0.0000000011

48. $-800, 000, 000, 000$

Practice Exercises

49. 9   **51.** $\frac{1}{81}$ **53.** 8   **55.** -10 **57.** 108   **59.** $-\frac{1}{27}$ **61.** $\frac{3}{x^2}$ **63.** $\frac{-2y^3}{x^2}$ **65.** $\frac{1}{5}$

67. $(x - 2)^8$ **69.** $\frac{5}{16}$ **71.** $\frac{4}{3}$ **73.** 6   **75.** $-64x^3$ **77.** $-\frac{x^6}{27y^3}$ **79.** $\frac{9x^2}{y^6}$

81. $\frac{1}{9x^2y^2}$ **83.** $\frac{1}{64}$ **85.** $\frac{1}{y^2}$ **87.** $\frac{64z}{xy^2}$ **89.** $\frac{4z^{10}}{9}$ **91.** $\frac{2}{x^3y^2}$ **93.** 8.35×10^8

95. Paper: 1,669,120; Metals: 3,408,330; Glass: 1,369,200; Yard: 459,008; Plastics: 143,440; Food: 0

97. 7.0×10^{-5} mm or 70 nanometers   **99.** 1.357×10^{17} mi   **101.** $-125x^6z^{12}$ **103.** $\frac{27x^6y^3}{8u^3v^9}$

105. $9x^6$ m^3 **107.** $8t^6$ cm^3

Challenge Problems

109. 3^{2n} **111.** 5^{2n} **113.** 81   **115.** x^{n+2}

PROBLEM SET 1.2

Warm-Ups

1. $x + 6$; 1   **2.** $2x^3y - 3x^2y^2 + 4xy^3 - 5$; 4   **3.** $x^2 - 3x + 4$; 2   **4.** $-x^3 + 9$; 3

5. $-3x^3 + 4x^2 + \frac{1}{2}x - 7$; 3   **6.** $-17x^7 + 4x^3 + \frac{2}{3}x + 89$; 7   **7.** 1   **8.** 21   **9.** -16 **10.** 6

11. 5   **12.** 3   **13.** -7 **14.** 2   **15.** $7x - 14$ **16.** $3y^2 + 6y - 12$ **17.** $5x^2 - 3xy - 3y^2$
18. $15x^3 + x^2 + 4x - 8$ **19.** $-5x^6 + 3x^4 - 3$ **20.** $4x^5 - 7x^4 + x^3 + 4$
21. $-15x^4y + 12x^3y^2 - 7x^2y^2 + 12$ **22.** $-3x^3y^2 + 5xy^2 - 6$ **23.** $-2x - 12$
24. $-z^2 - 5z - 8$ **25.** $-4x^2 - 8xy + 11y^2$ **26.** $-x^4 + 9x^3 + 3x^2 - 6x + 5$ **27.** $3x - 3$
28. $-v^4 - 3v^3 + 9v^2 - 7v - 6$ **29.** $4x^3y - 3x^2y^2 + 11xy + 12$ **30.** $-t^3 - 27$

Practice Exercises

31. $x - 4$; 1; binomial   **33.** $x^2 - x + 1$; 2; trinomial   **35.** $-2x^5 + 7x^4 + \frac{2}{5}x - 9$; 5

37. $x^5 + y^5$; 5; binomial   **39.** $x^7y^2 - x^4y^3 + y^7$; 9; trinomial   **41.** -2 **43.** 11   **45.** -8
47. 0   **49.** 1   **51.** -5 **53.** $24x + 4$ **55.** $4y + 1$ **57.** $-2x^2 - 11xy + 4y^2$

59. $15x^3 + 2x^2 + 4x - 8$ **61.** $8x^5 - 3x^4 + 13x^3 - x^2 + 7x + 12$
63. $24x^5 + 14x^4 - 4x^3 - 10x$ **65.** $18y + 18$ **67.** $w^2 - 6w + 5$ **69.** $-t^3 + 81$
71. $6x^2 - 4x + 8$ **73.** $3x^2 - 4x + 9$ **75.** $x^2 - x - 1$ **77.** $2y - 2x$ **79.** $6t^3 + 8t^2 - 2t - 8$
81. $s^3 + 3s^2 - 6$ units   **83.** $t^2 + 9t - 6$ yd   **85.** 48 ft; 0 ft   **87.** $230 - 4x$ degrees
89. $2s - 9$ ft   **91.** $120 - x$ degrees

Challenge Problems

93. -2 if n is odd; 0 if n is even   **95.** 0   **97.** $7x^{2n} + x^n + 4$ **99.** $2x^{2n} - 2x^n + 2$

PROBLEM SET 1.3

Warm-Ups

1. $x^2 + 3x + 2$ **2.** $x^2 + 9x + 20$ **3.** $\frac{1}{4}x^2 + \frac{3}{2}x + 2$ **4.** $x^2 + 8x + 7$

5. $x^2 - 13x + 42$ **6.** $x^2 - 13x + 40$ **7.** $x^2 + 2x - 35$ **8.** $x^2 - 2x - 63$
9. $-12a^3b^2$ **10.** $2x^5y^3$ **11.** rs^4 **12.** $3x^2 + 12x$ **13.** $-3x^2 + 3x^3$ **14.** $2x^2 + 5x$
15. $15x^2 - 61x + 56$ **16.** $12x^2 + 11x - 5$ **17.** $x^3 - 2x^2y^2 - x^2y + 3xy^3 - y^4$
18. $x^3 + 5x^2y + 7xy^2 + 2y^3$ **19.** $x^2 + 10x + 25$ **20.** $25x^2 + 120x + 144$
21. $9x^2 - 24xy + 16y^2$ **22.** $36x^2 + 84x + 49$ **23.** $x^2 - 4$ **24.** $1 - 16y^2$
25. $25x^2 - 49$ **26.** $64 - 81z^2$ **27.** $x^3 + 6x^2 + 12x + 8$ **28.** $x^3 - 12x^2 + 48x - 64$
29. $27a^3 - 27a^2b + 9ab^2 - b^3$ **30.** $27w^3 + 108w^2x + 144wx^2 + 64x^2$ **31.** $3x^2 - 12x - 3$
32. $-x^2 + 17x + 49$ **33.** $-8x^2 - 2x - 1$ **34.** $-5x^2 + 7x + 25$

Practice Exercises

35. $-8a^4b^2$ **37.** $2rs^5$ **39.** $-2^3x^5y^4$ **41.** $-3^7a^3b^8c^3$ **43.** $-r^{20}t^5$ **45.** $9r^{17}s^3t^{15}$

47. $a^{26}b^{16}c^{16}$ **49.** $3a^4bc + 5a^4b^4c^7 - 4a^3b^2c$ **51.** $-3x^3 + \frac{9}{5}x^2y + \frac{3}{5}x^2$

53. $x^{11} - 3x^{10} + \frac{1}{3}x^7 - x^4$ **55.** $x^2 + 10x + 9$ **57.** $\frac{1}{4}x^2 - 1$ **59.** $35 + 12x + x^2$

61. $x^4 + 2x^2 - 24$ **63.** $x^2 + \frac{1}{3}x - \frac{2}{9}$ **65.** $6 - 5x + x^2$ **67.** $28 + 3x^2 - x^4$

69. $8x^2 + 30x + 25$ **71.** $33x^2 + 26x + 5$ **73.** $20 - 21x + 4x^2$ **75.** $5x^4 - 41x^2 + 42$
77. $c^3 - 7c^2d + 6cd^2 + cd^3 - d^4$ **79.** $x^3 + 5x^2 - 29x - 105$ **81.** $x^2 - 9$
83. $x^2 + 12x + 36$ **85.** $81x^2 - 144x + 64$ **87.** $49x^2 - 25$ **89.** $25x^2 - 30xy + 9y^2$
91. $16x^2 + 56xy + 49y^2$ **93.** $x^3 + 9x^2 + 27x + 27$ **95.** $a^3 - 6a^2b + 12ab^2 - 8b^3$
97. $-x^2 + 25x - 5$ **99.** $7x^2 - 13x + 34$ **101.** $N^2 + 2N$ **103.** $4\pi x^2 - 20\pi x + 25\pi$ ft^2
105. $N^3 - 3N^2 + 3N - 1$ **107.** $3W^3$ yd^2

Challenge Problems

109. $x^{n+2} - x^{n+1} + x^n$ **111.** $x^{2n} + 2x^n + 1$ **113.** $x^{2n} + 2x^ny^n + y^{2n}$ **115.** $(x - y)^2 - 49$

PROBLEM SET 1.4

Warm-Ups

1. $x^3 + 3x^2 + 3x + 1$ **2.** $x^3 - 6x^2 + 12x - 8$ **3.** $x^4 + 8x^3 + 24x^2 + 32x + 16$

4. $243x^5 - 810x^4 + 1080x^3 - 720x^2 + 240x - 32$

5. $64x^6 + 576x^5y + 2160x^4y^2 + 4320x^3y^3 + 4860x^2y^4 + 2916xy^5 + 729y^6$

6. $x^6 + 6x^5y + 15x^4y^2 + 20x^3y^3 + 15x^2y^4 + 6xy^5 + y^6$

7. $u^6 - 12u^5w + 60u^4w^2 - 160u^3w^3 + 240u^2w^4 - 192uw^5 + 64w^6$

8. $x^8 + 16x^7 + 112x^6 + 448x^5 + 1120x^4 + 1792x^3 + 1792x^2 + 1024x + 256$

9. $x^{10} - 20x^9 + 180x^8 - 960x^7 + 3360x^6 - 8064x^5 + 13{,}440x^4 - 15{,}360x^3 + 11{,}520x^2 - 5120x$
   $+ 1024$

Practice Exercises

11. $x^3 - 3x^2 + 3x - 1$ **13.** $y^4 - 8y^3 + 24y^2 - 32y + 16$

15. $b^5 - 15b^4 + 90b^3 - 270b^2 + 405b - 243$

17. $16x^4 - 32x^3 + 24x^2 - 8x + 1$ **19.** $a^5 + 5a^4 + 10a^3 + 10a^2 + 5a + 1$

21. $16x^4 + 96x^3 + 216x^2 + 216x + 81$

23. $27x^3 - 54x^2 + 36x - 8$ **25.** $243x^5 + 810x^4 + 1080x^3 + 720x^2 + 240x + 32$

27. $x^6 - 6x^5y + 15x^4y^2 - 20x^3y^3 + 15x^2y^4 - 6xy^5 + y^6$

29. $x^8 - 16x^7 + 112x^6 - 448x^5 + 1120x^4 - 1792x^3 + 1792x^2 - 1024x + 256$

31. $-1{,}082{,}565x^8y^3$ **33.** $700{,}000z^4$

PROBLEM SET 1.5

Warm-Ups

1. 2^6 **2.** $2^3 \cdot 7$ **3.** $3^2 \cdot 5$ **4.** $2^2 \cdot 7^2$ **5.** $2 \cdot 3^4$ **6.** $2^3 \cdot 5^3$ **7.** 3^5 **8.** $2^2 \cdot 3^2 \cdot 5^2$

9. $2^3 \cdot 3^3$ **10.** $2^2 \cdot 3^2 \cdot 5$ **11.** 24 **12.** 25 **13.** 18 **14.** 18 **15.** x^3y^2 **16.** ab^2

17. st^2 **18.** $3x^2y^2$ **19.** $16x$ **20.** $20yz^4$ **21.** $24x^2y^3$ **22.** $xy(x - y)$ **23.** $16a^2b^2c(3b + 4a)$

24. $7ab(3a - 2b + 1)$ **25.** $4p^2q^3(5q^4 - 7p^3q^2 + 9p^6)$ **26.** $5(3 - a)$ **27.** $15t^2(t - 1)$

28. $2x^2y^3z^2(9z^2 - 8y - 7x^4y^3z^4)$ **29.** $3m^2n^3(5 - 3m^2n + 2mn^2)$ **30.** $2p^7(6p + 7 - 5p^2)$

31. $-3x(2x - 1)$ **32.** $-9x^4(2x + 3)$ **33.** $-3v(2u - 1)$ **34.** $(a + 2)(b + 5)$

35. $(p - q)(m + n)$ **36.** $2(x + 3)$ **37.** $(x + y)(x + y + 3)$ **38.** $(z + 2)^2(az + 2a - b)$

39. $(1 - r)^2[5(1 - r)^2 + 3(1 - r) - 1]$ **40.** $(5 + 2t)(5x + 2tx - 1)$ **41.** $(s - t)(r + u)$

42. $(1 - x)(2x + 3y)$ **43.** $(3u - v)[w(3u - v)^2 - 3u(3u - v) - v]$

44. $8(y - x)[2(y - x)^2 + 3(y - x) + 4]$ **45.** $(x + y)(a + 7)$ **46.** $(y + 1)(x + b)$

47. $(a - b)(a + 2)$ **48.** $(r - t)(6 - s)$ **49.** $(a + b)(a + 1)$ **50.** $(y - x)(y - 1)$

51. $(r - s)(5 - s)$ **52.** $(a + 2)(b + 5)$ **53.** $(u + w)(v - 1)$ **54.** $(z - 1)(z^2 + 1)$

55. $a(a - c)(a + b)$ **56.** $x^3(x + y)(x^2 + 1)$ **57.** $z(x - 1)(z + 1)$ **58.** $6t(s + 2)(5t - 4)$

59. $5a(x - y)(3b - 5c)$ **60.** $y(y - x)(y - 1)$

Practice Exercises

61. $6a^2b^2c(7a^2 + 9b^3)$ **63.** $24r^3s^3(rs - 2)$ **65.** $9ab(3a^3 - 2b^4 + 1)$ **67.** $-8x^3(2 + 3x^3)$

69. $12t^3(t - 1)$ **71.** $\sqrt{3}x(2 - x)$ **73.** $2\sqrt{5}x(1 + 2x^2)$ **75.** $(d - 2)(c + b)$ **77.** $(s - t)(a - u)$

79. $(x + y)^2(x + y - 5)$ **81.** $(1 - r)^2[3(1 - r) - 3(1 - r)^2 - 4]$ **83.** $3t(a - x)$
85. $(5 + b)^2[c^2(5 + b)^2 - d^2(5 + b)^3 - 2]$ **87.** $5(a - b)[3(a - b)^3 + 9(a - b) + 7]$
89. $(x + 6)(y + 1)$ **91.** $(a - b)(r - s)$ **93.** $(a - b)(1 - b)$ **95.** $(s + 3)(4 - b)$
97. $(t + 1)(t^4 + s^4)$ **99.** $(q^2 - t)(q^2 + 1)$ **101.** $x(a + b^2)(x - 1)$ **103.** $8c(r + 2s)(2c - 1)$
105. $w^2(w - t)(1 - w)$

Challenge Problems

107. $x^{n+2}(1 - x)$ **109.** $x^{2mn}(1 + x^{2mn})$

PROBLEM SET 1.6

Warm-Ups

1. $(x - 3)(x + 3)$ **2.** $(2x - 1)(2x + 1)$ **3.** $(3x - 2)(3x + 2)$ **4.** $(8y - 5)(8y + 5)$
5. $(x - 1)(x^2 + x + 1)$ **6.** $(2 + y)(4 - 2y + y^2)$ **7.** $(4 + t)(16 - 4t + t^2)$
8. $(3x - 2y)(9x^2 + 6xy + 4y)$ **9.** $4(x - 4)(x + 4)$ **10.** $3(2x - 1)(2x + 1)$
11. $3(t + 2s)(t^2 - 2st + 4s^2)$ **12.** $2x(x - 2)(x^2 + 2x + 4)$ **13.** $(x + 2)(x + 1)$
14. $(s + 1)(s + 6)$ **15.** $(x - 3)(x - 2)$ **16.** $(x - 1)(x - 9)$ **17.** $(z - 3)(z + 4)$
18. $(x + 3)(x - 8)$ **19.** $(x - 5)(x + 2)$ **20.** $(x + 5y)(x - 4y)$ **21.** $(2x + 1)(x + 1)$
22. $(3x - 4)(2x - 1)$ **23.** $(4s - 5t)(2s - 3t)$ **24.** $(2 - 7z)(1 - 3z)$ **25.** $(2x - 3)(x - 2)$
26. $(2t - 3)(t + 2)$ **27.** $(3a - 1)(2a + 3)$ **28.** $(2y - 3)(2y - 5)$ **29.** $4(x + 1)(x - 2)$
30. $5(x + 1)(x + 2)$ **31.** $-6y(y - 2)(y + 1)$ **32.** $3s(2s - 1)(s + 3)$

Practice Exercises

33. $(x - 9)(x + 9)$ **35.** $(x - 2)(x^2 + 2x + 4)$ **37.** $(x + 1)(x + 8)$ **39.** $(t + 5)(t - 4)$
41. $(z + 3)(z^2 - 3z + 9)$ **43.** $(3x + 5)(x - 2)$ **45.** $(4 - x)(4 + x)$ **47.** $(3 - 2v)(9 + 6v + 4v^2)$
49. $(6y + z)(y + 4z)$ **51.** $(5t - 3)(t - 5)$ **53.** $(x - 2y)(x^2 + 2xy + 4y^2)$ **55.** $2x(2x - 1)(2x + 1)$
57. $6y(y + 2)(y - 1)$ **59.** $2(x + 2y)(x + y)$ **61.** $3x(3x + 2)(9x^2 - 6x + 4)$ **63.** $7(t + 3)^2$
65. $(x^2 + 2)(x + 2)(x - 2)$ **67.** $(z^2 + 1)(z - 2)(z + 2)$ **69.** $(x^2 + 1)(x + 1)(x - 1)$
71. $(x^3 - 5)(x^3 - 3)$ **73.** $4(x^3 - 2)(x + 1)(x^2 - x + 1)$ **75.** Not factored; $16(x - 1)(x + 1)$
77. Not factored; $(b + x)(2a - y)$ **79.** Factored **81.** Not factored; $(3s + 5t)(2s - 3t)$ **83.** Prime
85. 16 **87.** ± 6 **89.** -42 **91.** $V = \pi h(x - y)(x + y)$ cubic units **93.** $\pi(y - x)(y + x)$ sq units

Challenge Problems

95. $(x + y + 2)(x + y - 2)$ **97.** $[5 - (s - t)][25 + 5(s - t) + (s - t)^2]$
99. $(1 - x)(1 + x + x^2)(1 + x^3 + x^6)$ **101.** $(x^p - 1)(x^p + 1)$ **103.** $(x^p + 1)(x^{2p} - x^p + 1)$
105. $(x - 1)(x^2 + x + 1)(x + 1)(x^2 - x + 1)$; $(x - 1)(x + 1)(x^4 + x^2 + 1)$
107. $(x^n - 2)(x^n + 4)$ **109.** $(x^n + y^m)^2$

PROBLEM SET 1.7

Warm-Ups

1. $25(x - 1)(x + 1)$ **2.** $x(x - 1)(x + 1)$ **3.** $2(w + 5x)(w^2 - 5wx + 25x^2)$
4. $(3x^2 + 4)(x - 1)(x + 1)$ **5.** $x^2(x + 2)(x - 1)$ **6.** $2x(x + 1)(x - 3)$ **7.** $2(z - 4)(y - 1)$

8. $a(1 - y)(z^2 + 4)$ **9.** prime   **10.** $(2a - y)(b + x)$ **11.** $(4x + 1)^2$
12. $(4 - y)(4 + y)(1 + y^2)$ **13.** factored   **14.** $(2x - 7)(x + 1)$ **15.** $9(2a^3 + 4b^3c^3 + 3b^2c^2)$
16. $(x^3 + 2)(x - 1)(x^2 + x + 1)$ **17.** $(x - 5)^2$ **18.** $x^3(x + 1)(x^2 + 1)$ **19.** $(a - 1)(a + 1)(x + z)$
20. factored   **21.** $(y - z)(y + 1)$ **22.** prime   **23.** $3(7x - 6)(x + 2)$ **24.** prime

Practice Exercises

25. $(x + 7)(x + 2)$ **27.** $45r^2t^4(3rt + 5)$ **29.** $(1 - 3ab)(1 + 3ab + 9a^2b^2)$ **31.** $(x + y)(x - y + 1)$
33. factored   **35.** $2(4 - x)(4 + x)$ **37.** $(x - t)(r - 1)(r^2 + r + 1)$ **39.** $(3r + 4)(9r^2 - 12r + 16)$
41. $(r + s)(x - y)$ **43.** $(x^2 - 2)(x^2 + 2)(x^4 + 4)$ **45.** $(x - y)(a + b)$ **47.** prime
49. $x(x - 1)(x^2 + 1)$ **51.** $(x^2 + 5)(x - 2)(x + 2)$ **53.** $(9 - x)^2$ **55.** $(a^3 + 2)(a^6 - 2a^3 + 4)$

Challenge Problems

57. $(x^n - 1)^2(x^n + 1)^2$ **59.** $x^{2n}(x^n - 1)(x^n + 1)$
61. (a) $(x^2 - x + 2)(x^2 + x + 2)$ (b) $(x^2 - 2x + 3)(x^2 + 2x + 3)$ (c) $(x^2 - x + 4)(x^2 + x + 4)$
    (d) $(x^2 - 2x - 2)(x^2 + 2x - 2)$ (e) $(x^2 - x - 5)(x^2 + x - 5)$

PROBLEM SET 1.8

Warm-Ups

1. a^4b **2.** $\dfrac{y^2}{x^2}$ **3.** $(x - y)^2z$ **4.** $\dfrac{2a}{b} + 1$ **5.** $x + \dfrac{2y}{x}$ **6.** $x - 1$ **7.** $z - 6$

8. $x - 2$ **9.** $x - 3$ **10.** $x + 1 + \dfrac{2}{x + 2}$ **11.** $x - 1 + \dfrac{-1}{x + 3}$

12. $2x^2 - x + 1 + \dfrac{-x + 3}{2x^2 + x - 2}$ **13.** $x^3 - x^2 - 4x - 7 + \dfrac{3x + 25}{x^2 - 2x + 3}$ **14.** $x + y + \dfrac{4y^2}{x - 3y}$

15. $x^2 - xy + y^2$ **16.** $x^2 - xy + y^2$ **17.** $x^2 - x + 2 + \dfrac{-x + 3}{x^2 - 1}$

Practice Exercises

19. $\dfrac{3a}{2b}$ **21.** $\dfrac{x^2}{2y} + \dfrac{x}{y} - \dfrac{3}{2y}$ **23.** $\dfrac{x}{y} - 1 + \dfrac{y}{x}$ **25.** $3x - 1$ **27.** $y + 3$ **29.** $x^2 - x - 1 + \dfrac{4}{2x + 1}$

31. $2x^2 - x + 1 + \dfrac{-7}{3x + 2}$ **33.** $2x + 1$ **35.** $3x + y$ **37.** $x^2 - x + 1 + \dfrac{-4x + 5}{3x^2 + x - 1}$

39. $3x^2 - x + 1 + \dfrac{4}{7x + 4}$ **41.** $x^3 - 3x^2 + 9x - 28 + \dfrac{86}{3 + x}$ **43.** $x^3 + 2 + \dfrac{-1}{x^3 - 1}$

45. $2x^3 - 4x + 1 + \dfrac{4x}{2x^2 + 1}$ **47.** $x^2 + 3x - 4$ **49.** 27

Challenge Problems

51. $x^n - 3 + \dfrac{7}{x^n + 1}$ **53.** 12

PROBLEM SET 1.9

Warm-Ups

1. $2x + 1$ **2.** $y + 3$ **3.** $x^2 + 2x + 1$ **4.** $x^2 - 2x + 4$ **5.** Remainder is 0
6. Remainder is -1

Practice Exercises

7. $x^3 + x^2 - x - 1$ **9.** $4x^2 - x + 3 + \dfrac{-1}{x + 2}$ **11.** $2x - 3$ **13.** $x^3 + 2x^2 + 2x - 1 + \dfrac{1}{x - 2}$

15. $2x^2 + 3x + 3 + \dfrac{2}{x - 2}$ **17.** $y^3 - y^2 + y - 1 + \dfrac{2}{y + 1}$ **19.** $x^3 + x^2 + 2x + 2$

21. $x^3 + 2x^2 + 3x + 4 + \dfrac{5}{x - 1}$ **23.** Remainder is 0

REVIEW PROBLEMS CHAPTER 1

1. -35 **3.** 25   **5.** $-\dfrac{4}{27}$ **7.** $\dfrac{y^{12}}{y^8}$ **9.** $(a + b)^5$ **11.** $\dfrac{4x^2}{y^2}$ **13.** $-\dfrac{1}{10}$ **15.** $-\dfrac{1}{6}$ **17.** $\dfrac{125}{16}$

19. $a^3b^4c^2 - a^2bc^5$ **21.** $6p^3 - 4p^2 - 6p - 2$ **23.** $\dfrac{y^2z}{x}$ **25.** $r^6s^3t^5$

27. $-x^4 - 3x^3y + 4x^2y^2 - 12xy + y^5$ **29.** $x^2 - 2x + 4 + \dfrac{-8}{x + 7}$ **31.** $t^3 - 15t^2 + 75t - 125$

33. $14r^2 - 11r + 2$ **35.** $c^2 - 9d^2$ **37.** $-2y$ **39.** $x^2y^3 - xy^2$ **41.** $4s^2 - 6s + 9$

43. $10x + 5$ **45.** $x^2 + 3x + \dfrac{-6x + 6}{x^2 - x + 2}$ **47.** $5r^2 - rt - t^2$ **49.** $-x^2 + 3x + 12$

51. $(x - 9)(x + 2)$ **53.** $(4y + 5)(16y^2 - 20y + 25)$ **55.** Prime   **57.** $(2x - 1)(x + 1)$
59. $2z(s - t)(2z - 1)$ **61.** Prime   **63.** $(12 - x)(12 + x)$ **65.** $6(u - v)(5u - 5v - 7)$
67. Prime   **69.** $(3y - 8)(y - 1)$ **71.** $(a^2 - 5)(a^4 + 5a^2 + 25)$ **73.** $(5 - 2x)(3 + x)$
75. $(4x + 1)^2$ **77.** $3a(3a - 1)(3a + 1)$ **79.** $(x - y - z)(x + y - z)$ **81.** x^2 ft^2
83. 9.98×10^{-4} **85.** 1.23456×10^2 **87.** $100,100,000,000$ **89.** 0.0000002006

Let's Not Forget . . .

90. 2   **91.** 3   **92.** 1   **93.** 4   **94.** $4x^2$ **95.** $2x^2 + 4x$ **96.** $x^2 + x - 2$
97. $x^2 - 4x + 4$ **98.** $x^3 + 3x^2 + 3x + 1$ **99.** -9 **100.** 9   **101.** -27
102. b^2 **103.** $-b^3$ **104.** factored   **105.** $(2x + 5)(4x^2 - 10x + 25)$
106. $(s + t)^2(s + t - a)$ **107.** factored   **108.** factored   **109.** $(1 - b)(a - 1)$
110. Two terms; $y + 7$ is a factor   **111.** One term; $y + 7$ is a factor
112. Three terms; $y + 7$ is not a factor   **113.** Three terms; $y + 7$ is a factor
114. Two terms; $y + 7$ is a factor

CHAPTER 1 TEST

1. A   **2.** D   **3.** A   **4.** B   **5.** B   **6.** D   **7.** $6 - 4x$ **8.** $10x^2 - x - 21$
9. $2x^3y^2 - 3x^2y - 4xy$ **10.** $16x^2 - 56x + 49$ **11.** $6x^2 - 8x + 9 + \dfrac{-11}{x + 1}$
12. $4x^3 - x^2 + 2x + 11$ **13.** $x^2 - 2x - 5$ **14.** $(2a - 3)(4a^2 + 6a + 9)$ **15.** $2(2x - 1)(x + 5)$
16. $(a - b)(x + 2)$ **17.** $(p - q)(x + y)$ **18.** $(a - 2)(a + 2)(a^2 + 2)$ **19.** $(2y - 5)(2y + 5)$
20. $9a^2b^2(3b - 4a + 5a^2b^2)$ **21.** 1.076×10^{-5} **22.** 9.009×10^6 **23.** $52,200$ **24.** 0.0007009

► Chapter 2

PROBLEM SET 2.1

Warm-Ups

1. $\{6\}$ **2.** $\{2\}$ **3.** $\{5\}$ **4.** $\{-3\}$ **5.** $\{3\}$ **6.** $\{3\}$ **7.** $\{3\}$ **8.** $\emptyset$ **9.** $\{5\}$ **10.** $\left\{\dfrac{1}{10}\right\}$

11. $\mathbb{R}$ **12.** $\{2\}$ **13.** $\{6\}$ **14.** $\{-26\}$ **15.** $\left\{-\dfrac{3}{2}\right\}$ **16.** $\left\{-\dfrac{5}{3}\right\}$

17. $\left\{-\dfrac{4}{3}\right\}$ **18.** $\mathbb{R}$ **19.** $\emptyset$ **20.** $\mathbb{R}$ **21.** $\{3\}$ **22.** $\emptyset$ **23.** $\{6\}$ **24.** $\{12\}$ **25.** $\left\{\dfrac{21}{2}\right\}$

26. $\mathbb{R}$ **27.** $\left\{-\dfrac{10}{3}\right\}$ **28.** $\{16\}$ **29.** $\{3\}$ **30.** $\{57\}$ **31.** $\{-1\}$ **32.** $\{-9\}$ **33.** $\emptyset$ **34.** $\{-3\}$

35. $\{1.1\}$ **36.** $\{3.84\}$ **37.** $\{-17\}$ **38.** $\{-3.\overline{6}\}$ **39.** $\{-3.8\overline{3}\}$ **40.** $\{0.3125\}$

41. The strawberry patch is 17 meters by 22 meters.   **42.** 11 m   **43.** The numbers are 2, 10, and 12.

44. They are 15 and 17.

Practice Exercises

45. $\{4\}$ **47.** $\left\{\dfrac{19}{2}\right\}$ **49.** $\{1\}$ **51.** $\left\{\dfrac{4}{3}\right\}$ **53.** $\emptyset$ **55.** $\{-3\}$ **57.** $\mathbb{R}$ **59.** $\left\{-\dfrac{1}{2}\right\}$ **61.** $\{-2\}$

63. $\{-1\}$ **65.** $\{2\}$ **67.** $\mathbb{R}$ **69.** $\left\{-\dfrac{5}{3}\right\}$ **71.** $\emptyset$ **73.** $\left\{\dfrac{3}{2}\right\}$ **75.** $\left\{\dfrac{1}{10}\right\}$

77. $\{14\}$ **79.** $\{18\}$ **81.** $\{14\}$ **83.** $\{22\}$ **85.** $\{-21\}$ **87.** $\{2\}$ **89.** $\{-1.2\}$ **91.** $\{-0.87\}$

93. $\{3.\overline{3}\}$ **95.** $\{-0.4\}$ **97.** Janet Jackson performed in 75 shows, while Madonna completed 37.

99. He completed 23 passes.   **101.** Each angle measures 60°.   **103.** The angles are 36° and 144°.

Challenge Problems

105. $\left\{\dfrac{4b + a}{2}\right\}$

PROBLEM SET 2.2

Warm-Ups

1. $r = \dfrac{C}{2\pi}$ **2.** $h = \dfrac{V}{\pi r^2}$ **3.** $x = 5 - a$ **4.** $x = \dfrac{a + 7}{2}$ **5.** $C = \dfrac{5F - 160}{9}$ **6.** $D = \dfrac{dm_2}{m_1}$

7. $x = \dfrac{1}{a}$ **8.** $x = \dfrac{b}{b - a}$ **9.** $x = \dfrac{ab + b^2}{a - b}$ **10.** $x = \dfrac{2a - 6b}{2b - 1}$ **11.** $x = -2a$ **12.** $x = \dfrac{3b}{7a}$

Practice Exercises

13. $l = \dfrac{A}{w}$ **15.** $w = \dfrac{V}{lh}$ **17.** $t = \dfrac{I}{Pr}$ **19.** $d = \dfrac{C}{\pi}$ **21.** $P = \dfrac{A}{1 + rt}$ **23.** $l = \dfrac{P - 2w}{2}$

25. $h = \dfrac{S}{2\pi r}$ **27.** $a = \dfrac{2S - n^2d + nd}{2n}$ **29.** The length is 7.0$\overline{75}$ ft.   **31.** His rate is $\dfrac{2}{11}$ km/min.

33. Its length is $27\dfrac{5}{8}$ ft.

35.

City	Fahrenheit	Celsius
Washington	$32°$	$0°$
Paris	$41°$	$5°$
Fairbanks	$-31°$	$-35°$
Nassau	$68°$	$20°$
Moscow	$14°$	$-10°$
Miami	$77°$	$25°$

37. Box 1 would be 8 inches high; box 2, 12 inches high; and box 3, 13 inches high.

39. $x = \dfrac{a - 5}{3}$ **41.** $x = \dfrac{b - ab}{a}$ **43.** $x = a + 6$ **45.** $x = -\dfrac{14}{3a}$ **47.** $x = \dfrac{2}{a - b}$

49. $x = \dfrac{3a + 2b}{a + 1}$ **51.** $x = \dfrac{a^2 - 4}{b - a}$ **53.** $x = \dfrac{2ab}{a - b}$ **55.** $x = \dfrac{15a + 15b}{6 - 10b}$ **57.** $x = \dfrac{10a - 6b^2}{a - 5b}$

Challenge Problems

59. $x = 2 - a^2; x \neq 1, x \neq a$

PROBLEM SET 2.3

Warm-Ups

1. $\{x \mid x \leq 7\}$ **2.** $\{x \mid x > -6\}$ **3.** $\{x \mid x \geq 10\}$

4. $\{x \mid x < 14\}$ **5.** $\{x \mid x < -3\}$ **6.** $\left\{x \mid x \geq \dfrac{5}{3}\right\}$

7. $\left\{x \mid x \leq -\dfrac{1}{2}\right\}$ **8.** $\{x \mid x \geq -3\}$ **9.** $\left\{x \mid x \geq \dfrac{5}{2}\right\}$ **10.** $\left\{x \mid x \leq -\dfrac{7}{5}\right\}$ **11.** $\{x \mid x > 1.25\}$

12. $\{t \mid t \geq 8\}$ **13.** $\left\{x \mid x < \dfrac{10}{7}\right\}$ **14.** $\left\{x \mid x \geq -\dfrac{11}{3}\right\}$ **15.** $\mathbb{R}$ **16.** $\emptyset$

17. (interval from 1 to 2)   **18.** (interval from 2 to 10)   **19.** (interval from −1 to 1)   **20.** (interval from 0 to 2)

21. (interval −7 to −2)   **22.** (interval 0 to 3)   **23.** (interval −1 to 4)   **24.** (interval −4 to 2)

25. $\{x \mid -3 < x < 1\}$ **26.** $\{x \mid 1 \leq x \leq 2\}$ **27.** $(2, 3)$ **28.** $(-\infty, -1] \cup [1, +\infty)$

29. $\left(-\dfrac{1}{2}, 0\right]$ **30.** $(-\infty, 0] \cup (1, +\infty)$ **31.** $(-\infty, 5) \cup (10, +\infty)$ **32.** $[-2, 2)$

33. $(5, +\infty)$ **34.** $(-\infty, -3]$ **35.** $2x + 1 \leq 7$; all numbers three or less.

36. $\dfrac{90 + 75 + 82 + x}{4} \geq 80$; Jim must make at least 73.

37. $2w + 2(w + 4) \leq 72$; The largest width is 16 ft.

38. $2(x + 3) \geq 12$; All numbers greater than or equal to 3.

Practice Exercises

39. $\{x \mid x \le 1\}$

41. $\left\{x \mid x < \dfrac{1}{2}\right\}$

43. $\{x \mid x > -10\}$

45. $\left\{x \mid x \le -\dfrac{5}{2}\right\}$

47. $\left\{x \mid x \ge -\dfrac{8}{27}\right\}$

49. $\left\{x \mid x < -\dfrac{3}{7}\right\}$

51. $\{x \mid x > -14\}$

53. $\left\{x \mid x > -\dfrac{7}{4}\right\}$

55. $\varnothing$

57. $\{x \mid x > 9\}$

59. $\{y \mid y > -5\}$

61. $\left\{x \mid x \le \dfrac{20}{9}\right\}$

63. $\left\{s \mid s \le \dfrac{16}{11}\right\}$

65. $\left\{y \mid y < -\dfrac{6}{5}\right\}$

67. $\mathbb{R}$

69. $\{x \mid x \ge 7\}$

71. $\{x \mid x > 7\}$

73. $\left\{r \mid r < \dfrac{13}{3}\right\}$

75. $\{x \mid x \le 24\}$

77. $\{x \mid x < 2\}$

79. $\varnothing$

81. $\mathbb{R}$

83.

85.

87.

89.

91. Nonsense **93.** Makes sense **95.** Makes sense **97.** $(-\infty, 0)$ **99.** $[4, 6]$

101. $(-11, -8)$

103. $(-\infty, -3] \cup (3, +\infty)$

105. $\{x \mid 1 \le x \le 2\}$

107. $\{x \mid -9 \le x \le -3\}$

109. $\{x \mid 14 < x \le 20\}$

111. $\left\{x \mid \dfrac{7}{4} < x < 3\right\}$

113. $4x - 2 \le 10$; all numbers 3 or less

115. $\dfrac{82 + 65 + 73 + x}{4} \ge 80$; she must score at least 97.

117. It would take at least 45 minutes of exercise on the rowing machine.

119. The team must win at least four more games.

Challenge Problems

121. $x < b + 5$

123. $x > \dfrac{b + 2a}{3}$

125. $x < \dfrac{5}{a}$

127. $\{x \mid x \le a \text{ or } x \ge b\}$

129. $\left\{x \mid \dfrac{3}{a} < x < \dfrac{1}{a}\right\}$

131. $\{x \mid x < -1 \text{ or } 1 < x < 2\}$

PROBLEM SET 2.4

Warm-Ups

1. 10%   **2.** 5 years   **3.** Approx. $11,376.39   **4.** $24,876.73   **5.** $15,500 at 5.5%; $9500 at 6.5%
6. $4000 at 5% and $6000 at 6%   **7.** $13,000   **8.** 7.75%   **9.** 3 cm   **10.** 5 m, 10 m
11. 7 cm by 14 cm   **12.** Yes   **13.** Between 11 and 12 years   **14.** Between 8 and 9 yrs   **15.** 19
16. 50   **17.** 243   **18.** −2   **19.** 7   **20.** $2n$

21.

Number of Days	Number of Bacteria Present
1	2,000
2	4,000
3	8,000
4	16,000
5	32,000
t	$1,000 \cdot 2^t$

22. Approximately 1,073,742 inches or about 17 miles!

Practice Exercises

23. (a) $2750.00   **(b)** $5427.43   **(c)** $5668.16

25. (a)

Birthday	Principal	Interest Earned during Previous Year	Value
7th	$10,000	$500	$10,500
8th	10,500	525	11,025
9th	11,025	551.25	11,576.25
10th	11,576.25	578.81	12,155.06
11th	12,155.06	607.75	12,762.81

   (b) $12,762.81   **(c)** $16,288.95
27. $8000 is invested at 5.5% and $4000 is invested at 6.5%.
29. They invested $70,000 at 10% and $30,000 at 8%.
31. He should invest $75,000 in AAAA bonds and $25,000 in AA bonds.
33. He invested $180,000 in the 20% deal and $60,000 in the 5% deal.
35. The side is 11 inches.
37. Glen gave away 10 tickets.
39. The number is 18.   **41.** The number is 30.
43. $1 + 3 + 5 + 7 + 9 = 25$ **45.** Between 12 and 13 years, and
   $1 + 3 + 5 + 7 + 9 + 11 = 36$ between 29 and 30 years

47. (a)

Number of Years after 1991	World Population
1	$5.4 + 5.4(0.015) = 5.4(1.015) \approx 5.48$ billion
2	$5.4(1.015)^2 \approx 5.56$ billion
3	$5.4(1.015)^3 \approx 5.65$ billion
4	$5.4(1.015)^4 \approx 5.73$ billion

   (b) Population $5.4(1.015)^t$ billion, where t is years after 1991

Challenge Problems

49. (a)

Years after 1992	Population	Food-Production Capacity
1	105,000	250,000
2	110,250	300,000
3	115,762	350,000
4	121,550	400,000
5	127,628	450,000
6	134,009	500,000
	$100,000(1.05)^t$	$200,000 + 50,000t$

(b) Between 2067 and 2068, population will overtake food-production capacity

PROBLEM SET 2.5

Warm-Ups

1. Christine should add 45 gal of 60% solution.   **2.** $2\frac{1}{2}$ gal

3. The chemist should add 5/3 gal of pure alcohol.   **4.** 10 gal   **5.** It will take 4/3 hr to fill the pool.

6. $\frac{12}{7}$ min   **7.** It should take 12 min.   **8.** 1495 km; 1656 km

9. They will meet in 4 hr. Frank walked 8 mi and Chuck walked 12 mi.   **10.** 40 mph

11. They will be 38 miles apart in 3 hours.

Practice Exercises

13. Virginia should add 5 oz of 50% alloy.   **15.** Jasinia should add 10 ml of water.

17. The glass manufacturer should add 28 kg of pure lead.   **19.** It will take $\frac{30}{11}$ hours.

21. It will take $7\frac{1}{2}$ minutes.   **23.** It will take $\frac{6}{7}$ minutes.   **25.** It will take 10 hours.

27. It will take them 35 minutes.   **29.** It is 150 miles.   **31.** The slow freight has traveled 70 miles.

33. It will take 9 min.

35. He can ride 8 8/9 miles. If he doubled his rental time, he could double his distance.

37. To finish the race in 50 min, he would have to increase his speed to 2 km/hr.

39. Mike will overtake Ray in 2 hr. They will have traveled 12 mi.

Challenge Problem

41. The speed of the current is 4/3 km/hr.

PROBLEM SET 2.6

Warm-Ups

1. $\{-3, 3\}$ **2.** $\{-10, 0\}$ **3.** $\left\{-\frac{5}{2}, \frac{5}{2}\right\}$ **4.** $\left\{-\frac{1}{2}, \frac{1}{2}\right\}$ **5.** $\left\{-\frac{3}{2}, \frac{5}{2}\right\}$ **6.** $\left\{\frac{1}{6}, \frac{1}{2}\right\}$ **7.** $\varnothing$ **8.** $\varnothing$

9. $\{0\}$ **10.** $\{7\}$ **11.** $\left\{\frac{5}{2}\right\}$ **12.** $\left\{\frac{1}{2}\right\}$ **13.** $\{-1, 4\}$ **14.** $\varnothing$ **15.** $\{1, 4\}$ **16.** $\{-4\}$ **17.** $\{-1\}$

18. $\{1\}$ **19.** $\left\{-\dfrac{2}{3}, 4\right\}$ **20.** $\left\{-\dfrac{1}{4}, \dfrac{1}{2}\right\}$

21. The smallest is 1500 calories; the largest is 2500. $|x - 2000| = 500$.

22. Shortest, $2\dfrac{15}{16}$; longest, $3\dfrac{1}{16}$; $|x - 3| = \dfrac{1}{16}$

Practice Exercises

23. $\{-6, 6\}$ **25.** $\{0\}$ **27.** $\{2, 6\}$ **29.** $\varnothing$ **31.** $\{-20, -6\}$ **33.** $\left\{-\dfrac{13}{4}, \dfrac{3}{4}\right\}$ **35.** $\left\{\dfrac{3}{10}\right\}$

37. $\left\{-4, -\dfrac{2}{5}\right\}$ **39.** $\{-0.8, 4\}$ **41.** $\{-2.2, 6.8\}$ **43.** $\{-1, 3\}$ **45.** $\left\{-\dfrac{3}{2}\right\}$ **47.** $\left\{-3, -\dfrac{1}{2}\right\}$

49. $\left\{0, -\dfrac{2}{3}\right\}$ **51.** $\left\{-4, \dfrac{1}{3}\right\}$ **53.** $\left\{-16, -\dfrac{2}{3}\right\}$ **55.** $\left\{\dfrac{1}{5}, 1\right\}$ **57.** $\left\{-2, -\dfrac{2}{3}\right\}$

59. $|x| = 3$; there are two numbers, ± 3.   **61.** $|4x - 3| = 1$; the numbers are $\dfrac{1}{2}$ and 1.

63. $|x - 40| = 1$; the lowest temperature is 39°F, and the highest is 41°F.   **65.** Yes

Challenge Problems

67. $\{a + 1, a - 1\}$ **69.** $\left\{-\dfrac{a}{2}\right\}$

71. $\varnothing$ **73.** $\left\{-\dfrac{1}{2}\right\}$

PROBLEM SET 2.7

Warm-Ups

1. $\{x \mid -4 < x < 6\}$ **2.** $\{x \mid x < -4 \text{ or } x > -2\}$ **3.** $\{x \mid x \le -10 \text{ or } x \ge 6\}$

4. $\{x \mid -2 \le x \le 16\}$ **5.** $\left\{x \mid 1 < x < \dfrac{9}{5}\right\}$ **6.** $\left\{x \mid x \le -\dfrac{5}{4} \text{ or } x \ge \dfrac{3}{4}\right\}$ **7.** $\left(-2, -\dfrac{4}{3}\right)$

8. $(-\infty, -2] \cup \left[\dfrac{2}{3}, +\infty\right)$ **9.** $(-\infty, 0) \cup (2, +\infty)$ **10.** $\left[-8, -\dfrac{2}{3}\right]$ **11.** $\mathbb{R}$

12. $\{x \mid x \ne -8\}$ **13.** $\varnothing$ **14.** $\mathbb{R}$ **15.** $2 \le x \le 12$ **16.** $x < -\dfrac{1}{2} \text{ or } x > \dfrac{11}{2}$

17. $x \le -\dfrac{15}{4} \text{ or } x \ge \dfrac{13}{4}$ **18.** $-4 < x < 8$ **19.** All numbers between -6 and -2, inclusive.

20. All numbers between 4 and 6.   **21.** All numbers between $\dfrac{1}{6}$ and $\dfrac{5}{6}$.

22. All numbers between -5 and 5, inclusive.   **23.** All numbers less than -7 or greater than 7.

24. All numbers greater than 10 or less than -10, inclusive.

25. All numbers between -3 and 3, inclusive.

26. All numbers less than $-\dfrac{7}{3}$ or greater than $-\dfrac{5}{3}$, inclusive.

Practice Exercises

27. $\{x \mid x \le -8 \text{ or } x \ge 8\}$ **29.** $\{x \mid -6 < x < 12\}$ **31.** $\{x \mid -8 \le x \le -2\}$

33. $\{x \mid x < -6 \text{ or } x > 1\}$ **35.** $\{x \mid 4 \le x \le 5\}$ **37.** $\left\{x \mid x < -\dfrac{1}{5} \text{ or } x > \dfrac{17}{5}\right\}$

39. $\left\{y \mid y \le -\dfrac{2}{5} \text{ or } y \ge -\dfrac{1}{5}\right\}$ **41.** $\{x \mid -1 < x < 3\}$ **43.** $\left\{x \mid -\dfrac{11}{3} \le x \le -\dfrac{7}{3}\right\}$

45. $(-3.8, 8.8)$ **47.** $(-\infty, -3.99) \cup (9.45, +\infty)$ **49.** $\left[-\dfrac{11}{2}, \dfrac{13}{2}\right]$

51. $(-\infty, 1] \cup [3, +\infty)$ **53.** $\mathbb{R}$ **55.** $\varnothing$ **57.** $\varnothing$ **59.** $(-\infty, -2] \cup [1, +\infty)$

61. $[-3, 15]$ **63.** $-4 \le x \le 2$ **65.** $x < -4.3 \text{ or } x > 4.3$ **67.** $-5 < x < -1$

69. $-\dfrac{5}{6} < x < \dfrac{7}{6}$ **71.** All numbers between -4 and 4.   **73.** All numbers between -8 and 8, inclusive.

75. All numbers between -1 and 3.   **77.** Richard will fly at a heading between $266°$ and $274°$, inclusive.

79. Between 17% and 27%, inclusive, of the total budget can be spent on advertising.

Challenge Problems

81. $\{x \mid a - 4 \le x \le a + 4\}$ **83.** $\{x \mid a - d < x < a + d\}$ **85.** $\{x \mid 0 < x < 2a\}$

87. $\{a\}$ **89.** $\{x \mid -2a < x < 0\}$ **91.** All numbers between $a - s$ and $a + s$, inclusive.

REVIEW PROBLEMS CHAPTER 2

1. $\left\{\dfrac{33}{5}\right\}$ **3.** $\left\{-\dfrac{1}{2}, \dfrac{7}{2}\right\}$ **5.** $\left(\dfrac{1}{2}, 2\right)$ **7.** $(-\infty, -24) \cup (-8, +\infty)$

9. $\left[\dfrac{1}{3}, +\infty\right)$ **11.** $\{-13, 7\}$ **13.** $\{-4\}$ **15.** $(-\infty, 7)$ **17.** $\left\{\dfrac{4}{3}\right\}$

19. $\mathbb{R}$ **21.** $\left\{\dfrac{13}{19}\right\}$ **23.** $\varnothing$ **25.** $(-\infty, 5]$

27. $\left[-\dfrac{5}{3}, 1\right]$ **29.** $\{-2\}$ **31.** $\left\{-\dfrac{11}{3}, \dfrac{1}{3}\right\}$ **33.** $\varnothing$ **35.** $\{-1, 3\}$ **37.** $\left\{\dfrac{5}{3}\right\}$

39. $\varnothing$ **41.** $\varnothing$ **43.** The largest picture frame would be 12 in. by 18 in.

45. The integers are 24 and 26.   **47.** The integers are 13, 15, and 17.   **49.** She should add 9 gal.

51. The dimensions are 15 m by 28 m.   **53.** All numbers less than or equal to 11

55. The length of a side would be 25 ft or less.   **57.** She must score at least 89.

Let's Not Forget

59. Factored   **60.** $(x - 3)(x^2 + 3x + 9)$ **61.** $(x + y)(1 - 4x - 4y)$ **62.** Factored   **63.** Factored

64. Two; $x + 2$ is a factor.   **65.** One; $x + 2$ is a factor.   **66.** Two; $x + 2$ is a factor.

67. One; $x + 2$ is a factor.   **68.** Two; $x + 2$ is a factor.   **69.** Three; $x + 2$ is a factor.

70. -64 **71.** 64 **72.** $x^2 - 4xy + 4y^2$ **73.** $a^3 + 6a^2b + 12ab^2 + 8b^3$ **74.** x^4y^2

75. $x^5y - x^4y$ **76.** Equation; $\left\{\dfrac{7}{2}\right\}$ **77.** Inequality; $\left\{x \mid x > \dfrac{7}{2}\right\}$ **78.** Expression; 4

79. Inequality; $\left\{x \mid -\dfrac{8}{3} \le x \le \dfrac{4}{3}\right\}$ **80.** Expression; $2x - 11$

CHAPTER 2 TEST

1. C   **2.** B   **3.** B   **4.** C   **5.** C   **6.** A   **7.** C   **8.** $\{-5, 2\}$ **9.** $\left\{x \mid x > -\dfrac{13}{3}\right\}$

10. $\left\{-\dfrac{5}{2}\right\}$ **11.** $\varnothing$ **12.** $\left\{x \mid x < -\dfrac{11}{3} \text{ or } x > 3\right\}$ **13.** The Honda traveled at a rate of 30 mph.

14. The largest frame will be 31 in. by 36 in.   **15.** She should add 8 quarts of 20% solution.

► **Chapter 3**

PROBLEM SET 3.1

Warm-Ups

1. 3   **2.** -1 **3.** 5   **4.** 0   **5.** None   **6.** 0   **7.** None   **8.** 9   **9.** 2   **10.** -8

11. $-3, 1$ **12.** 4, 6   **13.** 0, 2   **14.** 3, -2 **15.** ± 2 **16.** -3 **17.** $-\dfrac{3}{x}; \dfrac{3}{-x}$

18. $\dfrac{-x^2}{8}, -\dfrac{x^2}{8}$ **19.** $\dfrac{-7}{x}; \dfrac{7}{-x}$ **20.** $\dfrac{x-1}{-x}, -\dfrac{x-1}{x}$ **21.** $-\dfrac{2x}{x+5}; \dfrac{-2x}{x+5}$ **22.** $\dfrac{-x}{x+3}, \dfrac{x}{-(x+3)}$

23. $\dfrac{-(x-1)}{x^2}; -\dfrac{x-1}{x^2}$ **24.** $\dfrac{x+6}{-(x+3)}, -\dfrac{x+6}{x+3}$ **25.** x^5 **26.** $\dfrac{1}{y}$ **27.** $-\dfrac{2}{xy}$ **28.** $3a$

29. $\dfrac{x}{2(x+y)}$ **30.** $\dfrac{x+1}{(x-1)(x+3)}$ **31.** $\dfrac{2(x-1)}{3x}$ **32.** $x+4$ **33.** $\dfrac{1}{x^2-x+1}$

34. $\dfrac{x+2}{x+3}$ **35.** $\dfrac{4(x^2+x+1)}{x-2}$ **36.** $\dfrac{x+z}{x-z}$ **37.** -1 **38.** $-\dfrac{x+3}{x+1}$ **39.** $\dfrac{1-2x}{x+2}$ **40.** $-x$

41. $\dfrac{2-x}{x+3}$ **42.** $-\dfrac{m^2+mn+n^2}{m+n}$ **43.** $\dfrac{50}{10x}$ **44.** $\dfrac{75b^2c}{10bc^2}$ **45.** $\dfrac{r^2(m+n)}{6m+6n}$ **46.** $\dfrac{-3x(x-1)}{x^2-1}$

47. $\dfrac{-x(x+1)}{x-x^2}$ **48.** $\dfrac{2z(3-z)}{z^2-6z+9}$ **49.** $\dfrac{-x(x+3)}{x^2-9}$ **50.** $\dfrac{-3z(2z+y)}{4z^2-y^2}$

Practice Exercises

51. 0   **53.** 0   **55.** 7   **57.** 2, 5   **59.** $-1, 1$ **61.** x^5 **63.** $3t^2$

65. $\dfrac{x}{4}$ **67.** $\dfrac{-3}{x^2y^2}$ **69.** $\dfrac{5x}{7y^2}$ **71.** $-\dfrac{n}{(n-m)^3}$ **73.** $\dfrac{x+2}{(x-1)(x+5)}$ **75.** $-\dfrac{a+3}{a+1}$

77. $-a$ **79.** $\dfrac{1}{5-3x}$ **81.** $\dfrac{3(4y^3+1)}{5(3y^2-5)}$ **83.** $\dfrac{1}{x+2}$ **85.** $\dfrac{1}{x^2+x+1}$ **87.** $x-4$ **89.** $\dfrac{x+3}{x+1}$

91. $\dfrac{p^2-3p+9}{p+2}$ **93.** $-\dfrac{x+5}{x+4}$ **95.** $-\dfrac{s+1}{s^2+s+1}$ **97.** $\dfrac{x^2+x+1}{2(x+1)}$ **99.** $\dfrac{x^2+1}{x^2+3}$ **101.** $-\dfrac{2x+3}{x+1}$

103. $\dfrac{a+3}{c+6}$ **105.** $\dfrac{q-p}{q+p}$ **107.** $\dfrac{x+y}{x-y}$ **109.** $\dfrac{60x}{12x^2}$ **111.** $\dfrac{75b^2}{10bc^2}$ **113.** $\dfrac{2p}{2p-10}$

115. $\dfrac{-3x(x+1)}{x^2-1}$ **117.** $\dfrac{2z^2+6z}{z^2+6z+9}$ **119.** $\dfrac{-3}{y-x}$ **121.** $\dfrac{-x^2-2x}{x^2-4}$ **123.** $\dfrac{2x(x^2-xy+y^2)}{x^3+y^3}$

125. $\dfrac{2x(x^2+xy+y^2)}{x^3-y^3}$ **127.** $\dfrac{3a-15}{3}$ **129.** $\dfrac{x^2-7x}{x}$ **131.** $\dfrac{160x}{x+12}$ mph   **133.** $\dfrac{2N}{8-N}$ hr

135. $2, \$7.5, \$17.6, \$52.8, \$334.4$ million; impossible to estimate. The cost increases as the percent removed increases.

Challenge Problems

137. $x + 2$ **139.** $-3, -1$ **141.** $-2, 2$

PROBLEM SET 3.2

Warm-Ups

1. $\dfrac{yz}{x^3}$ **2.** $-2b^4c$ **3.** $-\dfrac{13}{3}$ **4.** q^2 **5.** $\dfrac{-4(a^2 + 2)}{a^2 - 3}$ **6.** $\dfrac{1 + s^4}{s^2 + 2st + 4t^2}$ **7.** $\dfrac{1}{2}$ **8.** 1.

9. $-\dfrac{1}{t}$ **10.** $-y(y + 2)$ **11.** $-\dfrac{y^2}{x}$ **12.** $-\dfrac{x + 4}{x + 1}$ **13.** $\dfrac{cy^3}{bz^2}$ **14.** $-\dfrac{a}{2bx^3y}$ **15.** $\dfrac{x + a}{4a^2}$

16. $\dfrac{(5 - x)(2 - x)}{2x}$ **17.** $\dfrac{(x + 4)(x + 1)}{(x + 2)(x + 3)}$ **18.** $\dfrac{(2x - 1)(x - 4)}{(x + 1)(x - 5)}$ **19.** $\dfrac{b^2 - 2bc + 4c^2}{a(x - 1)(x + 1)}$

20. $\dfrac{(c - d)(x - y)^2}{(c + d)(x + y)^2}$ **21.** $-y$ **22.** $\dfrac{-1}{r^2 + rs + s^2}$

Practice Exercises

23. $\dfrac{y^2z^3}{x}$ **25.** $-\dfrac{3}{m^3}$ **27.** $-\dfrac{y^2}{xz^5}$ **29.** $-\dfrac{2}{m^2}$ **31.** $\dfrac{-xy^2}{z^2}$ **33.** $-\dfrac{b^9c^5}{x^2}$ **35.** x **37.** $-\dfrac{11}{3}$

39. $\dfrac{a}{b^6}$ **41.** $\dfrac{z + 2}{z + 1}$ **43.** $-\dfrac{4}{c - 1}$ **45.** $-x^3(x^2 + 5x + 25)$ **47.** $\dfrac{(x + 2)^2}{2(x - 1)(x - 2)}$

49. $\dfrac{(a + b)^2}{(a - 2b)(a - b)}$ **51.** $\dfrac{3x - 1}{2x - 1}$ **53.** $\dfrac{c^2y^2}{bz^4}$ **55.** $\dfrac{m^2n^3r^2s}{pt}$ **57.** $-\dfrac{a^2}{5bxy}$ **59.** $\dfrac{5y}{2x}$ **61.** $-ab^2s^3$

63. $-\dfrac{b^2rs^{11}}{a^2}$ **65.** $-\dfrac{p}{q}$ **67.** $\dfrac{2x}{ay}$ **69.** -1 **71.** $\dfrac{4x + 1}{(x + 4)(16x^2 + 4x + 1)}$ **73.** $\dfrac{(x + 2)(x - 2)}{(x + 1)(x + 3)}$

75. $-\dfrac{(x + 2)^2}{(x + 1)^2}$ **77.** $\dfrac{(x + 3)(x - 1)}{x^2(x - 3)}$ **79.** $\dfrac{(x + 3)(x - 1)}{x^2(x + 1)}$ **81.** $\dfrac{(b - 2)(b + 3)}{(b + 2)(b^2 - 3)}$

83. $\dfrac{(a + b)(a + 2b)}{(a + d)(a - b)}$ **85.** $\dfrac{x - 2}{x + 1}$ **87.** $\dfrac{(2p + t)(p + t)}{(p - 2t)(2p - t)}$ **89.** $\dfrac{x^2(x^2 + 1)}{(2x - y)(x + 2y)}$

Challenge Problems

91. $\dfrac{(x^n - 3)(x^p + 1)}{(x^n + 2)(x^p + 3)}$

PROBLEM SET 3.3

Warm-Ups

1. $\dfrac{5 - x}{x^2}$ **2.** $\dfrac{a - b}{t}$ **3.** $\dfrac{2}{y}$ **4.** 0 **5.** 2 **6.** $\dfrac{1}{r - 2}$ **7.** $\dfrac{m + 1}{m - 2}$ **8.** 1 **9.** $\dfrac{5y}{x - 5}$

10. $\dfrac{x - 3}{z - t}$ **11.** $\dfrac{p^2 + q^2}{p^2q^2}$ **12.** $\dfrac{x^2 - 3y^2}{x^4y^3}$ **13.** $\dfrac{3r + 1}{(r + 2)(r - 3)}$ **14.** $\dfrac{a^2 - 4a - 2}{(a + 2)(a - 3)}$ **15.** $\dfrac{41}{15(a + 2)}$

16. $\dfrac{x^2 + 4x - 4}{(x - 2)^2(x + 2)}$ **17.** $\dfrac{1}{(x + 3)(x - 1)}$ **18.** $\dfrac{7y^2 + 20y + 28}{(2 + y)(2 - y)(4 + 2y + y^2)}$ **19.** $\dfrac{x^2 + 2}{x}$

20. $\dfrac{a^2 + ab + 3}{a + b}$ **21.** $\dfrac{1}{t - 1}$ **22.** $\dfrac{2}{a(a + 1)}$ **23.** $\dfrac{9}{2 - a}$ **24.** $\dfrac{3 - 7p}{p(p - 1)}$

25. $\dfrac{x^2 + 3x - 8}{(x - 4)(x - 2)(x + 1)}$ **26.** $\dfrac{5x + 3}{(x - 2)(x + 2)}$

Practice Exercises

27. $\dfrac{6 - x}{y^3}$ **29.** $\dfrac{ac + b^3}{b^3 c^2}$ **31.** $\dfrac{2z - 4x + xyz}{x^2 yz}$ **33.** $\dfrac{3y - 1}{y(y - 1)}$ **35.** $\dfrac{2s - 4}{(s + 1)(s - 1)}$

37. $\dfrac{2q^2 + 2q}{(q - 4)(q + 6)}$ **39.** $-\dfrac{3a}{x - 1}$ **41.** $\dfrac{23}{12(r + 2)}$ **43.** $\dfrac{-7}{5(m - n)}$ **45.** $\dfrac{t + 3}{t - 1}$ **47.** $\dfrac{d + 3}{d + 1}$

49. $\dfrac{x + 1}{x + 3}$ **51.** $\dfrac{x + 2}{(x - 8)(x - 3)}$ **53.** $\dfrac{z^2 - 16}{(2z - 1)(z - 3)(z - 10)}$ **55.** $\dfrac{y + 3}{(y^2 - 3y + 9)(y - 3)}$

57. $\dfrac{3(z + 3)}{(z - 1)(z + 2)}$ **59.** $\dfrac{9y + 5}{(y - 5)^2 (y + 5)}$ **61.** $\dfrac{3 - 4s}{(s - 8)(s + 3)(s - 2)}$ **63.** $\dfrac{u}{u - 1}$ **65.** $\dfrac{-a - 2b}{a + b}$

67. $\dfrac{-x + 8}{x(x + 2)}$ **69.** $-\dfrac{x^2 + 8x + 14}{(x + 4)(x + 3)(x + 2)}$ **71.** $\dfrac{1 + q}{q - t}$ **73.** $\dfrac{3z^2 - 9}{(z + 1)(z - 1)(z + 2)}$

75. $\dfrac{x - 42}{(x + 2)(x - 4)}$ **77.** $\dfrac{2x + 5}{x(x + 5)}$ **79.** $\dfrac{3x - 2}{(x + 1)(2x - 3)}$

Challenge Problems

81. $\dfrac{4x - 1}{x - 1}$ **83.** $\dfrac{12x}{(x - 3)^2 (x + 3)^2}$

PROBLEM SET 3.4

Warm-Ups

1. $\{1\}$ **2.** $\{1\}$ **3.** $\{2\}$ **4.** $\{3\}$ **5.** $\{3\}$ **6.** $\left\{-\dfrac{4}{3}\right\}$ **7.** $\left\{\dfrac{1}{3}\right\}$ **8.** $\left\{-\dfrac{1}{2}\right\}$ **9.** $\{2\}$ **10.** $\{1\}$

11. $\left\{\dfrac{1}{2}\right\}$ **12.** $\{-2\}$ **13.** $\{-2.5\}$ **14.** $\{3.91\overline{6}\}$ **15.** $\{5\}$ **16.** $\{1\}$ **17.** $\{10\}$ **18.** $\{-1\}$

19. $\{-3\}$ **20.** $\varnothing$ **21.** $\varnothing$ **22.** $\varnothing$ **23.** $\{-2\}$ **24.** $\left\{-\dfrac{1}{3}\right\}$ **25.** $\{-10\}$ **26.** $(-\infty, -5) \cup [5, +\infty)$

27. $(-\infty, 1) \cup [4, +\infty)$ **28.** $(5, 15]$ **29.** $(-6, -3]$ **30.** $(-1, 1) \cup (5, +\infty)$
31. $(-7, 1) \cup (9, +\infty)$ **32.** Jeri's rate is 25 mph.   **33.** His average speed is 16 mph.
34. The speed of the United jet is 250 mph.   **35.** There are 40 girls.   **36.** 180 are bluebirds.
37. There are 30.48 cm in a foot.   **38.** There are 11 pounds.   **39.** Any number between 0 and 3
40. Any number between -2 and -1, or any number greater than 0.

Practice Exercises

41. $\{-12\}$ **43.** $\dfrac{10 - y}{5y}$ **45.** $\left\{\dfrac{4}{3}\right\}$ **47.** $\{-2\}$ **49.** $\{1\}$ **51.** $\{6\}$ **53.** $\{11.5\overline{90}\}$ **55.** $\varnothing$

57. $\{-2\}$ **59.** $\{-3\}$ **61.** $\left\{\dfrac{2}{3}\right\}$ **63.** $\{-7\}$ **65.** $\varnothing$ **67.** $\varnothing$ **69.** $\{x \mid x < -1 \text{ or } x > 1\}$

71. $\{x \mid 4 \le x < 5\}$

73. $\{x \mid -2 < x < 0\}$ **75.** $(-\infty, -9] \cup (-2, +\infty)$ **77.** $(-\infty, -4) \cup [1, +\infty)$ **79.** $(-1, 1)$
81. Joyce's rate is 8 mph.   **83.** His speed going to work is 30 mph.   **85.** There are 120 Republicans.
87. There are 19 liters in 5 gallons.   **89.** The numbers are $\dfrac{16}{3}$ and 16.   **91.** The number is 1.
93. All numbers greater than 1

Challenge Problems

95. $\left\{\dfrac{2a + b}{2c}\right\}$ **97.** $\{x \mid b < x \le a\}$

PROBLEM SET 3.5

Warm-Ups

1. $\dfrac{a}{2}$ **2.** $\dfrac{z^2}{t^2 x^2}$ **3.** $-\dfrac{s^2}{6r}$ **4.** $\dfrac{-2a^2 n^2}{3bm}$ **5.** $\dfrac{s}{3}$ **6.** $-\dfrac{1}{2}$ **7.** $\dfrac{3}{2}$ **8.** $-\dfrac{2}{15}$ **9.** $\dfrac{1 - y}{1 + y}$
10. $-\dfrac{a + 1}{a + 3}$ **11.** $\dfrac{n + m}{n - m}$ **12.** $\dfrac{-1}{a + b}$ **13.** $\dfrac{2b - 1}{3b - 2}$ **14.** $\dfrac{1}{x + 1}$ **15.** $\dfrac{b + a^2}{a}$ **16.** $\dfrac{3x}{6 - 2x}$

Practice Exercises

17. $\dfrac{a^2}{3}$ **19.** $\dfrac{y}{a}$ **21.** $\dfrac{3}{4}$ **23.** $-\dfrac{2}{9}$ **25.** $\dfrac{1}{3}$ **27.** $-\dfrac{3(x^2 + 5x + 25)}{16}$ **29.** $\dfrac{r - 2}{r}$ **31.** $\dfrac{3 - 2y}{1 + 4y}$
33. 1   **35.** $t + s$ **37.** $\dfrac{2}{a^2 - ab + b^2}$ **39.** $\dfrac{1}{t - 4}$ **41.** $\dfrac{1}{x + 3}$ **43.** $\dfrac{1}{x - 1}$ **45.** 0   **47.** $\dfrac{1}{x - 1}$
49. $\dfrac{3yz - 2xz - 4xy}{yz - xz - xy}$ **51.** (a) 7 ft (b) $\dfrac{7}{13}$ ft

Challenge Problems

53. $\dfrac{5}{3}$ **55.** $\dfrac{2x^2 + x - 1}{x - 1}$

REVIEW PROBLEMS CHAPTER 3

1. 0   **3.** None   **5.** $-3, -1$ **7.** $\dfrac{8p^2 - 16p}{6p^2(p - 2)}$ **9.** $\dfrac{x^2 - x - 12}{x^2 - 8x + 16}$ **11.** $\dfrac{1}{m^2 n^2}$ **13.** $\dfrac{x - 5}{x^2 - x + 1}$
15. $t(s - 1)$ **17.** $\dfrac{-3(x + 1)(x + 2)}{(x + 3)(x - 2)}$ **19.** $\dfrac{a(a + b)}{(a - b)(a + 3)}$ **21.** $-\dfrac{rt^7 v^2}{u^3}$ **23.** $\dfrac{u - v}{(u + 2w)^2}$
25. $\dfrac{y + 4}{3y + 1}$ **27.** $\dfrac{4x + 1}{(x + y)(4x + 3)}$ **29.** $\dfrac{3t^3(2t - 3)}{(t^2 - 2)(t^4 + 4)(t - 3)}$ **31.** $-\dfrac{4t^3 s^5}{9}$ **33.** $\dfrac{-24}{a^5 b^2(a - b)}$
35. $\dfrac{-4(m + 4n)}{(2m^2 + 1)(16n^2 + 4mn + m^2)}$ **37.** xy **39.** $\dfrac{y - 2}{2y + 1}$ **41.** t **43.** $\dfrac{1}{2(x^2 + 1)}$ **45.** s
47. $(-\infty, -14] \cup (-7, +\infty)$ **49.** $\left\{\dfrac{4}{3}\right\}$ **51.** $[-3, -2)$ **53.** Ø   **55.** $(-\infty, -2) \cup (1, +\infty)$ **57.** $\left\{-\dfrac{9}{4}\right\}$
59. The speed of the truck is 54 mph.   **61.** The wall should be 10 inches by 16 inches.

63. The flag's dimensions must be $6\frac{2}{3}$ inches by $12\frac{2}{3}$ inches.   **65.** The number is $\frac{3}{2}$.

67. Any number between -1 and 0.

Let's Not Forget . . .

68. Factored   **69.** $(3x - 1)(9x^2 + 3x + 1)$ **70.** $(a - b)(x + y)$ **71.** Factored   **72.** Factored
73. Two terms; not a factor.   **74.** One term; $x - 1$ is a factor.   **75.** Three terms; not a factor.
76. Three terms; $x - 1$ is a factor.   **77.** Two terms; $x - 1$ is a factor.   **78.** Two terms; $x - 1$ is a factor.
79. $x^2 + 4xy + 4y^2$ **80.** $c^3 - 6c^2 + 12c - 8$ **81.** -4 **82.** 4 **83.** -8 **84.** a^4 **85.** $-a^5$
86. Won't reduce   **87.** $\dfrac{a - b}{y - z}$ **88.** $\dfrac{5x - 6}{(x - 2)(x + 2)}$ **89.** $\dfrac{(x + 2)^2}{(x - 2)^2}$ **90.** $\left\{-\dfrac{3}{2}\right\}$

91. Inequality; $(-1, 1) \cup [3, +\infty)$ **92.** Expression; $\dfrac{4xy}{(x - y)(x + y)}$ **93.** Equation; $\left\{\dfrac{8}{5}\right\}$

94. Inequality; $(-2.5, 0.5)$ **95.** Expression; $-\dfrac{80}{9}$ **96.** Expression; $\dfrac{1 + x}{1 - x}$

CHAPTER 3 TEST

1. C   **2.** B   **3.** A   **4.** C   **5.** B   **6.** $\dfrac{a^2 + b^2}{ab(a - b)}$ **7.** $\dfrac{x(x + 3)}{(x - 3)(x - 1)}$ **8.** $\dfrac{-1}{y + 3}$ **9.** $\dfrac{3x - 2}{x - 1}$

10. $\dfrac{3t + s}{3t - s}$ **11.** $[-5, -1)$ **12.** $\left\{-\dfrac{7}{2}\right\}$ **13.** $\{3\}$ **14.** All numbers between -1 and 1
15. Jane's speed is 7.5 mph, and Bob's speed is 4.5 mph.

► Chapter 4

PROBLEM SET 4.1

Warm-Ups

1. 14   **2.** -8 **3.** 5   **4.** -12 **5.** 8   **6.** $\dfrac{5}{7}$ **7.** $\dfrac{16}{25}$ **8.** $\dfrac{3}{4}$ **9.** $-2x$ **10.** $-2x^2y$

11. $10z^3$ **12.** $-12j^3l^2$ **13.** $\dfrac{11y^2}{xz}$ **14.** $\dfrac{6x^5}{y^4}$ **15.** $5\sqrt{2}$ **16.** $-6\sqrt{5}$ **17.** $6\sqrt{7}$ **18.** $4\sqrt{15}$

19. $2\sqrt[3]{5}$ **20.** $9\sqrt[3]{4}$ **21.** $3x^2y\sqrt[3]{x}$ **22.** $\dfrac{xy^2\sqrt[5]{xy^4}}{2}$ **23. (a)** $2y$ **(b)** Won't simplify   **(c)** $2 + y$

24. (a) $4k^2$ **(b)** Won't simplify   **(c)** $(2 + k)^2$

Practice Exercises

25. 18   **27.** -10 **29.** 6   **31.** $-4x^2y$ **33.** $6\sqrt{2}$ **35.** $12\sqrt{2}$ **37.** $3\sqrt[3]{2}$ **39.** $-3\sqrt[5]{2}$
41. $\dfrac{4}{9}$ **43.** $\dfrac{3\sqrt{3}}{7}$ **45.** $-2\sqrt{2}$ **47.** $3xy\sqrt{x}$ **49.** $\dfrac{4}{3}$ **51.** $\dfrac{5\sqrt{3}}{7}$ **53.** $\dfrac{2\sqrt[3]{7}}{3}$ **55.** $3xy\sqrt{7yz}$
57. $-15xy^4\sqrt{2xy}$ **59.** $\dfrac{14}{p^2q}$ **61.** $2z\sqrt[3]{2}$ **63.** $\dfrac{6a^2b^4c\sqrt[3]{c}}{d^9}$ **65.** $-2x^3z^5\sqrt[5]{3z}$ **67.** $13 + 2x$
69. $3\sqrt{4 + v^2}$

Challenge Problems

71. No   **73.** $|x|$ **75.** $-b$ **77.** $5 + a$ **79.** $|a + b|$

PROBLEM SET 4.2

Warm-Ups

1. $8\sqrt{13}$ **2.** $2\sqrt{3}$ **3.** $-8\sqrt{3}$ **4.** Not possible   **5.** $4\sqrt{6}$ **6.** $3\sqrt{3} - 3\sqrt{2}$ **7.** $-2 + 3\sqrt{3}$
8. 5   **9.** $5t\sqrt{2t}$ **10.** $11x^2\sqrt{2x}$ **11.** $11x\sqrt[3]{xy^2}$ **12.** $18xy^2\sqrt{3y}$

Practice Exercises

13. $11\sqrt{13}$ **15.** $-3\sqrt{2}$ **17.** $17\sqrt{3}$ **19.** $2\sqrt[3]{2}$ **21.** $12\sqrt{3x}$ **23.** $12\sqrt{6}$ **25.** $3t^2\sqrt{2t}$
27. $2x\sqrt{3x}$ **29.** $8\sqrt[3]{2}$ **31.** $6\sqrt[3]{3}$ **33.** $\frac{4}{3}\sqrt{2x}$ **35.** 0

Challenge Problems

37. $2x^{m+n}$ **39.** $2x^{m+n}$

PROBLEM SET 4.3

Warm-Ups

1. $2z$ **2.** $3y\sqrt{2xz}$ **3.** $6y\sqrt[3]{x^2}$ **4.** $-4x$ **5.** 63   **6.** 8   **7.** $\sqrt{6} - 3$ **8.** $6x\sqrt{2} + 6x\sqrt{3}$
9. $18 + 6\sqrt{2}$ **10.** $3 - 3\sqrt[3]{2x}$ **11.** $3 + 3\sqrt{3}$ **12.** $-5\sqrt{6}$ **13.** $30 - 12\sqrt{6}$ **14.** -3

Practice Exercises

15. $x\sqrt{x}$ **17.** $14xy^2\sqrt{2}$ **19.** $2\sqrt{14} - \sqrt{21}$ **21.** $3st\sqrt[3]{2t}$ **23.** 27   **25.** $7 - 4\sqrt{2}$
27. $s^2 - 2s\sqrt{2t} + 2t$ **29.** $4\sqrt{3} - 3\sqrt{2}$ **31.** $4\sqrt{15} + 24$ **33.** $2xy^2 + 4xy^2\sqrt[3]{3}$ **35.** $3k^2$
37. $6 + 2\sqrt{3}$ **39.** $2x^2 - 3y^2$ **41.** $8x + 4\sqrt{6x} + 3$ **43.** $-30 + 3\sqrt{2}$ **45.** $\sqrt{2} - 2$
47. $2 - \frac{9}{4}\sqrt{2}$ **49.** $3 + 3\sqrt[3]{4} + 3\sqrt[3]{2}$ **51.** $7 + 4\sqrt[3]{25} + 5\sqrt[3]{5}$ **53.** $3\sqrt[3]{18} - 3\sqrt[3]{12} - 1$
55. $\sqrt{2}(\sqrt{5} + \sqrt{3})$ **57.** $\sqrt{3}(\sqrt{2} - 1)$ **59.** $2(\sqrt{3} + \sqrt{5})$

Challenge Problems

61. 1   **63.** $a - b$; difference of two cubes

PROBLEM SET 4.4

Warm-Ups

1. $2\sqrt{2}$ **2.** -2 **3.** $3\sqrt{2}$ **4.** 3   **5.** $\sqrt{2} - \sqrt{5}$ **6.** $1 + 4\sqrt{7}$ **7.** $\frac{\sqrt{5}}{5}$ **8.** $\frac{2\sqrt{3}}{9}$

9. $\dfrac{\sqrt{6}}{2}$ **10.** $\dfrac{\sqrt{14}}{4}$ **11.** $\dfrac{\sqrt[3]{49}}{7}$ **12.** $\dfrac{\sqrt[3]{6}}{6}$ **13.** $\dfrac{\sqrt[4]{4}}{2}$ **14.** $\dfrac{x^2\sqrt[5]{4}}{2}$ **15.** $5\sqrt{5} + 10$

16. $-\dfrac{\sqrt{2} + \sqrt{10}}{2}$ **17.** $-2\sqrt{6} - 6$ **18.** $-\sqrt{2}$ **19.** $\dfrac{1}{1 + \sqrt{6}}$ **20.** $\dfrac{3}{7 + 2\sqrt{10}}$

Practice Exercises

21. $\sqrt{3}$ **23.** $2\sqrt[3]{2}$ **25.** $2\sqrt{3}$ **27.** $\sqrt{2} - 1$ **29.** 2 **31.** $\sqrt{5}$ **33.** $2 + 3\sqrt{3}$ **35.** $y\sqrt{2y}$

37. 3 **39.** $5 + 2\sqrt{5}$ **41.** $\dfrac{2\sqrt{11}}{11}$ **43.** $\dfrac{\sqrt{5}}{10}$ **45.** $\dfrac{\sqrt{15}}{5}$ **47.** $\dfrac{9 - 3\sqrt{2}}{7}$ **49.** $\dfrac{\sqrt{10}}{5}$

51. $2\sqrt{3} - 3$ **53.** $-3\sqrt{3} - 6$ **55.** $-\dfrac{3 - \sqrt{5}}{2}$ **57.** $\dfrac{\sqrt[3]{4}}{2}$ **59.** $2\sqrt[4]{2}$ **61.** $\dfrac{6}{5\sqrt{6}}$ **63.** $\dfrac{2}{\sqrt{2} - 2}$

65. $-\dfrac{5}{\sqrt{3} + \sqrt{2}}$ **67.** $\dfrac{1}{5 - 2\sqrt{5}}$ **69.** $\dfrac{5}{13 - 7\sqrt{6}}$

Challenge Problems

71. $\dfrac{1}{\sqrt{x + h} + \sqrt{x}}$ **73.** $\sqrt[3]{9} + \sqrt[3]{6} + \sqrt[3]{4}$

PROBLEM SET 4.5

Warm-Ups

1. $\{3\}$ **2.** $\varnothing$ **3.** $\{11\}$ **4.** $\left\{\dfrac{26}{3}\right\}$ **5.** $\{-7\}$ **6.** $\{0\}$ **7.** $\{0\}$ **8.** $\{-1\}$ **9.** $\varnothing$ **10.** $\left\{\dfrac{15}{2}\right\}$

11. $\{1\}$ **12.** $\{3\}$ **13.** $\varnothing$ **14.** $\{1\}$ **15.** $\{3\}$ **16.** $\{1\}$ **17.** $\varnothing$ **18.** $\{4\}$ **19.** $\{17\}$ **20.** $\{10\}$

21. $\{3\}$ **22.** $\varnothing$ **23.** $\{1\}$ **24.** $\left\{-\dfrac{3}{2}\right\}$ **25.** $\{0\}$ **26.** $\varnothing$ **27.** $\{0\}$ **28.** $\{3\}$ **29.** $\{-2\}$

30. $\{1\}$ **31.** $\varnothing$

Practice Exercises

33. $\varnothing$ **35.** $\{4\}$ **37.** $\{-24\}$ **39.** $\varnothing$ **41.** $\{2\}$ **43.** $\{5\}$ **45.** $\left\{-\dfrac{1}{2}\right\}$ **47.** $\varnothing$ **49.** $\varnothing$

51. $\{-10\}$ **53.** $\{1\}$ **55.** $\{1\}$ **57.** $\{3\}$ **59.** $\{5\}$ **61.** $\{-1\}$ **63.** 582.034 **65.** 6.036

67. 2.867 **69.** The height is $\dfrac{1}{2}$ foot.   **71.** The can would need to be 2.1 inches high.

PROBLEM SET 4.6

Warm-Ups

1. 7 **2.** 11 **3.** 13 **4.** -10 **5.** Not a real number   **6.** -10 **7.** 3 **8.** 2 **9.** 2 **10.** 9

11. 32 **12.** 1728 **13.** 128 **14.** $\dfrac{1}{3}$ **15.** $\dfrac{1}{9}$ **16.** $-\dfrac{1}{343}$ **17.** Not a real number   **18.** $\dfrac{1}{8}$

19. $\dfrac{4}{25}$ **20.** $\dfrac{27}{8}$ **21.** $\dfrac{2744}{2197}$ **22.** 32 **23.** $\dfrac{1331}{1000}$ **24.** $\dfrac{36}{49}$ **25.** 8 **26.** 49 **27.** 5 **28.** $\dfrac{8}{s^{12}}$

29. 4   **30.** 9   **31.** $6x$ **32.** $12xy^2$ **33.** $\dfrac{7x}{y}$ **34.** $2ab^{1/3}$ **35.** $-\dfrac{x^2}{4z^{1/3}}$ **36.** $-\dfrac{27x^3}{y^3}$

37. $\dfrac{2\sqrt{2}y^{1/2}}{x}$ **38.** $-\dfrac{11x}{y}$ **39.** $\dfrac{27a^2}{x}$ **40.** $x^{3/2} + x^{1/2}$ **41.** $6x^{13/6} - 10x^{7/6}$ **42.** $2x^{4/3} + 3$

43. $x + 2 + \dfrac{1}{x}$ **44.** $x - \dfrac{1}{x}$

Practice Exercises

45. 6   **47.** 15   **49.** -10 **51.** 9   **53.** -2 **55.** 6   **57.** Not a real number   **59.** 8   **61.** 128

63. 243   **65.** $\dfrac{1}{16}$ **67.** $\dfrac{1}{9}$ **69.** $\dfrac{16}{49}$ **71.** $\dfrac{4096}{3375}$ **73.** $\dfrac{5832}{2197}$ **75.** $5y$ **77.** $\dfrac{4x^2}{y^3}$ **79.** $\dfrac{x}{4z^{1/3}}$

81. $\dfrac{3\sqrt{2}x}{y}$ **83.** $-\dfrac{27a^3}{x}$ **85.** $\dfrac{5}{x^{1/3}y^{1/3}}$ **87.** $\dfrac{27z^2}{8xy^3}$ **89.** $\dfrac{xy}{125}$ **91.** $6x^{10/3} - 15x^{4/3}$

93. $6 + \dfrac{2b^{3/2}}{a^{1/2}}$ **95.** $x^3 - \dfrac{1}{x^3}$

Challenge Problems

97. $a + 3a^{2/3}b^{1/3} + 3a^{1/3}b^{2/3} + b$ **99.** $x - 3x^{1/3} + \dfrac{3}{x^{1/3}} - \dfrac{1}{x}$ **101.** $(x^{1/3} - 3)(x^{1/3} + 2)$

103. $a^{m/n}$ does *not* equal $\sqrt[n]{a^m}$ *unless* $\sqrt[n]{a}$ is a real number, and $\sqrt{-1}$ is not a real number.   **105.** $\sqrt{xy}$

PROBLEM SET 4.7

Warm-Ups

1. $0 + 3i$ **2.** $0 + 7i$ **3.** $0 + (-11)i$ **4.** $1 + \sqrt{3}i$ **5.** $3 + (-2)i$ **6.** $0 + \sqrt{5}i$

7. $5\sqrt{2} + 0i$ **8.** $0 + 3\sqrt{3}i$ **9.** $-2 + 0i$ **10.** -1 **11.** $-i$ **12.** i **13.** 1   **14.** $-i$

15. i **16.** $3 + 4i$ **17.** $1 + 2i$ **18.** $8 - 7i$ **19.** $-3 + 4i$ **20.** 7   **21.** -14 **22.** $-2 + 3i$

23. $15 - 6i$ **24.** $-1 + 5i$ **25.** $7 - 3i$ **26.** $2i$ **27.** $-5 - 12i$ **28.** $-1 - 2\sqrt{6}i$

29. $-4 - 8\sqrt{6}i$ **30.** 5   **31.** 29   **32.** 5   **33.** 12   **34.** $\dfrac{1}{2} - \dfrac{1}{2}i$ **35.** $\dfrac{2}{5} + \dfrac{1}{5}i$ **36.** $\dfrac{1}{5} + \dfrac{3}{5}i$

37. $\dfrac{7}{10} - \dfrac{1}{10}i$ **38.** $\dfrac{4}{13} - \dfrac{7}{13}i$ **39.** $\dfrac{3}{5} - \dfrac{4}{5}i$

Practice Exercises

41. $0 + 5i$ **43.** $2 + (-\sqrt{5})i$ **45.** $0 + \sqrt{7}i$ **47.** $0 + 2\sqrt{2}i$ **49.** $-i$ **51.** -1 **53.** i

55. $3 - 4i$ **57.** $8 + 7i$ **59.** $-2 - 2i$ **61.** -9 **63.** $3 + 2i$ **65.** $5 - i$ **67.** $-7 - 9i$

69. $-8 + i$ **71.** $-2i$ **73.** $-7 + 24i$ **75.** $-6 + 6\sqrt{3}i$ **77.** $1 + 2\sqrt{6}i$ **79.** 10   **81.** 149

83. 7   **85.** 5   **87.** $\dfrac{1}{2} + \dfrac{1}{2}i$ **89.** $\dfrac{3}{5} + \dfrac{6}{5}i$ **91.** $\dfrac{1}{8} + \dfrac{1}{8}i$ **93.** $-\dfrac{1}{10} + \dfrac{7}{10}i$ **95.** $\dfrac{12}{13} - \dfrac{5}{13}i$

97. $-46 + 9i$

Challenge Problems

99. LS: $i^2 + 1 = -1 + 1 = 0$; RS: 0 If x is $-i$, then LS: $(-i)^2 + 1 = i^2 + 1$ **101.** $(x + 3i)(x - 3i)$

REVIEW PROBLEMS CHAPTER 4

1. $5\sqrt{3}$ **3.** $2xz^2\sqrt{2xy}$ **5.** $\dfrac{-2}{xy}$ **7.** -7 **9.** $\dfrac{1}{4}$ **11.** $\dfrac{64x^3}{y^2z^2}$ **13.** $\dfrac{x}{y}$ **15.** $\dfrac{3}{4}$ **17.** $\sqrt{3}$

19. $2\sqrt{3}-\sqrt{2}$ **21.** $5-2\sqrt{6}$ **23.** $xy-1$ **25.** $4x-\dfrac{9}{y}$ **27.** $\dfrac{11}{7}$ **29.** $\dfrac{\sqrt[3]{36}}{6}$ **31.** $\sqrt[3]{2}+1$

33. $\dfrac{-1+\sqrt{3}}{2}$ **35.** $\left\{\dfrac{2}{3}\right\}$ **37.** $\{1\}$ **39.** $\varnothing$ **41.** $0+15i$ **43.** $2+2i$ **45.** $3+2i$

47. $11+7i$ **49.** 85 **51.** $-46-9i$

Let's Not Forget . . .

52. Factored   **53.** $(x-y)(r^2+t^2)$ **54.** Factored   **55.** Factored   **56.** $(2x-3)(4x^2+6x+9)$

57. $(2x-3y)^2$ **58.** $2;\ 1-x$ is a factor   **59.** $2;\ 1-x$ is a factor   **60.** $3;\ 1-x$ is a factor

61. $1;\ 1-x$ is a factor   **62.** $3;\ 1-x$ not a factor   **63.** $-\dfrac{4x^2+2x+1}{2x+1}$ **64.** $\dfrac{-5z^3}{xy^2}$ **65.** $-\dfrac{1}{2}$

66. $\dfrac{2a^2}{3b^3}$ **67.** $\dfrac{10}{3}$ **68.** $\sqrt{3}-i$ **69.** x^2+y^2 **70.** $2\sqrt{3}+6$ **71.** $x^2+2\sqrt{2}x+2$

72. $8-12x+6x^2-x^3$ **73.** $r-s$ **74.** $1+\dfrac{\sqrt{2}}{2}$ **75.** $\{1\}$ **76.** $\dfrac{3x-9}{(x+1)(x-2)}$

77. $\dfrac{x^2-2x}{x^2-1}$ **78.** Equation; $\{-2,7\}$ **79.** Equation; $\{1\}$ **80.** Expression; $\dfrac{2}{(x+3)(x-2)}$

81. Equation; $\left\{-\dfrac{1}{13}\right\}$ **82.** Expression; $16-24x+9x^2$ **83.** Expression; $5+i$

CHAPTER 4 TEST

1. A   **2.** B   **3.** C   **4.** B   **5.** B   **6.** D   **7.** A   **8.** C   **9.** D   **10.** $\{3\}$ **11.** $\varnothing$

12. $\dfrac{b^6}{3a^2c^3}$ **13.** $\dfrac{1}{2}+\dfrac{1}{2}i$ **14.** $\dfrac{\sqrt{6}}{4}$ **15.** $-9+8\sqrt{2}$

► Chapter 5

PROBLEM SET 5.1

Warm-Ups

1. $\{-7,-3\}$ **2.** $\{-11,5\}$ **3.** $\{\pm5\}$ **4.** $\{0,3\}$ **5.** $\{-5,0\}$ **6.** $\left\{-3,\dfrac{1}{2}\right\}$ **7.** $\left\{0,\dfrac{1}{3}\right\}$

8. $\left\{-\dfrac{7}{3},\dfrac{5}{2}\right\}$ **9.** $\{-4\}$ **10.** $\left\{\dfrac{9}{4}\right\}$ **11.** $\left\{3,\dfrac{9}{2}\right\}$ **12.** $\{\pm\sqrt{3}\}$ **13.** $\{\pm2i\}$ **14.** $\{\pm\sqrt{3}\,i\}$

15. $\{-1,2\}$ **16.** $\{-5,-1\}$ **17.** $\{1,3\}$ **18.** $\left\{\dfrac{2}{3},1\right\}$ **19.** $\left\{-1,-\dfrac{1}{4}\right\}$ **20.** $\left\{-3,\dfrac{1}{2}\right\}$ **21.** $\{0,2\}$

22. $\{0,7\}$ **23.** $\{-4,-1\}$ **24.** $\{-7,-1\}$ **25.** $\left\{-1,\dfrac{5}{2}\right\}$ **26.** $\left\{-\dfrac{1}{2},\dfrac{2}{3}\right\}$ **27.** $\left\{-2,\dfrac{1}{3}\right\}$

28. $\left\{-3,\dfrac{2}{3}\right\}$ **29.** $\{\pm2\}$ **30.** $\{\pm4\}$ **31.** $\{0,3\}$ **32.** $\left\{0,\dfrac{8}{3}\right\}$ **33.** $\left\{-\dfrac{1}{6},1\right\}$ **34.** $\left\{-1,\dfrac{3}{7}\right\}$

35. $\{8\}$ **36.** $\left\{\dfrac{1}{2}\right\}$ **37.** $\{x \mid -1 < x < 2\}$ **38.** $\{x \mid x < -4 \text{ or } x > 7\}$ **39.** $\{x \mid -1 \le x \le 3\}$
40. $\{x \mid -3 \le x \le 2\}$ **41.** $(1, 3)$ **42.** $(-\infty, 2) \cup (3, +\infty)$ **43.** $(-\infty, -5) \cup (-1, +\infty)$
44. $(-\infty, -1] \cup [3, +\infty)$ **45.** $(-2, 1)$ **46.** $[-2, 2]$

Practice Exercises

47. $\{-1, 6\}$ **49.** $\{-4, -3\}$ **51.** $\{-1, 7\}$ **53.** $\{-2, 8\}$ **55.** $\{5, 10\}$ **57.** $\{12\}$ **59.** $\left\{\dfrac{1}{5}, 2\right\}$
61. $\left\{-1, -\dfrac{2}{3}\right\}$ **63.** $\left\{\dfrac{3}{2}, 2\right\}$ **65.** $\left\{-\dfrac{4}{5}, 2\right\}$ **67.** $\{1, 8\}$ **69.** $\{3, 5\}$ **71.** $\left\{-\dfrac{3}{2}, \dfrac{4}{3}\right\}$
73. $\left\{-\dfrac{3}{2}, \dfrac{1}{2}\right\}$ **75.** $\{-1, 2\}$ **77.** $\{-3, 4\}$ **79.** $\{-1, 13\}$ **81.** $\left\{-\dfrac{2}{3}, \dfrac{3}{5}\right\}$ **83.** $\left\{-\dfrac{3}{4}, \dfrac{2}{5}\right\}$
85. $\left\{x \mid -\dfrac{1}{3} < x < 4\right\}$ **87.** $\varnothing$ **89.** $\{-5\}$ **91.** $\{x \mid x < 1 \text{ or } x > 6\}$ **93.** $\{x \mid x < -1 \text{ or } x > 8\}$
95. $\{x \mid x < -4 \text{ or } x > 4\}$ **97.** $\{t \mid -5 \le t \le 0\}$ **99.** $\{s \mid -3 < s < -1\}$ **101.** $\{x \mid -2 \le x \le 7\}$
103. $\left\{x \mid x < -\dfrac{3}{5} \text{ or } x > \dfrac{1}{2}\right\}$ **105.** $\left[-\dfrac{4}{3}, \dfrac{3}{7}\right]$ **107.** $\left(-\dfrac{2}{3}, \dfrac{1}{4}\right)$ **109.** $\left(-\infty, -\dfrac{7}{5}\right] \cup \left[\dfrac{7}{2}, +\infty\right)$

Challenge Problems

111. $\{\pm\sqrt{5}\}$ **113.** $\{\pm\sqrt{5}i\}$ **115.** $\left\{\pm\dfrac{5}{2}i\right\}$ **117.** $x^2 - 3x + 2 = 0$ **119.** $x^2 - 4 = 0$
121. $x^2 + 4 = 0$

PROBLEM SET 5.2

Warm-Ups

1. The girls are 18 and 16 years old.   **2.** They are -13 and -11.   **3.** The trails are 7 and 13 miles.
4. They are 4 and 17.   **5.** The integers are 4, 5, and 6.   **6.** They are -7, -6, and -5.
7. Between 2 and 3 seconds   **8.** After 20 yrs   **9.** The frame should be 1 ft wide.
10. It should be $2\dfrac{1}{2}$ ft.   **11.** The base is 16 ft and the altitude is 7 ft.   **12.** Height, 5 yd; base, 10 yd
13. The legs are 5 ft and 12 ft.   **14.** The sides are 3, 4, and 5 units.

Practice Exercises

15. The integers are -11 and -12.   **17.** The numbers are 4 and 5.   **19.** The integers are 1, 3, and 5.
21. The rectangle is $6\dfrac{1}{2}$ m by 10 m.   **23.** Between 0 and 10 minutes.   **25.** It should be $2\dfrac{1}{2}$ ft wide.
27. The sheet should be 14 in. by 14 in.   **29.** The poster would be 7 in. by 10 in.
31. The billboard is 10 m by 12 m.   **33.** Alice must sell more than 20 dresses.
35. The office is 20 ft by 30 ft. 2,764,800 bottles are needed.
37. The base is 14 yd, and the height is 7 yd.   **39.** The legs are 10 m and 24 m.

Challenge Problems

41. The southbound train is traveling at the rate of 120 mph and the other train is traveling at the rate of 50 mph.

PROBLEM SET 5.3

Warm-Ups

1. $\{\pm 5\}$ **2.** $\{\pm 3i\}$ **3.** $\{\pm\sqrt{5}\}$ **4.** $\{\pm\sqrt{5}i\}$ **5.** $\{\pm 4\sqrt{3}\}$ **6.** $\{\pm 3\sqrt{2}i\}$ **7.** $\{\pm 2\sqrt{2}i\}$

8. $\{\pm\sqrt{3}i\}$ **9.** $\{-5, -1\}$ **10.** $\{-3 \pm 2\sqrt{3}\}$ **11.** $\{5 \pm \sqrt{7}i\}$ **12.** $\{-8 \pm 3\sqrt{2}i\}$ **13.** 4   **14.** 9

15. $\dfrac{49}{4}$ **16.** $\dfrac{25}{4}$ **17.** $\dfrac{1}{4}$ **18.** $\dfrac{169}{4}$ **19.** $\{-1, 5\}$ **20.** $\{1 \pm i\}$ **21.** $\{-3, 2\}$ **22.** $\{-1 \pm \sqrt{5}\}$

23. $\{3 \pm 2\sqrt{2}\}$ **24.** $\left\{-\dfrac{3}{2} \pm \dfrac{\sqrt{7}}{2}i\right\}$ **25.** $\{-1, 3\}$ **26.** $\{1 \pm i\}$ **27.** $\left\{-\dfrac{3}{2} \pm \dfrac{\sqrt{3}}{2}\right\}$

28. $\left\{\dfrac{5}{4} \pm \dfrac{\sqrt{17}}{4}\right\}$

Practice Exercises

29. $\{-2, 4\}$ **31.** $\{1\}$ **33.** $\{1, 3\}$ **35.** $\{-1, 4\}$ **37.** $\{1, 4\}$ **39.** $\left\{\dfrac{1}{2} \pm \dfrac{\sqrt{5}}{2}\right\}$ **41.** $\left\{-\dfrac{5}{4} \pm \dfrac{\sqrt{57}}{4}\right\}$

43. $\left\{\dfrac{2}{3} \pm \dfrac{\sqrt{26}}{3}i\right\}$ **45.** $\left\{1, \dfrac{5}{2}\right\}$ **47.** $\left\{\dfrac{3}{4} \pm \dfrac{\sqrt{31}}{4}i\right\}$ **49.** ± 21.071 **51.** $\pm 2.449i$

53. ± 0.950

55. (a) The ad is 1.25 in. by 3.75 in. (b) The ad will cost \$130.78. (c) The cost is \$199.83. (d) No.

57. The distance is approximately 17.7 in.   **59.** One side would be approximately 209 ft.

61. The leg is $5\sqrt{2}$ m.   **63.** The width is approximately 3.1 ft or less.

Challenge Problems

65. $\{\pm\sqrt{a + b}\}$ **67.** $\{-1 \pm \sqrt{a}\}$ **69.** $\{-b \pm 2\sqrt{2}\}$ **71.** $\left\{-\dfrac{a}{2} \pm \dfrac{\sqrt{a^2 + a}}{2}\right\}$

73. $\{0.3 \pm \sqrt{0.0775}\}$ **75.** $\left\{\dfrac{-b \pm \sqrt{b^2 - 4ac}}{2a}, a \neq 0\right\}$

PROBLEM SET 5.4

Warm-Ups

1. $\{-4, 2\}$ **2.** $\{\pm 2\}$ **3.** $\{-2, 0\}$ **4.** $\left\{\dfrac{1}{2} \pm \dfrac{\sqrt{5}}{2}\right\}$ **5.** $\left\{-\dfrac{1}{2} \pm \dfrac{\sqrt{5}}{2}\right\}$ **6.** $\{-2\}$ **7.** $\left\{\dfrac{1}{2} \pm \dfrac{\sqrt{3}}{2}i\right\}$

8. $\left\{\dfrac{3}{2} \pm \dfrac{\sqrt{39}}{2}i\right\}$ **9.** $\{\pm i\}$ **10.** $\{\pm\sqrt{2}i\}$ **11.** $\left\{-\dfrac{5}{2} \pm \dfrac{\sqrt{11}}{2}i\right\}$ **12.** $\left\{\dfrac{1}{2} \pm \dfrac{\sqrt{11}}{2}i\right\}$

13. $\{1 \pm \sqrt{2}\}$ **14.** $\{-1 \pm \sqrt{2}\}$ **15.** $\{-2 \pm \sqrt{6}\}$ **16.** $\{-3 \pm i\}$ **17.** $\{-1 \pm i\}$

18. $\{-2 \pm \sqrt{6}i\}$ **19.** $\{1 \pm \sqrt{3}i\}$ **20.** $\{1 \pm \sqrt{5}i\}$ **21.** $\left\{-7, \dfrac{3}{2}\right\}$ **22.** $\left\{-\dfrac{2}{5}, \dfrac{3}{2}\right\}$

23. $\{4 \pm \sqrt{15}\}$ **24.** $\left\{\dfrac{1}{2} \pm \dfrac{\sqrt{5}}{2}\right\}$ **25.** $\left\{-2, \dfrac{3}{2}\right\}$ **26.** $\left\{-\dfrac{5}{2} \pm \dfrac{\sqrt{21}}{2}\right\}$

27. -15; two complex solutions   **28.** 8; two real solutions   **29.** 41; two real solutions
30. 0; one real solution   **31.** $[3 - \sqrt{5}, 3 + \sqrt{5}]$ **32.** $(-\infty, 3 - \sqrt{2}) \cup (3 + \sqrt{2}, +\infty)$
33. $\emptyset$ **34.** $(-\infty, +\infty)$

Practice Exercises

35. $\{-3\}$ **37.** $\left\{-\dfrac{1}{2}, 1\right\}$ **39.** $\{-2, 4\}$ **41.** $\left\{-\dfrac{2}{5}, 3\right\}$ **43.** $\left\{-\dfrac{2}{3}, \dfrac{1}{2}\right\}$ **45.** $\{0, 1\}$ **47.** $\{\pm 5i\}$

49. $\left\{-\dfrac{1}{2} \pm \dfrac{\sqrt{17}}{2}\right\}$ **51.** $\left\{\dfrac{3}{2} \pm \dfrac{\sqrt{11}}{2}i\right\}$ **53.** $\{1 \pm i\}$ **55.** $\{3 \pm i\}$ **57.** $\left\{-\dfrac{7}{4} \pm \dfrac{\sqrt{41}}{4}\right\}$

59. $\{-0.3, 0.2\}$ **61.** $\{-7, -1\}$ **63.** $\{\pm\sqrt{3}\}$ **65.** $\left\{-\dfrac{1}{3}, 1\right\}$ **67.** $\left\{0, \dfrac{1}{2}\right\}$ **69.** $\{\pm 2\}$

71. $\left\{\dfrac{1}{2} \pm \dfrac{1}{2}i\right\}$ **73.** $\left\{-\dfrac{2}{3}\right\}$ **75.** $[3 - \sqrt{3}, 3 + \sqrt{3}]$ **77.** $(-\infty, +\infty)$
79. $(-\infty, -2 - \sqrt{6}) \cup (-2 + \sqrt{6}, +\infty)$ **81.** $\emptyset$ **83.** b^2; real   **85.** $k > 1$ **87.** $k = 1$
89. Approximately 3 sec   **91.** Approximately 12.71 amps or approximately 1.89 amps

Challenge Problems

93. $x = \dfrac{-3 \pm \sqrt{9 + 8a}}{2a}$ **95.** $x = -n$ or $x = m$ **97.** $L = \dfrac{d \pm \sqrt{d^2 + 4D^2a^2}}{2D}$

99. $d = \pm\dfrac{\sqrt{2 - LR^2}}{R}$ **101.** $-\dfrac{b}{a}$ **103.** Sum $= -3/2$; product $= 2$
105. Sum $= 2/5$; product $= 17/5$

PROBLEM SET 5.5

Warm-Ups

1. $\{-4, -3, 5\}$ **2.** $\{\pm 1, 5\}$ **3.** $\left\{-17, -\dfrac{3}{2}, 0\right\}$ **4.** $\left\{0, \dfrac{17}{41}\right\}$ **5.** $\{\pm 1, \pm 3\}$ **6.** $\{\pm\sqrt{5}\}$

7. $\{\pm 2, \pm i\}$ **8.** $\{\pm\sqrt{3}, \pm\sqrt{2}i\}$ **9.** $\left\{3, -\dfrac{3}{2} \pm \dfrac{3\sqrt{3}}{2}i\right\}$ **10.** $\left\{0, 1, -\dfrac{1}{2}, \pm\dfrac{\sqrt{3}}{2}i\right\}$

11. $\{x \mid -2 < x < 0 \text{ or } x > 1\}$ **12.** $\{x \mid -5 < x < 0 \text{ or } x > 2\}$
13. $\{x \mid -4 < x < -3 \text{ or } x > 1\}$ **14.** $\{x \mid x \le 1 \text{ or } 2 \le x \le 3\}$

Practice Exercises

15. $\{\pm\sqrt{2}, \pm\sqrt{3}i\}$ **17.** $\{\pm 1, \pm 2\sqrt{3}\}$ **19.** $\{0, \pm\sqrt{2}i\}$ **21.** $\{\pm\sqrt{2}, \pm\sqrt{2}i\}$ **23.** $\left\{\pm\dfrac{\sqrt{3}}{2}, \pm\sqrt{2}i\right\}$

25. $\left\{\pm\dfrac{\sqrt{5}}{2}, \pm\dfrac{\sqrt{3}}{2}i\right\}$ **27.** $\{\pm\sqrt{2}, \pm\sqrt{2}i\}$ **29.** $\left\{-1, \dfrac{1}{2} \pm \dfrac{\sqrt{3}}{2}i\right\}$ **31.** $\left\{5, -\dfrac{5}{2} \pm \dfrac{5\sqrt{3}}{2}i\right\}$

33. $\left\{-3, 0, \dfrac{3}{2} \pm \dfrac{3\sqrt{3}}{2}i\right\}$ **35.** $\{\pm 1, \pm 6\}$ **37.** $\{0, \pm\sqrt{3}\}$ **39.** $\{\pm\sqrt{7}, \pm\sqrt{6}i\}$ **41.** $\{\pm 2, \pm\sqrt{3}i\}$

43. $\{\pm\sqrt{3}, \pm\sqrt{2}i\}$ **45.** $\left\{\pm 1, \pm\dfrac{\sqrt{3}}{2}i\right\}$ **47.** $\left\{\pm\dfrac{\sqrt{7}}{3}, \pm\sqrt{2}i\right\}$ **49.** $\{\pm 2, \pm 2i\}$

51. $\left\{-3, \dfrac{3}{2} \pm \dfrac{3\sqrt{3}}{2}i\right\}$ **53.** $\left\{-5, \dfrac{5}{2} \pm \dfrac{5\sqrt{3}}{2}i\right\}$ **55.** $\left\{0, 1, -\dfrac{1}{2} \pm \dfrac{\sqrt{3}}{2}i\right\}$ **57.** $\{\pm\sqrt{5}, \pm\sqrt{5}i\}$

59. $[-3, 1] \cup [7, +\infty)$ **61.** $(-1, 0) \cup (1, +\infty)$ **63.** $[-2, 0] \cup [2, +\infty)$

65. They are 5 ft by 25 ft.   **67.** The height is 48 ft.   **69.** At most 3 ft high with a 3 ft² base

71. The length should be greater than or equal to 3 ft.

Challenge Problems

73. $\{\pm 2, 1 \pm \sqrt{3}i, -1 \pm \sqrt{3}i\}$ **75.** $\{2, \pm 2i\}$ **77.** $\{-3, -2, 0, 1\}$ **79.** $(x - 4)(x - 2)(x + 1) = 0$

81. $\{x \mid x > -2\}$ **83.** $\{x \mid x \geq 2\}$

PROBLEM SET 5.6

Warm-Ups

1. $\{-1, 6\}$ **2.** $\left\{1, \dfrac{7}{2}\right\}$ **3.** $\{-6, 1\}$ **4.** $\{-2, 6\}$ **5.** $\left\{-3, -\dfrac{1}{3}\right\}$ **6.** $\{-5\}$ **7.** $\{5\}$ **8.** $\{-6\}$

9. $\{6\}$ **10.** $\varnothing$ **11.** $\varnothing$ **12.** $\varnothing$ **13.** $\varnothing$ **14.** $(-\infty, -1] \cup (0, 2]$ **15.** $[-4, 0) \cup [3, +\infty)$

16. $\left\{-1, \dfrac{1}{2}\right\}$ **17.** $\left\{-\dfrac{1}{6}, \dfrac{1}{4}\right\}$ **18.** $\left\{-\dfrac{1}{3}, \dfrac{1}{6}\right\}$ **19.** $\{\pm 1\}$ **20.** $\left\{\pm\dfrac{1}{2}, \pm i\right\}$

21. The ship's speed is 6 knots.   **22.** It traveled at 50 mph   **23.** His walking speed was 4 mph.

24. It was 3 mph   **25.** The larger pipe will take 6 hr and the smaller pipe 12 hr.

26. Alex 3 hr; Odette 6 hr

Practice Exercises

27. $\left\{-2, \dfrac{5}{2}\right\}$ **29.** $\left\{-1, -\dfrac{1}{6}\right\}$ **31.** $\{-3, -2\}$ **33.** $\{-2, 3\}$ **35.** $\varnothing$ **37.** $\{-4, 2\}$ **39.** $\{-2\}$

41. $\left\{-\dfrac{4}{5}, 1\right\}$ **43.** $\{-2\}$ **45.** $\left\{-3, 5\right\}$ **47.** $\{\pm 3\}$ **49.** $\{1\}$ **51.** $\{-2, 7\}$ **53.** $\{-6\}$

55. $\varnothing$ **57.** $\left\{-\dfrac{3}{2}, 0\right\}$ **59.** $\{-1\}$ **61.** $\{-2, 0\}$ **63.** $\{-1\}$ **65.** $\left\{-\dfrac{1}{2}, \dfrac{1}{5}\right\}$ **67.** $\left\{\dfrac{1}{7}, \dfrac{2}{3}\right\}$

69. $\left\{\dfrac{4}{9}, \dfrac{1}{3}\right\}$ **71.** $\left\{\dfrac{1}{5}\right\}$ **73.** $\left\{\pm 1, \pm\dfrac{\sqrt{2}}{3}\right\}$ **75.** $(-\infty, -1) \cup (0, 1) \cup (3, +\infty)$

77. $(-\infty, -5] \cup [-2, 2) \cup (5, +\infty)$ **79.** $[-2, -1] \cup (0, +\infty)$ **81.** His speed on his bike was 6 mph.

83. Her original speed was 16 mph.   **85.** It will take Kathy 9 hr and Jim 18 hr.

Challenge Problems

87. $\left\{\pm\dfrac{\sqrt{2}}{3}, \pm\dfrac{\sqrt{2}}{5}\right\}$ **89.** $\left\{\pm\dfrac{3\sqrt{2}}{2}, \pm i\right\}$ **91.** $\left\{\pm\dfrac{2\sqrt{5}}{5}, \pm\dfrac{2\sqrt{3}}{3}i\right\}$ **93.** $\{6\}$ **95.** $\{-3, 0\}$

97. $\{0\}$ **99.** $\{-2a, a\}$ **101.** $\left\{-2, -\dfrac{3}{2}, -\dfrac{1}{2}, 0\right\}$ **103.** $\{-1, 2\}$

105. His rate in still water is 10 mph.

PROBLEM SET 5.7

Warm-Ups

1. $\{1, 2\}$ **2.** $\{1\}$ **3.** $\{2\}$ **4.** $\{1, 3\}$ **5.** $\{-3, -1\}$ **6.** $\{0\}$ **7.** $\left\{\dfrac{1}{2}, 3\right\}$ **8.** $\{3\}$

9. $\left\{-\dfrac{1}{2}, -\dfrac{1}{3}\right\}$ **10.** $\{2\}$ **11.** $\emptyset$ **12.** $\{4\}$ **13.** $\{1, 5\}$ **14.** $\{3, 27\}$ **15.** $\{7, 15\}$ **16.** $\{3, 7\}$

17. $\emptyset$ **18.** $\{0\}$ **19.** $\{8\}$ **20.** $\{-27\}$ **21.** $\{14\}$ **22.** $\{\pm 3\}$

Practice Exercises

23. $\{2\}$ **25.** $\{1, 9\}$ **27.** $\{4\}$ **29.** $\{8\}$ **31.** $\{1, 3\}$ **33.** $\{-2\}$ **35.** $\{5\}$ **37.** $\{-3\}$ **39.** $\{0\}$

41. $\{2, 18\}$ **43.** $\{0\}$ **45.** $\{1\}$ **47.** $\left\{\dfrac{7}{3}\right\}$ **49.** The equilibrium price is \$5.50.

Challenge Problems

51. $\{0, 4\}$ **53.** $\left\{\pm\dfrac{1}{2}\right\}$ **55.** $\{-1, 8\}$ **57.** $\{-1, 243\}$

REVIEW PROBLEMS CHAPTER 5

1. $\left\{-\dfrac{2}{3}, \dfrac{1}{4}\right\}$ **3.** $(-\infty, 0) \cup (7, +\infty)$ **5.** $\left\{\dfrac{-2 \pm \sqrt{13}}{3}\right\}$ **7.** $\{-3, -1\}$ **9.** $\left\{-\dfrac{3}{2}\right\}$

11. $\left\{\dfrac{1}{8} \pm \dfrac{\sqrt{15}}{8}i\right\}$ **13.** $\left\{-\dfrac{5}{3}, \dfrac{1}{3}\right\}$ **15.** $\{4\}$ **17.** $\left\{-\dfrac{5}{6}, \dfrac{1}{3}\right\}$ **19.** $\left\{\pm\dfrac{3}{2}\right\}$ **21.** $\left\{-2, \dfrac{3}{2}\right\}$

23. $\{-3\}$ **25.** $\left\{\dfrac{1}{2}, -\dfrac{1}{4} \pm \dfrac{\sqrt{3}}{4}i\right\}$ **27.** $\left\{-2, -\dfrac{2}{3}\right\}$ **29.** $\{2\}$ **31.** $\left(-\infty, -\dfrac{2}{3}\right] \cup [1, +\infty)$

33. $\{\pm 7, \pm 1\}$ **35.** $\{0, 4\}$ **37.** $\{2, 10\}$ **39.** $\{-3\}$ **41.** $\left\{-\dfrac{1}{2} \pm \dfrac{\sqrt{15}}{10}i\right\}$ **43.** $\left\{\dfrac{1}{2}, -\dfrac{4}{3}\right\}$

45. $\{2 \pm \sqrt{6}\}$ **47.** $\{1, 9\}$ **49.** $\{\pm\sqrt{3}, \pm\sqrt{5}\}$ **51.** $\{-3 \pm \sqrt{5}i\}$ **53.** $\left\{\pm\dfrac{\sqrt{7}}{4}, \pm i\right\}$

55. $(-3 - 2] \cup (1, 7]$ **57.** $\{7 \pm 2\sqrt{2}i\}$ **59.** The numbers are -13 and 30. **61.** The leg is $\sqrt{14}$ m.

63. His speed was 24 mph. **65.** It will be greater after 30 years.

Let's Not Forget . . .

66. $(5 - 2x)(25 + 10x + 4x^2)$ **67.** Factored **68.** Factored **69.** $(4 - x - 1)(4 + x + 1)$

70. $(x + y)^2$ **71.** $(c - d)(y + x)$ **72.** Two; $x - y$ is a factor. **73.** Three; $x - y$ is not a factor.

74. Two; $x - y$ is a factor. **75.** Two; $x - y$ is not a factor. **76.** $2 \pm \sqrt{3}i$ **77.** $\dfrac{-3}{xy^3}$

78. $\dfrac{-9}{a^2 b^4}$ **79.** $\dfrac{b^2}{a^3} + \dfrac{b^3}{a}$ **80.** $-\dfrac{1}{8}$ **81.** Cannot be simplified. **82.** $x + 2x^{1/2}y^{1/2} + y$

83. $27y^3 - 27y^2 + 9y - 1$ **84.** Will not reduce **85.** $\dfrac{x - y}{b - c}$ **86.** $\{2\}$ **87.** $\dfrac{-x^2 - x + 9}{(x + 4)(x + 3)}$

88. $\dfrac{2x^2 + 11x + 12}{4x + 12}$ **89.** Expression; $3 - i$ **90.** Equation; $\{-5, 2\}$ **91.** Expression; $\dfrac{-2}{(x - 7)(x + 2)}$

92. Equation; {2, 10}   **93.** Equation; {−9, 15}   **94.** Equation; {−4, −1}   **95.** Equation; $\left\{\dfrac{1}{7}\right\}$

CHAPTER 5 TEST

1. B   **2.** A   **3.** D   **4.** C   **5.** C   **6.** $\left(-\dfrac{2}{3}, 2\right)$ **7.** $\left\{-\dfrac{5}{2}, 1\right\}$ **8.** {0, 6}

9. $(-\infty, 0) \cup (3, +\infty)$ **10.** $\{2, -1 \pm \sqrt{3}i\}$ **11.** $\left\{\pm 1, \pm\dfrac{1}{2}i\right\}$ **12.** {3}   **13.** $\{1 \pm \sqrt{2}\}$ **14.** 9; B

15. Her original speed was 20 mph.

► Chapter 6

PROBLEM SET 6.1

Warm-Ups

1. I   **2.** II   **3.** IV   **4.** III   **5.** x-axis   **6.** y-axis   **7.** IV   **8.** III   **9.** Both x and y-axes

10.
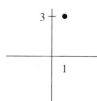

11.
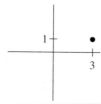

12.
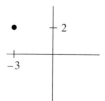

13.
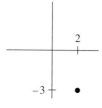

14.
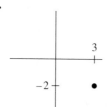

15.

16.
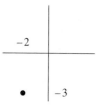

17.
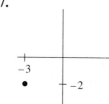

18.
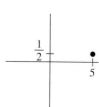

19.
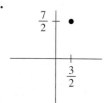

20.
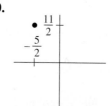

21.
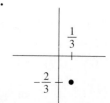

22.

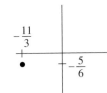

$-\frac{11}{3}$

$-\frac{5}{6}$

23.

4.6

−2.3

24.

−2.5

−0.1

25. 5   **26.** $3\sqrt{5}$ **27.** 10   **28.** 5   **29.** 13   **30.** 4   **31.** $2\sqrt{2}$ **32.** $\sqrt{82}$ **33.** $7\sqrt{2}$

34. $3\sqrt{2}$ **35.** $\sqrt{5}$ **36.** $\dfrac{\sqrt{85}}{3}$ **37.** $\dfrac{\sqrt{26}}{6}$ **38.** $(4, 1)$ **39.** $(1, 0)$ **40.** $(-4, 10)$

Practice Exercises

41. 10   **43.** 13   **45.** 8   **47.** $3\sqrt{5}$ **49.** 5   **51.** $\sqrt{2}$ **53.** 9   **55.** $\sqrt{82}$ **57.** $7\sqrt{2}$

59. $\sqrt{37}$ **61.** $\dfrac{5\sqrt{2}}{12}$ **63.** $(-3, -7)$ **65.** $\left(\dfrac{\sqrt{5}}{2}, \dfrac{3}{2}\right)$; 3.74   **67.** $\left(-\dfrac{5}{8}, \dfrac{\pi + 1}{2}\right)$; 5.21

Challenge Problems

69. I   **71.** II   **73.** III   **75.** III   **77.** I   **79.** *x*-axis

81. They are collinear because the distance between P_1 and P_2 is $2\sqrt{5}$ units, the distance between P_1 and P_3 is $\sqrt{5}$ units, and the distance between P_2 and P_3 is $\sqrt{5}$ units.

83. It is not a right triangle because the Pythagorean Theorem does not hold.

PROBLEM SET 6.2

Warm-Ups

1.

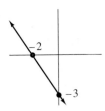

2.

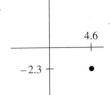

3.

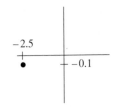

4.

5.

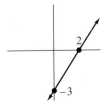

6.

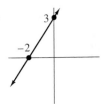

7.

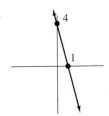

8.

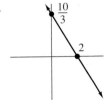

9.

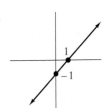

10.

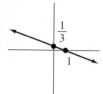

11.

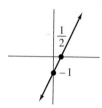

12.

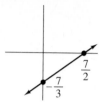

13.

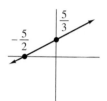

14.

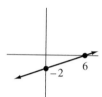

15.

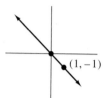

16.

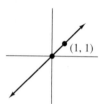

17.

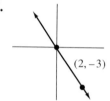

18.

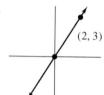

19.

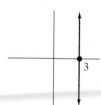

20.

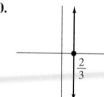

21.

22.

23.

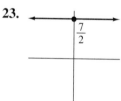

24.

25. (a)

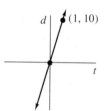

(b) It will take 7 hours.

26. (a)

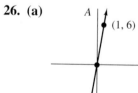

(b) The length is 16m.

27. (a)

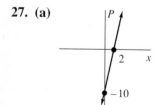

(b) Two shirts must be sold to break even.
(c) The profit is $490.
(d) The profit is a loss of $5.

28. (a)

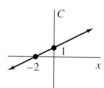

(b) The cost is $11.

Practice Exercises

29.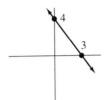

31.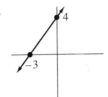

33.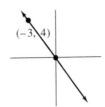

35.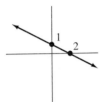

37.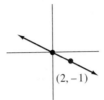

39.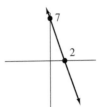

41.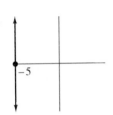

43.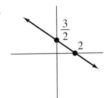

45.

47.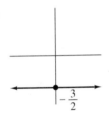

49.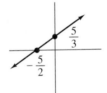

51.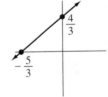

53.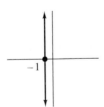

55.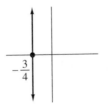

57. (a)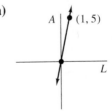

(b) The length will be 21 meters.

59. (a)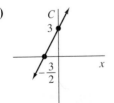

(b) The cost is $33.

61. (a)

(b) $13,000

Challenge Problems

63.

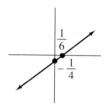

65.

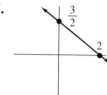

67.

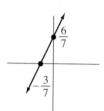

69.

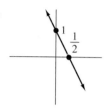

71.

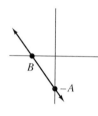

73.

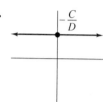

PROBLEM SET 6.3

Warm-Ups

1. 3   **2.** −3   **3.** $\frac{1}{3}$ **4.** $-\frac{1}{3}$ **5.** 2   **6.** 2   **7.** −2   **8.** $\frac{2}{5}$ **9.** $-\frac{1}{2}$ **10.** $\frac{3}{2}$ **11.** $-\frac{5}{2}$

12. $\frac{3}{4}$ **13.** $-\frac{9}{4}$ **14.** 2   **15.** 0   **16.** 0   **17.** 0   **18.** Undefined slope   **19.** Undefined slope

20. Undefined slope   **21.** $-\frac{2}{3}$ **22.** 5   **23.** $-\frac{1}{7}$ **24.** $-\frac{2}{3}$ **25.** 5   **26.** $-\frac{1}{7}$ **27.** $\frac{4}{7}$ **28.** $\frac{1}{2}$

29. 1   **30.** 0   **31.** Undefined slope   **32.** 0

33. $m = 1, b = -1$ **34.** $m = -1, b = 1$ **35.** $m = \frac{1}{2}, b = \frac{3}{2}$

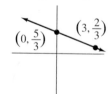   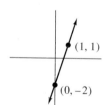

36. $m = -\frac{1}{3}, b = \frac{5}{3}$ **37.** $m = 3, b = -2$ **38.** $m = -4, b = 0$

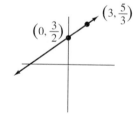   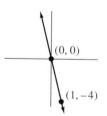

39. 8   **40.** $-\frac{4}{3}$ **41.** 0   **42.** Undefined slope   **43.** $\frac{3}{4}$ **44.** $\frac{2}{5}$

Practice Exercises

45. −2   **47.** $-\frac{5}{3}$ **49.** $\frac{-2}{3}$ **51.** −4/3   **53.** −3/4   **55.** 0   **57.** −1/2   **59.** 3/4

61. 1   **63.** Undefined slope   **65.** 0   **67.** −3/4   **69.** −1/5   **71.** 4   **73.** −3/7   **75.** 3/2

77. Undefined slope   **79.** $-\dfrac{3}{2}$ **81.** $m = 1, b = -2$ **83.** $m = \dfrac{1}{3}, b = \dfrac{2}{3}$ **85.** $m = -3, b = 2$

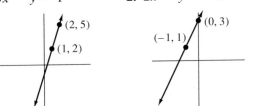

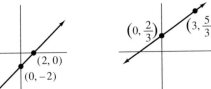

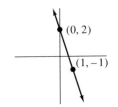

Challenge Problems

87. $y = -\dfrac{1}{2}x + 1$

PROBLEM SET 6.4

Warm-Ups

1. $3x - y = 1$ **2.** $2x - y = -3$ **3.** $2x + y = 11$ **4.** $x + y = -7$

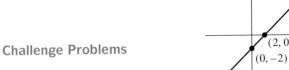

5. $x - 2y = 16$ **6.** $5x - 5y = -3$ **7.** $x - y = -1$ **8.** $x + 3y = 11$ **9.** $x - y = 5$

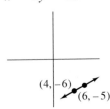

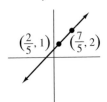

10. $2x - 5y = 10$ **11.** $5x - y = 0$ **12.** $7x - 5y = 3$ **13.** $2x - y = -14$ **14.** $6x - y = 11$

15. $3x - 2y = -7$ **16.** $x - 3y = 37$ **17.** $y = \dfrac{1}{2}x + 2$ **18.** $y = -4x + 3/2$ **19.** $y = -1$

20. $x = -4$ **21.** $\dfrac{5}{9}F - C = \dfrac{160}{9}$ **22.** $E - 60t = 240$; $840 million

Practice Exercises

23. $2x - y = 3$ **25.** $x + y = 7$ **27.** $x + 2y = -4$ **29.** $x - y = -1$ **31.** $3x - y = 9$
33. $2x - y = 0$ **35.** $x - 3y = -12$ **37.** $2x - 2y = -1$ **39.** $3x - 2y = -6$
41. $4x - y = -8$ **43.** $2x - y = -6$ **45.** $x = 2$ **47.** $x - y = 0$ **49.** $y = 0$
51. $x + 4y = 44$ **53.** $4t - 10S = -25$. Sales will be $6.5 billion.
55. (a) $C = 0.04P + 1.35$; (b) $2.95; (c) 50 lb; (d) The bill increases as weight increases.

Challenge Problems

57. $qx - py = 0$ **59.** 3; 2   **61.** $\dfrac{1}{2}; \dfrac{1}{3}$ **63.** $\dfrac{5}{3}$; no y-intercept

PROBLEM SET 6.5

Warm-Ups

1. Yes; (10, 100), (20, 400)   **2.** Yes; (3, 2), (2, 2)   **3.** Yes $(\sqrt{5}, 1)\left(\frac{1}{2}, 1\right)(10, \sqrt{7}), (12, \sqrt{7})$

4. No; (4, 5), (4, 6), (0, 7)   **5.** -1 **6.** -11 **7.** -5 **8.** -4 **9.** 2   **10.** 1   **11.** 0
12. 12   **13.** 6   **14.** 27   **15.** 3   **16.** 9   **17.** $2t^2 + t - 1$ **18.** $2(a + b)^2 + (a + b) - 1$
19. $2a^2 + a + 2b^2 + b - 2$ **20.** $8t^2 + 2t - 1$ **21.** $2a + \sqrt{a} - 1$ **22.** $\mathbb{R}$ **23.** $\mathbb{R}$ **24.** $\mathbb{R}$
25. $\mathbb{R}$ **26.** $\{x \mid x \neq -2\}$ **27.** $\{x \mid x \neq 0\}$ **28.** $\{x \mid x \neq 1\}$ **29.** $\{x \mid x \neq -2 \text{ and } x \neq 7\}$
30. $\{x \mid x \geq -2\}$ **31.** $\{x \mid x \geq 1\}$ **32.** $\{x \mid x \leq -2 \text{ or } x \geq 3\}$ **33.** $\{x \mid x \leq -2 \text{ or } x \geq 1\}$

Practice Exercises

35. $\{-6, -3, 0, 4, 7\}$ **37.** $\{y \mid y \text{ is an odd natural number}\}$ **39.** $\{11\}$ **41.** $\mathbb{R}$ **43.** $\mathbb{R}$
45. $\{x \mid x \neq 1\}$ **47.** $\{x \mid x \neq 3\}$ **49.** $\{x \mid x \neq -3\}$ **51.** $\{x \mid x \geq 2\}$ **53.** $\mathbb{R}$ **55.** $\mathbb{R}$ **57.** $\mathbb{R}$
59. $\mathbb{R}$ **61.** 7   **63.** 1   **65.** $3\sqrt{3} + 1$ **67.** $3(a - b) + 1$ **69.** $6t + 1$ **71.** $\sqrt{3a} + 1$
73. $\sqrt{5}$ **75.** 1   **77.** 3   **79.** $\sqrt{2k + 2}$ **81.** $a - 1$ **83.** $1 - 2(t^2 + 2th + h^2)$ **85.** $-4th - 2h^2$
87. $2 - 4t^2$ **89.** $k - 2kt^2$ **91.** $-10t^2$ **93.** -1 **95.** Undefined   **97.** 18   **99.** Undefined
101. 26   **103.** 144   **105.** 128
107. (a) $4.00 (b) $F(3)$ is $2.25, which is the fare for a 3-mile ride. (c) 6 miles
109. (a) 0 (b) 2 (c) $N(10)$ is 810, which is the number of ants after 10 minutes.

Challenge Problems

111. $\{x \mid x > 2\}$

PROBLEM SET 6.6

Warm-Ups

1.
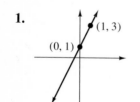

2.
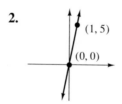

3.
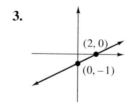

4.
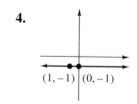

5. (a)
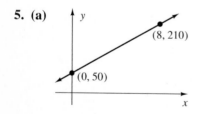

6. (a)
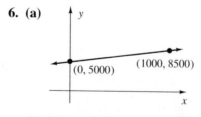

(b) The tank started with 50 gal of oil. (c) 8 min

(b) The start-up cost is $5000.

7. D: $\mathbb{R}$; R: $\{y \mid y \geq -1\}$ **8.** D: $\mathbb{R}$; R: $\mathbb{R}$
9. D: $\mathbb{R}$; R: $\{y \mid y \geq -2\}$ **10.** D: $\mathbb{R}$; R: $\{y \mid y \leq 4\}$

11. D: $\{x \mid x \geq 0\}$; R: $\{y \mid y \geq 0\}$ **12.** D: $\{x \mid -3 \leq x \leq 3\}$; R: $\{y \mid -3 \leq y \leq 0\}$
13. Yes   **14.** No   **15.** No   **16.** No   **17.** Yes   **18.** Yes

19.
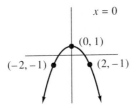

20.
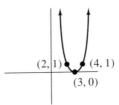

21.
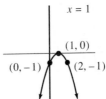

22.
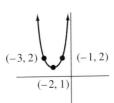

23.
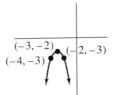

24.
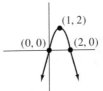

25.
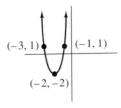

26.
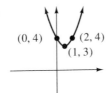

27.
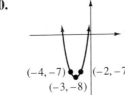

28.
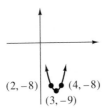

29.
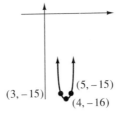

30.
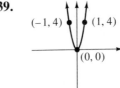

31.
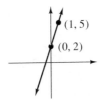

32.
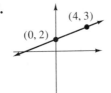

33. $\{-2, 4\}$ **34.** $\{-6, 5\}$

Practice Exercises

35.
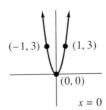

D: $\mathbb{R}$
R: $\mathbb{R}$

37.
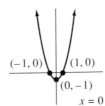

D: $\mathbb{R}$
R: $\mathbb{R}$

39.

D: $\mathbb{R}$
R: $\{y \mid y \geq 0\}$

41.

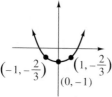

$\left(-1, -\frac{2}{3}\right)$ $\left(1, -\frac{2}{3}\right)$
$(0, -1)$

D: $\mathbb{R}$
R: $\{y \mid y \geq -1\}$

43.

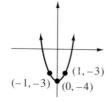

$(1, -3)$
$(-1, -3)$ $(0, -4)$

D: $\mathbb{R}$
R: $\{y \mid y \geq -4\}$

45.

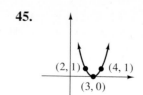

$(2, 1)$ $(4, 1)$
$(3, 0)$

D: $\mathbb{R}$
R: $\{y \mid y \geq 0\}$

47.

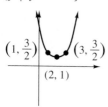

$\left(1, \frac{3}{2}\right)$ $\left(3, \frac{3}{2}\right)$
$(2, 1)$

D: $\mathbb{R}$
R: $\{y \mid y \geq 1\}$

49.

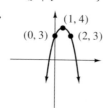

$(1, 4)$
$(0, 3)$ $(2, 3)$

D: $\mathbb{R}$
R: $\{y \mid y \leq 4\}$

51.

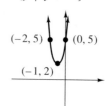

$(-2, 5)$ $(0, 5)$
$(-1, 2)$

D: $\mathbb{R}$
R: $\{y \mid y \geq 2\}$

53.

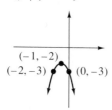

$(-1, -2)$
$(-2, -3)$ $(0, -3)$

D: $\mathbb{R}$
R: $\{y \mid y \leq -2\}$

55.

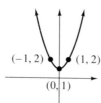

$(-1, 2)$ $(1, 2)$
$(0, 1)$

Minimum y value is 1.

57. (a) The baby would weigh approximately 19.32 pounds **(b)** No, it is not reasonable. A 20-year-old would weigh 9050.4 pounds.

59. (a) The revenue equation is $R(x) = -3x^2 + 600x$. **(b)** The maximum revenue is $30,000.   **61.** 244 ft

63. (a) _____

x	h(x)
20	24
60	48
100	56
140	48
180	24
200	6

(b) The javelin is 48 feet high when it is 60 feet from the foul line.
(c) When the javelin is 20 feet from the foul line, the javelin is 24 feet high.
(d) As the javelin goes up and then comes back down, at two different distances from the foul line it will be 48 feet high.

Challenge Problems

65.

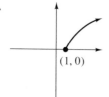

$(1, 0)$

D: $\{x \mid x \geq 1\}$
R: $\{y \mid y \geq 0\}$

67.

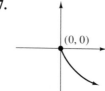

$(0, 0)$

D: $\{x \mid x \geq 0\}$
R: $\{y \mid y \leq 0\}$

69.

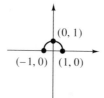

$(0, 1)$
$(-1, 0)$ $(1, 0)$

D: $\{x \mid -1 \leq x \leq 1\}$
R: $\{y \mid 0 \leq y \leq 1\}$

71.

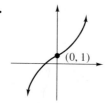

$(0, 1)$

D: $\mathbb{R}$
R: $\mathbb{R}$

73.

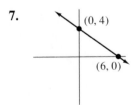

(0, 0)

D: $\mathbb{R}$
R: $\{y \mid y \geq 0\}$

75. (a) the length is $l \leq 300 - 2w$ **(b)** the area is $A \leq 300w - 2w^2$ **(c)** the largest possible area is 11,250 sq ft.

PROBLEM SET 6.7

Warm-Ups

1. $A = kr^2$ **2.** $V = kr^2h$ **3.** $V = Kr^3$ **4.** $F = \dfrac{k}{G}$ **5.** $u = \dfrac{kv}{w}$ **6.** $x = ky^2z/w$

7. x is 75 when y is 5.   **8.** 9   **9.** s is 193.2 when t is $\sqrt{2}$ and g is 32.2.
10. 10   **11.** $k = 2$. A force of 20 lb is required.   **12.** 20 amperes
13. The stone will have fallen 144.9 ft.   **14.** 30 candlepower

15. A 4-in. pendulum has a period of $\dfrac{1}{4}$ sec.

Practice Exercises

17. $s = kv$ **19.** $V = klwh$ **21.** $h = \dfrac{k\sqrt{g}}{d}$ **23.** x is $\dfrac{18}{25}$.

25. N is $\dfrac{32}{9}$ when l is $\dfrac{\sqrt{2}}{2}$ and M is $\dfrac{3}{2}$.   **27.** its slope is $\dfrac{4}{3}$.

29. a force of $\dfrac{5}{2}$ lb is required.

REVIEW PROBLEMS CHAPTER 6

1. $5, \left(3, \dfrac{5}{2}\right)$ **3.** $\sqrt{5}, \left(\dfrac{3}{2}, 2\right)$ **5.** $\sqrt{2}, \left(\dfrac{17}{2}, -\dfrac{5}{2}\right)$

In Problems 7 to 15 the x-intercept is listed first.

7.

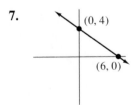

(0, 4)
(6, 0)

9.

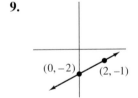

(0, −2)   (2, −1)

11.

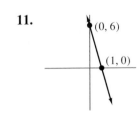

(0, 6)
(1, 0)

13.

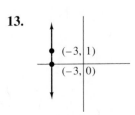

(−3, 1)
(−3, 0)

15.

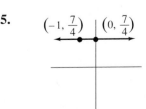

$\left(-1, \dfrac{7}{4}\right)$ $\left(0, \dfrac{7}{4}\right)$

17. -1 **19.** -2 **21.** $\dfrac{3}{5}$ **23.** $-\dfrac{9}{5}$ **25.** Slope $\dfrac{2}{3}$; y-intercept, -2 **27.** Slope 0; y-intercept, $-\dfrac{3}{2}$

29. $x + y = 2$ **31.** $5x - 7y = -3$ **33.** $2x + 3y = -19$ **35.** $4x + 3y = 15$

37. $\{y \mid y \text{ is a negative integer}\}$ **39.** $\mathbb{R}$ **41.** $\{x \mid x \le 0 \text{ or } x \ge 1\}$

43. $\{x \mid x \ne -1 \text{ and } x \ne 2\}$ **45.** Undefined   **47.** $\dfrac{6}{11}$ **49.** $\dfrac{3}{2}$ **51.** $\dfrac{3}{a^2 + 5}$

53.

D: $\mathbb{R}$
R: $\{y \mid y \ge 3\}$

55.

D: $\mathbb{R}$
R: $\{y \mid y \ge 3\}$

57.

D: $\mathbb{R}$
R: $\{y \mid y \ge -5\}$

59.

D: $\mathbb{R}$
R: $\{y \mid y \ge -1\}$

61. **(a)** $6x - y = 0$; **(b)** $(1, 6)$; **(c)** One fathom is equivalent to 6 ft;   **63.** u is 24

65. The force is 18 units.

Let's Not Forget . . .

67. $(3 + a)(9 - 3a + a^2)$ **68.** Factored   **69.** Factored   **70.** $a\left(\dfrac{1}{b} + ab\right)$

71. $(f - g - 3)(f - g + 3)$ **72.** $b(a^2c + d^2r)$ **73.** Two; not a factor.   **74.** Two; not a factor.

75. Three; is a factor.   **76.** Two; is a factor.   **77.** $\dfrac{1}{81}$ **78.** -81 **79.** 81 **80.** $\dfrac{1}{81}$

81. Will not simplify.   **82.** $2x\sqrt[3]{4 + 8x^2}$ **83.** a^4b^9 **84.** $\dfrac{-3b^4}{a^4}$ **85.** Will not reduce.

86. $\dfrac{x + y}{x^2 + xy + y^2}$ **87.** $\dfrac{x + 3}{x - 1}$ **88.** $\{-2, 5\}$ **89.** $-\dfrac{3}{2}$ **90.** Equation; $\{2, 8\}$

91. Expression; $\dfrac{-1}{2s - 12}$ **92.** Expression; $12 - 2\sqrt{35}$ **93.** Inequality; $\left\{x \mid -\dfrac{2}{3} < x < \dfrac{3}{2}\right\}$

94. Equation; $\{6\}$ **95.** Equation; $\{\pm 2, \pm i\}$

CHAPTER 6 TEST

1. B   **2.** C   **3.** A   **4.** B   **5.** C   **6.** C   **7.** C   **8.** B   **9.** A   **10.** A

11.   **12.**   **13.**   **14.**

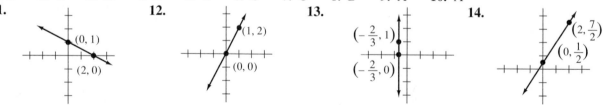

15. (a)   **(b)**

16. (a) $\{x \mid x \le -1 \text{ or } x \ge 2\}$ **(b)** $\{x \mid x \ne 7\}$

17. $y = x + 1$ **18.** $x = -2$

19. 13 **20.** $\left(-1, -\dfrac{7}{2}\right)$ **21.** $-\dfrac{2}{3}$

22. Slope $-\dfrac{5}{3}$; y-intercept $\dfrac{7}{3}$ **23.** x is 2.

► Chapter 7

PROBLEM SET 7.1

Warm-Ups

1. $\{(2, 1)\}$ **2.** $\{(4, -2)\}$ **3.** $\{(1, 2)\}$ **4.** $\{(2, 0)\}$ **5.** $\{(-3, -2)\}$ **6.** $\{(0, -1)\}$
7. $\{(x, y) \mid x + y = 4\}$ **8.** Ø   **9.** Ø   **10** $\{(3, 4), (3, -4)\}$ **11.** $\{(0, 1), (1, 2)\}$
12. $\left\{(-2, 0), \left(-\dfrac{10}{3}, -\dfrac{8}{3}\right)\right\}$ **13.** $\{(2, -1)\}$ **14.** $\{(2, 2)\}$ **15.** $\{(-1, 3)\}$ **16.** $\left\{\left(1, \dfrac{1}{2}\right)\right\}$
17. $\left\{\left(\dfrac{1}{3}, 2\right)\right\}$ **18.** $\left\{\left(-\dfrac{1}{4}, \dfrac{11}{4}\right)\right\}$ **19.** $\{(1, 1)\}$ **20.** $\{(3, 2)\}$ **21.** $\{(2, 5)\}$ **22.** $\{(12, 15)\}$
23. $\{(0, 0)\}$ **24.** $\{(-1, -1)\}$ **25.** Ø   **26.** $\{(x, y) \mid 2x - 9y = 0\}$ **27.** Ø   **28.** 12 and 18
29. 9 and -9 **30.** 16 and 48

Practice Exercises

31. $\{(1, 2)\}$ **33.** $\{(2, 1)\}$ **35.** $\{(0, -2)\}$ **37.** $\left\{\left(\dfrac{1}{2}, 2\right)\right\}$ **39.** Ø   **41.** $\left\{\left(\dfrac{3}{2}, 1\right)\right\}$ **43.** Ø
45. $\{(0, 0)\}$ **47.** $\{(1, 2), (2, 1)\}$ **49.** $\{(3, -1)\}$ **51.** $\{(2, -1)\}$ **53.** $\{(-1, 1)\}$ **55.** $\{(3, 3)\}$
57. $\left\{\left(-\dfrac{1}{3}, \dfrac{1}{3}\right)\right\}$ **59.** $\{(0, 0)\}$ **61.** $\{(0, 0)\}$ **63.** $\left\{\left(-\dfrac{2}{3}, -2\right)\right\}$ **65.** $\left\{\left(-\dfrac{1}{4}, -\dfrac{1}{5}\right)\right\}$
67. $\left\{\left(-\dfrac{1}{5}, -\dfrac{7}{5}\right)\right\}$ **69.** $\{(x, y) \mid 17x - 11y = -1\}$ **71.** $\left\{\left(\dfrac{1}{2}, \dfrac{1}{20}\right)\right\}$ **73.** $\left\{\left(0, \dfrac{15}{14}\right)\right\}$ **75.** $\{(0, 0)\}$
77. $\{(2, 2), (2, -2)\}$ **79.** $\{(0, -1), (1, 0), (-1, 0)\}$ **81.** 11 and 13   **83.** 7, 27, and 34

Challenge Problems

85. $\{(3, 1)\}$

PROBLEM SET 7.2

Warm-Ups

1. She should mix 12 liters of 17% solution and 14 liters of 30% solution.   **2.** 22 oz of 16% solution and 12 oz pure water.   **3.** A pound of tomatoes costs 99 cents and a head of lettuce costs 69 cents.
4. Snickers are 33¢ and Mr. Goodbar 35¢   **5.** The dimensions are 10 m by 17 m.   **6.** 4 yd by 10 yd.
7. The boat's rate is 5 mph and the current 3 mph.   **8.** Airplane, 225 km/hr; wind, 25 km/hr.   **9.** It takes Frank 6 hr and Jack 12 hr.   **10.** One pipe; 12 hr alone; other pipe 24 hr alone.
11. They are 8 and 11.

Practice Exercises

13. Lemons are 10 cents each and grapes are 99 cents per pound.   **15.** It takes Tom 2 hr and Randy 3 hr.
17. 120 by 480 ft   **19.** Carolyn's rate is 5 mph and Bill's is 3 mph.   **21.** The sides are 14″, 17″, and 17″.
23. 12 by 18 in.   **25.** A leotard sells for $45 and tights for $12.50.
27. They ran for $\frac{1}{2}$ hr and walked for $\frac{1}{3}$ hr.

Challenge Problems

29. It takes 40 min.

PROBLEMS SET 7.3

Warm-Ups

1. $\{(1, 2, 3)\}$ **2.** $\{(1, -1, 2)\}$ **3.** $\{(-1, 0, 2)\}$ **4.** $\{(5, -2, 2)\}$ **5.** $\left\{\left(0, 0, \dfrac{1}{3}\right)\right\}$ **6.** $\{(0, 0, 0)\}$
7. $\varnothing$ **8.** Dependent   **9.** $\varnothing$ **10.** Dependent

Practice Exercises

11. $\{(1, 3, 2)\}$ **13.** $\{(3, 1, 0)\}$ **15.** Dependent   **17.** $\varnothing$ **19.** Dependent   **21.** $\left\{\left(\dfrac{1}{2}, -\dfrac{1}{3}, -\dfrac{1}{5}\right)\right\}$

23. $\varnothing$ **25.** $\{(0, -2, 0)\}$ **27.** Dependent   **29.** $\left\{\left(\dfrac{2}{3}, \dfrac{1}{3}, -\dfrac{1}{3}\right)\right\}$ **31.** $\left\{\left(1, \dfrac{1}{2}, -2\right)\right\}$ **33.** 1, 5 and 6
35. -2, 3 and 5   **37.** Small drink costs 65¢ medium drink \$1.00; popcorn 95¢.   **39.** -1, 1, 4, 7

Challenge Problems

41. $\{(1, 3, 5, 4)\}$ **43.** $\left\{\left(0, 1, 3, \dfrac{1}{2}, 0\right)\right\}$

PROBLEM SET 7.4

Warm-Ups

1. 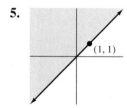   **2.** 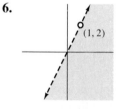   **3.** 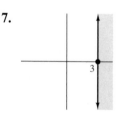   **4.**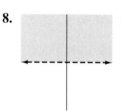

5. 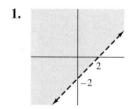   **6.** 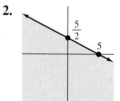   **7.**    **8.**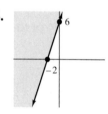

9. 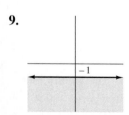   **10.**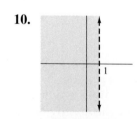

Practice Exercises

11.
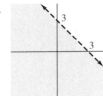

13.
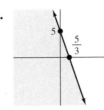

15.
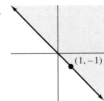

17.
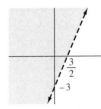

19.

21.
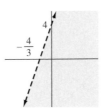

23.
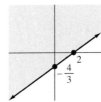

25.
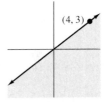

27.
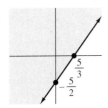

29.
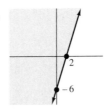

31.
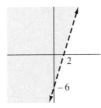

33.
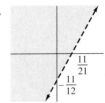

35.
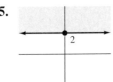

37.
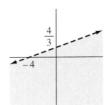

39.
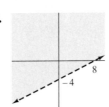

41.
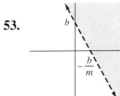

43.

45.
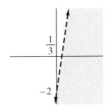

47. (a) $350g + 600v \leq 20{,}000$

(b)
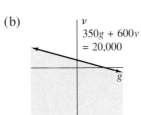

(c) She can buy no more than 40 violins.

Challenge Problems

49.
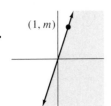

51.
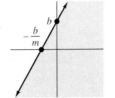

53.
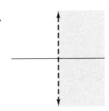

55.

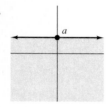

57.

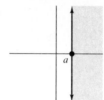

PROBLEM SET 7.5

Warm-Ups

1.

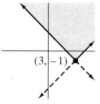

2.

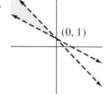

3.

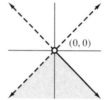

4.

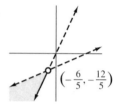

5.

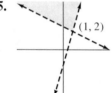

6.

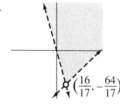

7.

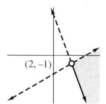

8.

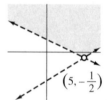

9.

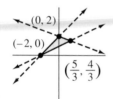

10.

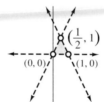

Practice Exercises

11.

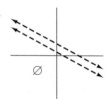

13.

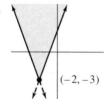

15.

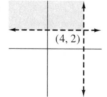

17.

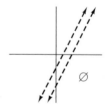

19.

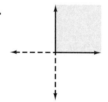

21.

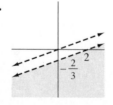

23.

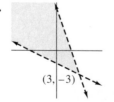

25.

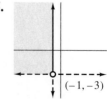

27.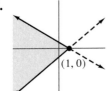

29.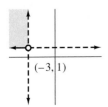

31.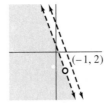

33.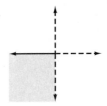

35.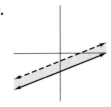

37.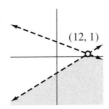

39.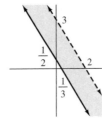

41.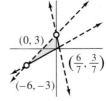

43. $D + 2P \leq 30$
$2D + P \leq 40$
$P \geq 0$
$D \geq 0$

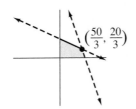

Challenge Problems

45.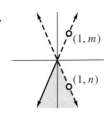

47.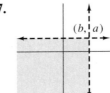

PROBLEM SET 7.6

Warm-Ups

1. 1   **2.** −2   **3.** 0   **4.** −16   **5.** 22   **6.** 1   **7.** −7   **8.** 15   **9.** −42   **10.** 0   **11.** 2
12. 303

Practice Exercises

23. 1   **25.** 0   **27.** 27   **29.** 10   **31.** 2   **33.** −6   **35.** 0   **37.** −117   **39.** −24

Challenge Problems

61. −27   **63.** 240

PROBLEM SET 7.7

Warm-Ups

1. $\begin{bmatrix} 2 & 3 & -4 \\ 5 & -6 & 1 \end{bmatrix}$ **2.** $\begin{bmatrix} 3 & -1 & 2 \\ 1 & 7 & -3 \end{bmatrix}$ **3.** $\begin{bmatrix} 2 & -3 & 4 & 1 \\ 5 & -1 & 1 & 0 \\ 7 & 6 & -2 & 5 \end{bmatrix}$ **4.** $\begin{bmatrix} 1 & 2 & 0 & -5 \\ 3 & 0 & -4 & 1 \\ 0 & 1 & 2 & 10 \end{bmatrix}$

5. $\begin{bmatrix} 1 & 2 & -1 & 0 \\ -1 & -1 & 0 & 11 \\ 0 & 1 & -4 & 13 \end{bmatrix}$ **6.** $\begin{bmatrix} 2 & 3 & -4 & 1 & 17 \\ 5 & 0 & 6 & -1 & 8 \\ 7 & -2 & 0 & -9 & 5 \\ 0 & 1 & 2 & 2 & -1 \end{bmatrix}$ **7.** $\begin{cases} x + 2y = 4 \\ 9x - 8y = -1 \end{cases}$ **8.** $\begin{cases} x - y = 7 \\ y + z = 1 \\ \phantom{y + {}} z = -4 \end{cases}$

9. $\begin{cases} x = 5 \\ y = -3 \\ z = -2 \end{cases}$ **10.** $\begin{cases} x - 2y + 3z - 4w = 5 \\ 4x - y + z \phantom{{}- 4w} = 2 \\ 2x + y \phantom{{}+ z - 4w} = 1 \\ -x + y - z + w = -1 \end{cases}$ **11.** $\{(1, 2)\}$ **12.** $\{(-1, 2)\}$ **13.** $\{(5, -2)\}$

14. $\{(7, -2)\}$ **15.** $\{(1, 2, -1)\}$ **16.** $\{(5, -1, 2)\}$

Practice Exercises

17. $\begin{bmatrix} 3 & 2 & 5 \\ 6 & -5 & 3 \end{bmatrix}$ **19.** $\begin{bmatrix} 3 & -2 & 5 & 4 \\ 2 & 3 & -1 & 0 \\ 1 & -1 & -2 & 1 \end{bmatrix}$ **21.** $\begin{bmatrix} 1 & -2 & 3 & 0 \\ -1 & 0 & -1 & 10 \\ -4 & 1 & 1 & -3 \end{bmatrix}$ **23.** $\begin{cases} 2x + y = 3 \\ 6x - 5y = 11 \end{cases}$

25. $\begin{cases} z = -3 \\ y = 4 \\ x = 2 \end{cases}$ **27.** $\{(1, 1)\}$ **29.** $\{(7, 2)\}$ **31.** $\{(1, 2, 3)\}$ **33.** $\{(4, 2, 1)\}$ **35.** $\{(3, -2, 1)\}$

CHAPTER 7 REVIEW PROBLEMS

1. $\{(1, 2)\}$ **3.** $\{(2, -3)\}$ **5.** $\{(3, 0), (5, 2)\}$ **7.** $\{(4, -1)\}$ **9.** $\{(6, 1)\}$ **11.** $\left\{\left(-1, \dfrac{1}{3}\right)\right\}$

13. $\left\{\left(\dfrac{1}{3}, \dfrac{1}{2}\right)\right\}$ **15.** $\left\{\left(\dfrac{1}{5}, \dfrac{1}{4}\right)\right\}$ **17.** $\{(5, 2, 1)\}$ **19.** $\{(0, 0, 0)\}$ **21.** $\{(0, 0, 6)\}$ **23.** $\{(-2, 0, 3)\}$

25. $\{(-1, 2, 3)\}$ **27.** Cantaloupes are \$1.19, honeydews are \$1.59, and watermelons are \$1.99.

29. 3 and -7, or 7 and -3

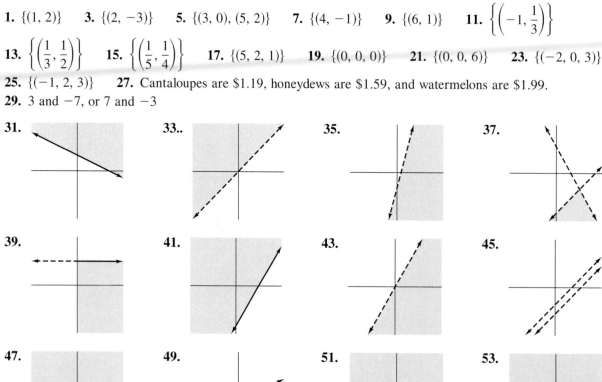

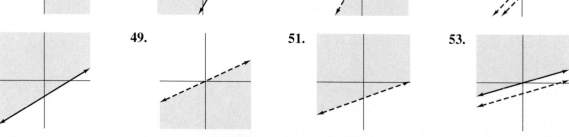

Let's Not Forget . . .

92. $5(5 + 3s)(5 - 3s)$ **93.** Factored   **94.** Factored   **95.** $(x + 2)(x^2 - 2x + 4)$
96. 2; $x + 2$ is a factor.   **97.** 2; not a factor.   **98.** 2; $x + 2$ is a factor.   **99.** 4; $x + 2$ is a factor.

100. $-\dfrac{1}{16}$ **101.** $-\dfrac{1}{16}$ **102.** $\dfrac{1}{9}$ **103.** $-\dfrac{1}{8}$ **104.** $\dfrac{-2c}{3b^3k^2}$ **105.** $\dfrac{3x^2}{2y^2}$ **106.** $2 \pm \sqrt{2}$

107. Won't reduce   **108.** $\dfrac{4 - x}{1 - x^2}$ **109.** $\{-2\}$ **110.** $-\dfrac{1}{x}$ **111.** Inequality; $\{x \,|\, 2 < x < 8\}$

112. Equation; $\{\pm 1\}$ **113.** Expression; $2x\sqrt[3]{1 + 8x^3}$ **114.** Expression; $\dfrac{2}{x^2 - 2}$

CHAPTER 7 TEST

1. D   **2.** B   **3.** D   **4.** B   **5.** C

6. 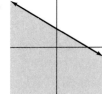   **7.** 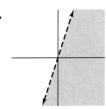   **8.** 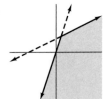   **9.**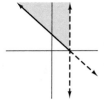

10. $\{(-1, 4), (4, -1)\}$ **11.** $\left\{\left(\dfrac{1}{3}, -1\right)\right\}$ **12.** $\{(1, -2)\}$ **13.** $\{(x, y) \,|\, y = 3x - 6\}$

14. $\{(1, -1, 3)\}$ **15.** Bananas cost 33¢ a pound and apples 99¢ a pound.   **16.** -3

17. $\left\{\left(\dfrac{4}{5}, -\dfrac{1}{5}\right)\right\}$ **18.** $\left\{\left(3, -\dfrac{4}{3}, -\dfrac{1}{3}\right)\right\}$ **19.** $\left\{\left(\dfrac{1}{3}, \dfrac{1}{3}\right)\right\}$ **20.** $\left\{\left(\dfrac{6}{5}, \dfrac{4}{5}, \dfrac{7}{5}\right)\right\}$

 Chapter 8

PROBLEM SET 8.1

Warm-Ups

In Problems 1 through 4, answers given in the order: $(f + g)(x)$, $(f - g)(x)$, $(fg)(x)$,
$\left(\dfrac{f}{g}\right)(x)$.

1. $3x - 5$, $x + 5$, $2x^2 - 10x$, $\dfrac{2x}{x - 5}$ **2.** $\dfrac{5}{3}x - 10$; $\dfrac{1}{3}x - 4$; $\dfrac{2}{3}x^2 - \dfrac{23}{3}x + 21$; $\dfrac{3(x - 7)}{2x - 9}$

3. $-x^2 + x + 4$, $-x^2 - x + 2$, $-x^3 - x^2 + 3x + 3$, $\dfrac{3 - x^2}{x + 1}$

4. $2x^2 - 2x - 3$; -3; $x^4 - 2x^3 - 2x^2 + 3x$; $\dfrac{x^2 - x - 3}{x^2 - x}$ **5.** 2 **6.** 0 **7.** $-\dfrac{1}{3}$

8. -6 **9.** $(f \circ g)(x) = 35x - 12$; $(g \circ f)(x) = 35x + 52$

10. $(f \circ g)(x) = \dfrac{3}{2}x + 9$; $(g \circ f)(x) = \dfrac{3}{2}x + \dfrac{9}{4}$ **11.** $(f \circ g)(x) = x^2 - 6x + 11$; $(g \circ f)(x) = x^2 - 1$

12. $(f \circ g)(x) = (x - 1)^3$; $(g \circ f)(x) = x^3 - 1$ **13.** -1 **14.** 0 **15.** -1 **16.** -1

Practice Exercises

Answers given in the order; $(f + g)(x), (f - g)(x), (fg)(x), \left(\dfrac{f}{g}\right)(x), (f \circ g)(x).$

17. $4x + 8, 2x - 8, 3x^2 + 24x, \dfrac{3x}{x + 8}, 3x + 24$ **19.** $7x - 7, x + 11, 12x^2 - 30x - 18, \dfrac{4x + 2}{3x - 9}, 12x - 34$

21. $\dfrac{4}{3}x, \dfrac{2}{3}x - 6, \dfrac{1}{3}x^2 + 2x - 9, \dfrac{3(x - 3)}{x + 9}, \dfrac{1}{3}x$

23. $2x^2 + 3x - 2, 2x^2 - x + 8, 4x^3 - 8x^2 + x - 15, \dfrac{2x^2 + x + 3}{2x - 5}, 8x^2 - 38x + 48$

25. $-x^2 + x + 4, -x^2 - x - 2, -x^3 - 3x^2 + x + 3, \dfrac{1 - x^2}{x + 3}, -x^2 - 6x - 8$

27. $2x^2 - 3x - 2, x - 2, x^4 - 3x^3 + 4x, \dfrac{x + 1}{x}; x \neq 2, x^4 - 4x^3 + 3x^2 + 2x - 2$ **29.** 9   **31.** -1
33. 1   **35.** -11

Challenge Problems

37. $(f \circ g)(x) = (g \circ f)(x) = x$ **39.** No

PROBLEM SET 8.2

Warm-Ups

1. One-to-one   **2.** Not one-to-one   **3.** One-to-one   **4.** Not one-to-one   **5.** Not one-to-one

6. Not one-to-one   **7.** One-to-one   **8.** One-to-one   **9.** $\dfrac{1}{3}(x + 1)$ **10.** $x + 4$ **11.** $2(x - 3)$

12. $\dfrac{3}{2}\left(x - \dfrac{1}{2}\right)$ **13.** $\sqrt[3]{x}$ **14.** $\sqrt[3]{x - 1}$

Practice Exercises

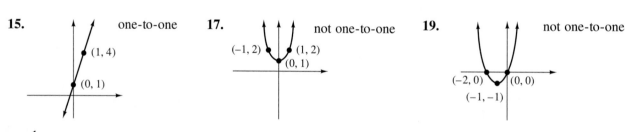

15.   one-to-one   **17.**   not one-to-one   **19.**   not one-to-one

21. $\dfrac{1}{5}(x + 1)$ **23.** $3(x + 1)$ **25.** $\sqrt[3]{x - 2}$

Challenge Problems

31.   The graph of f^{-1} is the graph of f reflected about the line $y = x$.

PROBLEM SET 8.3

Warm-Ups

1. 5.196   **2.** 1.308   **3.** 46.416   **4.** 4.729

5. 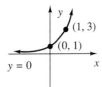   **6.** 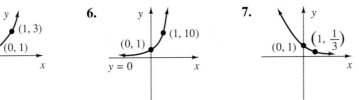   **7.** 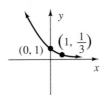   **8.**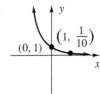

9. {4}   **10.** {−2}   **11.** {$\frac{1}{2}$}   **12.** {$\frac{5}{2}$}   **13.** 20.0855   **14.** 9.3565   **15.** 0.0183   **16.** 23.1407
17. The pressure is 11.9 lb/in.². **18.** There are 49 rabbits.   **19.** There will be 22,795 bacteria.
20. $46,480.98

Practice Exercises

21. 1.4422   **23.** 11.3924   **25.** 42.7494   **27.** 7.3891   **29.** 22.2740   **31.** 0.1353   **33.** 3.1318
35.   **37.** 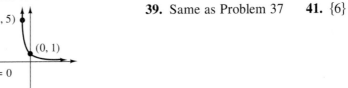   **39.** Same as Problem 37   **41.** {6}

43. {1}   **45.** {$\frac{3}{2}$}   **47.** {−3}   **49.** {$\frac{2}{3}$}   **51.** {−5}   **53.** {$\frac{-8}{5}$}   **55.** 3^5 **57.** $3^{0.5}$ **59.** $5^{\sqrt{3}}$

61. It will be 224,084.   **63.** About 26.37 grams will remain.   **65.** About 1.28/million curies remained.
67. (a)

x	$f(x)$
1	5
2	25
3	125
4	625
5	3125

(b) By the end of the eighth week, over 1 million people are involved.

69. $40,000   **71.** $162,000   **73.** f is 22% and g is 10.7%.
75. About $3\frac{1}{2}$ yr to double; about 7 yr to quadruple.   **77.** $27,000   **79.** $73,000

Challenge Problems

81.   **83.** 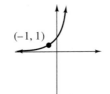   **85.**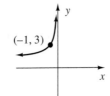

PROBLEM SET 8.4

Warm-Ups

1. $2^3 = 8$ **2.** $8 = \left(\frac{1}{2}\right)^{-3}$ **3.** $8^2 = 64$ **4.** $3 = 3^1$ **5.** $\log_2 8 = 3$ **6.** $\log_4 \frac{1}{16} = -2$
7. $\log_{1/3} \frac{1}{9} = 2$ **8.** $\log_3 1 = 0$ **9.** {16}   **10.** {4}   **11.** {125}   **12.** {1}   **13.** 3   **14.** 4

15. undefined   **16.** undefined   **17.** 2   **18.** 3   **19.** 2   **20.** 0   **21.** −2   **22.** 3   **23.** 0   **24.** 2
25. {1000}   **26.** {0.1}   **27.** D: $\{x \mid x > 0\}$
R: $\mathbb{R}$

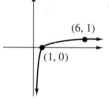

28.

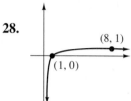

29. 10 dB   **30.** 140 dB
31. 65 dB   **32.** 125 dB

Practice Exercises

33. {25}   **35.** {3}   **37.** {16}   **39.** {2}   **41.** {1}   **43.** {2}   **45.** {1}   **47.** {9}   **49.** {4}   **51.** {−2}
53. {2}   **55.** $\left\{\frac{1}{2}\right\}$ **57.** {2}   **59.** {3}   **61.** {1}   **63.** {98}   **65.**

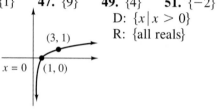

D: $\{x \mid x > 0\}$
R: {all reals}

67. D: $\{x \mid x > 0\}$ R: $\mathbb{R}$ **69.** 1.1761   **71.** 0.9031   **73.** 1.2041   **75.** −0.1938   **77.** −0.1938
79. 0.1505   **81.** $G \approx 27$ dB

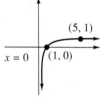

Challenge Problems

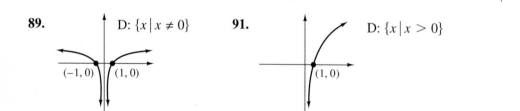

83.   (1, 0)

85.   (2, 0)   *x*

87.   (1, −1)

89.   D: $\{x \mid x \ne 0\}$ (−1, 0)   (1, 0)

91.   D: $\{x \mid x > 0\}$ (1, 0)

PROBLEM SET 8.5

Warm-Ups

1. 2.94   **2.** 0.30   **3.** 2.64   **4.** 0.81   **5.** −1.32   **6.** 4.56   **7.** 1.02   **8.** 1.47   **9.** $\left\{\frac{9}{2}\right\}$
10. {4}   **11.** $\left\{\frac{1}{2}\right\}$ **12.** {24}   **13.** {4}   **14.** $\left\{\frac{5}{2}\right\}$ **15.** {342}   **16.** {$e + 1$}   **17.** {3}   **18.** {e}
19. {5}   **20.** {−2, 5}   **21.** {±2}   **22.** {2}   **23.** 0.827   **25.** 4.107   **27.** About 14 years

Practice Exercises

29. 4.14   **31.** 5.04   **33.** -1.62 **35.** 0.72   **37.** T   **39.** F   **41.** F   **43.** F   **45.** $\left\{\dfrac{1}{4}\right\}$

47. $\{48\}$ **49.** $\{64\}$ **51.** $\{33\}$ **53.** $\left\{\dfrac{e-1}{2}\right\}$ **55.** $\{\pm3\}$ **57.** $\{-1, 4\}$ **59.** $\{5\}$ **61.** 8.827

63. -50.806 **65.** -0.549 **67.** 0.964   **69.** $\approx$11.6 yr   **71.** $\approx$14.7 yr   **73.** The two populations will be equal in about 13.2 years. There will be about 480,000 individuals each. Neither population will ever be zero.

Challenge Problems

54. $\log_b \dfrac{1}{x} = \log_b 1 - \log_b x = -\log_b x$

REVIEW PROBLEMS

1. $x^2 + 3x - 5$ **3.** $-x^2 + 3x - 5$ **5.** $3x^3 - 5x^2$ **7.** 13   **9.** -8 **11.** $-\dfrac{14}{3}$ **13.** -5

15. $-\dfrac{4}{9}$ **17.** $f^{-1}x = \dfrac{1-x}{2}$ **19.** $f^{-1}(x) = \sqrt[3]{x+7}$

21.

23.

25. $\{7\}$

27. $\{1\}$ **29.** $\{e^2 - e\}$ **31.** $\{1\}$ **33.** $\{2\}$ **35.** $\{2\}$ **37.** $\{7\}$ **39.** $\{2\}$ **41.** $\left\{\dfrac{1}{3}\right\}$ **43.** 0.671
45. 5.129   **47.** 0.805   **49.** -0.562 **51.** $\approx$11 yr   **53.** $\approx$10 yr

Let's Not Forget . . .

55. Factored   **56.** $(5 - 2a)(25 + 10a + 4a^2)$ **57.** $(2 + a + b)(2 - a - b)$ **58.** Factored
59. $b(a^3 + c^2)$ **60.** Factored   **61.** 2; Yes   **62.** 3; No   **63.** 2; Yes   **64.** 2; No

65. $2\sqrt[5]{2x^5 - 4y^5}$ **66.** $x^2 y^2$ **67.** $x - 3$ **68.** $\dfrac{y^8}{x^2}$ **69.** $\dfrac{4}{3}$ **70.** $2 \pm \sqrt{3}\,i$ **71.** $\dfrac{-5}{x^3 y}$

72. $\dfrac{3a^4 b^2}{4}$ **73.** $27x^3 - 27x^2 + 9x - 1$ **74.** $4x^2 - 12xy + 9y^2$ **75.** Won't reduce   **76.** $\dfrac{2 + a}{4 + a}$

77. $\dfrac{4x + 3}{(x + 3)^2}$ **78.** $\{-4\}$ **79.** $-(x + 3)$ **80.** Equation; $\{2\}$ **81.** Equation; $\emptyset$ **82.** Expression; 4

83. Expression; -4 **84.** Expression; $8 + 2\sqrt{15}$ **85.** Inequality; $\left\{x \mid x \le \dfrac{26}{5}\right\}$ **86.** Equation; $\{21\}$

87. Equation; $\{-11\}$ **88.** Inequality; $\{x \mid x < -2 \text{ or } x > 7\}$ **89.** Expression; $2\sqrt{3}$ **90.** Expression; 1

91. Expression; $\dfrac{b^3}{a^7}$ **92.** Inequality; $\{x \mid x < 3 \text{ or } x > 4\}$ **93.** Expression; $a^2 b - ab^2$ **94.** Equation; $\emptyset$

95. Inequality; $\{x \mid x \le -18 \text{ or } x \ge 4\}$

CHAPTER 8 TEST

1. B   **2.** B   **3.** D   **4.** A   **5.** C   **6.** 3   **7.** -6 **8.** $\dfrac{1}{x+1}$ **9.** $\dfrac{2}{x} - 3$ **10.** 3   **11.** $\{3\}$
12. $\{1\}$ **13.** $\{12\}$ **14.** -1.710 **15.** ≈ 11 yr

► **Chapter 9**

PROBLEM SET 9.1

Warm-Ups

1.

(0, 0)

$R = 2$

2.

(0, 0)

$R = \sqrt{5}$

3.

(0, 1)

$R = 3$

4.

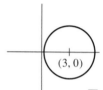

(3, 0)

$R = 2\sqrt{2}$

5.

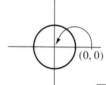

(−1, 2)

$R = \sqrt{10}$

6.

(1, −1)

$R = \sqrt{6}$

7.

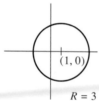

(1, 0)

$R = 3$

8.

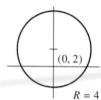

(0, 2)

$R = 4$

9.

$R = 2\sqrt{5}$

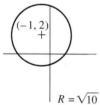

(0, −4)

10.

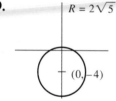

(−3, 0)

$R = 2\sqrt{2}$

11.

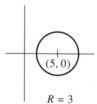

(5, 0)

$R = 3$

12.

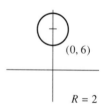

(0, 6)

$R = 2$

13.

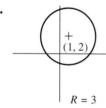

(1, 2)

$R = 3$

14.

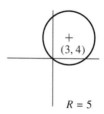

(3, 4)

$R = 5$

15.

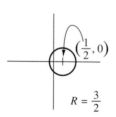

$\left(\dfrac{1}{2}, 0\right)$

$R = \dfrac{3}{2}$

16.

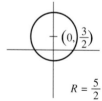

$\left(0, \dfrac{3}{2}\right)$

$R = \dfrac{5}{2}$

Practice Problems

17.

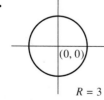

(0, 0)

$R = 3$

19.

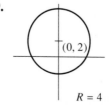

(0, 2)

$R = 4$

21.

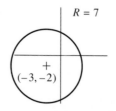

$R = 7$

(−3, −2)

23.

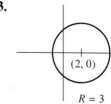

(2, 0)

$R = 3$

25.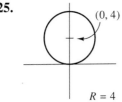
$(0, 4)$
$R = 4$

27.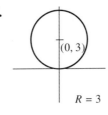
$(0, 3)$
$R = 3$

29.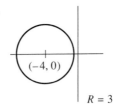
$(-4, 0)$
$R = 3$

31.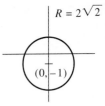
$R = 2\sqrt{2}$
$(0, -1)$

33.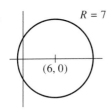
$R = 7$
$(6, 0)$

35.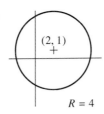
$(2, 1)$
$R = 4$

37.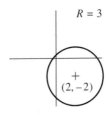
$R = 3$
$(2, -2)$

39.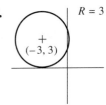
$R = 3$
$(-3, 3)$

41.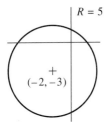
$R = 5$
$(-2, -3)$

43.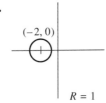
$(-2, 0)$
$R = 1$

45. $R = 2\sqrt{10}$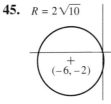
$(-6, -2)$

47.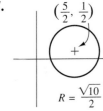
$\left(\frac{5}{2}, \frac{1}{2}\right)$
$R = \frac{\sqrt{10}}{2}$

49. $R = \frac{5\sqrt{2}}{2}$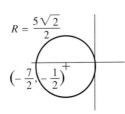
$\left(-\frac{7}{2}, -\frac{1}{2}\right)$

51.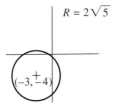
$R = 2\sqrt{5}$
$(-3, -4)$

53.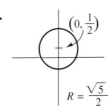
$\left(0, \frac{1}{2}\right)$
$R = \frac{\sqrt{5}}{2}$

55.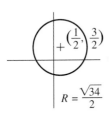
$\left(\frac{1}{2}, \frac{3}{2}\right)$
$R = \frac{\sqrt{34}}{2}$

57.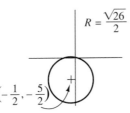
$R = \frac{\sqrt{26}}{2}$
$\left(-\frac{1}{2}, -\frac{5}{2}\right)$

Challenge Problems

61.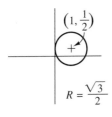
$\left(1, \frac{1}{2}\right)$
$R = \frac{\sqrt{3}}{2}$

63.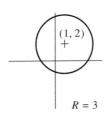
$(1, 2)$
$R = 3$

PROBLEM SET 9.2

Warm-Ups

1.

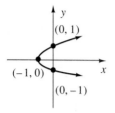

2.

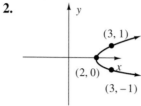

3.

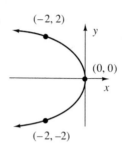

4.

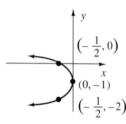

5.

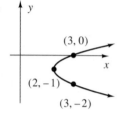

6.

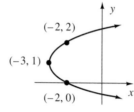

7.

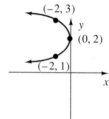

8.

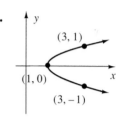

9.

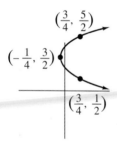

10.

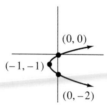

11.

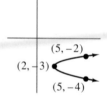

12.

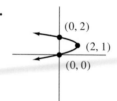

Practice Problems

13.

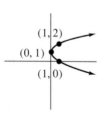

15.

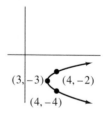

17.

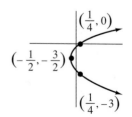

19.

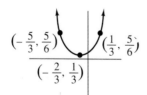

21.

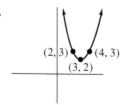

23.

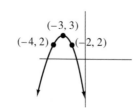

25.

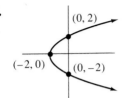

(0, 2)
(−2, 0)
(0, −2)

27.

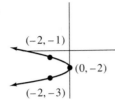

(−2, −1)
(0, −2)
(−2, −3)

29.

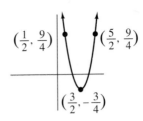

$\left(\frac{1}{2}, \frac{9}{4}\right)$ $\left(\frac{5}{2}, \frac{9}{4}\right)$
$\left(\frac{3}{2}, -\frac{3}{4}\right)$

31.

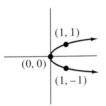

(1, 1)
(0, 0)
(1, −1)

33.

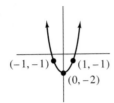

(−1, −1) (1, −1)
(0, −2)

35.

(3, 0) (6, 1)
(6, −1)

37.

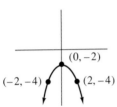

(0, −2)
(−2, −4) (2, −4)

39.

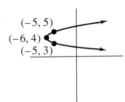

(−5, 5)
(−6, 4)
(−5, 3)

41.

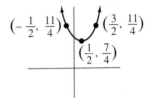

$\left(-\frac{1}{2}, \frac{11}{4}\right)$ $\left(\frac{3}{2}, \frac{11}{4}\right)$
$\left(\frac{1}{2}, \frac{7}{4}\right)$

43.

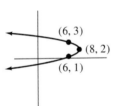

(6, 3)
(8, 2)
(6, 1)

45.

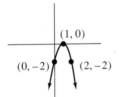

(1, 0)
(0, −2) (2, −2)

47.

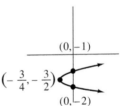

(0, −1)
$\left(-\frac{3}{4}, -\frac{3}{2}\right)$
(0, −2)

Challenge Problems

49.

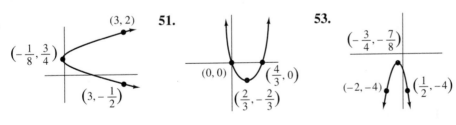

(3, 2)
$\left(-\frac{1}{8}, \frac{3}{4}\right)$
$\left(3, -\frac{1}{2}\right)$

51.

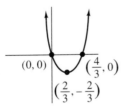

(0, 0) $\left(\frac{4}{3}, 0\right)$
$\left(\frac{2}{3}, -\frac{2}{3}\right)$

53.

$\left(-\frac{3}{4}, -\frac{7}{8}\right)$
(−2, −4) $\left(\frac{1}{2}, -4\right)$

55.

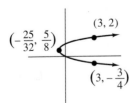

(3, 2)
$\left(-\frac{25}{32}, \frac{5}{8}\right)$
$\left(3, -\frac{3}{4}\right)$

PROBLEM SET 9.3

Warm-Ups

1.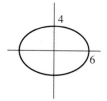

2.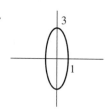

3.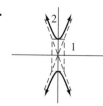

4.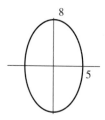

5.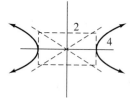

6.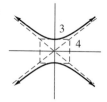

7.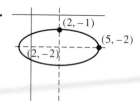

8.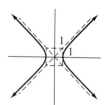

9.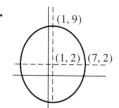

10.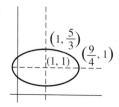

11.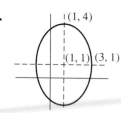

12.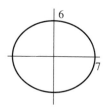

13.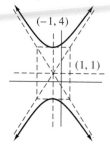

14.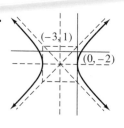

15.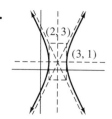

16.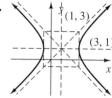

Practice Exercises

17.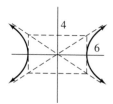

19.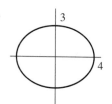

21.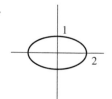

23.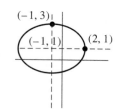

25.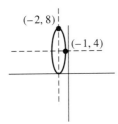

27.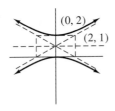

29.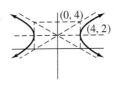

31.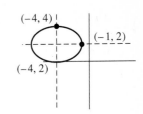

33.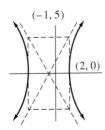

35.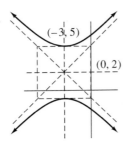

37.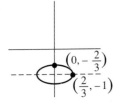

39.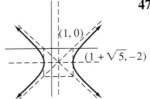

41.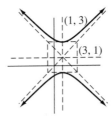

43.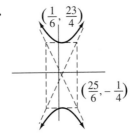

45.

47.

Challenge Problems

49.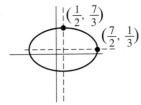

51.

PROBLEM SET 9.4

Warm-Ups

1. Hyperbola   **2.** Ellipse   **3.** Ellipse   **4.** Parabola   **5.** Circle   **6.** Hyperbola   **7.** Parabola
8. Circle

Practice Exercises

9. Parabola   **11.** Ellipse   **13.** Parabola   **15.** Hyperbola   **17.** Ellipse   **19.** Circle

Challenge Problems

21. Degenerate ellipse

● $(1, -1)$

REVIEW PROBLEMS

1. Hyperbola   **3.** Ellipse   **5.** Ellipse   **7.** Hyperbola   **9.** Parabola

11.

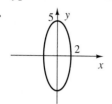

13.

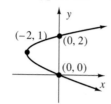

15.

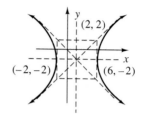

17.

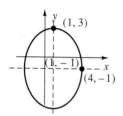

19.

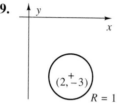

Let's Not Forget . . .

21. Factored   **22.** Factored   **23.** $(2 + z)(4 - 2z + z^2)$ **24.** Factored   **25.** $7(3x - 1)^2$
26. $5(1 - 2x)(1 + 2x)$ **27.** 2; $x - 3$ is a factor.   **28.** 3; not a factor   **29.** 3; $x - 3$ is a factor

30. 2; not a factor   **31.** $-\dfrac{1}{256}$ **32.** Not a real number   **33.** $-\dfrac{2x + 3}{4x^2 + 6x + 9}$ **34.** $\dfrac{x^2 + 4}{8}$

35. $\dfrac{-3y^3}{x^2 z^2}$ **36.** Can't be simplified.   **37.** $x + 4\sqrt{x} + 4$ **38.** $216 - 540x + 450x^2 - 125x^3$

39. $a^3 b^2 c$ **40.** $\dfrac{2 + b}{x + 7}$ **41.** Won't reduce.   **42.** $\dfrac{2 - x}{x - 5}$ **43.** $\left\{\dfrac{5}{4}\right\}$ **44.** $-\dfrac{x + 1}{x - 2}$

45. Expression; $x^2 - 14x + 49$ **46.** Equation; parabola opening up with vertex at (7, 0)

47. Equation; $\left\{-\dfrac{26}{5}, 8\right\}$ **48.** Expression; $\dfrac{1}{(2x - 5)(x + 1)(x + 2)}$ **49.** Expression; $\dfrac{1}{2} - \dfrac{5}{2}i$

50. Equation; {3}

CHAPTER 9 TEST

1. B   **2.** A   **3.** B   **4.** D   **5.** B   **6.** D   **7.** A   **8.** A   **9.** C
10. Center is (0, 2); radius is $\sqrt{5}$.   **11.** Vertex is $(-1, -1)$; opens down.

12.

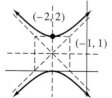

13.

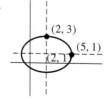

14.

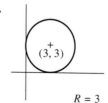

15.

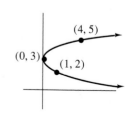

► Chapter 10

PROBLEM SET 10.1

Warm-Ups

1. 1, 2, 3, 4, 5   **2.** 3, 5, 7, 9, 11   **3.** 4, 6, 8, 10, 12   **4.** 11, 8, 5, 2, −1   **5.** 1, 4, 9, 16, 25
6. 0, 2, 8, 18, 32   **7.** 3, 3, 3, 3, 3   **8.** 1, −8, 27, −64, 125   **9.** 0, −1, 2, −3, 4   **10.** $2n$ **11.** n^2
12. $3n - 1$ **13.** $(-1)^n n$ **14.** $(2n - 1)^2$ **15.** $(-1)^{n+1} 2^n$ **16.** Increasing   **17.** Decreasing
18. Decreasing   **19.** Increasing   **20.** Neither   **21.** Increasing   **22.** $d = 2; a_n = 3 + (n - 1)2$
23. $d = 1; a_n = 2 + (n - 1)1$ **24.** $d = 4; a_n = 1 + (n - 1)4$ **25.** $d = 10; a_n = 5 + (n - 1)10$

26. $d = -2; a_n = 0 + (n - 1)(-2)$ **27.** $d = -\dfrac{1}{2}; a_n = \dfrac{1}{2} + (n - 1)\left(-\dfrac{1}{2}\right)$ **28.** $r = 2; a_n = 3 \cdot 2^{n-1}$

29. $r = 5; a_n = 1 \cdot 5^{n-1}$ **30.** $r = 3; a_n = 3 \cdot 3^{n-1}$ **31.** $r = -2; a_n = 2 \cdot (-2)^{n-1}$

32. $r = \dfrac{1}{2}; a_n = 4 \cdot \left(\dfrac{1}{2}\right)^{n-1}$ **33.** $r = -\dfrac{1}{4}; a_n = 1 \cdot \left(-\dfrac{1}{4}\right)^{n-1}$

Practice Exercises

35. 2, 5, 8, 11, 14   **37.** 6, 4, 2, 0, −2   **39.** 1, 3, 7, 13, 21   **41.** 1, −4, 9, −16, 25   **43.** $2n + 2$
45. $4n - 5$ **47.** $(n - 1)^3$ **49.** Increasing   **51.** Decreasing   **53.** Decreasing
55. $d = 3; a_n = 1 + (n - 1)3$ **57.** $d = -2; a_n = 10 + (n - 1)(-2)$

59. $d = \dfrac{1}{2}; a_n = -\dfrac{1}{2} + (n - 1)\dfrac{1}{2}$ **61.** $r = 3; a_n = 1 \cdot 3^{n-1}$ **63.** $r = -3; a_n = 2 \cdot (-3)^{n-1}$

65. $r = \dfrac{1}{3}; a_n = 27 \cdot \left(\dfrac{1}{3}\right)^{n-1}$

PROBLEM SET 10.2

Warm-Ups

1. $2 + 4 + 6 + 8 + 10$ **2.** $0 + 3 + 6 + 9 + 12 + 15$ **3.** $0 + 1 + 4 + 9 + 16 + 25 + 36 + 49$
4. $2 + 4 + 8 + 16 + 32 + 64 + 128$ **5.** $9 + 4 + 1 + 0 + 1$ **6.** $1 + 0 + 1 + 4 + 9 + 16$
7. $1 - 2 + 3 - 4 + 5 - 6$ **8.** $-2 + 4 - 8 + 16 - 32$ **9.** $0 - \dfrac{1}{3} + \dfrac{1}{2} - \dfrac{3}{5} + \dfrac{2}{3}$ **10.** $\displaystyle\sum_{j=1}^{8} j$

11. $\displaystyle\sum_{j=1}^{6} (-1)^{j+1}(2j - 1)$ **12.** $\displaystyle\sum_{j=1}^{8} (-1)^{j-1} \dfrac{j}{2j - 1}$ **13.** $\displaystyle\sum_{j=1}^{7} \dfrac{x^{j-1}}{j}$ **14.** $\displaystyle\sum_{j=1}^{100} A_j$ **15.** $\displaystyle\sum_{k=1}^{7} (-1)^{k+1}\sqrt{k}\, x^{2k-2}$

Practice Exercises

17. $-10 - 8 - 6 - 4 - 2$ **19.** $1 + 2 + 4 + 8 + 16 + 32 + 64$ **21.** $-1 + 0 + 1 + 8 + 27 + 64$
23. $-1 + 2 - 4 + 8 - 16$ **25.** $\displaystyle\sum_{k=3}^{14} k$ **27.** $\displaystyle\sum_{k=0}^{4} \dfrac{x^{2k}}{2k + 1}$ **29.** $\displaystyle\sum_{k=0}^{6} \dfrac{k - \pi}{10^k}$

Challenge Problems

31. $1 + 2 + 3 + 4 + 5$ All are the same. The index does not affect the sum.

33. They are the same. $\displaystyle\sum_{j=1}^{n} kA_j = k \sum_{j=1}^{n} A_j.$

PROBLEM SET 10.3

Warm-Ups

1. 720   **2.** 40,320   **3.** 479,001,600   **4.** 3,628,800   **5.** 14   **6.** $\dfrac{1}{930}$ **7.** 70   **8.** 56   **9.** 28
10. 8   **11.** 20   **12.** 190   **13.** 56   **14.** 10   **15.** 10   **16.** 1   **17.** 6   **18.** 15   **19.** 20
20. 15   **21.** 6   **22.** 1   **23.** 1   **24.** 1

Practice Exercises

25. 5040   **27.** 39,916,800   **29.** $\dfrac{1}{18}$ **31.** 126   **33.** 36   **35.** 30   **37.** 84   **39.** 4   **41.** 7
43. 35   **45.** 21   **47.** 1

Challenge Problems

49. $\dbinom{n}{r} = \dfrac{n!}{(n-r)!r!}$

$\dbinom{n}{n-r} = \dfrac{n!}{(n-(n-r))!(n-r)!} = \dfrac{n!}{r!(n-r)!}$

PROBLEM SET 10.4

Warm-Ups

1. $a^{12} + 12a^{11}b + 66a^{10}b^2$ **2.** $x^{20} + 20x^{19}y + 190x^{18}y^2$ **3.** $s^{15} + 30s^{14} + 420s^{13}$
4. $512 + 2304t + 4608t^2$ **5.** $64a^6 + 576a^5b + 2160a^4b^2$ **6.** $x^{50} + 25x^{48} + 300x^{46}$ **7.** $715a^9b^4$
8. $495x^4$ **9.** $\dfrac{1,082,565}{8}x^8$ **10.** $\dfrac{7}{216}x^5$

Practice Exercises

11. $a^6 + 6s^5b + 15a^4b^2 + 20a^3b^3 + 15a^2b^4 + 6ab^5 + b^6$
13. $x^9 + 9x^8 + 36x^7 + 84x^6 + 126x^5 + 126x^4 + 84x^3 + 36x^2 + 9x + 1$
15. $64x^6 + 192x^5y + 240x^4y^2 + 160x^3y^3 + 60x^2y^4 + 12xy^5 + y^6$
17. $a^9 + 9a^8b + 36a^7b^2 + 84a^6b^3 + 126a^5b^4 + 126a^4b^5 + 84a^3b^6 + 36a^2b^7 + 9ab^8 + b^9$
19. $x^9 + 9x^8 + 36x^7 + 84x^6 + 126x^5 + 126x^4 + 84x^3 + 36x^2 + 9x + 1$
21. $128x^7 + 448x^6y + 672x^5y^2 + 560x^4y^3 + 280x^3y^4 + 84x^2y^5 + 14xy^6 + y^7$
23. $78a^2b^{11} + 13ab^{12} + b^{13}$ **25.** $270s^2 + 405s + 243$ **27.** $6048x^2y^5 + 1344xy^6 + 128y^7$
29. $66a^2b^{10} + 12ab^{11} + b^{12}$ **31.** $1215s^2 + 1458s + 729$ **33.** $20{,}412x^2y^5 + 10{,}206xy^6 + 2187y^7$
35. $a^{50} + 50a^{49}b$ **37.** $s^{500} + 1500s^{499}$ **39.** $1 + 100x$ **41.** $\dfrac{55}{3456}x^9$

Challenge Problems

43. $\left(\dfrac{1}{2}\right)^6 + 6\left(\dfrac{1}{2}\right)^5\left(\dfrac{1}{2}\right) + 15\left(\dfrac{1}{2}\right)^4\left(\dfrac{1}{2}\right)^2 + 20\left(\dfrac{1}{2}\right)^3\left(\dfrac{1}{2}\right)^3 + 15\left(\dfrac{1}{2}\right)^2\left(\dfrac{1}{2}\right)^4 + 6\left(\dfrac{1}{2}\right)\left(\dfrac{1}{2}\right)^5 + \left(\dfrac{1}{2}\right)^6$

$= \dfrac{1}{64} + \dfrac{6}{64} + \dfrac{15}{64} + \dfrac{20}{64} + \dfrac{15}{64} + \dfrac{6}{64} + \dfrac{1}{64} = \dfrac{64}{64} = 1$

45. $\displaystyle\sum_{j=0}^{n}(-1)^j\binom{n}{j}a^{n-j}b^j$ **47.** $x^6 - 6x^5y + 15x^4y^2 - 20x^3y^3 + 15x^2y^4 - 6xy^5 + y^6$

49. $q^8 - 16q^7 + 112q^6 - 448q^5 + 1120q^4 + 1792q^3 + 1792q^2 - 1024q + 256$

51. $81x^8 - 216x^6y^2 + 216x^4y^4 - 96x^2y^6 + 16y^8$

PROBLEM SET 10.5

Warm-Ups

1. 12   **2.** 1   **3.** 5   **4.** 20   **5.** 60   **6.** 120   **7.** 120   **8.** 6   **9.** 1   **10.** 9   **11.** 36   **12.** 84

13. 1   **14.** 9   **15.** 120   **16.** 7920   **17.** 1716   **18.** 1001

Practice Exercises

19. 1680   **21.** 1   **23.** 30   **25.** 360   **27.** 720   **29.** 21   **31.** 1   **33.** 28   **35.** 1   **37.** 28

39. 24,024   **41.** 1001   **43.** 1680   **45.** 11,440

Challenge Problems

47. $N!$ **49.** $\dfrac{N!}{2}$ **51.** 1   **53.** $\dfrac{(N + 2)(N + 1)}{2}$ **55.** $\dfrac{n(n + 1)}{2}$

REVIEW PROBLEMS

1. 4, 7, 10, 13, 16   **3.** 3, 15, 75, 375, 1875   **5.** $-3, 6, -9, 12, -15$ **17.** Increasing   **9.** Increasing

11. $d = 2; 4$ **13.** $r = 4; 64$ **15.** $r = \dfrac{3}{2}; 135$ **17.** $0 + 1 + 4 + 9 + 16 + 25$

19. $0 + 4 + 16 + 36 + 64 + 100$ **21.** $1 - 4 + 27 - 256$ **23.** $\displaystyle\sum_{j=0}^{6}(j + 1)k^j$ **25.** 362,880

27. 12   **29.** 1640   **31.** 84   **33.** 252   **35.** $x^7 + 7x^6 + 21x^5 + 35x^4 + 35x^3 + 21x^2 + 7x + 1$

37. $64 + 576y + 2160y^2 + 4320y^3 + 4860y^4 + 2916y^5 + 729y^6$

39. $1 + 10s + 45s^2 + 120s^3 + 210s^4 + 252s^5 + 210s^6 + 120s^7 + 45s^8 + 10s^9 + s^{10}$

41. $256 + 5120t + 44800r^2$ **43.** $3432a^7b^7$ **45.** 210   **47.** 332,640

Let's Not Forget . . .

48. Factored   **49.** $(2x - 3)(4x^2 + 6x + 9)$ **50.** $(12x - 5)(8x + 3)$ **51.** Factored   **52.** $7(9 + x^3)$

53. $(x - 1)^3$ **54.** 2; factor   **55.** 3; factor   **56.** 3; not a factor   **57.** $-\dfrac{1}{16}$ **58.** $\dfrac{-2z^2}{x^3y^2}$

59. $-\dfrac{4x + 5}{16x^2 + 20x + 25}$ **60.** Won't simplify.   **61.** $\dfrac{65x^3}{2x^3 + 8}$ **62.** $(a + b)^2$

63. $27x^3 + 54x^2y + 36xy^2 + 8y^3$ **64.** $11 + 4\sqrt{6}$ **65.** $-1 + \dfrac{\sqrt{7}}{7}i$ **66.** $-1 - r$ **67.** $\dfrac{x + 1}{x^2}$

68. $\left\{\dfrac{1}{4}\right\}$ **69.** $\dfrac{x}{x + 1}$

70. Equation; {3}   **71.** Expression; $\dfrac{1}{x+2}$ **72.** Expression; $4x^2 + 4xy + y^2$ **73.** Equation; Ø

74. Expression; $\dfrac{x-3}{(2x-1)^2(x+2)}$ **75.** Expression; $-\dfrac{3}{13} + \dfrac{2}{13}i$

CHAPTER 10 TEST

1. D   **2.** B   **3.** D   **4.** B   **5.** D   **6.** B   **7.** A   **8.** C   **9.** D   **10.** $a_n = 4 + (n-1)3$
11. $2, -2^2, 2^3, -2^4, 2^5$ **12.** 25   **13.** 6720   **14.** $81 + 108t + 54t^2 + 12t^3 + t^4$
15. $153a^2b^{16} + 18ab^{17} + b^{18}$

Photo Credits

Index

| on/off |

| $2^{nd}F$ | $\sqrt{x}$ x^2 | log 10^x | ln e^x | CE/C |

| DRG $\to$ DRG | hyp^{-1} hyp | sin^{-1} sin | cos^{-1} cos | tan^{-1} tan |

| M$-$ M+ | [(|)] | 1/x | y^x |

| 7 | 8 | 9 | STO | RCL |

| 4 | 5 | 6 | × | ÷ |

| 1 | 2 | 3 | + | − |

| 0 | . | π EXP | +/− | = |